Mathematical
Surveys
and
Monographs

Volume 244

Nonlinear Dirac Equation

Spectral Stability of Solitary Waves

Nabile Boussaïd
Andrew Comech

American Mathematical Society
Providence, Rhode Island

2010 *Mathematics Subject Classification.* Primary 35B32, 35B35, 35B40, 35C08, 35P05, 35P30, 35Q41, 37K40, 47J10, 81Q05.

For additional information and updates on this book, visit
www.ams.org/bookpages/surv-244

Library of Congress Cataloging-in-Publication Data

Cataloging-in-Publication Data has been applied for by the AMS.
See http://www.loc.gov/publish/cip/.

∞ The paper used in this book is acid-free and falls within the guidelines established to ensure permanence and durability.
Visit the AMS home page at `https://www.ams.org/`

10 9 8 7 6 5 4 3 2 1 24 23 22 21 20 19

To Rasmey, Norinn, and Léa,
N.B.

To Mary and Natalie,
A.C.

Contents

CHAPTER I

Introduction

The main aim of this monograph is to investigate the *spectral stability*, that is, the absence of eigenvalues with positive real part for the linearization at the solitary wave solutions to the nonlinear Dirac equation

$$\mathrm{i}\partial_t\psi = D_m\psi - f(\psi^*\beta\psi)\beta\psi, \qquad \psi(t,x)\in\mathbb{C}^N, \quad x\in\mathbb{R}^n, \tag{I.1}$$

in the case when the nonlinearity is represented by

$$f\in C^1(\mathbb{R}\setminus\{0\})\cap C(\mathbb{R}), \qquad f(\tau)=|\tau|^\kappa+O(|\tau|^K), \qquad 0<\kappa<K.$$

We prove the spectral stability of weakly relativistic solitary wave solutions, $\omega\lessapprox m$, in the following cases:

$$\kappa\lessapprox 2/n, \qquad K>\kappa, \qquad n\in\mathbb{N}; \tag{I.2}$$

$$\kappa=2/n, \qquad K>\frac{4}{n}, \qquad n\in\mathbb{N}. \tag{I.3}$$

We use the notation $\omega\lessapprox m$ to indicate that there is $\varepsilon>0$ sufficiently small such that $\omega\in(m-\varepsilon,m)$.

In particular, we prove the spectral stability for the quintic nonlinear Dirac equation in (1+1)D (that is, in one spatial dimension). We cannot prove the spectral stability for all subcritical values $\kappa\in(0,2/n)$: the point spectrum of the nonlinear Schrödinger equation of order $1+2\kappa$ linearized at a solitary wave becomes rich for small $\kappa>0$; such cases would require a more detailed analysis. Many of our results for the nonlinear Dirac equation have already appeared in [**BC16**, **BC17**, **BC18**, **BC19**].

Let us give the plan of the monograph. The introductory chapters are elementary and provide some background on Functional Analysis, Spectral Theory, Quantum Mechanics, and linear stability theory. In particular, we describe the Derrick theorem and the Kolokolov stability criterion. We think that these results already give the feeling of the material and also emphasize the importance of the Spectral Theory approach to the stability of solitary waves. Then we develop the tools needed for the Dirac operator: the Dirac–Pauli Theorem on the choice of Dirac matrices (which we do in arbitrary dimension), the limiting absorption principle, and the Carleman estimates. In the final chapters, we prove the spectral stability of weakly relativistic solitary waves in the charge-subcritical cases and in the "charge-critical" case. More precisely, we show that the presence of eigenvalues with nonzero real part in the spectrum of the linearization at a solitary wave in the nonrelativistic limit, $\omega\lessapprox m$, is essentially described by the Kolokolov stability criterion [**Kol73**], which initially appeared in the context of the nonlinear Schrödinger equation. The spectral stability result opens the way to the proofs of asymptotic stability.

Subjective historical review of the electron theory. In the laboratory conditions, electrons are observed in London in 1838, when Michael Faraday attaches high voltage to a vacuum tube in his laboratory at the Royal Institution and observes cathode rays. New tubes are being constructed and the vacuum is improving. Further experiments with vacuum tubes are done in Bonn by Heinrich Geissler in 1857 (with the pressure around 100 Pa), Julius Plucker, Johann Wilhelm Hittorf, and then by Sir William Crookes in London and Arthur Schuster in Manchester (now with the pressure below 0.1 Pa), who study the deflection of the *cathode rays* in magnetic and electric fields. The charge-to-mass ratio of the *cathode rays* is estimated. This name, *Kathodenstrahlen*, is coined in 1876 by a German physicist Eugen Goldstein.

In 1874, Irish physicist George Johnstone Stoney, while working on electrolysis, suggests that there is a "single definite quantity of electricity", estimates its charge, and in 1881 coins the term *electrolion* (and then going to *electron*).

While developing the light bulb, Thomas Edison inserts the third wire into a vacuum tube and files a patent for the "Electrical indicator" (1884), a vacuum tube which is to become a triode in twenty years, several lines below.

In 1896, at the Cavendish Laboratory, Cambridge, Joseph John Thomson accurately measures the mass-to-charge ratio of the cathode rays and suggests the existence of the electrons [**Tho97**]. In [**Tho04**], taking into account Earnshaw's theorem on instability of stationary configurations of point charges [**Ear42**], Thomson comes up with the plum-pudding model, in which point electrons sit (or, rather, move) in the pudding-like atom. Partial differential equations are near: J.J. Thomson was a PhD student of John William Strutt, 3rd Baron Rayleigh, the author of the celebrated two-volume *The Theory of Sound* [**Str77**, **Str78**], an outstanding PDE monograph (Richard Courant admired this book and, according to Kurt Friedrichs, strongly recommended to his students). Rayleigh's ideas were later used by Schrödinger as the basis for what is now the Rayleigh–Schrödinger perturbation theory.

In 1904, John Fleming uses Edison's "electrical indicator" for amplifying the radio signal for applications to transatlantic communications. This idea is developed by Lee DeForest in 1906 into the "Audion", or a triode: a vacuum tube with a grid that is used to control the current through the tube, which allows one to amplify the electric signal; the electronic age begins.

J.J. Thomson's influence on the electron theory continues: one of his PhD students is Ernest Rutherford, under whose direction the Geiger–Marsden gold foil experiment in 1909 leads to the Bohr–Rutherford model of an atom, putting an end to Thomson's plum-pudding model. In 1911, Niels Bohr receives his doctorate from the University of Copenhagen and sets off to Cambridge, where he works as a postdoc under J.J. Thomson. Two years later, Bohr formulates his famous postulates [**Boh13**] aimed at describing electron's behaviour in atoms.

In 1911, Louis de Broglie accompanies his older brother, Maurice, to the First Solvay Conference on Physics. In 1923, following discussions of Planck's and Einstein's research on wave-particle duality in the context of photons (the Second Solvay Conference, 1913), he puts forward the idea that the electrons could be described as waves, published in his PhD Thesis [**Bro25**] and presented at the Fourth Solvay Conference (1924). In November 1925, Peter Debye, a colloquium organizer at ETH, prompts Erwin Schrödinger to give a colloquium on de Broglie's *phase* *waves*. Shortly after his talk, again having been prompted by Debye and spending Christmas of 1925 in Arosa, Switzerland, Schrödinger writes down a relativistically invariant equation for the wave function of the electron, now known as the Klein–Gordon equation. Introduction of the Coulomb potential into this equation leads to the energy values that agree—in the leading order—with the electron energy levels $E_n = -\frac{m\,e^4}{2\hbar^2 n^2}$, $n \in \mathbb{N}$ (with m and e electron's mass and charge), from the "Old Quantum Theory" by Niels Bohr and Arnold Sommerfeld; that expression explains the empirical formula for the frequencies of Hydrogen spectral lines obtained by Johannes Rydberg back in 1888. To find E_n, one writes the wave function as $e^{-ar}F(r)$, with an appropriate $a > 0$ and $F(r) = F_0 + F_1 r + \dots$ a polynomial whose coefficients are computed recurrently; the values of E_n are to be such that the polynomial $F(r)$ has finitely many terms, or else the series representing $F(r)$ could be shown to converge to a function which grows like e^{2ar} (see, e.g., **[Sch49]**). Noticing the inconsistency of the higher order "relativistic corrections" to the values of E_n with the accurately measured wavelengths of the emitted light, Schrödinger shelves his relativistic *H-Atom. Eigenschwingungen* draft and retreats to the non-relativistic limit,

$$E = \sqrt{m^2c^4 + \mathbf{p}^2c^2} \approx mc^2 + \frac{\mathbf{p}^2}{2m},$$

arriving at what we now know as the Schrödinger equation **[Sch26]**; see **[Sch49, Moo94, Meh01]**. Now of course it is accepted that, under spatial rotations, electron's wave function transforms according to the four-dimensional "spinorial" representation of $\mathbf{SO}(3)$, and thus could not be described by a scalar-valued solution to the relativistic Schrödinger (or Klein–Gordon) equation.

In the summer of 1926, Clinton Davisson attends the Oxford meeting of the British Association for the Advancement of Science on recent advances in quantum mechanics, and learns there at Max Born's lecture that the plots with peaks from his and Lester Germer's 1923 experiment are used as a confirmation of the de Broglie hypothesis about the wave nature of electrons. In 1927, diffraction patterns from electron scattering on crystals are independently obtained by George Paget Thomson, J.J. Thomson's son.

The relativistically invariant equation for electrons is invented in December 1927, in Cambridge, by Paul Dirac, who notices that

$$(\sigma_1 p_1 + \sigma_2 p_2 + \sigma_3 p_3)^2 = p_1^2 + p_2^2 + p_3^2, \qquad \forall\, \mathbf{p} = (p_1, p_2, p_3) \in \mathbb{R}^3,$$

with σ_i the Pauli matrices, doubles the matrix size to be able to extract the square root of the energy-momentum relation $E^2 = m^2c^4 + \mathbf{p}^2c^2$, arriving at the first-order relation $E = c\boldsymbol{\alpha} \cdot \mathbf{p} + \beta mc^2$ where $\boldsymbol{\alpha} = (\alpha^1, \alpha^2, \alpha^3)$, with α^i, $1 \le i \le 3$, and β the 4×4 Dirac matrices, and then uses Schrödinger's substitution of E and $\mathbf{p}$ by the operators $\mathrm{i}\hbar\partial_t$ and $-\mathrm{i}\hbar\nabla$ to write down the evolution equation for the electron wave function **[Dir28]** (for the historical details, see **[Meh01]**). The matrix form of the Dirac equation results in the spinor-valued wave function and triumphally yields the accurate values of the relativistic corrections to the energy levels of the Hydrogen atom.

Self-interacting spinor fields. Let us point out that the Lamb shift **[LR47]** could not be explained in the framework of the linear Dirac equation in the external potential, thus illustrating the need for a more accurate description of the nonlinear effects in the Dirac–Maxwell system, which are known to physicists as the interaction of an electron with the *vacuum energy fluctuations*. As a matter of fact, neither can the linear theory explain electron's need for "quantum jumps" from one state to another, postulated by Bohr, since in a linear system any superposition of solutions is also a solution. Speaking of the *Copenhagen interpretation of quantum mechanics*, Steven Weinberg says: "According to Bohr, in a measurement the state of a system such as a spin collapses to one result or another in a way that cannot itself be described by quantum mechanics, and is truly unpredictable. This answer is now widely felt to be unacceptable" [**Wei17**]. At the same time, *quantum jumps* can be rigorously explained in the framework of the global attractors of *nonlinear* dispersive Hamiltonian systems [**Kom03, KK07, Kom12**]: the energy leaks via a weak, higher order self-interaction (caused by a nonlinear nature of the system) into the essential spectrum and disperses, until the system arrives at a one-frequency state (*Bohr's quantum orbit*) which no longer radiates the energy. To summarize, it is possible that, in the words of the Russian writer Sergey Dovlatov, the nonlinear effects in Quantum Physics could be *nice, but small; or, rather, small, but nice* (he was speaking about his salary).

The idea to employ nonlinear models in Quantum Physics could be traced back to the last paragraph in the article of Dmitri Ivanenko [**Iva38**] (remarkably, in the context of the nonlinear Dirac equation). It is followed up in particular in [**FLR51, FFK56, Hei57**]. Widely known models of self-interacting spinor fields are the massive Thirring model [**Thi58**] (spinor field with the vector self-interaction) and the Soler model [**Iva38, Sol70**] (spinor field with the scalar self-interaction); the one-dimensional analogue of the Soler model is known as the (massive) Gross–Neveu model of quark confinement [**GN74, LG75**]. Let us also mention that nonlinear equations of Dirac type also appear under the name of coupled-mode equations in nonlinear optics [**SS94**] and in the description of matter-wave Bose–Einstein condensates trapped in an optical lattice [**PSK04**].

In the Klein–Gordon context, self-interacting fields appear in Leonard Schiff's nonlinear meson theory [**Sch51a, Sch51b**] and start receiving mathematical attention since the articles by Konrad Jörgens and Irving Segal [**Jör61, Seg63**] in the study of the well-posedness of the nonlinear Klein–Gordon equation in the energy space. This is followed by the research on nonlinear scattering by Irving Segal, Walter Strauss, and Cathleen Morawetz [**Seg66, Str68, MS72**], the work on the existence of solitary waves [**Str77a, BL79, BL83a**], and an extensive research on their linear and orbital stability, in the nonlinear Schrödinger and Klein–Gordon context; see, e.g., [**Kol73, BC81, CL82, Sha83, Wei85, Sha85, GSS87, Gri88, DBGRN15, DBRN19**], which is followed by the research on the asymptotic stability of solitary waves by Avy Soffer and Michael Weinstein [**SW90, SW92**], Vladimir Buslaev and Galina Perelman [**BP92b, BP95**], which is further developed in [**PW97, SW99, Cuc01, BS03, Cuc03**]. The next aim of the theory could be the Soliton Resolution Conjecture [**Kom03, Sof06, Tao07, KK07, Tao09, KLLS15, DJKM17**].

Although the research on the Dirac-based systems follows much slower, there is an increasing interest to this subject. In particular, the existence of standing waves in the nonlinear Dirac equation is studied in [**Sol70, CV86, Mer88, ES95**]. The discussion of the applications of classical self-interacting spinor fields in Quantum Theory is in [**Rañ83a**]. The relation of nonlinear theory and the Pauli exclusion principle is considered in [**Rañ83b**]. Local and global well-posedness of the nonlinear Dirac equation is further addressed in [**EV97**] (semilinear Dirac equation in (3+1)D) and in [**MNNO05**] (nonlinear Dirac equation in (3+1)D). There are many results on the local and global well-posedness: in one spatial dimension, we mention [**ST10, MNT10, Huh11, Can11, Pel11, Huh13a, Huh13b, HM15, ZZ15**]; the higher-dimensional setting is considered in [**Bou04, Bou08a, BC14, Huh14, BH15, BH16**].

Stability. Now we move to the question of stability of solitary waves. The stability is understood in many cases for the nonlinear Schrödinger, Klein–Gordon, and Korteweg–de Vries equations (see, e.g., the review [**Str89**]). In these systems, at the points of the functional space corresponding to solitary waves, the Hamiltonian function is of finite Morse index. In simpler cases, the Morse index is equal to one, and the perturbations in the corresponding direction are prohibited by one of the conservation laws when the Kolokolov stability condition [**Kol73**] is satisfied (also known as Vakhitov–Kolokolov stability criterion). In other words, the solitary waves could be demonstrated to correspond to conditional minimizers of the energy under the charge constraint; this results not only in spectral stability but also in orbital stability [**CL82, GSS87**]. The nature of stability of solitary wave solutions of the nonlinear Dirac equation, observed in the early numerical studies of dynamics [**AC81**], seems different from this picture [**Rañ83a**, Section V]. The Hamiltonian function is not bounded from below, and is of infinite Morse index at the points corresponding to solitary waves; the NLS-style approach to stability fails. As a consequence, we do not know how to prove the *orbital stability* [**CL82, GSS87**] except via proving the asymptotic stability first. The only known exception is the completely integrable massive Thirring model in (1+1)D, where the orbital stability was proved by means of a coercive conservation law [**PS14, CPS16**] coming from higher order integrals of motion.

On the way to proving the asymptotic stability, one starts with the linear (also known as *spectral*) stability. The purely imaginary essential spectrum of the linearization operator is readily available via Weyl's theorem on the essential spectrum; the discrete spectrum is more delicate. While the linear instability of ground states in the nonlinear Schrödinger equation can only come from a positive eigenvalue, whose presence is conveniently controlled by the Kolokolov stability criterion [**Kol73**], the Dirac equation presents a less comfortable situation: the linearization at the solitary waves in the nonlinear Dirac equation can possess eigenvalues anywhere in the complex plane. In particular, the linear instability in nonlinear equations of Dirac type may develop due to the bifurcations from the embedded thresholds at $\pm\mathrm{i}(m+|\omega|)$ as in [**BPZ98**], in the context of the one-dimensional coupled-mode equation; from the collision of the thresholds $\pm\mathrm{i}(m\pm|\omega|)$ at $z=\pm m\mathrm{i}$ when $\omega=0$, as in [**KS02**], in the context of the massive Thirring model; "in the nonrelativistic limit" $\omega\lessapprox m$ (by this we mean that there is $\varepsilon>0$ small enough so that $\omega\in(m-\varepsilon,m)$), when linear instability of Schrödinger equation is inherited by weakly relativistic solitary waves in the nonlinear Dirac equation with supercritical nonlinearity [**CGG14**]. (It was discovered numerically that this latter instability disappears when $\omega\in(0,m)$ becomes sufficiently small [**CMKS$^+$16**], although it later reappears for $\omega\gtrapprox 0$; that is, for $\omega\in(0,\varepsilon)$ with $\varepsilon>0$ small enough.) The birth of "unstable" eigenvalues with positive real part can also happen from the collision of purely imaginary eigenvalues in the spectral gap (away from the origin), like in the Soler model in two spatial dimensions [**CMKS⁺16**].

The spectral stability of solitary waves to the cubic nonlinear Dirac equation in (1+1)D (the Gross–Neveu model) was demonstrated in [**BC12a**], where the spectrum of the linearization at solitary waves was computed via the Evans function technique; no nonzero-real-part eigenvalues have been detected. This (1+1)D stability result was later confirmed in [**Lak18**] via numerical simulations of the evolution. Subsequent numerical computations of the spectral stability in the nonlinear Dirac in one and two spatial dimensions were performed in [**CMKS⁺16**].

Our main goal is to prove the asymptotic stability in the general situation, in systems with the translational invariance, without any simplifying restrictions (like "radially symmetric") on the perturbations. This is to be achieved by the combination of the description of the structure of the solitary wave manifold and the spectral stability results (which are the main subject of this monograph) with the decay estimates. In particular, we mention the local decay estimates, L^p decay estimates, and the Strichartz estimates. In the context of the Schrödinger operator, the classical results are [**JK79, KY89, KY89, JSS91, GJY04, RS04, Bec11**] and [**Str77b, Yaj87, GV92, KT98**]. A concise exposition of the stationary scattering theory of Agmon–Jensen–Kato can be found in [**KK12**]. In the context of the Dirac operator, the related results are in [**MNO03, MNNO05, Bou06, DF07, Bou08b**]. Given the amount of the recent material and the rate of the progress in the field, we choose not to include the description of these results in the present monograph.

Some results on asymptotic stability in the context of the nonlinear Dirac equation were obtained in the three-dimensional case with the external potential [**Bou06, Bou08b**]. There, the spectrum of the linear part of the operator $D_m + V$, besides the essential spectrum $\mathbb{R}\setminus(-m, m)$, is assumed to contain two simple eigenvalues; let us denote them by λ_0 and λ_1, with $\lambda_0 < \lambda_1$. There is a bifurcation of small amplitude solitary waves for the nonlinear equation from the associated eigenspaces. The corresponding linearized operators are small exponentially localized perturbations of $D_m + V$, so that the perturbation theory allows for a precise knowledge of the resulting spectral stability. Depending on the distance from λ_0 to λ_1 compared to the distance from λ_0 to the essential spectrum, the resulting point spectrum for the linearized operator may have nonzero-real-part eigenvalues, or instead it may be discrete and purely imaginary and hence spectrally stable if a "Fermi Golden Rule" assumption, at the linearization level, is satisfied (similarly to the Schrödinger case; we refer to [**Sig93, BP95, SW90, SW92, TY02a, TY02c, TY02d, TY02b, Miz08, GNT04, CM08**]). In the former case, linear and dynamical instabilities take place. In the latter case, the linear stability follows from the spectral one via the perturbation theory. Using the dispersive properties of perturbations of D_m, one concludes that there is a stable manifold of real codimension 2. Due to the presence of nonzero discrete modes, even in the linearly stable case, the dynamical stability is not guaranteed. Before considering the results on the dynamics outside this manifold, for perturbations along the remaining two real directions, one could ask what might happen if $D_m + V$ had only one eigenvalue. The answer follows quite immediately with the ideas from [**Bou06, Bou08b**]. In this case, there is only one family of solitary waves and it is asymptotically stable. Notice that the asymptotic profile is possibly another solitary wave which is close to the perturbed one. In the one-dimensional case, this was studied properly in [**PS14**].

Note that the one-dimensional framework suffers from relatively weak dispersion which makes the analysis of the stabilization process rather delicate. As for the dynamics outside the above-mentioned stable manifold, the techniques rely on the analysis of nonlinear resonances between discrete isolated modes and the essential spectrum where the dispersion takes place. This requires the normal form analysis in order to isolate the leading resonant interactions. The former is possible due to the "Fermi Golden Rule" assumption. Such an analysis was done in **[BC12b]** but in a slightly different framework: instead of considering the perturbative framework, the authors chose the translation-invariant case, imposing a series of assumptions that led to the spectral stability of solitary waves. These assumptions are verified in some perturbative context with $V \neq 0$; this case is analyzed in **[CT16]**.

The asymptotic stability approach from **[BC12b, PS12, CPS17]** is developed under important restrictions on the types of admissible perturbations. Such restrictions are needed to avoid the translational invariance and, most importantly, to prohibit the perturbations in the direction of exceptional eigenvalues $\pm 2\omega\mathrm{i}$ of the linearization operator at a solitary wave $\phi_\omega(x)e^{-\mathrm{i}\omega t}$. These eigenvalues are a feature of the Soler model and are present in the spectrum for any nonlinearity f in the Soler model (I.1); see **[BC12a, Gal77, DR79]**. These eigenvalues violate the "Fermi Golden Rule": they do not "interact" (that is, do not resonate) with the essential spectrum; the energy from the corresponding modes does not disperse to infinity. To proceed with the proof of the asymptotic stability for this case, one would need to consider the set of bi-frequency solitary waves and to include modulation equations for the corresponding parameters.

There are several open questions which we would like to mention.

(1) Formally, the Hamiltonian of the Soler model (nonlinear Dirac equation with the scalar self-interaction) has the "infinite Morse index" at the points of the functional space corresponding to the solitary wave solutions. At the same time, this model seems to be even "more stable" than the nonlinear Schrödinger equation. For example, in dimension two, the cubic nonlinear Schrödinger equation is critical (the solitary waves $\phi_\omega(x)e^{-\mathrm{i}\omega t}$ with $\phi_\omega(x) = |\omega|^{1/2}\Phi(|\omega|^{1/2}x)$ have the same charge for all $\omega < 0$), and as a consequence the linearization at any solitary wave has a larger size Jordan block at $z = 0$, resulting in an instability of solitary waves **[CP03]**. On the contrary, by **[CMKS$^+$16, BC19]**, the nonlinear Dirac equation with cubic nonlinearity in two dimensions is spectrally stable, without a higher order Jordan block at $z = 0$. We also mention that there is a blow-up phenomenon in the charge-critical as well as in the charge-supercritical nonlinear Schrödinger equation, see in particular **[ZSS71, ZS75, Gla77, Wei83, Mer90]**; at the same time, we are not aware of the blow-up results in the models based on the nonlinear Dirac equation with $\mathbf{U}(1)$-invariance. Overall, the nonlinear Dirac equation seems surprisingly stable. Does it possess a stability mechanism which we do not know yet?

We point out that the stability in the nonlinear Dirac equation *does not* need the standard *Dirac sea* assumption that all the negative energy states are filled and that the Pauli principle prohibits more than one particle in a state.

(2) While the nonlinear Dirac equation is only a simplified model of self-interacting spinor fields, we expect that our approach can be modified to tackle the spectral stability of solitary waves in the Dirac–Maxwell system **[Gro66, Wak66]** and in similar Dirac-based models. (Let us mention that the local well-posedness of the

Dirac–Maxwell system was proved in [**Bou96**]. The existence of standing waves in the Dirac–Maxwell system is proved in [**EGS96**] (for $\omega \in (-m, 0)$) and [**Abe98**] (for $\omega \in (-m, m)$); see also the review [**ES02**].) Are there stable localized solutions in Dirac–Maxwell and similar systems? Are there any interesting implications in Physics?

(3) Presently, the Dirac equation gives a consistent description of the electron in the Hydrogen atom, but even for a two-electron Helium one faces problems [**KOY15**]. What is an appropriate model for the many-body problem based on the nonlinear Dirac equation? Are there localized solutions? How does one approach the linear stability in such models? Let us mention [**BES05, ES05, Der12, Lev14**].

ACKNOWLEDGMENTS

We would like to thank our families for their patience and support.

Both authors are most grateful to Dr. Yulia Meshkova for many valuable suggestions and for pointing out numeruos errors and typos in Chapters II and III, and to Dr. Sergei Gelfand for the encouragement and advice.

Support from the grant ANR-10-BLAN-0101 of the French Ministry of Research is gratefully acknowledged by the first author. During the work on the project, the second author was partially supported by Université Bourgogne Franche–Comté, by the Metchnikov fellowship from the French Embassy, and by the Mittag-Leffler Institute. Both authors acknowledge the financial support of the région Bourgogne Franche–Comté.

CHAPTER II

Distributions and function spaces

In this chapter, we introduce the background information on distributions and related spaces. A hands-on theory of the Sobolev spaces is presented in books by Peter Gilkey [**Gil84**], Chris Sogge [**Sog93**], and Haïm Brezis [**Bre10**], while a comprehensive reference on the theory of distributions is the monograph by Lars Hörmander [**Hör83**].

Unless stated otherwise, we assume that the functions are complex-valued.

II.1 Distributions

II.1.1 Test functions and distributions. The topological space $\mathscr{D}(\mathbb{R}^n)$, $n \geq 1$, denotes the space of smooth compactly supported functions, called *test functions*. As a set, it coincides with $C^\infty_{\mathrm{comp}}(\mathbb{R}^n)$. This space is equipped with an inductive limit topology; for instance, the one corresponding to the sequence of spaces $C^\infty_{\mathrm{comp}}(\mathbb{B}^n_R)$, $R \to \infty$ (the set of smooth functions on $\mathbb{R}^n$ with support in $\mathbb{B}^n_R$), equipped with the topology generated by the seminorms

$$\|\varphi\|_{\alpha,R} = \sup_{x\in\mathbb{B}^n_R} (|\partial^\alpha_x \varphi(x)|), \qquad \alpha \in \mathbb{N}^n_0.$$

More precisely, for $k \in \mathbb{N}_0$ and $R > 0$, we equip $C^k_{\mathrm{comp}}(\mathbb{B}^n_R)$ with the topology generated by the norm

$$\|\varphi\|_{C^k(\mathbb{B}^n_R)} = \sum_{\alpha\in\mathbb{N}^n_0,\,|\alpha|\leq k} \|\varphi\|_{\alpha,R} = \sum_{\alpha\in\mathbb{N}^n_0,\,|\alpha|\leq k} \sup_{x\in\mathbb{B}^n_R} |\partial^\alpha_x \varphi(x)|,$$

then we equip $C^\infty_{\mathrm{comp}}(\mathbb{B}^n_R)$, $R > 0$, with the coarsest topology which makes the injections

$$C^\infty_{\mathrm{comp}}(\mathbb{B}^n_R) \subset C^k_{\mathrm{comp}}(\mathbb{B}^n_R)$$

continuous for any $k \in \mathbb{N}$, and then the topology on $\mathscr{D}(\mathbb{R}^n)$ is defined as the finest topology which makes the injections

$$C^\infty_{\mathrm{comp}}(\mathbb{B}^n_R) \subset \mathscr{D}(\mathbb{R}^n)$$

continuous for any $R > 0$.

Problem II.1 *(1)* Show that $C^\infty_{\mathrm{comp}}(\mathbb{B}^n_R)$ equipped with the topology corresponding to the seminorms $\|\varphi\|_\alpha$ is metrizable, not complete, and not normable.

(2) Show that $\mathscr{D}(\mathbb{R}^n)$ is complete but not metrizable.

Since $\mathscr{D}(\mathbb{R}^n)$ is not metrizable, one needs to elaborate on the convergence of sequences and the characterization of continuity. A sequence $(\varphi_j)_{j\in\mathbb{N}}$ converges in

$\mathscr{D}(\mathbb{R}^n)$ if and only if

(1) There exists $R > 0$ such that $\operatorname{supp} \varphi_j \subset \mathbb{B}^n_R$;
(2) For each $\alpha \in \mathbb{N}^n_0$, $(\partial^\alpha_x \varphi_j)_{j\in\mathbb{N}}$ is convergent uniformly on $\mathbb{B}^n_R$.

Similarly, a linear map T from $\mathscr{D}(\mathbb{R}^n)$ to a Banach space E is continuous if and only if for any $R > 0$ there exist $C_R > 0$ and $k \in \mathbb{N}_0$ such that

$$\|T(\varphi)\|_E \leq C_R \sum_{\alpha\in\mathbb{N}^n_0,\, |\alpha|\leq k} \|\varphi\|_{\alpha,R}, \qquad \forall\varphi \in C^\infty_{\text{comp}}(\mathbb{B}^n_R)\,.$$

The dual space $\mathscr{D}'(\mathbb{R}^n)$ is called the space of distributions. This corresponds to $E = \mathbb{C}$ in the previous property. Note that k in this property depends on R. If a finite k can be chosen uniformly in $R > 0$ then the distribution is said to have a finite order.

The space $\mathscr{D}(\mathbb{R}^n)$ is considered as a subspace of $\mathscr{D}'(\mathbb{R}^n)$, with the following natural embedding: given $\psi \in \mathscr{D}(\mathbb{R}^n)$, we define the corresponding element ψ of $\mathscr{D}'(\mathbb{R}^n)$ (which we again denote by ψ) by

$$\psi:\ \mathscr{D}(\mathbb{R}^n) \to \mathbb{C}, \qquad \varphi \mapsto \langle\psi, \varphi\rangle = \int_{\mathbb{R}^n} \bar{\psi}(x)\varphi(x)\, dx.$$

By duality, any bounded linear map from $\mathscr{D}(\mathbb{R}^n)$ to itself extends to $\mathscr{D}'(\mathbb{R}^n)$. Fundamental examples of such transformations are given by multiplications by smooth functions or partial derivatives of any order.

II.1.2 Tempered distributions and Fourier transforms. The Schwartz class of test functions is denoted by $\mathscr{S}(\mathbb{R}^n)$ and is equipped with the topology determined by the seminorms

$$\|u\|_{\alpha,\beta} := \sup_{x\in\mathbb{R}^n} |x^\alpha \partial^\beta_x u(x)|, \qquad \alpha,\, \beta \in \mathbb{N}^n_0. \tag{II.1}$$

Problem II.2 Show that the space $\mathscr{S}(\mathbb{R}^n)$ is metrizable.

The dual space $\mathscr{S}'(\mathbb{R}^n)$ is called the space of tempered distributions.

Problem II.3 Show that the Dirac delta-function

$$\delta:\ u \mapsto u(0)$$

defines δ as a tempered distribution.

Problem II.4 Show that for any T in $\mathscr{S}'(\mathbb{R}^n)$ there exist $m,\, k \in \mathbb{N}_0$ and $C > 0$ such that

$$|T(u)| \leq C \sum_{|\alpha|\leq k, |\beta|\leq m} \|u\|_{\alpha,\beta}, \qquad \forall u \in \mathscr{S}(\mathbb{R}^n),$$

with $\|\cdot\|_{\alpha,\beta}$ from (II.1). This shows that tempered distributions have a finite order.

The Fourier transform $\mathcal{F}: \mathscr{S}(\mathbb{R}^n) \to C(\mathbb{R}^n)$ of $u \in \mathscr{S}(\mathbb{R}^n)$ is defined by

$$\hat{u}(\xi) = (\mathcal{F}u)(\xi) = \int_{\mathbb{R}^n} e^{-\mathrm{i}\xi\cdot x} u(x)\, dx, \qquad \xi \in \mathbb{R}^n.$$

The Fourier transform is not invariant in the space $\mathscr{D}(\mathbb{R}^n)$, hence there is no extension of the Fourier transform to distributions.

Problem II.5 Show that the Fourier transform of a compactly supported measurable function which is not identically zero has only isolated zeros and thus can not have compact support.

At the same time, the Fourier transform defines a continuous linear map from $\mathscr{S}(\mathbb{R}^n)$ to itself. The integration by parts shows that, as the matter of fact, the mapping $\mathcal{F} : \mathscr{S}(\mathbb{R}^n) \to \mathscr{S}(\mathbb{R}^n)$ is continuous (in the topology determined by seminorms $\|\cdot\|_{\alpha,\beta}$). One can readily show that the inverse of the Fourier transform is given by

$$(\mathcal{F}^{-1}v)(x) = \int_{\mathbb{R}^n} e^{\mathrm{i}\xi\cdot x} v(\xi)\,\frac{d\xi}{(2\pi)^n}, \qquad v \in \mathscr{S}(\mathbb{R}^n).$$

Indeed, the above transformation is continuous in $\mathscr{S}(\mathbb{R}^n)$ (just like the Fourier transform), while $e^{-\epsilon\xi^2/2}\hat{u}(\xi)$ converges to $\hat{u}(\xi)$ in $\mathscr{S}(\mathbb{R}^n)$ as $\epsilon \to 0+$, so we may first apply what is to be the inverse Fourier transform to $e^{-\epsilon\xi^2/2}\hat{u}(\xi)$, and send $\epsilon \to 0+$ later. We have:

$$\text{(II.2)} \qquad \int_{\mathbb{R}^n} e^{\mathrm{i}\xi\cdot x} e^{-\epsilon\xi^2/2}\hat{u}(\xi)\,\frac{d\xi}{(2\pi)^n} = \int_{\mathbb{R}^n}\left(\int_{\mathbb{R}^n} e^{\mathrm{i}\xi\cdot(x-y)} e^{-\epsilon\xi^2/2}\,\frac{d\xi}{(2\pi)^n}\right) u(y)\,dy,$$

where the order of integration could be interchanged due to the absolute convergence. The integral in the brackets can be explicitly evaluated, giving $(2\pi\epsilon)^{-n/2}e^{-|x-y|^2/(2\epsilon)}$, which converges to $\delta_x = \delta(\cdot - x)$ (in $\mathscr{S}'(\mathbb{R}^n)$) as $\epsilon \to 0+$. Thus, (II.2) converges to $u(x)$.

The space $\mathscr{S}(\mathbb{R}^n)$ identifies as a subspace of $\mathscr{S}'(\mathbb{R}^n)$ and the Fourier transform extends by duality as a continuous linear map from $\mathscr{S}'(\mathbb{R}^n)$ to itself.

According to the Plancherel theorem,

$$(2\pi)^{-n}\|\mathcal{F}u\|^2_{L^2} = \|u\|^2_{L^2}, \qquad u \in \mathscr{S}(\mathbb{R}^n);$$

so, using the density of $\mathscr{S}(\mathbb{R}^n)$ in $L^2(\mathbb{R}^n)$, one extends the Fourier transform to the continuous mapping $\mathcal{F} : L^2(\mathbb{R}^n) \to L^2(\mathbb{R}^n)$.

Problem II.6 Show that the Fourier transform of a function u in $L^2(\mathbb{R}^n)$ (obtained by the extension from $\mathscr{S}(\mathbb{R}^n)$) coincides with the Fourier transform of u considered as a tempered distribution.

Remark II.7 For consistency with the Quantum Mechanics notations (the plane wave being $\psi(t,x) = e^{\mathrm{i}(\xi\cdot x - \omega t)}$, with particular $\xi \in \mathbb{R}^3$ and $\omega \in \mathbb{R}$), the Fourier transform with respect to time is usually introduced with the opposite sign in the exponent:

$$\text{(II.3)} \qquad \hat{f}(\omega) = (\mathcal{F}f)(\omega) = \int_{\mathbb{R}} e^{\mathrm{i}\omega t} f(t)\,dt, \qquad \omega \in \mathbb{R};$$

$$\text{(II.4)} \qquad f(t) = (\mathcal{F}^{-1}\hat{f})(t) = \int_{\mathbb{R}} e^{-\mathrm{i}\omega t}\hat{f}(\omega)\,\frac{d\omega}{2\pi}, \qquad t \in \mathbb{R}.$$

II.2 Sobolev spaces

Given an open subset $\Omega \subset \mathbb{R}^n$ and $k \in \mathbb{N}_0$, we denote the standard L^2-based Sobolev spaces (of complex valued functions) by

$$H^k(\Omega) = H^k(\Omega,\mathbb{C}) = \left\{\, u \in L^2(\Omega,\mathbb{C}):\ \ \partial_x^\alpha u \in L^2(\Omega,\mathbb{C}), \quad \forall\alpha \in \mathbb{N}_0^n,\ |\alpha| \le k \,\right\},$$

with the derivatives in the sense of distributions and with the norm defined by

$$\|u\|^2_{H^k(\Omega)} = \sum_{|\alpha|\le k} \|\partial_x^\alpha u\|^2_{L^2(\Omega)}.$$

We denote $H^\infty(\Omega) = \cap_{k\in\mathbb{N}_0} H^k(\Omega)$.

For $\alpha \in \mathbb{R}$, we introduce fractional powers of the regularized Laplace operator via the Fourier transform:

$$(I-\Delta)^{\alpha/2} : \mathscr{S}(\mathbb{R}^n) \to \mathscr{S}(\mathbb{R}^n), \qquad (I-\Delta)^{\alpha/2} u(x) = \int_{\mathbb{R}^n} e^{\mathrm{i}\xi\cdot x} (1+\xi^2)^{\alpha/2} \hat{u}(\xi) \frac{d\xi}{(2\pi)^n},$$

which is extended by duality to $(I-\Delta)^{\alpha/2} : \mathscr{S}'(\mathbb{R}^n) \to \mathscr{S}'(\mathbb{R}^n)$. For $\alpha \in \mathbb{R}$, we denote

$$\langle \mathrm{i}\nabla\rangle^{\alpha} = (I-\Delta)^{\alpha/2},$$

with the right-hand side understood in the sense of the Fourier multiplier $(1+\xi^2)^{\alpha/2}$. For α, $s \in \mathbb{R}$, we define the weighted Sobolev spaces:

$$H^\alpha_s(\mathbb{R}^n) = \big\{ u \in \mathscr{S}'(\mathbb{R}^n) \colon \ \ \|u\|_{H^\alpha_s} < \infty \big\}, \qquad \|u\|_{H^\alpha_s} = \|\langle x\rangle^s \langle -\mathrm{i}\nabla\rangle^\alpha u\|_{L^2},$$

with $\langle x\rangle$ being (the operator of multiplication by) $(1+x^2)^{1/2}$, $x \in \mathbb{R}^n$. For $s \in \mathbb{R}$, we write $L^2_s(\mathbb{R}^n)$ for $H^0_s(\mathbb{R}^n)$:

$$L^2_s(\mathbb{R}^n) = H^0_s(\mathbb{R}^n) = \big\{ u \in L^2_{\mathrm{loc}}(\mathbb{R}^n) \colon \ \ \langle x\rangle^s u \in L^2(\mathbb{R}^n) \big\}, \quad \|u\|_{L^2_s} = \|\langle x\rangle^s u\|_{L^2}.$$

Above, we use the standard notation L^2_{loc}:

$$\text{(II.5)} \qquad L^2_{\mathrm{loc}}(\mathbb{R}^n) = \big\{ u \in \mathscr{D}'(\mathbb{R}^n) \colon \ \ \varphi u \in L^2(\mathbb{R}^n) \ \ \forall \varphi \in \mathscr{D}(\mathbb{R}^n) \big\}.$$

One similarly defines L^∞_{loc} and other spaces. The space $H^\alpha_s(\mathbb{R}^n)$ with $\alpha \in \mathbb{R}$ and $s = 0$ is denoted by $H^\alpha(\mathbb{R}^n)$. In the case when $\alpha = k \in \mathbb{N}_0$ and $\Omega = \mathbb{R}^n$, this definition coincides with the earlier definition of $H^k(\mathbb{R}^n)$.

Given an open set $\Omega \subset \mathbb{R}^n$ and $\alpha \in \mathbb{R}$, we use the notation $H^\alpha_{\mathrm{comp}}(\Omega)$ for the subset of $H^\alpha(\mathbb{R}^n)$ made up by functions with compact support in Ω. The set $H^\alpha_{\mathrm{comp}}(\mathbb{R}^n)$ is dense in $H^\alpha(\mathbb{R}^n)$. If $\Omega \subsetneq \mathbb{R}^n$, the closure of $H^\alpha_{\mathrm{comp}}(\Omega)$ in $H^\alpha(\Omega)$ is denoted by $H^\alpha_0(\Omega)$. The notation $H^\alpha_{\mathrm{loc}}(\Omega)$ is used for the set of functions $u \in \mathscr{S}'(\mathbb{R}^n)$ such that $\eta u \in H^\alpha_{\mathrm{comp}}(\Omega)$ for any $\eta \in C^\infty_{\mathrm{comp}}(\Omega)$.

Problem II.8 Show that $H^k_0(\mathbb{R}^n) = H^k(\mathbb{R}^n)$ for any integer k.

For $s \in \mathbb{R}$ and $1 \le p < \infty$, we denote by $W^{s,p}(\mathbb{R}^n)$ the space of all $u \in \mathscr{S}'(\mathbb{R}^n)$ such that $\langle \mathrm{i}\nabla\rangle^s u \in L^p(\mathbb{R}^n)$, with the norm

$$\|u\|_{W^{s,p}} = \|\langle \mathrm{i}\nabla\rangle^s u\|_{L^p}.$$

The space $W^{s,p}_0(\Omega)$ is defined as the closure of $C^\infty_{\mathrm{comp}}(\Omega)$ with respect to the norm $\|\cdot\|_{W^{s,p}}$.

Theorem II.9 (The Sobolev embedding theorem) *Let $1 \le p \le q < \infty$. Then*

$$W^{s,p}(\mathbb{R}^n) \subset L^q(\mathbb{R}^n) \qquad \textit{as long as} \quad \frac{s}{n} \ge \frac{1}{p} - \frac{1}{q}, \quad p > 1$$

and

$$W^{s,p}(\mathbb{R}^n) \subset L^\infty(\mathbb{R}^n) \qquad \textit{as long as} \quad \frac{s}{n} > \frac{1}{p}, \quad p \ge 1.$$

Proof. The proof is given in [**Sog93**, Theorem 0.3.7]. Here we give a short proof of the non-sharp version with $p = 2$:

$$H^s(\mathbb{R}^n) \subset L^q(\mathbb{R}^n), \qquad \frac{s}{n} > \frac{1}{2} - \frac{1}{q}, \qquad 2 \le q < \infty.$$

Let $n \in \mathbb{N}$, $s/n > 1/2 - 1/q$, and assume that $u \in H^s(\mathbb{R}^n)$. Then, since $u(x) = \mathcal{F}^{-1}[\hat{u}(\xi)](x)$, we need to show that $\hat{u} \in L^{q'}(\mathbb{R}^n)$, where $1 < q' < 2$ is such that $\frac{1}{q} + \frac{1}{q'} = 1$. By the Hölder's inequality,

$$\int_{\mathbb{R}^n} |\hat{u}(\xi)|^{q'}\,d\xi = \Big(\int_{\mathbb{R}^n} |(1+|\xi|)^s \hat{u}(\xi)|^{q'A}\,d\xi\Big)^{1/A}\Big(\int_{\mathbb{R}^n} |(1+|\xi|)^s|^{-q'A'}\,d\xi\Big)^{1/A'},$$

where A' is such that $\frac{1}{A} + \frac{1}{A'} = 1$. We set $A = 2/q'$; then the first term on the right is finite (since $u \in H^s(\mathbb{R})$). The second term on the right is finite if $sq'A' > n$, or $q' > \frac{n}{sA'} = \frac{n}{s}(1 - \frac{1}{A}) = \frac{n}{s}(1 - \frac{q'}{2})$, or $q'(1 + \frac{n}{2s}) > \frac{n}{s}$, $\frac{s}{n} + \frac{1}{2} > \frac{1}{q'} = 1 - \frac{1}{q}$, which is satisfied since

$$\frac{s}{n} > \frac{1}{2} - \frac{1}{q}.$$

The inclusion $H^s(\mathbb{R}^n) \subset L^\infty(\mathbb{R}^n)$ for $s > n/2$ follows from the fact that if $u \in H^s(\mathbb{R}^n)$ with $s > n/2$, then $\hat{u}$ is integrable. Its inverse Fourier transform is then continuous and tends to zero at infinity by the Riemann–Lebesgue theorem. $\square$

As a consequence, if $s > n/2$, the Sobolev space $H^s(\mathbb{R}^n)$ is a Banach algebra, as provided by the following lemma.

Lemma II.10 *Let $n \in \mathbb{N}$, $s > n/2$. For any $u, v \in H^s(\mathbb{R}^n)$, one has $uv \in H^s(\mathbb{R}^n)$, and moreover*

$$\|uv\|_{H^s} \le \big(2\max\{1, 2^{s-1}\}c_s\big)^{1/2}\|u\|_{H^s}\|v\|_{H^s},$$

with $c_s := \int_{\mathbb{R}^n}(1+\xi^2)^{-s}\,d\xi$.

PROOF. Since $s > n/2$, by Theorem II.9,

$$H^s(\mathbb{R}^n) \subset L^q(\mathbb{R}^n), \qquad \forall q \in [2, \infty].$$

So if u and v are in $H^s(\mathbb{R}^n)$, u and v are in $L^4(\mathbb{R}^n)$ and thus uv is in $L^2(\mathbb{R}^n)$. We can thus write

$$(1+\xi^2)^{s/2}\,|\widehat{uv}(\xi)| = (1+\xi^2)^{s/2}\,|\hat{u} * \hat{v}(\xi)| \le (1+\xi^2)^{s/2}|\hat{u}| * |\hat{v}|(\xi).$$

Using the inequality

$$(1+\xi^2)^{s/2} \le \max\{1, 2^{s-1}\}\big((1+\eta^2)^{s/2} + (1+(\xi-\eta)^2)^{s/2}\big),$$

we deduce:

$$\begin{aligned}(1+\xi^2)^{s/2}|\hat{u}| * |\hat{v}|(\xi) &= (1+\xi^2)^{s/2}\int_{\mathbb{R}^n} |\hat{u}(\eta)||\hat{v}(\xi-\eta)|\,d\eta \\ &\le \max\{1, 2^{s-1}\}\int_{\mathbb{R}^n}(1+\eta^2)^{s/2}|\hat{u}(\eta)||\hat{v}(\xi-\eta)|\,d\eta \\ &+ \max\{1, 2^{s-1}\}\int_{\mathbb{R}^n}|\hat{u}(\eta)|(1+(\xi-\eta)^2)^{s/2}|\hat{v}(\xi-\eta)|\,d\eta.\end{aligned}$$

Using Young's inequality, we have:

$$\begin{aligned}&\int_{\mathbb{R}^n}(1+\xi^2)^s\,(|\hat{u}|*|\hat{v}|(\xi))^2\,d\xi\\&\le\max\{1,2^{s-1}\}\|\hat{v}\|^2_{L^1(\mathbb{R}^n)}\int_{\mathbb{R}^n}(1+\eta^2)^s|\hat{u}(\eta)|^2\,d\eta\\&+\max\{1,2^{s-1}\}\|\hat{u}\|^2_{L^1(\mathbb{R}^n)}\int_{\mathbb{R}^n}(1+\eta^2)^s|\hat{v}(\eta)|^2\,d\eta\\&\le 2\max\{1,2^{s-1}\}c_s\int_{\mathbb{R}^n}(1+\eta^2)^s|\hat{u}(\eta)|^2\,d\eta\int_{\mathbb{R}^n}(1+\eta^2)^s|\hat{v}(\eta)|^2\,d\eta.\end{aligned}$$

This shows that $uv\in H^s(\mathbb{R}^n)$, with the desired estimates on the norm. □

Theorem II.11 (Rellich–Kondrashov Theorem) *Let $\Omega\subset\mathbb{R}^n$ be an open bounded domain. Then the embedding*

$$W_0^{s,p}(\Omega)\subset W^{t,q}(\mathbb{R}^n),\qquad p<q,\qquad \frac{s-t}{n}>\frac{1}{p}-\frac{1}{q},$$

is compact.

See e.g. [**Eva98**, Section 5.7] (for the particular case $s=1$, $t=0$).

Remark II.12 In the left-hand side, one either needs to consider $W_0^{s,p}(\Omega)$ and a bounded open neighborhood $\Omega\subset\mathbb{R}^n$, or $W^{s,p}(\Omega)$ and a bounded open neighborhood $\Omega\subset\mathbb{R}^n$ with smooth (sufficiently regular) boundary, see [**GT01**, **AF03**].

Lemma II.13 (Gagliardo–Nirenberg–Sobolev inequalities) *Let $n\in\mathbb{N}$, $n\ge 2$.*

(1) Let $1\le p<n$. There is $C_{n,p}>0$ such that

$$\|u\|_{L^{p^*}(\mathbb{R}^n)}\le C_{n,p}\|\nabla u\|_{L^p(\mathbb{R}^n)},\qquad p^*:=\frac{np}{n-p},\tag{II.6}$$

for any $u\in W^{1,p}(\mathbb{R}^n)$.

(2) Let $1\le r\le\infty$, $1\le q\le\infty$, $0<\alpha<1$, and let p_α be defined by

$$\frac{1}{p_\alpha}=\Big(\frac{1}{r}-\frac{1}{n}\Big)\alpha+\frac{1-\alpha}{q}.\tag{II.7}$$

Then there is $C_{n,q,r,\alpha}>0$ such that

$$\|u\|_{L^{p_\alpha}(\mathbb{R}^n)}\le C_{n,q,r,\alpha}\|\nabla u\|^{\alpha}_{L^r(\mathbb{R}^n)}\|u\|^{1-\alpha}_{L^q(\mathbb{R}^n)},\tag{II.8}$$

for any $u\in L^q(\mathbb{R}^n)\cap W^{1,r}(\mathbb{R}^n)$ if both q and r are finite; if $q=\infty$, then for any $u\in C(\mathbb{R}^n)\cap L^\infty(\mathbb{R}^n)\cap W^{1,r}(\mathbb{R}^n)$ with

$$\lim_{R\to\infty}\|\mathbb{1}_{|x|>R}\,u\|\|_{L^\infty(\mathbb{R}^n)}=0;$$

if $r=\infty$, then for any $u\in C^1(\mathbb{R}^n)\cap L^q(\mathbb{R}^n)\cap W^{1,\infty}(\mathbb{R}^n)$ with

$$\lim_{R\to\infty}\|(|u|+|\nabla u|)\mathbb{1}_{|x|>R}\|_{L^\infty(\mathbb{R}^n)}=0.$$

(3) Let $1\le r\le\infty$, $1\le q\le\infty$, $m\in\mathbb{N}$, $j\in\mathbb{N}_0$, $0\le j\le m$, $\frac{j}{m}\le\alpha<1$, and let p_α be defined by

$$\frac{1}{p_\alpha}=\frac{j}{n}+\Big(\frac{1}{r}-\frac{m}{n}\Big)\alpha+\frac{1-\alpha}{q}.\tag{II.9}$$

Then there is $C_{n,q,r,m,j,\alpha} > 0$ such that

$$\sum_{|\gamma|=j} \|\partial_x^\gamma u\|_{L^{p_\alpha}(\mathbb{R}^n)} \le C_{n,q,r,m,j,\alpha} \sum_{|\gamma|=m} \|\partial_x^\gamma u\|^\alpha_{L^r(\mathbb{R}^n)} \|u\|^{1-\alpha}_{L^q(\mathbb{R}^n)}, \tag{II.10}$$

for any $u \in L^q(\mathbb{R}^n) \cap W^{m,r}(\mathbb{R}^n)$ if q and r are both finite or both infinite; if $q = \infty$ and r is finite, then for any $u \in C(\mathbb{R}^n) \cap L^\infty(\mathbb{R}^n) \cap W^{m,r}(\mathbb{R}^n)$ with

$$\lim_{R\to\infty} \|\mathbb{1}_{|x|>R}\, u\|_{L^\infty(\mathbb{R}^n)} = 0;$$

if q is finite and $r = \infty$, then for any $u \in C^m(\mathbb{R}^n) \cap L^q(\mathbb{R}^n) \cap W^{m,\infty}(\mathbb{R}^n)$ with

$$\lim_{R\to\infty} \||\mathbb{1}_{|x|>R}\, \partial_x^\gamma u\|_{L^\infty(\mathbb{R}^n)} = 0 \qquad \forall \gamma \in \mathbb{N}_0^n, \quad |\gamma| \le m.$$

Similar inequalities were published by Ladyzhenskaya, Gagliardo, and Nirenberg during 1957–1959. For more general cases of the Gagliardo–Nirenberg inequalities (and also for the case $n = 1$), see e.g. [**Bre10, BM18**].

Proof. Let us give the proof for $u \in C^1_{\mathrm{comp}}(\mathbb{R}^n)$.

We first consider the case $p = 1$. From $|u(x)| \le \int_{-\infty}^{x^i} |\partial_i u(x^1, ..., y^i, ..., x^n)|\, dy^i$, with $\partial_i = \frac{\partial}{\partial x^i}$, we have:

$$|u(x)|^{\frac{n}{n-1}} \le \prod_{i=1}^n \Big(\int_{\mathbb{R}} |\partial_i u(x^1, ..., y^i, ..., x^n)|\, dy^i \Big)^{\frac{1}{n-1}}. \tag{II.11}$$

We integrate this inequality in x^1 and apply the Hölder inequality in the form

$$\int |I_1|^{\frac{1}{n-1}} \dots |I_{n-1}|^{\frac{1}{n-1}}\, dt \le \Big(\int |I_1|\, dt \dots \int |I_{n-1}|\, dt \Big)^{\frac{1}{n-1}}$$

to the terms from (II.11) with $2 \le i \le n$ (the terms which contain x^1):

$$\int_{\mathbb{R}} |u(x)|^{\frac{n}{n-1}}\, dx^1 \le \int_{\mathbb{R}} \prod_{i=1}^n \Big(\int_{\mathbb{R}} |\partial_i u(x^1, ..., y^i, ..., x^n)|\, dy^i \Big)^{\frac{1}{n-1}} dx^1$$
$$\le \Big(\int_{\mathbb{R}} |\partial_1 u(y^1, x^2, ...)|\, dy^1 \Big)^{\frac{1}{n-1}} \prod_{i=2}^n \Big(\int_{\mathbb{R}^2} |\partial_i u(x^1, ..., y^i, ..., x^n)|\, dx^1\, dy^i \Big)^{\frac{1}{n-1}}.$$

Then, integrating similarly in x^2, ..., x^n, we arrive at

$$\int_{\mathbb{R}^n} |u(x)|^{\frac{n}{n-1}}\, dx \le \prod_{i=1}^n \Big(\int_{\mathbb{R}^n} |\partial_i u(x^1, ..., y^i, ..., x^n)|\, dx^1 \dots dy^i \dots dx^n \Big)^{\frac{1}{n-1}},$$

which (after renaming the integration variables) takes the desired form

$$\int_{\mathbb{R}^n} |u(x)|^{\frac{n}{n-1}}\, dx \le \Big(\int_{\mathbb{R}^n} |\nabla u(x)|\, dx \Big)^{\frac{n}{n-1}}.$$

Let us deal with the case $p > 1$. Again, let $u \in C^1_{\mathrm{comp}}(\mathbb{R}^n)$. Assume that $1 < M < \infty$. Using (II.6) with $p = 1$ which we just proved and the Hölder inequality, we have:

$$\|u\|^M_{L^{\frac{Mn}{n-1}}} = \||u|^M\|_{L^{\frac{n}{n-1}}} \le M \||u|^{M-1} |\nabla u|\|_{L^1} \le M \|\nabla u\|_{L^p} \||u|^{M-1}\|_{L^{p'}}, \tag{II.12}$$

for any p, $p' \in [1, +\infty]$ satisfy $1/p + 1/p' = 1$.

Assume that $1 < p < n$. We choose M such that

$$\frac{Mn}{n-1} = (M-1)p', \qquad \frac{1}{1-n^{-1}} = \frac{1-M^{-1}}{1-p^{-1}}, \qquad M^{-1} = 1 - \frac{1-p^{-1}}{1-n^{-1}} = \frac{n/p-1}{n-1};$$

note that this consistent with our assumption on M. Let $p^* := Mn/(n-1) = np/(n-p)$; then we arrive at

$$\|u\|_{L^{p^*}}^{M} \le M\|\nabla u\|_{L^p}\|u\|_{L^{p^*}}^{M-1}. \tag{II.13}$$

Canceling $\|u\|_{L^{p^*}}^{M-1}$ (which is nonzero or else there is nothing to prove), we obtain

$$\|u\|_{L^{p^*}} \le M\|\nabla u\|_{L^p}. \tag{II.14}$$

for $u \in C^1_{\mathrm{comp}}(\mathbb{R}^n)$.

For the general case, we conclude the proof by the density argument since $C^1_{\mathrm{comp}}(\mathbb{R}^n)$ is dense in $W^{1,p}(\mathbb{R}^n)$. This proves *Part (1)*.

To prove *Part (2)*, we again assume that $u \in C^1_{\mathrm{comp}}(\mathbb{R}^n)$, that

$$r,\, r' \in [1,+\infty], \qquad \frac{1}{r}+\frac{1}{r'} = 1, \qquad 1 < M < \infty,$$

and write (II.12) in the following form:

$$\|u\|_{L^{\frac{Mn}{n-1}}}^{M} \le M\|\nabla u\|_{L^r}\||u|^{M-1}\|_{L^{r'}}.$$

With an arbitrary $A \in (1,M)$, the above inequality yields

$$\begin{aligned} \|u\|_{L^{\frac{Mn}{n-1}}}^{M} &\le M\|\nabla u\|_{L^r}\||u|^{A-1}|u|^{M-A}\|_{L^{r'}} \\ &\le M\|\nabla u\|_{L^r}\||u|^{A-1}\|_{L^Q}\||u|^{M-A}\|_{L^S} \\ &= M\|\nabla u\|_{L^r}\|u\|_{L^{Q(A-1)}}^{A-1}\|u\|_{L^{S(M-A)}}^{M-A}, \qquad 1 < A < M, \end{aligned} \tag{II.15}$$

where

$$r,\, Q,\, S \in [1,+\infty], \qquad \frac{1}{r}+\frac{1}{Q}+\frac{1}{S} = 1; \tag{II.16}$$

we choose S so that

$$\frac{Mn}{n-1} = S(M-A), \qquad S = \frac{Mn}{(M-A)(n-1)} \in (1,+\infty). \tag{II.17}$$

We divide (II.15) by $\|u\|_{L^{Mn/(n-1)}}^{M-A}$ (which is nonzero or else there is nothing to prove), arriving at

$$\|u\|_{L^{Mn/(n-1)}}^{A} \le A\|\nabla u\|_{L^r}\|u\|_{L^{Q(A-1)}}^{A-1}, \qquad 1 < A < M.$$

We denote $\alpha = 1/A \in (M^{-1},1)$, set

$$Q = \frac{q}{\alpha^{-1}-1}, \tag{II.18}$$

and arrive at

$$\|u\|_{L^{p_\alpha}} \le \alpha^{-\alpha}\|\nabla u\|_{L^r}^{\alpha}\|u\|_{L^q}^{1-\alpha},$$

with $p_\alpha = Mn/(n-1)$. Using in (II.16) the values of S and Q from (II.17) and (II.18), we have:

$$\frac{1}{r}+\frac{\alpha^{-1}-1}{q}+\frac{(M-\alpha^{-1})(n-1)}{Mn} = 1, \qquad \frac{1}{r}+\frac{\alpha^{-1}-1}{q}-\frac{1}{n}-\frac{\alpha^{-1}}{p_\alpha} = 0,$$

which is in agreement with (II.7). Since $\alpha \in (M^{-1},1)$ with M arbitrarily large, this proves (II.8) for compactly supported functions.

The extension to the general case is done as in the proof of *Part (1)*. We leave to the reader the proof of the density of the set of smooth compactly supported functions in the space $L^q(\mathbb{R}^n) \cap W^{1,r}(\mathbb{R}^n)$ endowed with its natural norm when both q and r are finite. If q is infinite while r is finite, then the closure of the set of smooth compactly supported functions in $L^\infty(\mathbb{R}^n) \cap W^{1,r}(\mathbb{R}^n)$ is

$$\{\, u \in C(\mathbb{R}^n) \cap L^\infty(\mathbb{R}^n) \cap W^{1,r}(\mathbb{R}^n) : \quad \|\mathbb{1}_{|x|>R}\, u\|_{L^\infty(\mathbb{R}^n)} = 0 \,\}.$$

Indeed, the above set contains the closure of the set of smooth compactly supported functions in the topology of $L^\infty(\mathbb{R}^n) \cap W^{1,r}(\mathbb{R}^n)$ since the uniform limit of a sequence of smooth compactly supported functions in the corresponding norm is continuous and vanishes at infinity; the opposite inclusion follows as usual by truncation and mollification. By a similar argument, if q is finite and r is infinite, then the closure of the set of smooth compactly supported functions in the topology of $L^q(\mathbb{R}^n) \cap W^{1,\infty}(\mathbb{R}^n)$ is

$$\{\, u \in C^1(\mathbb{R}^n) \cap L^q(\mathbb{R}^n) \cap W^{1,\infty}(\mathbb{R}^n) : \quad \lim_{R\to\infty} \|(|u| + |\nabla u|)\mathbb{1}_{|x|>R}\|_{L^\infty(\mathbb{R}^n)} = 0 \,\}.$$

Notice that both q and r cannot be infinite at the same time due to (II.7).

For *Part (3)*, one picks $u \in C^2_{\rm comp}(\mathbb{R}^n)$ (which for brevity is assumed to be real-valued), assumes that $p \ge 2$, and considers the identity

$$\|\partial_{x^i} u\|_{L^p}^p = \int_{\mathbb{R}^n} \partial_{x^i} u \cdot \partial_{x^i} u |\partial_{x^i} u|^{p-2}\, dx = -\int_{\mathbb{R}^n} u \partial_{x^i}\left(\partial_{x^i} u |\partial_{x^i} u|^{p-2}\right) dx$$

for a particular $1 \le i \le n$, which leads to

$$\|\partial_{x^i} u\|_{L^p}^p \le (p-1) \int_{\mathbb{R}^n} |u||\partial_{x^i}^2 u||\partial_{x^i} u|^{p-2}\, dx \le (p-1)\|u\|_{L^q} \|\partial_{x^i}^2 u\|_{L^r} \||\partial_{x^i} u|^{p-2}\|_{L^s}$$

$$\le (p-1)\|u\|_{L^q} \|\partial_{x^i}^2 u\|_{L^r} \||\partial_{x^i} u|^{p-2-M}\|_{L^a} \||\partial_{x^i} u|^M\|_{L^b},$$

for arbitrary $M \in [0, p-2]$. We leave the details to the reader. This proves *Part (3)* for $u \in C^1_{\rm comp}(\mathbb{R}^n)$. The extension to the general case is done as in the proof of *Part (2)*, except for the case $q = r = \infty$ which is immediate since in that case $\alpha = j/m$ and $p_\alpha = \infty$. □

II.3 Compactness of H_r^1 in L^q

The L^2-subspace of spherically symmetric functions is denoted by

$$L_r^2(\mathbb{R}^n) \subset L^2(\mathbb{R}^n).$$

Similarly for other spaces.

We will need the following compactness result due to Strauss **[Str77a]**.

Theorem II.14 *Assume $n \ge 2$. There are $C_n > 0$ and $\alpha_n > 0$ which depend only on n such that for any $u \in H_r^1(\mathbb{R}^n)$ one has:*

$$|u(x)| \le C_n |x|^{(1-n)/2} \|u\|_{H^1}, \quad x \in \mathbb{R}^n \setminus \mathbb{B}^n_{\alpha_n} \ (a.e.).$$

Moreover, the following inclusion map is compact:

$$H_r^1(\mathbb{R}^n) \hookrightarrow L^q(\mathbb{R}^n), \qquad 2 < q < 2^* := \frac{2n}{n-2}.$$

Here $H_r^1(\mathbb{R}^n)$ denotes the subspace of spherically symmetric functions in $H^1(\mathbb{R}^n)$.

PROOF. We start with the following lemma.

Lemma II.15 ([Str77a, Radial Lemma 1]) *Let $n \ge 2$. Every function $u \in H^1(\mathbb{R}^n)$ which is spherically symmetric is almost everywhere equal to a function $U(x)$, continuous for $x \ne 0$, such that $|U(x)| \le c|x|^{(1-n)/2}$ for $|x| \ge \epsilon > 0$, where c depends only on n and ϵ.*

PROOF. Let $\rho \in C^\infty(\mathbb{R}_+)$ be such that $0 \le \rho \le 1$, $\rho(r) = 0$ for $r \le 1/2$, $\rho(r) = 1$ for $r \ge 1$. We may assume that $0 \le \rho' \le 3$. Denote $\rho_\epsilon(r) = \rho(r/\epsilon)$ and consider $f(r) = \rho_\epsilon(r) r^{\frac{n-1}{2}} u(r)$. Let us show that $f \in H^1(\mathbb{R}_+)$. Indeed,

$$\begin{aligned}\int_{\mathbb{R}_+} ((f')^2 + f^2)\,dr &= \int_{\mathbb{R}_+} \Big(\big((\rho_\epsilon r^{\frac{n-1}{2}})' u + \rho_\epsilon r^{\frac{n-1}{2}} u' \big)^2 + (\rho_\epsilon r^{\frac{n-1}{2}} u)^2 \Big)\,dr \\ &\le C_{n,\epsilon} \int_{\mathbb{R}_+} (u'(r)^2 + u(r)^2) r^{n-1}\,dr = \frac{C_{n,\epsilon}}{\operatorname{vol}(\mathbb{S}^{n-1})} \|u\|^2_{H^1(\mathbb{R}^n)},\end{aligned}$$

where $C_{n,\epsilon} > 0$ depends only on n and ϵ. Since $f \in H^1(\mathbb{R}_+) \subset C_{\mathrm{b}}(\mathbb{R}_+)$, we have $|f(r)| \le C\|f\|_{H^1(\mathbb{R}_+)} \le C'\|u\|_{H^1(\mathbb{R}^n)}$ for all $r \ge 0$. Hence $u(r)$ is continuous and there is some $c > 0$ which depends only on n and ϵ such that $|u(r)| \le c r^{-\frac{n-1}{2}} \|u\|_{H^1(\mathbb{R}^n)}$ for $r \ge \epsilon$. □

Problem II.16 Prove that for any spherically symmetric function $u \in H^1_r(\mathbb{R}^n)$ and any $\alpha \in (0,1)$ and $R > 1$ there is $c > 0$ depending only on α, R, and $\|u\|_{H^1(\mathbb{R}^n)}$ such that

$$|u(x) - u(y)| \le c|x-y|^\alpha, \qquad x,\, y \in \mathbb{R}^n, \quad |x|,\ |y| \in [R^{-1}, R]. \tag{II.19}$$

Hint: Show that a function from $H^1_r(\mathbb{R}^n)$ considered as a function of $|x|$ belongs to $H^1([1/R, R])$.

Now we study the compactness of the embedding $H^1_r \hookrightarrow L^q$. Let $u_j \in H^1_r(\mathbb{R}^n)$ be a bounded sequence. Let us show that there is a subsequence $\{u_{j_k} \colon\ k \in \mathbb{N}\}$ such that $u_{j_k} \to v$ (strongly) in L^q for $2 < q < 2^* := \frac{2n}{n-2}$ as $k \to \infty$. By (II.19), u_j and v are Hölder functions of $r = |x|$ on any interval $[N^{-1}, N]$. By the Ascoli-Arzelà lemma, we can choose a subsequence of u_j, which we will also denote u_j, such that for any $N \in \mathbb{N}$ the sequence u_j converges uniformly in $|x| \in [N^{-1}, N]$ to v. By the Fatou lemma, we have $\|v\|_{H^1} \le \sup_j \|u_j\|_{H^1}$.

We claim that for any $\epsilon > 0$, there exists N large enough so that for any $u \in H^1_r(\mathbb{R}^n)$,

$$\|u\|_{L^q(|x| \le \frac{1}{N})} + \|u\|_{L^q(|x| \ge N)} < \epsilon \|u\|_{H^1_r(\mathbb{R}^n)}.$$

Indeed, for $r < 1/N$, we choose Q such that $q < Q < 2^*$ and, using the embedding $H^1(\mathbb{R}^n) \subset L^Q(\mathbb{R}^n)$, we get:

$$\begin{aligned}\|u\|^q_{L^q(|x| \le \frac{1}{N})} = \int_{|x| < \frac{1}{N}} |u(x)|^q\,dx &\le \Big(\int_{|x|<\frac{1}{N}} |u(x)|^{q\cdot\frac{Q}{q}}\,dx \Big)^{\frac{q}{Q}} \Big(\int_{|x|<\frac{1}{N}} dx \Big)^{1-\frac{q}{Q}} \\ &\le C_{n,q,Q} \|u\|^q_{H^1(\mathbb{R}^n)} (N^{-n})^{1-\frac{q}{Q}}.\end{aligned}$$

For large r, by Lemma II.15, we can use the estimate $|u(x)| \le c|x|^{\frac{1-n}{2}} \|u\|_{H^1_r(\mathbb{R}^n)}$, getting

$$\|u\|_{L^q(|x|>N)} \le \mathrm{const}\, \|u\|_{H^1_r(\mathbb{R}^n)} \int_{r>N} \frac{r^{n-1}\,dr}{\big(r^{\frac{n-1}{2}}\big)^q} \le \mathrm{const}\, \|u\|_{H^1_r(\mathbb{R}^n)} N^{1+(n-1)(1-\frac{q}{2})}.$$

Now we check that $\|u_j - v\|_{L^q(\mathbb{R}^n)} \to 0$. For any $\epsilon > 0$, we choose $N \in \mathbb{N}$ large enough so that

$$\|u_j - v\|_{L^q(|x| \le \frac{1}{N})} + \|u_j - v\|_{L^q(|x| \ge N)} < \frac{\epsilon}{2}.$$

Then, since u_j converge to v uniformly in $|x|$ on $[N^{-1}, N]$, there is $j_0 \in \mathbb{N}$ such that for any $j \ge j_0$

$$\|u_j - v\|_{L^q(\frac{1}{N} \le |x| \le N)} < \frac{\epsilon}{2}.$$

Therefore, $\|u_j - v\|_{L^q} < \epsilon$ for $j \ge j_0$. □

II.4 Cotlar–Stein almost orthogonality lemma

We will need the almost orthogonality principle of M. Cotlar [**Cot55**]. This result is classical; we present the proof here in order to remind the convergence in the weak, strong, and uniform operator topology.

Definition II.17 (Weak, strong, and uniform operator topology) Let $\mathbf{X}$, $\mathbf{Y}$ be Banach spaces, let $A : \mathbf{X} \to \mathbf{Y}$, and let there be a sequence $A_j : \mathbf{X} \to \mathbf{Y}$, $j \in \mathbb{N}$.

(1) The operators A_j converge to A in the *weak operator topology* if for any $u \in \mathbf{X}$ and $v \in \mathbf{Y}^*$ one has $\langle v, A_j u\rangle \to \langle v, Au\rangle$ as $j \to \infty$;
(2) The operators A_j converge to A in the *strong operator topology* if for any $u \in \mathbf{X}$ one has $A_j u \to Au$ in $\mathbf{Y}$ as $j \to \infty$;
(3) The operators A_j converge to A *in the uniform operator topology* if $\|A_j - A\| \to 0$ as $j \to \infty$.

An often situation is that one can decompose the operator T into an infinite sum of operators $T = \sum_{i\in\mathbb{N}} T_i$, which satisfy certain estimates, and the question is, under which assumptions on T_i one can deduce an adequate estimate on T.

Lemma II.18 (Cotlar–Stein almost orthogonality lemma) *Let* $\mathbf{H}_1$ *and* $\mathbf{H}_2$ *be Hilbert spaces. The operators*

$$T_i : \mathbf{H}_1 \to \mathbf{H}_2, \qquad i \in \mathbb{N},$$

are called almost orthogonal *if*

$$\text{(II.20)} \quad A := \sup_{i\in\mathbb{N}} \sum_{j\in\mathbb{N}} \|T_i T_j^*\|^{1/2}_{\mathbf{H}_2\to\mathbf{H}_2} < \infty, \qquad B := \sup_{i\in\mathbb{N}} \sum_{j\in\mathbb{N}} \|T_i^* T_j\|^{1/2}_{\mathbf{H}_1\to\mathbf{H}_1} < \infty.$$

In this case, the series $\sum_{i\in\mathbb{N}} T_i$ *converges in the strong operator topology (but not necessarily in the uniform operator topology) to the operator* T *bounded by*

$$\|T\| \le \sqrt{AB}.$$

Proof. The proof is split into two steps. In Lemma II.19 below, we prove that the norm of $\sum_{i\in I} T_i$ is uniformly bounded for any finite $I \subset \mathbb{N}$. An immediate consequence (Corollary II.20) is that the series $\sum_{i\in\mathbb{N}} T_i$ converges in the weak operator topology. Then, in Lemmata II.21 and II.22, we prove that the series $\sum_{i\in\mathbb{N}} T_i$ converges in the strong operator topology.

In the operator norms below, we will not indicate explicitly in which spaces the norm is considered.

Lemma II.19 *For any finite subset $I \subset \mathbb{N}$, the operator*

$$T_I := \sum_{i\in I} T_i : \ \mathbf{H}_1 \to \mathbf{H}_2$$

is bounded by $\|T_I\| \le \sqrt{AB}$, with A, B defined in (II.20).

PROOF. We follow the proof from the book of E.M. Stein [**Ste93**, Theorem VII.2.1]. For brevity, we denote

$$a_{ij} = \|T_i^* T_j\|, \qquad b_{ij} = \|T_i T_j^*\|. \tag{II.21}$$

For any $N \in \mathbb{N}$, we have:

$$\|(T_I^* T_I)^N\| \le \sum_{i_1\in I,\, j_1\in I,\, \dots\, i_N\in I,\, j_N\in I} \|T_{i_1}^* T_{j_1} \dots T_{i_N}^* T_{j_N}\|. \tag{II.22}$$

For each term $T_{i_1}^* T_{j_1} \dots T_{i_N}^* T_{j_N}$, we have two following bounds:

$$\|(T_{i_1}^* T_{j_1}) \dots (T_{i_N}^* T_{j_N})\| \le a_{i_1 j_1} \dots a_{i_N j_N}, \tag{II.23}$$

$$\|T_{i_1}^* (T_{j_1} T_{i_2}^*) \dots (T_{j_{N-1}} T_{i_N}^*) T_{j_N}\| \le b_{i_1 i_1}^{1/2} b_{j_1 i_2} \dots b_{j_{N-1} i_N} b_{j_N j_N}^{1/2}. \tag{II.24}$$

Taking the weighted geometric mean of (II.23) and (II.24) and noting that $b_{i_1 i_1}^{1/4} \le B^{1/2}$ and $b_{i_N i_N}^{1/4} \le B^{1/2}$ (see (II.20)), we bound $\|T_{i_1}^* T_{j_1} \dots T_{i_N}^* T_{j_N}\|$ by

$$\|T_{i_1}^* T_{j_1} \dots T_{i_N}^* T_{j_N}\| \le B^{1/2} a_{i_1 j_1}^{1/2} b_{j_1 i_2}^{1/2} \dots b_{j_{N-1} i_N}^{1/2} a_{i_N j_N}^{1/2} B^{1/2}.$$

We first sum up in i_1; according to (II.20), $\sup_{j_1} \sum_{i_1} a_{i_1 j_1}^{1/2} \le A$. Similarly, we sum up in $j_1, \dots, i_N$. Finally, summation in j_N yields the factor of $|I| < \infty$ (the number of elements in I), resulting in the following bound on $\|(T_I^* T_I)^N\|$:

$$\begin{aligned} \sum_{i_1\in I,\, j_1\in I,\, \dots\, i_N\in I,\, j_N\in I} \|T_{i_1}^* T_{j_1} \dots T_{i_N}^* T_{j_N}\| &\le B^{1/2} A^N B^{N-1} \sum_{j_N\in I} B^{1/2} \\ &= A^N B^N \sum_{j_N\in I} 1 = |I| A^N B^N. \end{aligned} \tag{II.25}$$

Now we assume that $N = 2^n$, for some $n \in \mathbb{N}$. Using (II.22) and (II.25), we get:

$$\|T_I\|^{2N} = \|T_I^* T_I\|^N = \|(T_I^* T_I)^2\|^{N/2} = \dots = \|(T_I^* T_I)^N\| \le |I| A^N B^N, \tag{II.26}$$

or $\|T_I\| \le |I|^{\frac{1}{2N}} A^{1/2} B^{1/2}$. Since $N = 2^n$ is arbitrarily large, $\|T_I\| \le A^{1/2} B^{1/2}$. □

Corollary II.20 *The series $\sum_{i\in\mathbb{N}} T_i$ converges in the weak operator topology.*

PROOF. To prove the convergence of $\sum_{i\in\mathbb{N}} T_i$ in the weak operator topology, we need to show that for any $u \in \mathbf{H}_1$, $v \in \mathbf{H}_2$, the series of complex numbers

$$\sum_{i\in\mathbb{N}} \langle v, T_i u\rangle \tag{II.27}$$

converges. If it were not the case, then for any $c > 0$ there would exist a finite subset $I \subset \mathbb{N}$ such that $|\sum_{i\in I} \langle v, T_i u\rangle| > c$. In particular, for some finite subset $I \subset \mathbb{N}$,

$$\Big|\sum_{i\in I} \langle v, T_i u\rangle\Big| > \sqrt{AB}\|u\|\|v\|.$$

On the other hand, by Lemma II.19, $\Big|\sum_{i\in I} \langle v, T_i u\rangle\Big| = |\langle v, T_I u\rangle| \le \sqrt{AB}\|u\|\|v\|$. This contradiction finishes the proof. □

In the next two lemmata we prove that, as the matter of fact, $\sum_{i\in\mathbb{N}} T_i$ converges not only in the weak operator topology, but also in the strong operator topology.

Lemma II.21 *Let $m \in \mathbb{N}$, and let there be finite non-intersecting subsets $I_\alpha \subset \mathbb{N}$, $1 \le \alpha \le m$. Then*

$$\Big\| \sum_{\alpha=1}^m T_{I_\alpha}^* T_{I_\alpha} \Big\| \le AB, \qquad \text{where} \quad T_{I_\alpha} = \sum_{i\in I_\alpha} T_i.$$

PROOF. We denote

$$\varDelta := \bigcup_{\alpha=1}^m I_\alpha \times I_\alpha \subset \mathbb{N}\times\mathbb{N},$$

so that $\sum_{\alpha=1}^m T_{I_\alpha}^* T_{I_\alpha} = \sum_{(i,j)\in\varDelta} T_i^* T_j$. Then we have:

$$\begin{aligned} \Big\| \Big(\sum_{\alpha=1}^m T_{I_\alpha}^* T_{I_\alpha} \Big)^N \Big\| &\le \sum_{(i_1,j_1)\in\varDelta,\dots(i_N,j_N)\in\varDelta} \|T_{i_1}^* T_{j_1} \dots T_{i_N}^* T_{j_N}\| \\ &\le \sum_{i_1\in I,\, j_1\in I,\dots i_N\in I,\, j_N\in I} \|T_{i_1}^* T_{j_1} \dots T_{i_N}^* T_{j_N}\| \le |I| A^N B^N, \end{aligned}$$

where $I = \bigcup_{\alpha=1}^m I_\alpha$; in the last inequality, we used (II.25). We note that $\sum_{\alpha=1}^m T_{I_\alpha}^* T_{I_\alpha}$ is self-adjoint. Therefore, as in (II.26), we assume that $N = 2^n$, $n \in \mathbb{N}$, and arrive at

$$\Big\| \sum_{\alpha=1}^m T_{I_\alpha}^* T_{I_\alpha} \Big\|^N = \Big\| \Big(\sum_{\alpha=1}^m T_{I_\alpha}^* T_{I_\alpha} \Big)^2 \Big\|^{N/2} = \dots = \Big\| \Big(\sum_{\alpha=1}^m T_{I_\alpha}^* T_{I_\alpha} \Big)^N \Big\| \le |I| A^N B^N.$$

Since N could be arbitrarily large, the conclusion follows. □

Lemma II.22 *The series $\sum_{i\in\mathbb{N}} T_i$ converges in the strong operator topology.*

PROOF. Pick a vector $u \in \mathbf{H}_1$ and denote $v_i = T_i u$, $i \in \mathbb{N}$. We need to show that the series $\sum_{i\in\mathbb{N}} v_i$ converges in $\mathbf{H}_2$. Let us assume that this is not the case. Then there would be $\epsilon > 0$ and infinitely many non-intersecting finite subsets $I_\alpha \in \mathbb{N}$ such that $\|\sum_{i\in I_\alpha} v_i\| \ge \epsilon$. Therefore, there would exist $m \in \mathbb{N}$ such that

$$\sum_{\alpha=1}^m \Big\| \sum_{i\in I_\alpha} v_i \Big\|^2 > AB\|u\|^2. \tag{II.28}$$

On the other hand,

$$\begin{aligned} \sum_{\alpha=1}^m \Big\| \sum_{i\in I_\alpha} v_i \Big\|^2 &= \sum_{\alpha=1}^m \Big\langle \sum_{i\in I_\alpha} T_i u, \sum_{j\in I_\alpha} T_j u \Big\rangle = \sum_{\alpha=1}^m \Big\langle T_{I_\alpha} u, T_{I_\alpha} u \Big\rangle \\ &= \Big\langle u, \sum_{\alpha=1}^m T_{I_\alpha}^* T_{I_\alpha} u \Big\rangle \le AB\|u\|^2, \end{aligned}$$

with the last inequality due to Lemma II.21. This contradicts (II.28), allowing us to conclude that for any $u \in \mathbf{H}_1$ the series $\sum_{i\in\mathbb{N}} T_i u$ converges. □

Remark II.23 The sum of almost orthogonal operators may not converge in the uniform operator topology. Indeed, we have

$$I_{l^2(\mathbb{N})} = \sum_{i\in\mathbb{N}} \boldsymbol{e}_i \otimes \boldsymbol{e}_i^*, \tag{II.29}$$

with $\boldsymbol{e}_i$ the standard basis in $l^2(\mathbb{N})$ and $\boldsymbol{e}_i \otimes \boldsymbol{e}_i^*$ being almost orthogonal (as the matter of fact, "orthogonal", in the sense that in (II.21) $a_{ij} = 0$ and $b_{ij} = 0$ if $i \neq j$). Yet, the series in (II.29) does not converge in the uniform operator topology since each of the summands is of norm one.

This concludes the proof of the Cotlar–Stein Lemma. □

Problem II.24 Use the Cotlar–Stein lemma to prove that the Fourier transform is bounded in $L^2(\mathbb{R})$.
Hint: Use the partition of unity

$$1 = \sum_{X\in\mathbb{Z}} \rho(x - X), \qquad \forall x \in \mathbb{R},$$

for appropriate real-valued $\rho \in C^\infty_{\mathrm{comp}}([-1, 1])$, *to define the operators*

$$(\mathcal{F}_{\Xi X})v(\xi) = \int_{\mathbb{R}} \rho(\xi - \Xi)e^{-\mathrm{i}\xi x}\rho(x - X)v(x)\,dx, \qquad X,\, \Xi \in \mathbb{Z}.$$

Prove that for each $X,\ \Xi \in \mathbb{Z}$, *the operator* $\mathcal{F}_{\Xi X}$ *is bounded in* $L^2(\mathbb{R})$. *Prove that the operators* $\mathcal{F}_{\Xi X}$ *and* $\mathcal{F}_{\Xi' X'}$ *are almost orthogonal in the sense of Lemma* II.18 *(the bounds* (II.20) *are satisfied). In particular, note that* $\mathcal{F}^*_{\Xi' X'}\mathcal{F}_{\Xi X} = 0$ *if* $|\Xi - \Xi'| \geq 2$, *while for* $|X - X'| \geq 3$ *one can integrate by parts in the expression for the integral kernel of* $\mathcal{F}^*_{\Xi' X'}\mathcal{F}_{\Xi X}$,

$$K(\mathcal{F}^*_{\Xi' X'}\mathcal{F}_{\Xi X})(x, y) = \int_{\mathbb{R}} e^{\mathrm{i}\xi(x-y)}\rho(\xi - \Xi)\rho(\xi - \Xi')\rho(x - X)\rho(y - X')\,d\xi,$$

obtaining the bound $\|\mathcal{F}^*_{\Xi' X'}\mathcal{F}_{\Xi X}\| \leq C_N|X - X'|^{-N}$, *with some* $C_N > 0$, *for any desired value of* N, *and then the bounds* (II.20) *follow. (What is the smallest value of* $N \in \mathbb{N}$ *so that the almost orthogonality conditions* (II.20) *are satisfied?)*

Problem II.25 Show that the solution to Problem II.24 can be applied without any changes to prove that the pseudodifferential operator

$$Tu(x) = \int_{\mathbb{R}} e^{\mathrm{i}\xi x}a(x, \xi)\hat{u}(\xi)\,\frac{d\xi}{2\pi}, \qquad u \in L^2(\mathbb{R}), \tag{II.30}$$

with the *symbol* $a(x, \xi) \in C^N_{\mathrm{b}}(\mathbb{R} \times \mathbb{R})$, is bounded in $L^2(\mathbb{R})$ if $N \in \mathbb{N}$ is sufficiently large. The space $C^N_{\mathrm{b}}(\mathbb{R} \times \mathbb{R})$ is defined by

$$\{f \in C^N(\mathbb{R} \times \mathbb{R}):\ \ \forall (\alpha, \beta) \in \mathbb{N}_0 \times \mathbb{N}_0,\ \alpha + \beta \leq N,\ \ \sup_{(x,\xi)\in\mathbb{R}\times\mathbb{R}} |\partial_x^\alpha \partial_\xi^\beta a(x, \xi)| < \infty\}.$$

Problem II.26 Formulate and prove the generalization of Problem II.25 for operators in $L^2(\mathbb{R}^n)$, $n \in \mathbb{N}$.

II.5 The Pólya–Szegő inequality

Definition II.27 (Schwarz symmetrization) For a measurable function u in $\mathbb{R}^n$, its Schwarz symmetrization u^* is defined by

$$u^*(x) = \alpha, \qquad \text{where} \quad \alpha = \inf \left\{ a \ge 0, \quad \left|\mathbb{B}^n_{|x|}\right| \ge \left|\{y\colon \quad |u(y)| \ge a\}\right| \right\}.$$

The definition implies that $u^*(x)$ is measurable, spherically symmetric, nonnegative, and nonincreasing in $|x|$.

Lemma II.28 (The Pólya–Szegő inequality) *For any $p \ge 1$ and any $u \in W^{1,p}(\mathbb{R}^n)$, its Schwarz symmetrization u^* satisfies*

$$\int_{\mathbb{R}^n} |\nabla u^*|^p \, dx \le \int_{\mathbb{R}^n} |\nabla u|^p \, dx.$$

For a proof, see e.g. [**BS00**].

II.6 The Paley–Wiener theorem

Theorem II.29 (Paley–Wiener) *(1) Let $\varphi \in \mathscr{S}(\mathbb{R}^n)$. If $\operatorname{supp} \varphi \subset \overline{\mathbb{B}^n_R}$, with some $R > 0$, then $\hat\varphi(\xi)$ is an entire function of $\xi \in \mathbb{C}^n$ (analytic function in the whole space $\mathbb{C}^n$) and for any $N \in \mathbb{N}$ there is $C_N > 0$ such that*

$$|\hat\varphi(\xi)| \le C_N \langle \xi \rangle^{-N} e^{R|\operatorname{Im}\xi|}. \tag{II.31}$$

(2) Conversely, if $\hat\varphi \in \mathscr{S}(\mathbb{R}^n)$ has a holomorphic extension to $\mathbb{C}^n$ (also denoted $\hat\varphi$) which satisfies (II.31) with some $R > 0$, for any $N \in \mathbb{N}$, then $\varphi \in C^\infty(\mathbb{R}^n)$, $\operatorname{supp} \varphi \subset \overline{\mathbb{B}^n_R}$.

Above, $\operatorname{Im}\xi = (\operatorname{Im}\xi_1, \dots, \operatorname{Im}\xi_n) \in \mathbb{R}^n$.

PROOF. To prove the first part, one needs to integrate by parts in x in the integral

$$\hat\varphi(\xi) = \int_{\mathbb{R}^n} e^{-\mathrm{i}x\cdot\xi} \varphi(x) \, dx.$$

For the second part, we pick $x \ne 0$ and define $\omega = x/|x|$. Then, due to analyticity of $\hat\varphi$,

$$\varphi(x) = \int_{\mathbb{R}^n} \hat\varphi(\xi) e^{\mathrm{i}\xi\cdot x} \frac{d\xi}{(2\pi)^n} = \int_{\mathbb{R}^n} \hat\varphi(\xi + \mathrm{i}\tau\omega) e^{\mathrm{i}(\xi+\mathrm{i}\tau\omega)\cdot x} \frac{d\xi}{(2\pi)^n},$$

$$|\varphi(x)| \le C_N \int_{\mathbb{R}^n} \langle \xi \rangle^{-N} e^{R\tau} e^{-\tau|x|} \, d\xi.$$

Taking $N = n + 1$ and sending τ to $+\infty$, we see that for $|x| > R$ the integral is arbitrarily small, hence $\varphi(x) = 0$ for $|x| > R$. □

Problem II.30 Let $u \in H^2(\mathbb{R})$ satisfy

$$(-\Delta - \lambda^2) u(x) = f(x) u(x),$$

where $\lambda > 0$ and $f \in C(\mathbb{R})$ (not necessarily real-valued), and there is $\epsilon > 0$ such that $|f(x)| < e^{-\epsilon|x|}$ for all $x \in \mathbb{R}$. Prove that for any $N > 0$ there is $C_N > 0$ such that $|u(x)| < C_N e^{-N|x|}$, $x \in \mathbb{R}$.

Hint: $|f(x)u(x)| \le ce^{-\epsilon|x|}$, hence $\widehat{fu}(\xi)$ is analytic for $|\operatorname{Im}\xi| < \epsilon$.

Hint: So is $\hat u(\xi) = \frac{\widehat{fu}(\xi)}{\xi^2 - \lambda^2}$ since $\widehat{fu}(\pm\lambda) = 0$, or else $\hat u = \frac{\widehat{fu}(\xi)}{\xi^2-\lambda^2} \notin L^2(\mathbb{R})$.

Hint: By the Paley–Wiener theorem, $|u(x)| \le ce^{-\epsilon|x|}$; then $|f(x)u(x)| \le ce^{-2\epsilon|x|}$.

CHAPTER III

Spectral theory of nonselfadjoint operators

Here we provide the basic information on the spectral theory, presenting the approach for closed linear operators in Banach spaces. Our main references on the spectral theory of linear operators in Banach spaces are Gohberg and Krein [**GK57**, **GK69**], Goldberg [**Gol66**], and Edmunds and Evans [**EE87**]. Other standard references are Dunford and Schwartz [**DS58**, **DS63**], Kato [**Kat76**], Yosida [**Yos80**], Reed and Simon [**RS80**], and Davies [**Dav95**].

After standard definitions and examples (closed operators in Section III.1, the adjoint operator in Section III.2, the spectrum of an operator in Section III.3, and the Fredholm theory in Section III.4), we give a detailed description of normal eigenvalues (Section III.5). The Hilbert space case (symmetric, normal, and selfadjoint operators) is briefly mentioned in Section III.6. Then, in Section III.7, we discuss different subsets of the essential spectrum and give the proof of the Weyl theorem on the stability of the essential spectrum under relatively compact perturbations (nonselfadjoint case, Banach spaces). The Schur complement is considered in Section III.8. The theory of characteristic roots developed by M. Keldysh is briefly covered in Section III.9. Some Quantum Mechanics examples are in Section III.10. The spectrum of the Dirac operator is considered in Section III.11.

III.1 Basic theory of unbounded operators

Below, $\mathbf{X}$ and $\mathbf{Y}$ are complex Banach spaces.[1]

III.1.1 Closed operators. We will say that A is a *linear operator* (or, for brevity, an *operator*) from $\mathbf{X}$ to $\mathbf{Y}$,

$$A:\ \mathbf{X}\to\mathbf{Y},$$

if there is a subspace $\mathfrak{D}(A)\subset\mathbf{X}$, called the *domain* of A, such that A is a $\mathbb{C}$-linear map from $\mathfrak{D}(A)$ to $\mathbf{Y}$. We equip $\mathfrak{D}(A)$ with the *graph norm*

$$\|x\|_A=\|x\|+\|Ax\|,\qquad x\in\mathfrak{D}(A). \tag{III.1}$$

The set of values of A, called the *range of a linear operator*, is denoted by $\mathfrak{R}(A)\subset\mathbf{Y}$. Its dimension is called the *rank of a linear operator*:

$$\operatorname{rank}A=\dim\mathfrak{R}(A).$$

We say that a linear operator $A:\mathbf{X}\to\mathbf{Y}$ is *bounded* if $\mathfrak{D}(A)=\mathbf{X}$ and

$$\|A\|:=\sup_{x\in\mathbf{X},\,x\neq0}\frac{\|Ax\|_{\mathbf{Y}}}{\|x\|_{\mathbf{X}}}<\infty.$$

The set of bounded linear operators $\mathbf{X}\to\mathbf{Y}$ is a Banach space denoted by $\mathscr{B}(\mathbf{X},\mathbf{Y})$.

[1]Unless stated otherwise, all Banach spaces and Hilbert spaces are considered over $\mathbb{C}$.

We say that a linear operator $K : \mathbf{X} \to \mathbf{Y}$ is *compact* if $\mathfrak{D}(K) = \mathbf{X}$ and if it maps any bounded set of $\mathbf{X}$ to a *precompact* set of $\mathbf{Y}$ (a set with a compact closure). That is, if the image of any bounded sequence contains a Cauchy subsequence. The Banach space of compact linear operators $\mathbf{X} \to \mathbf{Y}$ is denoted by $\mathscr{B}_0(\mathbf{X}, \mathbf{Y})$.

Problem III.1 Show that $\mathscr{B}_0(\mathbf{X}, \mathbf{Y})$ is closed in $\mathscr{B}(\mathbf{X}, \mathbf{Y})$ in the uniform operator topology.

Problem III.2 Show that any operator with finite-dimensional range is compact.

We say that an operator A is *closed* if its graph

$$\mathcal{G}(A) := \{ (x, Ax) \in \mathbf{X} \times \mathbf{Y},\ x \in \mathfrak{D}(A) \} \tag{III.2}$$

is closed in $\mathbf{X} \times \mathbf{Y}$. That is, from $x_j \to x \in \mathbf{X}$ and $Ax_j \to y \in \mathbf{Y}$ it follows that $x \in \mathfrak{D}(A)$ and $Ax = y$. The set of closed linear operators $\mathbf{X} \to \mathbf{Y}$ is denoted by $\mathscr{C}(\mathbf{X}, \mathbf{Y})$.

Problem III.3 Construct a linear operator of rank one which is not closed.

Problem III.4 Construct a linear operator with zero kernel and dense range which is not closed.

Lemma III.5 *A linear operator $A : \mathbf{X} \to \mathbf{Y}$ is closed if and only if its domain $\mathfrak{D}(A)$ equipped with the graph norm* (III.1) *is a Banach space.*

Problem III.6 Prove Lemma III.5.

Given two linear operators $A : \mathbf{X} \to \mathbf{Y}$ and $B : \mathbf{X} \to \mathbf{Y}$ such that $\mathfrak{D}(A) \subset \mathfrak{D}(B)$ and $B|_{\mathfrak{D}(A)} = A$, we say that B is an extension of A and write $A \subset B$. We say that an operator $A : \mathbf{X} \to \mathbf{Y}$ is *closable* if it has a closed extension $\hat{A} : \mathbf{X} \to \mathbf{Y}$; that is, $\mathfrak{D}(A) \subset \mathfrak{D}(\hat{A})$, $\hat{A}|_{\mathfrak{D}(A)} = A$, and $\hat{A} \in \mathscr{C}(\mathbf{X}, \mathbf{Y})$. For a closable operator there is a minimal closed extension, which is called the *closure* of A and denoted $\overline{A}$; it could be characterized by its graph, which is the closure of the graph of A in $\mathbf{X} \times \mathbf{Y}$:

$$\mathcal{G}\left(\overline{A}\right) = \overline{\mathcal{G}(A)} \subset \mathbf{X} \times \mathbf{Y}.$$

Problem III.7 Show that a linear operator $A : \mathbf{X} \to \mathbf{Y}$ with domain $\mathfrak{D}(A)$ is closable if and only if $x_j \to 0$, $x_j \in \mathfrak{D}(A)$, and $Ax_j \to y$ imply that $y = 0$.

Problem III.8 Give an example of a linear operator $A : \mathbf{X} \to \mathbf{X}$ which is not closable but has closed range.

Problem III.9 Show that the closure of a subspace of $\mathscr{C}(\mathbf{X}, \mathbf{Y})$ consisting of operators with finite-dimensional range in the uniform operator topology is a subspace of $\mathscr{B}_0(\mathbf{X}, \mathbf{Y})$.

The equality of $\mathscr{B}_0(\mathbf{X}, \mathbf{Y})$ and the closure of operators with finite-dimensional range in the uniform operator topology is not necessarily true for general Banach spaces.

Problem III.10 Let $\mathbf{X}$ be a Banach space, $\mathbf{H}$ be a Hilbert space (each of them not necessarily separable), and let $K \in \mathscr{B}_0(\mathbf{X}, \mathbf{H})$.

(1) Show that the range of K is separable. *Hint: For any $n \in \mathbb{N}$ there is $k \in \mathbb{N}$ and a sequence $\left(x_i^{(n)}\right)_{1 \le i \le k}$ in $\mathbf{H}$ such that*

$$\mathfrak{R}(K|_{\mathbb{B}_1(\mathbf{X})}) \subset \cup_{1 \le i \le k} \mathbb{B}_{1/n}(x_i^{(n)}).$$

(2) Prove that the closure of finite rank operators from $\mathscr{B}(\mathbf{X}, \mathbf{H})$ in the uniform operator topology is $\mathscr{B}_0(\mathbf{X}, \mathbf{H})$.

Remark III.11 In general, if a Banach space $\mathbf{X}$ is such that any compact operator from $\mathscr{B}_0(\mathbf{X}, \mathbf{X})$ could be approximated with any precision by finite rank operators in the uniform operator topology, one says that $\mathbf{X}$ has a *compact approximation property.* If a Banach space $\mathbf{X}$ is such that the identity could be approximated with any precision by finite rank operators in the strong operator topology, one says that $\mathbf{X}$ has a *bounded approximation property.* There are examples of reflexive separable Banach spaces for which the compact approximation property and the bounded approximation property do not hold [**Enf73**].

For a linear operator $A : \mathbf{X} \to \mathbf{X}$, we say that A acts *in* $\mathbf{X}$. We denote

$$\mathscr{B}(\mathbf{X}) = \mathscr{B}(\mathbf{X}, \mathbf{X}), \qquad \mathscr{C}(\mathbf{X}) = \mathscr{C}(\mathbf{X}, \mathbf{X}), \qquad \mathscr{B}_0(\mathbf{X}) = \mathscr{B}_0(\mathbf{X}, \mathbf{X}).$$

The *kernel* (or the *null space*) of A is

$$\mathbf{ker}(A) = \big\{ u \in \mathfrak{D}(A): \quad Au = 0 \big\}.$$

Problem III.12 If $A : \mathbf{X} \to \mathbf{Y}$ is closed, then the kernel of A is a closed vector subspace.

The *inverse* A^{-1} of a linear operator $A : \mathbf{X} \to \mathbf{Y}$ with domain $\mathfrak{D}(A)$ is defined if and only if A is one to one, which is the case if and only if $Ax = 0$ implies that $x = 0$. A^{-1} is by definition the operator from $\mathbf{Y}$ to $\mathbf{X}$ that sends Ax to x for each $x \in \mathfrak{D}(A)$. Thus,

$$\mathfrak{D}(A^{-1}) = \mathfrak{R}(A), \qquad \mathfrak{R}(A^{-1}) = \mathfrak{D}(A);$$
$$A^{-1}(Ax) = x \quad \forall x \in \mathfrak{D}(A), \qquad A(A^{-1}y) = y \quad \forall y \in \mathfrak{R}(A).$$

We will say that the linear operator $A : \mathbf{X} \to \mathbf{Y}$ with domain $\mathfrak{D}(A)$ is *invertible* if A^{-1} exists.

Problem III.13 If $A : \mathbf{X} \to \mathbf{Y}$ is invertible, then it is closed if and only if so is A^{-1}.
Hint: Notice that the graphs of A and A^{-1} are related as follows:

$$\mathcal{G}(A^{-1}) = \mathcal{G}(A)^T \subset \mathbf{Y} \times \mathbf{X},$$

where the mapping $(\cdot, \cdot)^T : \mathbf{X} \times \mathbf{Y} \to \mathbf{Y} \times \mathbf{X}$ is defined by $(x, y)^T = (y, x)$.

Problem III.14 Prove that a bounded linear operator $A : \mathbf{X} \to \mathbf{Y}$ is closed.

Theorem III.15 (Closed Graph Theorem [Kat76, Theorem III.5.20**])** *If $A \in \mathscr{C}(\mathbf{X}, \mathbf{Y})$ and $\mathfrak{D}(A) = \mathbf{X}$, then A is bounded.*

We say that a linear operator $A : \mathbf{X} \to \mathbf{Y}$ is *bounded from below* if the following is satisfied:

$$\exists c > 0 \quad \text{so that} \quad \|Ax\| \ge c\|x\| \qquad \forall x \in \mathfrak{D}(A). \tag{III.3}$$

Problem III.16 Let $A : \mathbf{X} \to \mathbf{Y}$ be a linear operator with domain $\mathfrak{D}(A)$ and assume that A is bounded from below (see (III.3)). Prove that $\mathfrak{R}(A)$ is closed if and only if A is closed.
Hint: The bound from below, $\|Ax\| \ge c\|x\|$, implies that if $\big(Ax_j\big)_{j\in\mathbb{N}}$ is a Cauchy sequence, then so is $\big(x_j\big)_{j\in\mathbb{N}}$.

III.1.2 Generalized eigenvectors and root lineal. Let $\mathbf{X}$ be a Banach space and let $A : \mathbf{X} \to \mathbf{X}$ be a linear operator with domain $\mathfrak{D}(A)$. Let $\lambda \in \sigma_{\rm p}(A)$. Nonzero vectors from $\mathbf{ker}(A - \lambda I)$ are called *eigenvectors* of A corresponding to the eigenvalue λ. The dimension of $\mathbf{ker}(A - \lambda I)$ is called the *geometric multiplicity* of the eigenvalue λ.

The *root lineal*[2] of A corresponding to $\lambda \in \sigma_{\rm p}(A)$ is defined by

$$\mathfrak{L}_\lambda(A) = \big\{\, x \in \mathfrak{D}(A): \quad \exists k \in \mathbb{N},\ (A - \lambda I)^j x \in \mathfrak{D}(A)\ \forall j < k, \quad (A - \lambda I)^k x = 0 \,\big\}.$$

We denote

$$\mathfrak{L}(A) = \mathfrak{L}_0(A).$$

Nonzero vectors from $\mathfrak{L}_\lambda(A)$ are called *generalized eigenvectors* (*root vectors*) of A corresponding to the eigenvalue λ. The *algebraic multiplicity* of the eigenvalue λ is defined by

$$\nu_\lambda = \dim \mathfrak{L}_\lambda(A).$$

If there is the smallest value $k \in \mathbb{N}$ such that $(A - \lambda I)^j x = 0$ for all $j \geq k$ and for all $x \in \mathfrak{L}_\lambda$, and if $k \geq 2$, then we will say that λ corresponds to Jordan blocks of finite size (at most k).

If the root lineal $\mathfrak{L}_\lambda(A)$ is closed (for example, if the algebraic multiplicity ν_λ is finite), then it is called the *root subspace* (*generalized eigenspace*) of A corresponding to the eigenvalue $\lambda \in \mathbb{C}$.

If $\mathfrak{L}(A)$ is closed, it is called the *generalized null space* of A. While the null space of a closed linear operator is closed (see Problem III.12), it is not necessarily the case for the generalized null space:

Problem III.17 Construct a bounded linear operator A with the root lineal $\mathfrak{L}(A)$ which is not closed.
Hint: Construct A such that $Ax_1 = 0$, $Ax_{j+1} = x_j$ for $j \geq 2$, $x_j \to y \neq 0$, $Ay = y$.

III.1.3 Relatively bounded and relatively compact operators. The operator B is called *relatively bounded* with respect to A (or simply A-bounded) if $\mathfrak{D}(A) \subset \mathfrak{D}(B)$ and there are constants $a \geq 0$ and $b \geq 0$ such that

$$\|Bx\| \leq a\|x\| + b\|Ax\| \qquad \forall x \in \mathfrak{D}(A). \tag{III.4}$$

The infimum b_0 of possible values of b in (III.4) is called A-bound on B. Further, the operator B is called *A-compact* if $\mathfrak{R}\big(B|_{\overline{\mathbb{B}_1(\mathfrak{D}(A))}}\big)$ is compact, where

$$\overline{\mathbb{B}_1(\mathfrak{D}(A))} = \big\{\, x \in \mathfrak{D}(A): \quad \|x\|^2 + \|Ax\|^2 \leq 1 \,\big\}.$$

Problem III.18 Assume that the resolvent set $\rho(A)$ is nonempty. Prove that B is A-compact if and only if $B(A - zI)^{-1}$ is compact for any (and hence for all) $z \in \rho(A)$.

The following problem shows that the A-bound is not necessarily attained:

Problem III.19 Prove that there exist $a, b \geq 0$ such that for any $u \in H^2(\mathbb{R})$ one has

$$\|\partial_x u\|_{L^2(\mathbb{R})} \leq a\|u\|_{L^2(\mathbb{R})} + b\|\partial_x^2 u\|_{L^2(\mathbb{R})}.$$

Show that the infimum over all possible values of b equals 0, although the corresponding inequality with $b = 0$ is false.

---

[2]We recall that a *lineal* is a linear manifold which is not necessarily closed.

The sum of a closed operator A and an A-compact operator B is not guaranteed to be closed. Here is a sufficient condition for this to be the case.

Theorem III.20 (**[Kat58**, Theorem 2a]) *Let $A : \mathbf{X} \to \mathbf{Y}$ be a closed linear operator with finite-dimensional kernel and closed range. If the linear operator $B : \mathbf{X} \to \mathbf{Y}$ is closed and A-compact, then $A + B$ is closed with finite-dimensional kernel and closed range.*

Theorem III.21 (**[Kat76**, Theorem IV.1.1]) *Let $A : \mathbf{X} \to \mathbf{Y}$ be a linear operator. If B is A-bounded with A-bound $b \in [0, 1)$, then it is also $(A + B)$-bounded with relative $(A + B)$-bound not larger than $b/(1 - b)$. Then $A + B$ is closable if and only if A is closable; in this case the closures of A and $A + B$ have the same domain. Moreover, if A is closed, then so is $A + B$.*

PROOF. Let B be A-bounded with a relative bound $b \in [0, 1)$, so that there is $a \geq 0$ such that $\|Bx\| \leq a\|x\| + b\|Ax\|$, for any $x \in \mathfrak{D}(A)$. Then

$$\|Bx\| \leq a\|x\| + b\|(A + B)x\| + b\|Bx\|, \qquad \forall x \in \mathfrak{D}(A),$$

which yields the desired relation $(1-b)\|Bx\| \leq a\|x\| + b\|(A+B)x\|$. It follows that the graph norm $\|x\|_{A+B} = \|x\| + \|(A + B)x\|$ on $\mathfrak{D}(A)$ is equivalent to the graph norm $\|x\|_A = \|x\| + \|Ax\|$.

Since $\mathfrak{D}(B) \supset \mathfrak{D}(A)$, the domain of $A + B$ coincides with $\mathfrak{D}(A)$. If A is closed, then, by Lemma III.5, $\mathfrak{D}(A)$ with the graph norm $\|\cdot\|_A$ is a Banach space. Since the graph norms of $A + B$ and A are equivalent, it follows that $\mathfrak{D}(A + B)$ with the graph norm $\|x\|_{A+B}$ is a Banach space and hence $A + B$ is closed (again by Lemma III.5).

The case when A (or $A + B$) is closable is left to the reader. □

Theorem III.22 (**[Kat76**, Theorem IV.1.11]) *Let A and B be linear operators from $\mathbf{X}$ to $\mathbf{Y}$, $\mathfrak{D}(B) \supset \mathfrak{D}(A)$, and let B be A-compact. If A is closable, $A + B$ is also closable, the closures of A and $A + B$ have the same domain, and B is $(A + B)$-compact. In particular, if A is closed, then so is $A + B$.*

PROOF. Let us show that B is $(A + B)$-compact. Assume that x_j and $(A + B)x_j$, $j \in \mathbb{N}$ are uniformly bounded; we need to show that Bx_j contains a Cauchy sequence. Assume that the sequence Ax_j is not bounded; then we may assume (going to a subsequence) that $Ax_j \neq 0$, $\|Ax_j\| \to \infty$, and consider $y_j = x_j/\|Ax_j\|$. Then $y_j \to 0$, $(A + B)y_j \to 0$, and Ay_j are uniformly bounded. Since B is A-compact, we may assume that By_j converges to some $w \in \mathbf{Y}$. Then $Ay_j = (A + B)y_j - By_j \to 0 - w = -w$, and due to A being closed and $y_j \to 0$ one has $w = 0$; on the other hand, one has $Ay_j \to -w$, $\|Ay_j\| = 1$. This contradiction shows that the sequence Ax_j is uniformly bounded; hence Bx_j contains a Cauchy sequence due to B being A-compact.

Let us show that if A is closable, then so is $A+B$. If $x_j \to 0$ and $(A+B)x_j \to v \in \mathbf{Y}$, then we may assume that Ax_j is bounded (repeating the above argument); then, passing to a subsequence, we may assume that Bx_j converges to some $w \in \mathbf{Y}$. Then

$$Ax_j = (A+B)x_j - Bx_j \to v - w.$$

Since A is closable while $x_j \to 0$, $Ax_j \to v - w = 0$. From $x_j \to 0$ and $Ax_j \to 0$ we conclude that $Bx_j \to 0$; then $(A+B)x_j \to 0$, so $A+B$ is closable.

If A is closed, then so is $A+B$. Indeed, let $x_j \to x \in \mathfrak{D}(A)$ and $(A+B)x_j \to v \in \mathbf{Y}$. We only need to notice that $x \in \mathfrak{D}(A) = \mathfrak{D}(A) \cap \mathfrak{D}(B) = \mathfrak{D}(A+B)$; since $A+B$ is closable, one has $(A+B)x_j \to (A+B)x$, showing that $(A+B)x = v$.

We leave as an exercise the statement that the closures of A and $A+B$ have the same domain. □

Let A and B be linear operators and assume that B is A-compact. By Theorem III.22, if A is closed, then so is $A+B$; the converse is not true, as the following problem shows.

Problem III.23 Construct A and B such that B is A-compact, $A+B$ is closed, but A is not closed.
Hint: One can take A to be an unbounded rank one operator and $B = -A$.

According to the folklore, *if the operator B is A-compact, then it is also relatively bounded with A-bound zero*, but [**BH00**, Example 1] shows that this is not necessarily true in the case of non-reflexive Banach spaces. The following theorem addresses this issue.

Theorem III.24 ([KO90, Theorem 1.1], [BH00, Theorem 2]) *Let $A : \mathbf{X} \to \mathbf{Y}$ be an operator with dense domain $\mathfrak{D}(A)$. If B is A-compact, then B has A-bound zero if one of the following statements is true:*
(1) A is closed and $\mathbf{X}$, $\mathbf{Y}$ are reflexive;
(2) A is closed and $\mathfrak{D}(A)$ is reflexive;
(3) B is closable.

By [**BH00**], *(1)* implies *(2)* since the graph $(x, Ax) \in \mathbf{X} \times \mathbf{Y}$, being a closed subspace of a reflexive space, is reflexive.

Related results are in [**Hes69, Hes70, Sch86**], and [**EE87**, Corollary III.7.7].

Problem III.25 Let $A \in \mathscr{C}(\mathbf{X}, \mathbf{Y})$. Show that if B is A-bounded with A-bound zero, it is not necessarily A-compact.

III.2 Adjoint operators

Let $\mathbf{X}$, $\mathbf{Y}$ be Banach spaces. We will denote the pairing of a Banach space $\mathbf{X}$ and its dual $\mathbf{X}^*$ by

$$\langle\,\cdot\,,\,\cdot\,\rangle_{\mathbf{X}} : \mathbf{X}^* \times \mathbf{X} \to \mathbb{C},$$

and similarly for $\mathbf{Y}$. Whenever possible, in the notation of the pairing we will omit the subscript which refers to the space. Let $A : \mathbf{X} \to \mathbf{Y}$ be a linear operator which is not necessarily closed, with domain $\mathfrak{D}(A)$ which is not necessarily dense. Denote

$$\mathfrak{D}(A^*) = \left\{ \eta \in \mathbf{Y}^* : \ \exists c \ge 0 \text{ such that } |\langle \eta, Ax \rangle_{\mathbf{Y}}| \le c\|x\|_{\mathbf{X}} \quad \forall x \in \mathfrak{D}(A) \right\}. \qquad \text{(III.5)}$$

We say that a linear operator

$$B : \mathbf{Y}^* \to \mathbf{X}^*, \qquad \mathfrak{D}(B) \supset \mathfrak{D}(A^*)$$

is an *adjoint* of A if one has

$$\langle \eta, Ax \rangle_{\mathbf{Y}} = \langle B\eta, x \rangle_{\mathbf{X}}, \qquad \forall x \in \mathbf{X}, \quad \forall \eta \in \mathfrak{D}(A^*).$$

Such an operator always exists. Indeed, for any $\eta \in \mathfrak{D}(A^*)$, let us define $\xi_0 \in \mathfrak{D}(A)^*$ by

$$\xi_0 : \mathfrak{D}(A) \to \mathbb{C}, \qquad x \mapsto \langle \eta, Ax \rangle_{\mathbf{Y}};$$

note that $\|\xi_0\|_{\mathbf{X}^*} \le c$, with $c \ge 0$ from (III.5) corresponding to η. By the Hahn–Banach theorem, there is a bounded extension of ξ_0 from $\mathfrak{D}(A)$ onto $\mathbf{X}$, which we denote by $\xi : \mathbf{X} \to \mathbb{C}$. We define

$$B : \mathbf{Y}^* \to \mathbf{X}^* \quad \text{by} \quad B : \eta \mapsto \xi,$$

so that $\langle \eta, Ax \rangle_{\mathbf{Y}} = \langle B\eta, x \rangle_{\mathbf{X}}$.

Example III.26 Let $\{\boldsymbol{e}_j\}_{j \in \mathbb{N}_0}$ be the standard basis in $l^2(\mathbb{N}_0)$. Let $A \in \mathscr{C}(l^2(\mathbb{N}_0))$, $\mathfrak{D}(A) = l^2(\mathbb{N})$ (which is considered as a subspace of $l^2(\mathbb{N}_0)$), be defined by

$$A : \boldsymbol{e}_j \mapsto \boldsymbol{e}_j, \qquad j \in \mathbb{N}.$$

Then for any $b \in \mathbb{C}$ the operator $B \in \mathscr{B}(l^2(\mathbb{N}_0))$ defined by

$$B : \boldsymbol{e}_j \mapsto \boldsymbol{e}_j, \quad j \in \mathbb{N}; \qquad B : \boldsymbol{e}_0 \mapsto b\boldsymbol{e}_0$$

is an adjoint of A. Note that $\sigma(A) = \{1\}$, $\|A\| = 1$; $\sigma(B) = \{1, b\}$, $\|B\| = \max(1, |b|)$.

Example III.27 Let $A : l^2(\mathbb{N}) \to l^2(\mathbb{N})$ be an unbounded linear operator defined by $A : \boldsymbol{e}_j \mapsto j\boldsymbol{e}_1$, $j \in \mathbb{N}$. One can see that A is not closable, and also that the domain $\mathfrak{D}(A^*)$ from (III.5) contains zero vector only. The operator $B = 0$ with $\mathfrak{D}(B) = \{0\}$ is the adjoint of A. In particular, B is not densely defined. Note that A is unbounded, while $\|B\| = 0$.

If a linear operator $A : \mathbf{X} \to \mathbf{Y}$ has dense domain $\mathfrak{D}(A)$, then its adjoint, denoted by A^*, is uniquely defined on domain $\mathfrak{D}(A^*)$ from (III.5):

$$A^* : \mathbf{Y}^* \to \mathbf{X}^*, \qquad \langle A^*\eta, x \rangle_{\mathbf{X}} = \langle \eta, Ax \rangle_{\mathbf{Y}} \quad \forall x \in \mathfrak{D}(A), \quad \forall \eta \in \mathfrak{D}(A^*),$$

with $\mathfrak{D}(A^*)$ from (III.5).

Problem III.28 Show that the adjoint of $A : \mathbf{X} \to \mathbf{Y}$ is unique if and only if A is densely defined.

Problem III.29 Describe A^* for $A = \frac{d}{dx} : L^2(\mathcal{I}) \to L^2(\mathcal{I})$, $\mathcal{I} = (0, 1)$, with

(1) $\mathfrak{D}(A) = C^1_{\mathrm{comp}}(\mathcal{I}) = \{\, u \in C^1(\mathbb{R}): \ \operatorname{supp} u \subset \mathcal{I} \,\}$;
(2) $\mathfrak{D}(A) = \{\, u \in C^1(\mathcal{I}): \ \lim_{x\to 0+} u(x) = 0 \,\}$;
(3) $\mathfrak{D}(A) = \{\, u \in C^1(\mathcal{I}): \ \lim_{x\to 0+} u(x) = 0 = \lim_{x\to 1-} u(x) \,\}$;
(4) $\mathfrak{D}(A) = C^1(\mathcal{I})$.

Let $\mathbf{X}$ be a Banach space. For $E \subset \mathbf{X}$, let $E^\perp$ denote the vector subspace of $\mathbf{X}^*$ consisting of elements which vanish on E:

$$E^\perp = \{\xi \in \mathbf{X}^*: \ \langle \xi, x \rangle = 0 \ \ \forall x \in E\}. \tag{III.6}$$

Problem III.30 Let $A : \mathbf{X} \to \mathbf{Y}$ be densely defined. Show that

$$\mathcal{G}(A^*) = \left(\mathcal{G}(A)^\Sigma\right)^\perp,$$

where the isomorphism $(\cdot, \cdot)^\Sigma : \mathbf{X} \times \mathbf{Y} \to \mathbf{Y} \times \mathbf{X}$ is defined by $(x, y)^\Sigma = (y, -x)$.

Problem III.31 Prove that if A is densely defined, then A^* is closed.

Problem III.32 ([Kat76, Theorem III.5.28]) Prove that if A is closable, then A^* is densely defined.
Hint: Apply Problem III.31 *to* A^* *and show that if* $\mathbf{X}$ *and* $\mathbf{Y}$ *are considered as subspaces of* $(\mathbf{X}^*)^*$ *and* $(\mathbf{Y}^*)^*$*, respectively, then* $A \subset (A^*)^*$.

Problem III.33 Show that if $E \subset \mathbf{X}$, then $(E^\perp)^\perp \cap \mathbf{X} = \overline{E}$ (where $\mathbf{X}$ is considered as a subspace of $(\mathbf{X}^*)^*$). Show that if $\mathbf{X}$ is reflexive, then $(E^\perp)^\perp = \overline{E}$.
Hint: The inclusion $(E^\perp)^\perp \cap \mathbf{X} \supset \overline{E}$ *is immediate. Use the Hahn–Banach theorem to show that if* $x \in \mathbf{X}$ *is not in* $\overline{E}$*, then there exists* $\xi \in E^\perp$ *such that* $\langle \xi, x\rangle \neq 0$ *(that is,* $x \notin (E^\perp)^\perp$*).*

Problem III.34 ([Kat76, Theorem III.5.29]) Assume that the Banach spaces $\mathbf{X}$ and $\mathbf{Y}$ are reflexive. Prove that if $A : \mathbf{X} \to \mathbf{Y}$ is densely defined and closable, then A^* is densely defined and closed, and $(A^*)^* = \overline{A}$.
Hint: Use Problems III.30 *and* III.33.

Problem III.35 If A is bounded, prove that A^* is bounded with the same norm.

Problem III.36 Let $\mathfrak{D}(A)$ be dense. If A is a bijection (as a map from $\mathfrak{D}(A) \subset \mathbf{X}$ to $\mathbf{Y}$), prove that

$$A^* : \mathbf{Y}^* \to \mathfrak{D}(A)^*$$

is invertible with bounded inverse given by $(A^{-1})^*$.

Problem III.37 Let $\mathfrak{D}(A)$ be dense. Prove that if A^* is a bijection, then so is A.
Hint: Use the inclusion $A \subset (A^*)^*$.

Problem III.38 Prove that the map $A \mapsto A^*$ is continuous from $\mathscr{B}(\mathbf{X}, \mathbf{Y})$ to $\mathscr{B}(\mathbf{Y}^*, \mathbf{X}^*)$ in the uniform operator topology (the norm topology).

Problem III.39 If $A : \mathbf{X} \to \mathbf{Y}$ is compact, prove that A^* is compact.

Solution. Denote $\mathbf{U} = \mathfrak{R}\big(A|_{\mathbb{B}_1(\mathbf{X})}\big)$. Since A is a compact operator, $\overline{\mathbf{U}}$ is a compact subset of $\mathbf{Y}$. Let $(\eta_j)_{j\in\mathbb{N}}$ be a sequence in $\mathbb{B}_1(\mathbf{Y}^*)$. Then the sequence $f_j = \eta_j|_{\overline{\mathbf{U}}} \in \big(\overline{\mathbf{U}}\big)^*$ is bounded and equicontinuous, and by the Ascoli–Arzelà theorem, there is a convergent subsequence; thus, we may assume that f_j is a Cauchy sequence. One has:

$$\begin{aligned}\|A^*\eta_i - A^*\eta_j\|_{\mathbf{X}^*} &= \sup_{x\in\mathbb{B}_1(\mathbf{X})} |\langle A^*\eta_i - A^*\eta_j,\ x\rangle| = \sup_{x\in\mathbb{B}_1(\mathbf{X})} |\langle \eta_i - \eta_j,\ Ax\rangle| \\ &= \sup_{y\in\mathbf{U}} |\langle f_i - f_j,\ y\rangle| = \sup_{y\in\overline{\mathbf{U}}} |\langle f_i - f_j,\ y\rangle|.\end{aligned}$$

It follows that $A^*\eta_j$ is a Cauchy sequence in $\mathbf{X}^*$, finishing the proof.

Problem III.40 If $A : \mathbf{X} \to \mathbf{Y}$ is of finite rank, prove that A^* is of finite rank.
Hint: Consider the linear operators A, A^* *as compositions*

$$A : \mathbf{X} \to \mathfrak{R}(A) \hookrightarrow \mathbf{Y}, \qquad A^* : \mathbf{Y}^* \hookrightarrow \mathfrak{R}(A)^* \to \mathbf{X}^*$$

and use the fact that the dual of a finite-dimensional vector space is finite-dimensional.

III.3 Spectrum of a linear operator

Let $\mathbf{X}$ be a Banach space and let $A : \mathbf{X} \to \mathbf{X}$ be a linear operator with domain $\mathfrak{D}(A) \subset \mathbf{X}$.

Definition III.41 (Regular points, the resolvent set, and the spectrum) A point $z \in \mathbb{C}$ is called a *regular point* of a linear operator $A : \mathbf{X} \to \mathbf{X}$ if $A - zI : \mathbf{X} \to \mathbf{X}$ is invertible, and its inverse is bounded:

$$(A - zI)^{-1} \in \mathscr{B}(\mathbf{X}).$$

It follows that $A - zI : \mathfrak{D}(A) \to \mathbf{X}$ is a bijection: it is one to one (injective) and onto (surjective).

Remark III.42 In the case when $A \in \mathscr{C}(\mathbf{X})$, the condition that $A - zI : \mathfrak{D}(A) \to \mathbf{X}$ is a bijection is equivalent to z being a regular point. Indeed, by the closed graph theorem (Theorem III.15), if $A - zI : \mathfrak{D}(A) \to \mathbf{X}$ is a bijection, then $(A - zI)^{-1}$ is bounded.

The *resolvent set* $\rho(A) \subset \mathbb{C}$ of a linear operator A is defined as a set of all its regular points. In particular, for any $z \in \rho(A)$,

$$\mathfrak{D}\big((A - zI)^{-1}\big) = \mathbf{X}, \qquad \mathfrak{R}\big((A - zI)^{-1}\big) = \mathfrak{D}(A).$$

The spectrum of a linear operator is defined as the complement of its resolvent set:

$$\sigma(A) = \mathbb{C} \setminus \rho(A).$$

Problem III.43 Give an example of an operator with domain which is not dense and has a nonempty resolvent set.

Lemma III.44 *The spectrum of a compact operator consists of nonzero eigenvalues of finite multiplicity and a possible accumulation point at $\lambda = 0$.*

Problem III.45 Show that if $\mathbf{X}$ is infinite-dimensional and $K : \mathbf{X} \to \mathbf{X}$ is a compact operator, then either the kernel is infinite-dimensional or 0 is an accumulation point of $\sigma(A) \setminus \{0\}$.

Lemma III.46 *Let $A : \mathbf{X} \to \mathbf{X}$ be a linear operator such that its resolvent set is not empty. Then A is closed.*

Proof. Let $x_j \to x$, $x_j \in \mathfrak{D}(A)$; $y_j := Ax_j \to y$. By the assumption, $\rho(A)$ is not empty. We pick some $z \in \rho(A)$. Then $(A - zI)x_j = y_j - zx_j$, and by continuity of $(A - zI)^{-1} : \mathbf{X} \to \mathfrak{D}(A)$ one has

$$\begin{aligned} x &= \lim_{j\to\infty} x_j = \lim_{j\to\infty} (A - zI)^{-1}(y_j - zx_j) \\ &= (A - zI)^{-1} \lim_{j\to\infty} (y_j - zx_j) = (A - zI)^{-1}(y - zx). \end{aligned}$$

Thus, $x \in \mathfrak{D}(A)$ and $(A - zI)x = y - zx$, hence $Ax = y$. □

Example III.47 Let $A \in \mathscr{B}(\mathbf{X})$. Then $\sigma(A)$ is nonempty and $\sigma(A) \subset \overline{\mathbb{D}_{\|A\|}}$. Indeed, for $|z| > \|A\|$, one can see that the resolvent $(A - zI)^{-1}$ is given explicitly by the power series:

$$z^{-1}I + z^{-2}A + z^{-3}A^2 + \dots, \qquad z \in \mathbb{C}, \quad |z| > \|A\|.$$

In particular, one can see that $\|(A - zI)^{-1}\| \to 0$ as $|z| \to \infty$. If $\sigma(A)$ were empty, then for any $u, v \in \mathbf{X}$, $\langle u, (A - zI)^{-1}v\rangle$ would be an analytic function tending to zero as $|z| \to \infty$, hence would have to be identically zero by the Liouville theorem, leading to a contradiction.

Problem III.48 Use Problems III.13 and III.14 for an alternative proof of Lemma III.46.

Lemma III.49 *Let $A \in \mathscr{C}(\mathbf{X})$. Then for any $z \in \rho(A)$*

$$\|(A - zI)^{-1}\| \geq \frac{1}{\operatorname{dist}(z, \sigma(A))} \qquad \forall z \in \rho(A). \tag{III.7}$$

Proof. For any $z \in \rho(A)$ and $\zeta \in \mathbb{C}$,

$$A - \zeta I = (A - zI)\big(I + (A - zI)^{-1}(z - \zeta)\big). \tag{III.8}$$

If $|z-\zeta| < 1/\|(A-zI)^{-1}\|$, then $I+(A-zI)^{-1}(z-\zeta)$ is invertible with bounded inverse; then, by (III.8), so is $A-\zeta I$, hence $\zeta \in \rho(A)$. We conclude that $\operatorname{dist}(z, \sigma(A)) \geq 1/\|(A - zI)^{-1}\|$. □

Corollary III.50 *The set $\rho(A)$ is open in $\mathbb{C}$, hence $\sigma(A) = \mathbb{C} \setminus \rho(A)$ is closed.*

Example III.51 Let $M = \begin{bmatrix} 0 & 1 \\ 0 & 0 \end{bmatrix} : \mathbb{C}^2 \to \mathbb{C}^2$. Then one has $M^2 = 0$, $\sigma(M) = \{0\}$, $\mathfrak{L}(M) = \mathbb{C}^2$, and $(M - z)^{-1} = \begin{bmatrix} -\frac{1}{z} & -\frac{1}{z^2} \\ 0 & -\frac{1}{z} \end{bmatrix}$, so that

$$\|(M - z)^{-1}\| \geq \left\|(M - z)^{-1} \begin{bmatrix} 0 \\ 1 \end{bmatrix}\right\| = \Big(\frac{1}{|z|^4} + \frac{1}{|z|^2}\Big)^{1/2} > \frac{1}{|z|} = \frac{1}{\operatorname{dist}(z, 0)},$$

in agreement with Lemma III.49.

Lemma III.52 *Let $A \in \mathscr{C}(\mathbf{X})$ with $\mathfrak{D}(A)$ dense (so that A^* is uniquely defined according to Problem* III.28*). Then*

$$\sigma(A^*) = \{\lambda \in \mathbb{C}: \quad \bar{\lambda} \in \sigma(A)\}.$$

The assumption that $\mathfrak{D}(A)$ is dense in $\mathbf{X}$ is important in view of Example III.26.

Proof. By Problem III.31 and Problem III.32, $A^* : \mathbf{X}^* \to \mathbf{X}^*$ is densely defined and closed. We need to show that $\lambda \in \rho(A)$ if and only if $\bar{\lambda} \in \rho(A^*)$. Without loss of generality, we may assume that $\lambda = 0$.

Let us assume that $0 \in \rho(A)$. By Problem III.36, since A is invertible with bounded inverse, so is A^*, with $(A^*)^{-1} = (A^{-1})^*$. Since A^{-1} is closable (it is since it is bounded) and densely defined, so is $(A^{-1})^*$. Then, by Problem III.35, $(A^*)^{-1}$ is bounded and then (being densely defined) it is defined on all of $\mathbf{X}^*$. Thus, $0 \in \rho(A^*)$.

Conversely, if we assume that $0 \in \rho(A^*)$, then, using Problem III.37, we deduce that A is invertible with bounded inverse. □

For a bounded operator $A \in \mathscr{B}(\mathbf{X})$ the *spectral radius* is defined by

$$r(A) = \sup_{z \in \sigma(A)} |z|. \tag{III.9}$$

Lemma III.53 (Gelfand's formula) *Let $A \in \mathscr{B}(\mathbf{X})$. Then*

$$\lim_{n \to \infty} \|A^n\|^{1/n} = r(A).$$

PROOF. We follow the proof of [**DS63**, Chapter IX.1, Lemma 8], which is given for Banach algebras. For $z \in \mathbb{C}$ such that $|z| > \|A\|$, the resolvent of A is given by the power series

$$(A - zI)^{-1} = -z^{-1}(I - A/z)^{-1} = -z^{-1}\Big(I + \frac{A}{z} + \frac{A^2}{z^2} + \dots\Big),$$

allowing one to write for any $u \in \mathbf{X}$, $v \in \mathbf{X}^*$ the power series expansion of the function $\langle v, (A - zI)^{-1}u \rangle$ for $|z| > \|A\|$:

$$\langle v, (A - zI)^{-1}u \rangle = -z^{-1}\Big(\langle v, u \rangle + \frac{\langle v, Au \rangle}{z} + \frac{\langle v, A^2u \rangle}{z^2} + \dots\Big). \tag{III.10}$$

Since the function $\langle v, (A - zI)^{-1}u \rangle$ is analytic in the region $\mathbb{C} \setminus \overline{\mathbb{D}_{r(A)}} \subset \rho(A)$, the power series (III.10) is absolutely convergent for each $z \in \mathbb{C}$ with $|z| > r(A)$, and then

$$\sup_{N \in \mathbb{N}} \Big|\Big\langle v, \frac{A^n}{z^n} u \Big\rangle\Big| < \infty.$$

With the above being true for any $u \in \mathbf{X}$ and $v \in \mathbf{X}^*$, the Banach–Steinhaus theorem on the uniform boundedness yields

$$C := \sup_{n \in \mathbb{N}} \frac{\|A^n\|}{|z|^n} < \infty,$$

leading to $\|A^n\|^{1/n} \le C^{1/n}|z|$ and then to $\limsup_{n \to \infty} \|A^n\|^{1/n} \le |z|$. Since this inequality is true for each $z \in \mathbb{C}$ with $|z| > r(A)$, one has

$$\limsup_{n \to \infty} \|A^n\|^{1/n} \le r(A). \tag{III.11}$$

Now we note that if $A^n - z^n I = (A - zI)(A^{n-1} + \dots + z^{n-1}I)$ (with some $z \in \mathbb{C}$ and $n \in \mathbb{N}$) has a bounded inverse, then so does $A - zI$; therefore, if $z \in \sigma(A)$, then $z^n \in \sigma(A^n)$ for each $n \in \mathbb{N}$, and then $\|A^n\| \ge |z|^n$, implying that

$$\liminf_{n \to \infty} \|A^n\|^{1/n} \ge r(A). \tag{III.12}$$

The inequalities (III.11) and (III.12) show that $\lim_{n \to \infty} \|A^n\|^{1/n}$ exists and equals $r(A)$, finishing the proof. □

III.3.1 Point spectrum, continuous spectrum, residual spectrum. Let us assume that

$$A \in \mathscr{C}(\mathbf{X}).$$

We decompose the spectrum of A into the following three non-intersecting subsets:

(1) Point spectrum $\sigma_{\rm p}(A)$ consists of eigenvalues of A:

$$\sigma_{\rm p}(A) = \{\, \lambda \in \sigma(A): \quad \exists x \in \mathfrak{D}(A), \quad x \ne 0, \quad Ax = \lambda x \,\}.$$

Note that the point spectrum could contain isolated eigenvalues of infinite multiplicity and also eigenvalues embedded into the spectrum.

(2) Continuous spectrum $\sigma_{\rm cont}(A)$ consists of $\lambda \in \mathbb{C}$ which are not eigenvalues, $\mathfrak{R}(A - \lambda I)$ is dense in $\mathbf{X}$, but is not closed. That is, $\mathbf{ker}(A - \lambda I) = \{0\}$, but $(A - \lambda I)^{-1}$ is not bounded on the range of $A - \lambda I$.

(3) Residual spectrum $\sigma_{\rm res}(A)$ consists of $\lambda \in \mathbb{C}$ which are not eigenvalues and such that the range of $A - \lambda I$ is not dense in $\mathbf{X}$ (whether the inverse of $A - \lambda I$ on the range of $A - \lambda I$ is bounded or not):

$$\sigma_{\rm res}(A) = \{\, \lambda \in \mathbb{C}: \quad \mathbf{ker}(A - \lambda I) = \{0\}, \quad \overline{\mathfrak{R}(A - \lambda I)} \subsetneq \mathbf{X} \,\}.$$

So, for $\lambda \in \sigma_{\rm res}(A)$, the operator $(A - \lambda I)^{-1}$ can not be uniquely extended to an operator on $\mathbf{X}$.

Thus, there is a decomposition of the spectrum into three non-intersecting subsets:

$$\sigma(A) = \sigma_{\rm p}(A) \cup \sigma_{\rm cont}(A) \cup \sigma_{\rm res}(A). \tag{III.13}$$

Problem III.54 Show that for $A \in \mathscr{C}(\mathbf{X})$ with dense domain $\mathfrak{D}(A)$ one has

$$\sigma_{\rm res}(A) \subset \{\, \lambda \in \mathbb{C}: \quad \bar\lambda \in \sigma_{\rm p}(A^*) \,\} \subset \sigma_{\rm res}(A) \cup \sigma_{\rm p}(A).$$

Hint: Show that if λ is in $\sigma_{\rm res}(A)$, then there exists $x \ne 0$ in $\mathfrak{R}(A - \lambda I)^\perp \subset \mathbf{X}^*$ which belongs to $\mathfrak{D}((A - \lambda I)^*)$ and satisfies $(A - \lambda I)^* x = 0$.

Example III.55 Consider the left shift operator:

$$L : l^2(\mathbb{N}) \to l^2(\mathbb{N}), \qquad L : (a_1, a_2, a_3, \dots) \mapsto (a_2, a_3, a_4, \dots). \tag{III.14}$$

This operator is bounded. Since $\|L\| = 1$, $\sigma(L) \subset \{|z| \le 1\}$. For any $|z| < 1$, $(1, z, z^2, \dots)$ is a corresponding eigenvector; one can show that $\sigma_{\rm p}(L) = \{z: \ |z| < 1\}$.

One has $\sigma_{\rm res}(L) = \emptyset$ (see Problem III.56 below).

Since the spectrum is a closed set, it follows that $\sigma_{\rm cont}(L) = \{|z| = 1\}$.

Problem III.56 For the left shift operator L from (III.14), check that, for any $\theta \in \mathbb{R}$, one has $\overline{\mathfrak{R}(L - e^{\mathrm{i}\theta})} = l^2(\mathbb{N})$, so that

$$\sigma_{\rm res}(L) \cap \partial\mathbb{D}_1 = \emptyset.$$

Example III.57 Consider $R = L^*$:

$$R : l^2(\mathbb{N}) \to l^2(\mathbb{N}), \qquad R : (a_1, a_2, a_3, \dots) \mapsto (0, a_1, a_2, \dots).$$

Just like L, the operator R is bounded; it is also bounded from below (see (III.3); in fact, R is norm-preserving). One easily checks that $\sigma_{\rm p}(R) = \emptyset$. For $z \in \mathbb{C}$, $|z| < 1$, for any $v \in l^2(\mathbb{N})$, $(R - zI)v$ is orthogonal to $(1, \bar z, \bar z^2, \bar z^3, \dots) \in l^2(\mathbb{N})$, hence $\mathfrak{R}(R - zI)$ is closed but not dense: $\mathfrak{R}(R - zI) = \overline{\mathfrak{R}(R - zI)} \subsetneq \mathbf{X}$. Therefore,

$$\sigma_{\rm res}(R) = \{z \in \mathbb{C}: \quad |z| < 1\}, \qquad \sigma_{\rm cont}(R) = \{z \in \mathbb{C}, |z| = 1\}.$$

Example III.58 Consider the operator $Q : L^2(\mathbb{R}) \to L^2(\mathbb{R})$, $f(x) \mapsto x f(x)$, with the domain $\mathfrak{D}(Q) = \{\, f \in L^2(\mathbb{R}): \quad x f(x) \in L^2(\mathbb{R}) \,\}$. Then $\sigma(Q) = \mathbb{R}$,

$$\sigma_{\rm p}(Q) = \emptyset, \qquad \sigma_{\rm cont}(Q) = \mathbb{R}, \qquad \text{and} \quad \sigma_{\rm res}(Q) = \emptyset.$$

Problem III.59 Give examples of operators such that *(1)* $\lambda \in \sigma_p(A)$ and the range of $A - \lambda I$ is dense; *(2)* $\lambda \in \sigma_p(B)$ and the range of $B - \lambda I$ is not dense.

Problem III.60 Find the spectrum in $L^2(\mathcal{I})$, $\mathcal{I} = (0, 1)$, of the operator $A = \frac{d}{dx}$ with the following domains:

(1) $\mathfrak{D}(A) = H_0(\mathcal{I}) = \{ u \in H^1(\mathcal{I}) : \quad u(0) = u(1) = 0 \}$;

(2) $\mathfrak{D}(A) = \{ u \in H^1(\mathcal{I}) : \quad u(0) = 0 \}$;

(3) $\mathfrak{D}(A) = H^1(\mathcal{I})$.

III.3.2 Approximate point spectrum. Let $A \in \mathscr{C}(\mathbf{X})$. We say that $\lambda \in \mathbb{C}$ belongs to the *approximate point spectrum* (the set of *approximate eigenvalues*), denoted by $\sigma_{ap}(A)$, if $A - \lambda I$ is not bounded from below (see (III.3)):

$$\sigma_{ap}(A) = \{ \lambda \in \sigma(A) : \quad \exists \, (v_j)_{j \in \mathbb{N}}, \quad \|v_j\| = 1, \quad \|(A - \lambda I)v_j\| \to 0 \}.$$

Clearly, one has

$$\sigma_p(A) \cup \sigma_{cont}(A) \subset \sigma_{ap}(A).$$

Problem III.61 Show that $\sigma_{ap}(A) \subset \sigma(A)$.

Problem III.62 Show that if $\sigma_{res}(A) = \emptyset$, then $\sigma_{ap}(A) = \sigma(A)$.

We say that A is *quasinilpotent* if A is bounded and $\|A^j\|^{1/j} \to 0$ as $j \to \infty$.

Problem III.63 The operator A is quasinilpotent if and only if $\sigma(A) = \{0\}$. *Hint: For the* only if *part, notice that if A is quasinilpotent, then for any $z \in \mathbb{C}$ the series $\sum_{j \in \mathbb{N}} z^j A^j$ is convergent.*

Example III.64 In general, $\sigma_p(A) \cup \sigma_{cont}(A) \subsetneq \sigma_{ap}(A)$. Let

$$M : l^2(\mathbb{N}) \to l^2(\mathbb{N}), \qquad M : (a_1, a_2, a_3, \dots) \mapsto (0, 2^{-1}a_1, 2^{-2}a_2, 2^{-3}a_3, \dots).$$

Then M is quasinilpotent since $\|M^j\|^{1/j} = (2^{-1-2-\dots-j})^{1/j} \to 0$; hence $\sigma(M) = \{0\}$. One can see that $0 \notin \sigma_p(M)$. Moreover, $\mathfrak{R}(M)$ is not dense in l^2 since $\boldsymbol{e}_1 = (1, 0, \dots) \notin \overline{\mathfrak{R}(A)}$, hence $0 \notin \sigma_{cont}(M)$. Hence, $\sigma(M) = \sigma_{res}(M) = \{0\}$. On the other hand, using the standard basis $\{\boldsymbol{e}_j\}_{j \in \mathbb{N}}$ of $l^2(\mathbb{N})$, one sees that $\|M\boldsymbol{e}_j\| \to 0$ as $j \to \infty$; so, M is not bounded from below, and $\lambda = 0$ is an approximate eigenvalue.

Problem III.65 Give an example of a quasinilpotent operator which is not compact.

Problem III.66 Identify which of the following examples could provide solutions to some of the earlier problems:

$$A : \ \boldsymbol{e}_j \to j\boldsymbol{e}_j, \qquad j \in \mathbb{N}.$$

$$B; \ \boldsymbol{e}_j \to j\boldsymbol{e}_1, \qquad j \in \mathbb{N}.$$

$$C : \ \boldsymbol{e}_j \to \boldsymbol{e}_j + j\boldsymbol{e}_1, \qquad j \in \mathbb{N}.$$

$$D : \ \boldsymbol{e}_j \to \begin{cases} \boldsymbol{e}_{j+1}, & j \ \text{odd}; \\ \boldsymbol{e}_{j+2}/2^j, & j \ \text{even}. \end{cases}$$

$$E : \ \boldsymbol{e}_j \to \boldsymbol{e}_{j+1}/j, \qquad j \in \mathbb{N}.$$

III.4 Fredholm operators

Here we present several standard results from the theory of Fredholm operators which we will use for the proof of the Weyl theorem on the stability of the essential spectrum.

Since the definition of a Fredholm operator only applies to closed linear densely defined operators, in this section we assume:

$$\mathbf{X},\ \mathbf{Y} \text{ are Banach spaces}, \qquad A \in \mathscr{C}(\mathbf{X}, \mathbf{Y}), \qquad \mathfrak{D}(A) \text{ is dense in } \mathbf{X}.$$

We remind that the *kernel* (or the *null space*) of A is

$$\mathbf{ker}(A) = \left\{ x \in \mathfrak{D}(A): \quad Ax = 0 \right\}.$$

The *cokernel* of A, also called the *defect subspace* of A, is denoted by $\mathbf{coker}(A)$:

$$\mathbf{coker}(A) = \mathbf{Y}/\mathfrak{R}(A).$$

III.4.1 Fredholm and semi-Fredholm operators.

Definition III.67 The linear operator $A \in \mathscr{C}(\mathbf{X}, \mathbf{Y})$ is called *Fredholm* if its domain $\mathfrak{D}(A)$ is dense in $\mathbf{X}$, its range $\mathfrak{R}(A)$ is closed, and both $\mathbf{ker}(A)$ and $\mathbf{coker}(A)$ are finite-dimensional. We denote the *Fredholm domain* of A by

$$\Phi_A = \left\{ z \in \mathbb{C}: \quad A - zI \quad \text{is Fredholm} \right\}.$$

If $\mathfrak{D}(A)$ is dense, $\mathfrak{R}(A)$ is closed, and either $\mathbf{ker}(A)$ or $\mathbf{coker}(A)$ is finite-dimensional, we say that A is *semi-Fredholm* and denote the *semi-Fredholm domain* of A by

$$\Psi_A = \left\{ z \in \mathbb{C}: \quad A - zI \ \text{ is semi-Fredholm} \right\}.$$

We also denote

$$\begin{aligned} \Phi_A^+ &= \left\{ z \in \mathbb{C}: \quad \mathfrak{R}(A - zI) \text{ is closed}, \quad \dim \mathbf{ker}(A - zI) < \infty \right\}, \\ \Phi_A^- &= \left\{ z \in \mathbb{C}: \quad \mathfrak{R}(A - zI) \text{ is closed}, \quad \dim \mathbf{coker}(A - zI) < \infty \right\}, \end{aligned}$$

so that

$$\Phi_A = \Phi_A^+ \cap \Phi_A^-, \qquad \Psi_A = \Phi_A^+ \cup \Phi_A^-.$$

We denote the *index* of a Fredholm or semi-Fredholm operator A by

$$\operatorname{ind} A = \dim \mathbf{ker}(A) - \dim \mathbf{coker}(A) \in \{-\infty\} \sqcup \mathbb{Z} \sqcup \{+\infty\}; \tag{III.15}$$

above, we use the symbol $\sqcup$ to denote a disjoint union. That is, we write $\operatorname{ind} A = +\infty$ if $\mathbf{coker}(A)$ is finite-dimensional while $\mathbf{ker}(A)$ is not, and similarly $\operatorname{ind} A = -\infty$ if $\mathbf{ker}(A)$ is finite-dimensional while $\mathbf{coker}(A)$ is not.

Problem III.68 Prove that $A \in \mathscr{C}(\mathbf{X}, \mathbf{Y})$ with the dense domain $\mathfrak{D}(A)$ is Fredholm if and only if there exist $B \in \mathscr{B}(\mathbf{Y}, \mathbf{X})$ and compact operators $K_1 \in \mathscr{B}_0(\mathbf{Y})$ and $K_2 \in \mathscr{B}_0(\mathbf{X})$ such that

$$AB + K_1 = I_{\mathbf{Y}}, \qquad BA + K_2 = I_{\mathfrak{D}(A)}. \tag{III.16}$$

Remark III.69 If one of the Banach spaces $\mathbf{X}$, $\mathbf{Y}$ is finite-dimensional while the other is infinite-dimensional, then clearly there are no Fredholm operators $\mathbf{X} \to \mathbf{Y}$. If both $\mathbf{X}$ and $\mathbf{Y}$ are infinite-dimensional, then $K \in \mathscr{B}_0(\mathbf{X}, \mathbf{Y})$ can not be Fredholm; by Problem III.68, there should be K_1 and K_2 such that $KB + K_1 = I_{\mathbf{Y}}$, $BK + K_2 = I_{\mathbf{X}}$, with left-hand sides being compact operators.

Problem III.70 ([Gol66, Lemma V.1.5]) Let A be a semi-Fredholm operator and $\hat{A}$ an extension of A such that $n := \dim(\mathfrak{D}(\hat{A})/\mathfrak{D}(A))$ is finite. Show that $\hat{A}$ is closed with closed range and

$$\operatorname{ind} \hat{A} = \operatorname{ind} A + n.$$

Hint: First show that $\mathfrak{D}(\hat{A}) = \mathfrak{D}(A) \oplus \mathbf{N}$ *where* $\mathbf{N}$ *has dimension* n*. For the index, make the proof for* $n = 1$ *and then iterate.*

We point out that the assumption in Definition III.67 that $\mathfrak{R}(A)$ is closed is redundant as long as $\dim \mathbf{coker}(A) < \infty$:

Theorem III.71 ([EE87, Theorem I.3.2]) *Let* $A \in \mathscr{C}(\mathbf{X}, \mathbf{Y})$ *(with* $\mathfrak{D}(A)$ *not necessarily dense in* $\mathbf{X}$*) and assume that*

$$\mathbf{coker}(A) = \mathbf{Y}/\mathfrak{R}(A)$$

is finite-dimensional. Then $\mathfrak{R}(A)$ *is closed.*

PROOF. Assume that $A \in \mathscr{B}(\mathbf{X}, \mathbf{Y})$ and that $\mathbf{ker}(A) = \{0\}$. (If $\mathbf{ker}(A) \neq \{0\}$, consider A as a map from $\mathbf{X}/\mathbf{ker}(A)$ which is a Banach space since $\mathbf{ker}(A)$ is closed.) We extend A to a linear operator

$$\hat{A} : \mathbf{coker}(A) \oplus \mathbf{X} \to \mathbf{Y}, \qquad (y, x) \mapsto Ax + y,$$

which is bounded if we equip $\mathbf{coker}(A) \oplus \mathbf{X}$ with the norm

$$\|(y, x)\|_{\mathbf{coker}(A) \oplus \mathbf{X}} := \|y\|_{\mathbf{Y}} + \|x\|_{\mathbf{X}}.$$

Then, since $\hat{A}$ is bounded and onto (by construction), it is bi-continuous by the Schauder–Banach open mapping theorem. Therefore, it is bounded from below and then it has closed range by Problem III.16, so $\mathfrak{R}(A) = \hat{A}(\{0\} \times \mathbf{X})$ is closed.

In the general case $A \in \mathscr{C}(\mathbf{X}, \mathbf{Y})$, we consider A as a bounded linear operator from $\mathfrak{D}(A)$ to $\mathbf{Y}$, where $\mathfrak{D}(A)$ is a Banach space equipped with the graph norm $\|x\|_A = \|x\| + \|Ax\|$ (see Lemma III.5); this substitution does not change the cokernel. □

Lemma III.72 ([EE87, Lemma I.3.12]) *Let* $A \in \mathscr{C}(\mathbf{X}, \mathbf{Y})$*. Then* A *has closed range and* $\dim \mathbf{ker}(A) < \infty$ *if and only if there exist a Banach space* $\mathbf{Y}_1$*, a compact operator* $K \in \mathscr{B}_0(\mathbf{X}, \mathbf{Y}_1)$*, and* $c > 0$ *such that*

$$\|x\|_{\mathbf{X}} \le c\,(\|Ax\|_{\mathbf{Y}} + \|Kx\|_{\mathbf{Y}_1}), \qquad \forall x \in \mathfrak{D}(A). \tag{III.17}$$

Moreover, the operator K *can be chosen so that* $\operatorname{rank} K = \dim \mathbf{ker}(A)$*.*

PROOF. If the inequality holds, then for all $x \in \mathbf{ker}(A)$ one has $\|x\| \le c\|Kx\|$; this implies that $\mathbb{B}_1(\mathbf{X}) \cap \mathbf{ker}(A)$ is compact (it suffices to notice that if $(Kx_j)_{j\in\mathbb{N}}$ is a Cauchy subsequence and $\|x_j - x_k\| \le c\|Kx_j - Kx_k\|$ for all $j, k \in \mathbb{N}$, then $(x_j)_{j\in\mathbb{N}}$ is also a Cauchy sequence). This implies that $\mathbf{ker}(A)$ is finite-dimensional.

Let us prove that $\mathfrak{R}(A)$ is closed. Let $y_j \in \mathfrak{R}(A)$, $y_j \to y \in \mathbf{Y}$. There are $x_j \in \mathfrak{D}(A)$ such that $y_j = Ax_j$. Since $\mathbf{ker}(A)$ is finite-dimensional, we may write

$$\mathbf{X} = \mathbf{ker}(A) \oplus \mathbf{Z}, \tag{III.18}$$

with $\mathbf{Z}$ a closed vector subspace of $\mathbf{X}$, and we may assume that $x_j \in \mathbf{Z}$.

Let us assume that the sequence x_j is unbounded; passing to a subsequence, we may assume that $x_j \neq 0\ \forall j \in \mathbb{N}$ and that $\|x_j\| \to \infty$. Denote $z_j = x_j/\|x_j\| \in \mathbf{Z}$; then $Az_j \to 0$. Again passing to a subsequence, we may assume that $(Kz_j)_{j\in\mathbb{N}}$

converges. In view of (III.17), $(z_j)_{j\in\mathbb{N}}$ is a Cauchy sequence, converging to some $z \in \mathbf{Z}$. Since A is closed, one has $Az = 0$; then $z \in \mathbf{ker}(A) \cap \mathbf{Z}$, and by (III.18) this implies that $z = 0$, in contradiction to it being the limit of $(z_j)_{j\in\mathbb{N}}$ with $\|z_j\| = 1$.

Thus, the sequence $(x_j)_{j\in\mathbb{N}} \in \mathbf{Z}$ is bounded. Due to compactness of K and convergence of $(Ax_j)_{j\in\mathbb{N}}$, (III.17) implies that there is a subsequence of $(x_j)_{j\in\mathbb{N}}$ which converges to some $x \in \mathbf{Z}$. Since A is closed, $x \in \mathfrak{D}(A)$ and $y = Ax \in \mathfrak{R}(A)$, proving that $\mathfrak{R}(A)$ is closed.

Conversely, assume that $\mathfrak{R}(A)$ is closed and $\dim \mathbf{ker}(A) < \infty$. Let P be the projection from $\mathbf{X} = \mathbf{ker}(A) \oplus \mathbf{Z}$ (see (III.18)) onto $\mathbf{ker}(A)$, and let A_1 be the restriction of A onto $\mathbf{Z}$. Since A_1 is a closed linear operator with zero kernel, A_1^{-1}, with its domain the Banach space $\mathfrak{R}(A)$ and the range $\mathbf{Z} \cap \mathfrak{D}(A)$, is also closed. It is continuous by the closed graph theorem (Theorem III.15); thus, there is $c > 0$ such that

$$\|z\|_{\mathbf{X}} \le c\|A_1 z\|_{\mathbf{Y}}, \qquad \forall z \in \mathbf{Z} \cap \mathfrak{D}(A). \tag{III.19}$$

Now let $x \in \mathfrak{D}(A)$. Then, using (III.19),

$$\|x\| \le \|Px\| + \|(I - P)x\| \le \|Px\| + c\|A_1(I - P)x\| = \|Px\| + c\|Ax\|,$$

giving (III.17) with $K = P$ bounded and of finite rank. □

Theorem III.73 ([**Kat76**, Theorem IV.5.13]) *Let $A \in \mathscr{C}(\mathbf{X}, \mathbf{Y})$ have a dense domain $\mathfrak{D}(A)$. Then $\mathfrak{R}(A)$ is closed if and only if $\mathfrak{R}(A^*)$ is closed. In this case,*

$$\mathfrak{R}(A)^\perp = \mathbf{ker}(A^*), \qquad \mathbf{ker}(A)^\perp = \mathfrak{R}(A^*),$$

$$\dim \mathbf{ker}(A^*) = \dim \mathbf{coker}(A), \qquad \dim \mathbf{coker}(A^*)^\perp = \dim \mathbf{ker}(A).$$

A is Fredholm (semi-Fredholm) if so is A^*. In this case, $\operatorname{ind} A = -\operatorname{ind} A^*$.

Above, e.g. $\mathfrak{R}(A)^\perp \subset \mathbf{Y}^*$ denotes the vector subspace of $\mathbf{Y}^*$ consisting of elements $\eta \in \mathbf{Y}^*$ which vanish on $\mathfrak{R}(A)$; see (III.6). See also [**Gol66**, Theorem IV.2.3] and [**EE87**, Theorem I.3.7].

Problem III.74 Prove that the adjoint of a Fredholm (respectively, semi-Fredholm) operator is Fredholm (respectively, semi-Fredholm).

The following example illustrates the importance of the assumption in Theorem III.73 that $\mathfrak{R}(A)$ is closed; it also shows that the finiteness of $\dim \mathbf{ker}(A)$ does not imply that $\mathfrak{R}(A)$ is closed (cf. Theorem III.71).

Example III.75 Consider

$$K : l^2(\mathbb{N}) \to l^2(\mathbb{N}), \qquad (u_1, u_2, u_3, u_4, \dots) \mapsto (u_1, u_2/2, u_3/3, u_4/4, \dots).$$

Clearly $K = K^*$, $\mathbf{ker}(K) = \mathbf{ker}(K^*) = \{0\}$, $\overline{\mathfrak{R}(K)} = l^2(\mathbb{N})$. The range $\mathfrak{R}(K)$ is not closed since it does not include, for example, $(a_1, a_2/2, a_3/3, \dots) \in l^2(\mathbb{N})$, where all a_j, $j \in \mathbb{N}$, are either equal to 1 or to 0, with infinitely many of a_j equal to one, so that $(a_1, a_2, \dots) \notin l^2(\mathbb{N})$. As a consequence, there will be infinitely many linearly independent classes in $l^2(\mathbb{N})/\mathfrak{R}(K)$, so that the cokernel

$$\mathbf{coker}(K) = l^2(\mathbb{N})/\mathfrak{R}(K)$$

is infinite-dimensional (while $\mathbf{ker}(K^*) = \{0\}$). Also, $\mathbf{ker}(K)^\perp$ (a closed vector subspace) can not be equal to $\mathfrak{R}(K^*) = \mathfrak{R}(K)$ (which is not closed). Since $\mathfrak{R}(K)$ is not closed, K is neither Fredholm nor semi-Fredholm. (Of course, since K is compact, it is not Fredholm; see Remark III.69.)

The following generalizes Problem III.16.

Problem III.76 Let $A \in \mathscr{C}(\mathbf{X}, \mathbf{Y})$. Prove that A has closed range if and only if there is $c > 0$ such that

$$\|[x]\|_{\mathbf{X}/\operatorname{\mathbf{ker}}(A)} \le c\|Ax\|_{\mathbf{Y}}, \qquad \forall x \in \mathfrak{D}(A), \tag{III.20}$$

where

$$\|[x]\|_{\mathbf{X}/\operatorname{\mathbf{ker}}(A)} = \inf_{x_0 \in \operatorname{\mathbf{ker}}(A)} \|x + x_0\|_{\mathbf{X}}.$$

When A has closed range, show that the smallest possible constant in (III.20) is the norm of the inverse of $A_1 : \mathbf{X}/\operatorname{\mathbf{ker}}(A) \to \mathfrak{R}(A)$.

Theorem III.77 ([Gol66, Theorem IV.2.7]) *Let $\mathbf{X}_0$, $\mathbf{X}$, $\mathbf{Y}$ be Banach spaces. Let $A \in \mathscr{C}(\mathbf{X},\mathbf{Y})$ be an operator with closed range $\mathfrak{R}(A)$, and assume that* $\dim \operatorname{\mathbf{ker}}(A) < \infty$. *Let $B : \mathbf{X}_0 \to \mathbf{X}$ be a linear operator.*

(1) If B is closed, then so is AB.
(2) If B is densely defined and has closed range, then so is AB.
(3) If $\mathfrak{D}(A)$ is dense in $\mathbf{X}$ and both A and B are Fredholm, then so is AB and

$$\operatorname{ind} AB = \operatorname{ind} A + \operatorname{ind} B.$$

(4) If $\mathfrak{D}(A)$ and $\mathfrak{D}(B)$ are dense (in $\mathbf{X}$ and $\mathbf{X}_0$, respectively) and $\dim \operatorname{\mathbf{coker}}(B) < \infty$, *then $\mathfrak{D}(AB)$ is dense in $\mathbf{X}_0$.*

Theorem III.78 (First stability theorem) *Let $A \in \mathscr{C}(\mathbf{X},\mathbf{Y})$ be semi-Fredholm. There is $\delta > 0$ such that if $B \in \mathscr{B}(\mathbf{X}, \mathbf{Y})$ and $\|B\| < \delta$, then $A+B$ is semi-Fredholm and moreover*

$$\dim \operatorname{\mathbf{ker}}(A+B) \le \dim \operatorname{\mathbf{ker}}(A), \qquad \dim \operatorname{\mathbf{coker}}(A+B) \le \dim \operatorname{\mathbf{coker}}(A);$$

$$\operatorname{ind}(A+B) = \operatorname{ind} A.$$

For the case of Fredholm operators, see [**GK57**, Theorem 2.2]. The version for semi-Fredholm operators is proved in [**Gol66**, Theorem V.1.6] and [**EE87**, Theorem I.3.22].

We remind that, by Definition III.67, the linear operator is semi-Fredholm if it is closed, densely defined, has closed range, and either its kernel or cokernel (or both) is of finite dimension.

Theorem III.79 (Second stability theorem) *Let $A \in \mathscr{C}(\mathbf{X}, \mathbf{Y})$ with dense domain $\mathfrak{D}(A)$, and assume that B is A-compact. Then A is semi-Fredholm if and only if $A + B$ is semi-Fredholm. If this is the case, then:*

(1) $\dim \operatorname{\mathbf{ker}}(A) < \infty$ *if and only if* $\dim \operatorname{\mathbf{ker}}(A+B) < \infty$.
(2) $\dim \operatorname{\mathbf{coker}}(A) < \infty$ *if and only if* $\dim \operatorname{\mathbf{coker}}(A+B) < \infty$.

In both cases,

$$\operatorname{ind}(A+B) = \operatorname{ind} A.$$

The case of Fredholm operators is covered in [**GK57**, Theorem 2.3]. For the case of semi-Fredholm operators, see [**Gol66**, Corollary V.2.2] and [**EE87**, Theorem I.3.21].

PROOF. We notice that since A is closed and B is A-compact, $A+B$ is closed by Theorem III.22.

It is enough to give a proof for a *bounded* Fredholm operator A and for compact B. Indeed, following [**EE87**, Remark I.3.27], we notice that $A \in \mathscr{C}(\mathbf{X}, \mathbf{Y})$ and B being A-compact can be considered as $A \in \mathscr{B}(\mathfrak{D}(A), \mathbf{Y})$, $B \in \mathscr{B}_0(\mathfrak{D}(A), \mathbf{Y})$, where $\mathfrak{D}(A)$ is considered as a Banach space with the graph norm $\|x\|_A = \|x\| + \|Ax\|$ (see Lemma III.5); we note that $\mathbf{ker}(A)$ and $\mathbf{coker}(A)$ remain unchanged after this substitution. So, we assume that

$$A \in \mathscr{B}(\mathbf{X}, \mathbf{Y}), \qquad B \in \mathscr{B}_0(\mathbf{X}, \mathbf{Y}).$$

Let us consider the case when A has finite-dimensional kernel and $\mathfrak{R}(A)$ is closed. Then, by Lemma III.72, there exist $c > 0$, a Banach space $\mathbf{Y}_1$, and $K \in \mathscr{B}_0(\mathbf{X}, \mathbf{Y}_1)$ such that for any $x \in \mathbf{X}$ one has

$$\|x\|_{\mathbf{X}} \le c(\|Ax\|_{\mathbf{Y}} + \|Kx\|_{\mathbf{Y}_1}) \le c(\|(A+B)x\|_{\mathbf{Y}} + \|Bx\|_{\mathbf{Y}} + \|Kx\|_{\mathbf{Y}_1}),$$

and, again by Lemma III.72 (treating $\|Bx\|_{\mathbf{Y}} + \|Kx\|_{\mathbf{Y}_1}$ as the norm of the compact operator $B \oplus K : \mathbf{X} \to \mathbf{Y} \oplus \mathbf{Y}_1$ applied to x), $A+B$ has closed range and finite-dimensional kernel. The equality of indices follows from Theorem III.78 which provides that the mapping

$$[0,1] \to \{-\infty\} \sqcup \mathbb{Z} \sqcup \{+\infty\}, \qquad s \mapsto \operatorname{ind}(A+sB)$$

is continuous (we use the discrete topology in $\{-\infty\} \sqcup \mathbb{Z} \sqcup \{+\infty\}$). Notice that the equality in the case when indices are finite also implies the equality when indices are infinite: $\operatorname{ind} A = -\infty$ if and only if $\operatorname{ind}(A+B) = -\infty$.

As we already pointed out, the above argument also yields the results for the case when A is closed (but not necessarily bounded), $\mathbf{ker}(A)$ is finite-dimensional, $\mathfrak{R}(A)$ is closed, and B is A-compact: in this case, we conclude that $\mathbf{ker}(A+B)$ is finite-dimensional and $\mathfrak{R}(A+B)$ is closed.

Let now $A \in \mathscr{C}(\mathbf{X}, \mathbf{Y})$ and assume that the cokernel of A is finite-dimensional. Then $\mathfrak{R}(A)$ is closed (by Theorem III.71), so is $\mathfrak{R}(A^*)$ (by Theorem III.73), and then the conclusion follows if one applies the above argument to the adjoints.

If A is closed, $\mathbf{ker}(A+B)$ (or $\mathbf{coker}(A+B)$) is finite-dimensional, $\mathfrak{R}(A+B)$ is closed, and B is A-compact, then we notice that $A+B$ is closed and $-B$ is $(A+B)$-compact (by Theorem III.22) and apply the above argument to conclude that $\mathbf{ker}(A)$ (or, respectively, $\mathbf{coker}(A)$) is finite-dimensional and $\mathfrak{R}(A)$ is closed. □

Problem III.80 (Fredholm alternative) Let $K : \mathbf{X} \to \mathbf{X}$ be a compact operator. Then $\mathbf{ker}(I-K)$ is finite-dimensional, $\mathfrak{R}(I-K)$ is closed, and $I-K$ is invertible with bounded inverse if and only if $\mathbf{ker}(I-K^*) = \{0\}$.
Hint: Prove that if K is compact, then $I-K$ is Fredholm of index zero. See Theorem III.79.

Problem III.81 Let $A \in \mathscr{C}(\mathbf{X}, \mathbf{Y})$ be semi-Fredholm. Show that there is $\delta > 0$ such that if $B \in \mathscr{B}(\mathbf{X}, \mathbf{Y})$ and $\|B\| < \delta$, then $A+B$ is semi-Fredholm. Complete the proof of Theorem III.78.
Hint: Mimic the proof of Theorem III.79 *using*

$$\|x\|_{\mathbf{X}} \le C_A(\|(A+B)x\|_{\mathbf{Y}} + \|B\|\|x\|_{\mathbf{X}} + \|Kx\|_{\mathbf{Y}_1}),$$

and estimating $\|x\|$ in terms of $\|(A+B)x\|_{\mathbf{Y}}$ and $\|Kx\|_{\mathbf{Y}_1}$ as long as $C_A\|B\| <$ 1. *Take K to be a projector on* $\mathbf{ker}(A)$ *to show that $A+B$ has closed range* *and* $\dim \mathbf{ker}(A+B) \le \dim \mathbf{ker}(A)$ *when the latter is finite. Use the adjoints for* $\dim \mathbf{coker}(A+B)$.

If $\dim \mathbf{ker}(A)$ *is finite, consider* A_0*, which is a restriction of* A *to some supplement of* $\mathbf{ker}(A)$*, and use Problem* III.70 *to complete the proof.*

Theorem III.82 ([**GK57**, Theorem 3.3]) *Let* $A \in \mathscr{C}(\mathbf{X})$ *and let* Ω *be a connected component of* Φ_A*. Then at all* $z \in \Omega$ *except perhaps at some isolated points* λ_j *the value of* $\dim \mathbf{ker}(A - zI)$ *is constant. At these isolated points,* $\dim \mathbf{ker}(A - \lambda_j I)$ *takes larger values.*

III.5 Normal eigenvalues and the discrete spectrum

In this section, we consider isolated points of the spectrum of a linear operator A. This implies that the resolvent set of A is nonempty, and then, by Lemma III.46, A is closed; so we assume in this section that

$$A \in \mathscr{C}(\mathbf{X}).$$

When we deal with the Fredholm operators, by Definition III.67, we will also have to assume that they are densely defined.

III.5.1 The Riesz projector.

Definition III.83 (The Riesz projector) Let Γ be a rectifiable (simple or composite) path which lies entirely in $\mathbb{C} \setminus \sigma(A)$ and encloses an open set $G_\Gamma \subset \mathbb{C}$. Hence $\mathrm{Ind}_\Gamma(\lambda) := \frac{1}{2\pi \mathrm{i}} \oint_\Gamma \frac{dz}{z-\lambda} = 1$ for $\lambda \in G_\Gamma$, $\mathrm{Ind}_\Gamma(\lambda') = 0$ for any $\lambda' \in \mathbb{C} \setminus \overline{G_\Gamma}$. The Riesz projector corresponding to G_Γ is defined by

$$P_\Gamma = -\frac{1}{2\pi \mathrm{i}} \oint_\Gamma (A - zI)^{-1}\, dz. \tag{III.21}$$

If $G_\Gamma \cap \sigma(A) = \{\lambda\}$ for some $\lambda \in \mathbb{C}$, then we denote P_Γ by P_λ.

Remark III.84 Let us point out that the integral in (III.21) is well-defined since the integrand is continuous in the uniform operator topology $\mathscr{B}(\mathbf{X})$ (in fact, even $\mathscr{B}(\mathbf{X}, \mathfrak{D}(A))$). The integral converges strongly, as a Bochner integral, to a bounded operator. Since the operator-valued function $z \mapsto (A - zI)^{-1}$ is analytic on $\mathbb{C} \setminus \sigma(A)$, this integral does not change under smooth deformations of Γ.

Problem III.85 Show that $P_\Gamma^2 = P_\Gamma$; that is, P_Γ is a projector.
Hint: Use a double integration and the Cauchy theorem for analytic maps.

The following results were obtained by F. Riesz in 1912 for bounded operators (see [**RSN56**, Chapter XI, §148]), which generalize to closed operators; see [**GK57**, §4].

Lemma III.86 *Let* $A \in \mathscr{C}(\mathbf{X})$*, with* $\mathfrak{D}(A)$ *being dense. Then* A *is invariant on* $\mathfrak{R}(P_\Gamma)$ *and on* $\mathfrak{R}(I - P_\Gamma)$*;*

$$\mathfrak{D}(A|_{\mathfrak{R}(P_\Gamma)}) = \mathfrak{R}(P_\Gamma), \qquad \mathfrak{D}(A|_{\mathfrak{R}(I-P_\Gamma)}) = \mathfrak{D}(A) \cap \mathfrak{R}(I - P_\Gamma);$$

$$\sigma(A|_{\mathfrak{R}(P_\Gamma)}) = \sigma(A) \cap G_\Gamma, \qquad \sigma(A|_{\mathfrak{R}(I-P_\Gamma)}) = \sigma(A) \cap (\mathbb{C} \setminus G_\Gamma).$$

III.5.2 Normal eigenvalues.

Definition III.87 (Normal eigenvalues [GK57]) A point $\lambda \in \sigma(A)$ is called a *normal eigenvalue* of a linear operator $A \in \mathscr{C}(\mathbf{X})$ if the following two conditions are satisfied:

(1) The algebraic multiplicity of λ is finite:

$$\nu_\lambda = \dim \mathfrak{L}_\lambda(A) < \infty;$$

(2) The space $\mathbf{X}$ could be decomposed into a direct sum

$$\mathbf{X} = \mathfrak{L}_\lambda(A) \oplus \mathfrak{N}_\lambda, \tag{III.22}$$

where $\mathfrak{N}_\lambda$ is an invariant subspace of A in which $A - \lambda I$ has a bounded inverse. That is, the restriction A_2 of A onto $\mathfrak{N}_\lambda$ is an operator with domain $\mathfrak{D}(A_2) = \mathfrak{N}_\lambda \cap \mathfrak{D}(A)$ and $\mathfrak{R}(A_2 - \lambda I) \subset \mathfrak{N}_\lambda$, which has a bounded inverse.

The following equivalence result is based on [**GK57**, Theorems 4.1, 4.2] and on [**GK69**, Theorem 2.1].

Theorem III.88 *Let $A : \mathbf{X} \to \mathbf{X}$ be a closed densely defined linear operator. The following statements are equivalent:*

(1) $\lambda \in \sigma(A)$ is a normal eigenvalue *(in the sense of Definition* III.87*);*
(2) λ is an isolated point in $\sigma(A)$ and $A - \lambda I$ is semi-Fredholm;
(3) λ is an isolated point in $\sigma(A)$ and $A - \lambda I$ is Fredholm of index zero;
(4) λ is an isolated point in $\sigma(A)$ and $\operatorname{rank} P_\lambda < \infty$*;*
(5) λ is an isolated point in $\sigma(A)$, $\dim \mathfrak{L}_\lambda(A) < \infty$, and $\mathfrak{R}(A - \lambda I)$ is closed.

Moreover, if $\lambda \in \sigma(A)$ is a normal eigenvalue, then

$$\mathfrak{L}_\lambda(A) = \mathfrak{R}(P_\lambda). \tag{III.23}$$

PROOF. *(1) ⇒ (2), (3).* Assume that λ is a normal eigenvalue of the operator A, in the sense of Definition III.87. Let A_1 and A_2 be the restrictions of A onto $\mathfrak{L}_\lambda(A)$ and $\mathfrak{N}_\lambda(A)$, respectively. The operator $B_1 := A_1 - \lambda I$ is nilpotent since its spectrum consists of $z = 0$ only and $\dim \mathfrak{L}_\lambda < \infty$. So there is $n \in \mathbb{N}$ such that $B_1^n = 0$. Then there is the identity

$$-(z-\lambda)^n I = B_1^n - (z-\lambda)^n I = (A_1 - zI)\big[(z-\lambda)^{n-1} I + (z-\lambda)^{n-2} B_1 + \ldots + B_1^{n-1}\big],$$

hence

$$-(A_1 - \lambda)^{-1} = (z-\lambda)^{-1} I + (z-\lambda)^{-2} B_1 + \ldots + (z-\lambda)^{-n} B_1^{n-1}. \tag{III.24}$$

According to the definition of a normal eigenvalue, the operator $A_2 - \lambda I$ has a bounded inverse in $\mathfrak{N}_\lambda$; denoting $R_0 = (A_2 - \lambda I)^{-1}$, for z close to λ, one obtains:

$$(A_2 - zI)^{-1} = R_0 (I - (z-\lambda) R_0)^{-1} = \sum_{j \in \mathbb{N}_0} (z-\lambda)^j R_0^{1+j}. \tag{III.25}$$

Therefore, for $z \neq \lambda$, denoting by P the projection from $\mathbf{X}$ onto $\mathfrak{L}_\lambda$ parallel to $\mathfrak{N}_\lambda$ and using (III.24) and (III.25), we have:

(III.26)

$$\begin{aligned}(A - zI)^{-1} &= (A_1 - zI)^{-1}P + (A_2 - zI)^{-1}(I - P)\\ &= -(z-\lambda)^{-1}P - \sum_{j=1}^{n-1}(z-\lambda)^{-j-1}B_1^j P + \sum_{j\in\mathbb{N}_0}(z-\lambda)^j R_0^{1+j}(I-P),\end{aligned}$$

valid for $0 < |z - \lambda| < \epsilon$, with some $\epsilon > 0$. This shows that λ is an isolated point of the spectrum of A:

$$\overline{\mathbb{D}_\epsilon(\lambda)} \cap \sigma(A) = \{\lambda\}.$$

Let Γ denote a circle of radius $\epsilon/2 > 0$ around λ. Integrating (III.26) over Γ, we have:

$$-\frac{1}{2\pi \mathrm{i}}\oint_\Gamma (A - zI)^{-1}\,dz = \frac{1}{2\pi\mathrm{i}}\oint_\Gamma \frac{P\,dz}{z-\lambda} = P.$$

This shows that P_λ (the left-hand side of the above relation) coincides with the projection P of $\mathbf{X}$ onto $\mathfrak{L}_\lambda$ parallel to $\mathfrak{N}_\lambda$, proving (III.23).

To prove that $A - \lambda I$ is Fredholm, we notice that both

$$\mathbf{ker}(A - \lambda I) = \mathbf{ker}\left((A-\lambda I)|_{\mathfrak{L}_\lambda}\right) \oplus \mathbf{ker}\left((A-\lambda I)|_{\mathfrak{N}_\lambda}\right)$$

and

$$\begin{aligned}\mathbf{coker}(A - \lambda I) &= \mathbf{X}/\mathfrak{R}(A-\lambda I)\\ &= (\mathfrak{L}_\lambda \oplus \mathfrak{N}_\lambda)/\Big(\mathfrak{R}\big((A-\lambda I)|_{\mathfrak{L}_\lambda}\big) \oplus \mathfrak{R}\big((A-\lambda I)|_{\mathfrak{N}_\lambda}\big)\Big)\\ &= \Big(\mathfrak{L}_\lambda/\mathfrak{R}\big((A-\lambda I)|_{\mathfrak{L}_\lambda}\big)\Big) \oplus \Big(\mathfrak{N}_\lambda/\mathfrak{R}\big((A-\lambda I)|_{\mathfrak{N}_\lambda}\big)\Big)\end{aligned}$$

are finite-dimensional since $\mathfrak{L}_\lambda$ is finite-dimensional while $\mathbf{ker}(A - \lambda I)|_{\mathfrak{N}_\lambda} = \{0\}$ and $\mathfrak{R}(A - \lambda I)|_{\mathfrak{N}_\lambda} = \mathfrak{N}_\lambda$.

The conclusion that $\mathfrak{R}(A - \lambda I)$ is closed follows from Theorem III.71 and the finiteness of $\dim \mathbf{coker}(A - \lambda I)$.

Thus, $A - \lambda I$ is Fredholm. Since the operator $A - zI$, $0 < |z - \lambda| < \epsilon$ (with $\epsilon > 0$ small), is invertible with bounded inverse, one has $\operatorname{ind}(A - zI) = 0$, and then $A - \lambda I$ is a Fredholm operator of index zero by Theorem III.78. Theorem III.88 *(3)* follows.

(2) ⇒ (3). Assume that $A - \lambda I$ is semi-Fredholm and that λ is an isolated point of the spectrum. Then there is $\epsilon > 0$ such that $A - zI$ has a bounded inverse for all z such that $0 < |z - \lambda| < \epsilon$, so z belongs to a component of Φ_A where $\operatorname{ind}(A - zI) = 0$; by continuity of the index (Theorem III.78), $\operatorname{ind}(A - \lambda I) = 0$, so $\dim \mathbf{coker}(A - \lambda I) = \dim \mathbf{ker}(A - \lambda I)$ is also finite, showing that $A - \lambda I$ is Fredholm of index zero.

(3) ⇒ (4). Let $A - \lambda I$ be Fredholm of index zero. Let P_λ be the Riesz projector corresponding to the isolated point $\lambda \in \sigma(A)$; we need to prove that $\mathfrak{R}(P_\lambda)$ is finite-dimensional. We denote by A_1 the restriction of A onto $\mathfrak{R}(P_\lambda)$; then $\sigma(A_1) = \{\lambda\}$. $\mathfrak{R}(A_1 - \lambda I)$ is closed since it coincides with $P_\lambda \mathfrak{R}(A - \lambda I)$, which is a closed subspace of $\mathbf{X}$ since it is the kernel of the restriction of $I - P_\lambda$ to $\mathfrak{R}(A - \lambda I)$. Since $\mathbf{ker}(A_1 - \lambda I) = \mathbf{ker}(A - \lambda I)$ is finite-dimensional, while all points $z \neq \lambda$ from an open neighborhood of λ are regular points of A_1 (since $\mathfrak{R}(P_\lambda)$ is an invariant subspace of

the operator $A - zI$, which is invertible with bounded inverse for $0 < |z - \lambda| < \epsilon$), $A_1 - \lambda I$ is Fredholm of index zero. Thus, for each $z \in \mathbb{C}$, the restriction $A_1 - zI$ of $A - zI$ onto $\mathfrak{R}(P_\lambda)$ is Fredholm of index zero.

By [**GK57**, Theorem 3.2] (see Theorem III.91 below), since A_1 is a bounded linear operator acting in the Banach space $\mathfrak{M} = \mathfrak{R}(P_\lambda)$ such that $A_1 - zI$ is Fredholm for each $z \in \mathbb{C}$, this Banach space is finite-dimensional:

$$\dim \mathfrak{R}(P_\lambda) < \infty.$$

(4) ⇒ (1). If $\lambda \in \mathbb{C}$ is an isolated point of $\sigma(A)$ and $\operatorname{rank} P_\lambda < \infty$, then we consider A in $\mathfrak{M} = P_\lambda \mathbf{X}$ (which is finite-dimensional) and in $\mathfrak{N} = (I - P_\lambda)\mathbf{X}$, which are invariant subspaces of A. The restriction of A onto $\mathfrak{M}$ has as its spectrum the only point λ (see Lemma III.86), hence $(A - \lambda I)^\nu \mathfrak{M} = 0$, where $\nu = \operatorname{rank} P_\lambda < \infty$.

The point λ is a regular point of the restriction of A onto $\mathfrak{N}$. Then $\mathfrak{M}$ is a root subspace of A corresponding to λ, hence $\mathbf{X} = \mathfrak{M} \oplus \mathfrak{N} = \mathfrak{L}_\lambda(A) \oplus \mathfrak{N}$ is the required invariant decomposition, so that λ is a normal eigenvalue.

We have shown the equivalence of *(1)*, *(2)*, *(3)*, and *(4)*.

(5) ⇒ (2) follows from Definition III.67.

Finally, *(5)* follows from *(2)* (the conclusion that $\mathfrak{R}(A - \lambda I)$ is closed) and *(4)* and (III.23) ($\dim \mathfrak{L}_\lambda(A) = \operatorname{rank} P_\lambda < \infty$). This completes the proof of Theorem III.88. □

Problem III.89 Show that the requirement that $\mathfrak{R}(A - \lambda I)$ is closed in Theorem III.88 *(5)* is necessary: Give an example of an operator A such that $0 \in \sigma(A)$ is an isolated point in the spectrum, $\dim \mathfrak{L}(A) < \infty$, but $\operatorname{rank} P_0 = \infty$ (hence $\lambda = 0$ is not a normal eigenvalue).

Remark III.90 The requirement that $\dim \mathfrak{L}_\lambda(A) < \infty$ in Theorem III.88 *(5)* is also necessary. For simplicity, let us consider the situation when $\sigma(A) = \{0\}$; we recall that, by Problem III.63, this condition is equivalent to A being quasinilpotent. There are examples of quasinilpotent, non-nilpotent, bounded operators such that $\mathfrak{R}(A)$ is closed (and so is $\mathfrak{R}(A^n)$ for any $n \in \mathbb{N}$); see [**Apo76**, Theorem 3] and also [**Bur05**]. Such operators satisfy $\dim \mathbf{ker}(A) = \dim \mathbf{coker}(A) = \infty$ and thus are not semi-Fredholm; $\lambda = 0 \in \sigma(A)$ is not a normal eigenvalue.

Theorem III.91 *Let A be a bounded operator in $\mathbf{X}$ such that $\Phi_A = \mathbb{C}$ (that is, $A - zI$ is Fredholm for each $z \in \mathbb{C}$). Then $\mathbf{X}$ is finite-dimensional.*

PROOF. We reproduce the proof from [**GK57**, Theorem 3.2]. Given $A \in \mathscr{B}(\mathbf{X})$, it is invertible with bounded inverse for $\lambda \in \mathbb{C}$ larger than $\|A\|$; since $A - zI$ is Fredholm for all $z \in \mathbb{C}$, it then follows by Theorem III.78 that $\operatorname{ind}(A - zI) = 0$ for all $z \in \mathbb{C}$. Consider $\hat{A}$, the class of A in the Calkin algebra $\hat{\Omega} := \mathscr{B}(\mathbf{X})/\mathscr{B}_0(\mathbf{X})$, equipped with the norm

$$\|\hat{A}\|_{\hat{\Omega}} = \inf_{K \in \mathscr{B}_0(\mathbf{X})} \|A + K\|.$$

If we assume that $\mathbf{X}$ is infinite-dimensional, then for each $z \in \mathbb{C}$, $A - zI$ is not compact (since it is Fredholm; see Remark III.69) hence $\hat{A} - z\hat{I}$ is nonzero and is invertible in $\hat{\Omega}$ (this follows from the relations (III.16) in Problem III.68). On the other hand, by [**GRS46**] (see also [**Nai72**, Theorem II.8.1]), each element $\boldsymbol{a}$ of a unital Banach algebra $\mathcal{R}$ has a spectrum: that is, there is $\lambda \in \mathbb{C}$ such that $\boldsymbol{a} - \lambda \boldsymbol{e}$, with $\boldsymbol{e}$ the unit element of $\mathcal{R}$, is not invertible in the algebra, as the following lemma shows.

Lemma III.92 (Non-void spectrum) *Let $\mathcal{R}$ be a Banach algebra over the field of complex numbers, with the unit element which we denote $\boldsymbol{e} \in \mathcal{R}$, $\|\boldsymbol{e}\| = 1$. Then for each $\boldsymbol{a} \in \mathcal{R}$, its spectrum is non-void: that is, there is $\lambda \in \mathbb{C}$ such that $\boldsymbol{a} - \lambda\boldsymbol{e}$ does not have an inverse in $\mathcal{R}$.*

PROOF. We essentially reproduce the proof from [**Nai72**, Theorem II.8.1] noting that one does not need to assume that the algebra $\mathcal{R}$ is commutative. Pick $\boldsymbol{a} \in \mathcal{R}$. There is $z_0 \in \mathbb{C}$ such that $\boldsymbol{a} - z_0\boldsymbol{e}$ has the inverse (or else we are done); we denote this inverse by $\boldsymbol{r} \in \mathcal{R}$. Consider the $\mathcal{R}$-valued function defined by

$$\boldsymbol{f}(z) = (\boldsymbol{a}-z\boldsymbol{e})^{-1} = (\boldsymbol{a}-z_0\boldsymbol{e}-(z-z_0)\boldsymbol{e})^{-1} = (\boldsymbol{e}-(z-z_0)\boldsymbol{r})^{-1}\boldsymbol{r} = \sum_{j=0}^{\infty}(z-z_0)^j\boldsymbol{r}^{j+1},$$

with $z \in \mathbb{C}$, which is convergent for $|z - z_0| < 1/\|\boldsymbol{r}\|$, defining an analytic $\mathcal{R}$-valued function. If all the points $z \in \mathbb{C}$ were regular points of $\boldsymbol{f}(z)$, we could consider the Cauchy integral representation

$$\boldsymbol{f}(z) = \frac{1}{2\pi \mathrm{i}} \oint_{|\zeta - z|=R} \frac{\boldsymbol{f}(\zeta)}{\zeta - z}\, d\zeta, \qquad \forall z \in \mathbb{C}, \qquad R > 0. \tag{III.27}$$

We note that

$$\boldsymbol{f}(z) = (\boldsymbol{a} - z\boldsymbol{e})^{-1} = -z^{-1}(\boldsymbol{e} - z^{-1}\boldsymbol{a})^{-1} = -\sum_{j=0}^{\infty} z^{-j-1}\boldsymbol{a}^j,$$

which is absolutely convergent for $|z| > \|\boldsymbol{a}\|$, with $\|\boldsymbol{f}(z)\| \to 0$, uniformly as $|z| \to \infty$. Similarly to the proof of the Liouville theorem, sending R to infinity, we conclude from (III.27) that $\boldsymbol{f}(z) = 0$, at any given $z \in \mathbb{C}$, in contradiction to the assumption that it is the inverse to $\boldsymbol{a} - z\boldsymbol{e}$. □

This completes the proof of Theorem III.91. □

Problem III.93 Show that an unbounded operator can have a void spectrum. *Hint: Consider derivatives on bounded intervals.*

Problem III.94 ([Rud73, Exercise 5.9]) Let $\{\boldsymbol{e}_n\}_{n\in\mathbb{Z}}$ be the canonical basis in $l^2(\mathbb{Z})$. Let $E = \operatorname{Span}\{\boldsymbol{e}_n,\, n \in \mathbb{N}\}$ and $F = \operatorname{Span}\{\boldsymbol{e}_{-n} + n\boldsymbol{e}_n,\, n \in \mathbb{N}\}$. Show that

(1) $E \cap F = \{0\}$;
(2) $E + F$ is dense in $l^2(\mathbb{Z})$ but not closed.

III.5.3 The discrete spectrum.

Definition III.95 (Discrete spectrum) Let $A \in \mathscr{C}(\mathbf{X})$, with $\mathfrak{D}(A)$ being dense. The *discrete spectrum* $\sigma_{\mathrm{d}}(A)$ is the set of *normal eigenvalues* of A (in the sense of Definition III.87).

By Theorem III.88, this means that the discrete spectrum could be characterized as a set of isolated points λ of the spectrum with the corresponding Riesz projectors P_λ of finite rank, or, equivalently, as a set of isolated points λ of the spectrum with $A - \lambda I$ being Fredholm.

Problem III.96 Prove that $\lambda \in \sigma_{\mathrm{d}}(A)$ if and only if $\bar{\lambda} \in \sigma_{\mathrm{d}}(A^*)$.

III.6 Operators in the Hilbert space: symmetric, normal, self-adjoint

In this section, we assume that $\mathbf{H}$ is a complex Hilbert space and that $A : \mathbf{H} \to \mathbf{H}$ is a linear operator with dense domain $\mathfrak{D}(A)$.

III.6.1 Symmetric operators. A linear operator $A : \mathbf{H} \to \mathbf{H}$ with dense domain $\mathfrak{D}(A)$ is *symmetric* if

$$\langle Ax, y\rangle = \langle x, Ay\rangle, \qquad \forall x,\, y \in \mathfrak{D}(A). \tag{III.28}$$

Problem III.97 Prove that if A is symmetric, then $A \subset A^*$, and A is closable.

Problem III.98 Prove that if A is symmetric, then $\sigma_{\mathrm{p}}(A) \subset \mathbb{R}$ and that the eigenvectors corresponding to different eigenvalues are mutually orthogonal.

Problem III.99 Prove that if A is closed and symmetric, then $\mathfrak{R}(A-zI)$ is closed for any z in $\mathbb{C} \setminus \mathbb{R}$.
Hint: Use $\|(A - zI)x\| \ge |\operatorname{Im} z|\|x\|$, *for all* $x \in \mathfrak{D}(A)$.

Lemma III.100 *If A is symmetric, then $\sigma(A)\setminus\sigma_{\mathrm{res}}(A) \subset \mathbb{R}$.*

PROOF. For any $x \in \mathfrak{D}(A)$, the equality $\operatorname{Im}\langle x, Ax\rangle = 0$ leads to

$$\|(A - zI)x\|^2 = \|(A - \operatorname{Re} z)x\|^2 + |\operatorname{Im} z|^2\|x\|^2,$$

$$\|(A - zI)x\| \ge |\operatorname{Im} z|\|x\|, \qquad \forall z \in \mathbb{C},$$

so that A has a bounded inverse as an operator $\mathfrak{D}(A) \to \mathfrak{R}(A-zI)$. Let us assume that $z \notin \sigma_{\mathrm{res}}(A)$ and $\operatorname{Im} z \ne 0$. Then we have $\overline{\mathfrak{R}(A - zI)} = \mathbf{H}$, and it follows that $A-zI$ has a bounded inverse hence z is in the resolvent set of A. We conclude that $\sigma(A) \setminus \sigma_{\mathrm{res}}(A) \subset \mathbb{R}$. □

Problem III.101 Find the mistake in the following reasoning:

> *Let us try to prove that, for A symmetric, the range of $A - zI$ is dense in $\mathbf{H}$ as long as $\operatorname{Im} z \ne 0$. If this were not the case, there would be nonzero $y \in \mathfrak{D}(A)$ orthogonal to $\mathfrak{R}(A - zI)$; hence for all $x \in \mathfrak{D}(A)$,*
>
> $$0 = \langle y, (A - zI)x\rangle = \langle (A - \bar{z}I)y, x\rangle,$$
>
> *which is impossible (with $x = y$) since* $\operatorname{Im} z \ne 0$ *while* $\langle Ay, y\rangle \in \mathbb{R}$.

III.6.2 Normal operators. A linear operator $A : \mathbf{H} \to \mathbf{H}$ is called *normal* if

$$\mathfrak{D}(A) = \mathfrak{D}(A^*), \qquad \|Ax\| = \|A^*x\| \quad \forall x \in \mathfrak{D}(A).$$

Heuristically, we say that A is normal if it commutes with A^*, although such a statement presents difficulties for unbounded operators due to domain considerations.

Problem III.102 Show that if $A : \mathbf{H} \to \mathbf{H}$ is bounded, then A is normal if and only if it commutes with A^*.
Hint: Consider the identity $\langle A^*Ax, x\rangle = \|Ax\|^2 = \|A^*x\|^2 = \langle AA^*x, x\rangle$.

Problem III.103 Prove that if $A : \mathbf{H} \to \mathbf{H}$ is normal, then

$$\mathfrak{L}_\lambda(A) = \mathbf{ker}(A - \lambda I) \qquad \forall \lambda \in \sigma(A).$$

Solution. Without loss of generality, we assume that $\lambda = 0$. We need to show that A does not have Jordan blocks. Assume that

$$u \in \mathfrak{D}(A), \quad Au = v \in \mathfrak{D}(A), \qquad Av = 0.$$

Then $0 = \|Av\|^2 = \|A^*v\|^2$, hence $\|v\|^2 = \langle Au, v\rangle = \langle u, A^*v\rangle = 0$, thus $u \in \mathbf{ker}(A)$.

Lemma III.104 *If A is a normal bounded operator, then its norm equals its spectral radius $r(A)$ (see* (III.9)*):*

$$\|A\| = r(A).$$

PROOF. If a normal bounded operator T is also symmetric, so that $T = T^*$, then the result holds due to Gelfand's formula $r(T) = \lim_{j\to\infty} \|T^j\|^{1/j}$ (see Lemma III.53) and the relation

$$\|T\| = \|T^2\|^{1/2} = \|^{1/4} = \dots .$$

For a bounded normal operator A, Gelfand's formula and the relation $\|(A^*A)^j\| = \|A^j\|^2$, $j \in \mathbb{N}$, yield

$$r(A)^2 = \lim_{j\to\infty} \|A^j\|2/j = \lim_{j\to\infty} \|(A^*A)^j\|^{1/j} = r(A^*A),$$

and then $r(A)^2 = r(A^*A) = \|A^*A\| = \|A\|^2$. □

According to Lemma III.104, in the case of normal operators, the inequality (III.7) turns into equality:

Lemma III.105 *If A is a normal linear operator, then*

$$\|(A - zI)^{-1}\| = \frac{1}{\operatorname{dist}(z, \sigma(A))}, \qquad \forall z \in \rho(A).$$

Problem III.106 Show that if A is normal, then it is closed and has no strict normal extension (namely, prove that if N is normal and $A \subset N$, then $A = N$). *Hint: Show that $\mathfrak{D}(A) \subset \mathfrak{D}(N) = \mathfrak{D}(N^*) \subset \mathfrak{D}(A^*) = \mathfrak{D}(A)$.*

Problem III.107 Prove that if A is normal, then so is A^*.

Problem III.108 Prove that if A is normal, then it is closed and $A = (A^*)^*$.

Lemma III.109 *If A is normal, then $\sigma_{\mathrm{res}}(A) = \emptyset$ and $\sigma(A) = \sigma_{\mathrm{ap}}(A)$.*

PROOF. We first note that for $z \notin \sigma_{\mathrm{ap}}(A)$, there is $c > 0$ such that

$$c\|x\| \le \|(A - zI)x\| = \|(A^* - \bar{z}I)x\|, \qquad \forall x \in \mathfrak{D}(A); \tag{III.29}$$

the last equality is due to $A - zI$ being normal. By Problem III.62, it suffices to show that $\sigma_{\mathrm{res}}(A) = \emptyset$; if we had $z \in \sigma_{\mathrm{res}}(A)$, so that $\mathfrak{R}(A - zI)$ were not dense in $\mathbf{H}$, there would be nonzero $y \in \mathbf{H}$ such that $\langle y, (A - zI)x\rangle = 0$ for all $x \in \mathfrak{D}(A)$; hence $y \in \mathfrak{D}(A^*)$ and $(A^* - \bar{z}I)y = 0$. By (III.29), one has $y = 0$, which leads to a contradiction. □

For normal operators, the set of normal eigenvalues coincides with the common understanding of the discrete spectrum as the set of isolated eigenvalues of finite multiplicity:

Lemma III.110 *Let* $\mathbf{H}$ *be a Hilbert space and let* $A : \mathbf{H} \to \mathbf{H}$ *be a normal operator. Then* $\lambda \in \sigma(A)$ *is a normal eigenvalue of* A *if and only if it is an isolated point of the spectrum of* A *of finite geometric multiplicity.*

PROOF. If $\lambda \in \sigma(A)$ is a normal eigenvalue, then by Theorem III.88 *(5)* it is of finite algebraic (and hence geometric) multiplicity.

Going in the opposite direction, we assume that $\lambda \in \sigma(A)$ is an isolated point of the spectrum of finite geometric multiplicity. Due to Problem III.103, its algebraic multiplicity is also finite:

$$\nu_\lambda = \dim \mathfrak{L}_\lambda(A) < \infty.$$

According to Lemma III.86, we know that $\mathfrak{R}(I - P_\lambda)$ is an invariant subspace of A in which $A - \lambda I$ has a bounded inverse; here P_λ is the Riesz projector (III.21) with $\Gamma = \partial\mathbb{D}_\epsilon(\lambda)$ and with $\epsilon > 0$ sufficiently small (so that $\sigma(A) \cap \overline{\mathbb{D}_\epsilon(\lambda)} = \{\lambda\}$). Let us show that $\mathfrak{R}(P_\lambda) = \mathfrak{L}_\lambda(A)$. Without loss of generality, we can assume that $\lambda = 0$ (or else we consider $A - \lambda I$ which is also normal). Then, instead of A, we consider its restriction onto $\mathfrak{R}(P_0)$, which is the invariant subspace of A; this restriction, which we denote by A_1, is also normal since it is bounded and commutes with its adjoint. So A_1 is now a bounded normal operator such that $\sigma(A_1) = 0$; by Lemma III.104, $A_1 = 0$. This means that $A|_{\mathfrak{R}(P_0)} \equiv 0$, so $\mathfrak{R}(P_0) \subset \mathbf{ker}(A)$ and then $\mathfrak{R}(P_0) = \mathfrak{L}(A)$. Taking into account that $A|_{\mathfrak{R}(I-P_0)}$ has a bounded inverse, the decomposition $\mathbf{X} = \mathfrak{L}(A) \oplus \mathfrak{R}(I - P_0)$ shows that $\lambda = 0$ is a normal eigenvalue of A. This concludes the proof. □

III.6.3 Self-adjoint operators. We say that a linear operator $A : \mathbf{H} \to \mathbf{H}$ is *self-adjoint* if $A^* = A$; this means that a self-adjoint operator is both symmetric and normal. Conversely, if A is both symmetric (hence $A \subset A^*$; see Problem III.97) and normal (so that $\mathfrak{D}(A^*) = \mathfrak{D}(A)$), then A is self-adjoint.

Theorem III.111 (Kato–Rellich [Kat76, Theorem V.4.3**])** *Let* A *be self-adjoint. If* B *is symmetric and* A*-bounded with an* A*-bound less than one, then* $A+B$ *is also self-adjoint on* $\mathfrak{D}(A)$*. In particular,* $A + B$ *is self-adjoint if* B *is bounded and symmetric.*

III.7 Essential spectra and the Weyl theorem

III.7.1 The essential spectra. In the case of nonselfadjoint operators, there are many different, non-equivalent definitions of the essential spectrum; see [**ELZ83**, §1.5] and [**EE87**, §I.4]. All the definitions of the essential spectrum use the concept of the Fredholm operator; we assume in this section that

$$\mathbf{X} \text{ is a Banach space}, \qquad A \in \mathscr{C}(\mathbf{X}), \qquad \mathfrak{D}(A) \text{ is dense.}$$

Definition III.112 (Essential spectra) Let $\mathbf{X}$ be a Banach space and let A be a closed linear densely defined operator.

(1) $\lambda \in \sigma_{\mathrm{ess},1}(A)$ if $A - \lambda I$ is not semi-Fredholm: either $\mathfrak{R}(A - \lambda I)$ is not closed, or both $\mathbf{ker}(A - \lambda I)$ and $\mathbf{coker}(A - \lambda I)$ are infinite-dimensional;

(2) $\lambda \in \sigma_{\mathrm{ess},2}(A)$ if either $\mathfrak{R}(A - \lambda I)$ is not closed or $\dim \mathbf{ker}(A - \lambda I) = \infty$;

(3) $\lambda \in \sigma_{\mathrm{ess},3}(A)$ if $A - \lambda I$ is not Fredholm: either $\mathfrak{R}(A - \lambda I)$ is not closed or at least one of $\mathbf{ker}(A - \lambda I)$, $\mathbf{coker}(A - \lambda I)$ is infinite-dimensional;

(4) $\lambda \in \sigma_{\mathrm{ess},4}(A)$ if $A - \lambda I$ is not Fredholm of index zero;

(5) The essential spectrum $\sigma_{\mathrm{ess},5}(A)$ is the union of $\sigma_{\mathrm{ess},1}(A)$ with all components of $\mathbb{C} \setminus \sigma_{\mathrm{ess},1}(A)$ that do not intersect with the resolvent set of A.

We note that there are the following relations between the spectra and the Fredholm domains of A (see Definition III.67):

$$\text{(III.30)}\qquad \begin{aligned} &\sigma_{\mathrm{ess},1}(A) = \mathbb{C} \setminus (\Phi_A^+ \cup \Phi_A^-) = \sigma_{\mathrm{ess},2}(A) \cap \{\, z\colon \quad \bar{z} \in \sigma_{\mathrm{ess},2}(A^*) \,\}, \\ &\sigma_{\mathrm{ess},2}(A) = \mathbb{C} \setminus \Phi_A^+, \qquad \sigma_{\mathrm{ess},2}(A^*) = \mathbb{C} \setminus \Phi_A^-, \\ &\sigma_{\mathrm{ess},3}(A) = \mathbb{C} \setminus (\Phi_A^+ \cap \Phi_A^-) = \sigma_{\mathrm{ess},2}(A) \cup \{\, z\colon \quad \bar{z} \in \sigma_{\mathrm{ess},2}(A^*) \,\}. \end{aligned}$$

Remark III.113 Given $A \in \mathscr{C}(\mathbf{X})$, the set $\sigma_{\mathrm{ess},3}(A)$ is sometimes called the *Fredholm spectrum* of A (the set of λ such that $A - \lambda I$ is not Fredholm) while $\sigma_{\mathrm{ess},4}(A)$ is sometimes called the *Weyl spectrum* ($A - \lambda I$ is not Fredholm of index zero).

Lemma III.114 $\sigma_{\mathrm{ess},1}(A) \subset \sigma_{\mathrm{ess},2}(A) \subset \sigma_{\mathrm{ess},3}(A) \subset \sigma_{\mathrm{ess},4}(A) \subset \sigma_{\mathrm{ess},5}(A)$.

PROOF. The first three inclusions are immediate; we need to justify the inclusion $\sigma_{\mathrm{ess},4}(A) \subset \sigma_{\mathrm{ess},5}(A)$. If $\lambda \in \sigma_{\mathrm{ess},1}(A)$, then by definition $\lambda \in \sigma_{\mathrm{ess},5}(A)$. Now assume that $\lambda \in \sigma_{\mathrm{ess},4}(A) \setminus \sigma_{\mathrm{ess},1}(A)$. Since on the set $z \in \Psi_A$ where $A - zI$ is semi-Fredholm both $\dim \mathbf{ker}(A - zI)$ and $\dim \mathbf{coker}(A - zI)$ are semicontinuous from below (see Theorem III.78 or e.g. [**EE87**, Theorem I.3.22]), there are several types of components of $\mathbb{C} \setminus \sigma_{\mathrm{ess},1}(A)$:

(1) Both $\mathbf{ker}(A - zI)$ and $\mathbf{coker}(A - zI)$ are finite-dimensional and index is zero;
(2) Both $\mathbf{ker}(A - zI)$ and $\mathbf{coker}(A - zI)$ are finite-dimensional, index is nonzero;
(3) At least one of $\mathbf{ker}(A - zI)$ and $\mathbf{coker}(A - zI)$ is infinite-dimensional.

Only the first of the three types of components can have intersections with the resolvent set of A (where both kernel and cokernel have dimension zero), while the assumption $\lambda \in \sigma_{\mathrm{ess},4}(A) \setminus \sigma_{\mathrm{ess},1}(A)$ shows that λ can only belong to the last two types. Thus, λ belongs to a component of $\mathbb{C} \setminus \sigma_{\mathrm{ess},1}(A)$ which has no intersection with the resolvent set and so by definition belongs to $\sigma_{\mathrm{ess},5}(A)$. □

Problem III.115 Show that for the left shift operator L on $l^2(\mathbb{N})$ one has $0 \in \sigma_{\mathrm{ess},4}(L)$, while $0 \notin \sigma_{\mathrm{ess},3}(L)$. Prove the same for the right shift R on $l^2(\mathbb{N})$.

Example III.116 Consider $J : l^2(\mathbb{N}) \to l^2(\mathbb{N})$, $\boldsymbol{e}_j \mapsto \boldsymbol{e}_{2j}$ for each $j \in \mathbb{N}$. Then $\mathbf{ker}(J) = \{0\}$, $\dim \mathbf{coker}(J) = \infty$, hence $0 \in \sigma_{\mathrm{ess},2}(J)$ while $0 \notin \sigma_{\mathrm{ess},1}(J)$.

Problem III.117 Check that $\mathfrak{R}(J)$ is closed. Find $\sigma(J)$.

Problem III.118 Check that $0 \in \sigma_{\mathrm{ess},3}(J)$ but $0 \notin \sigma_{\mathrm{ess},2}(J)$.

Example III.119 Consider the right shift operator R on $l^2(\mathbb{N})$. Recall that

$$\sigma(R) = \sigma_{\mathrm{res}}(R) = \overline{\mathbb{D}_1}.$$

For any $z \in \partial\mathbb{D}_1$, the range of $R - zI$ is dense but not closed. (For simplicity, let $z = 1$; one can check that $\mathfrak{R}(R - zI)$ is dense in $l^2(\mathbb{N})$, but one can show that, for example, $\phi = (1, 1/2, 1/3, \dots) \in l^2(\mathbb{N}) \setminus l^1(\mathbb{N})$ is not in the range of $R - zI$. Alternatively, if $\mathfrak{R}(R - zI)$ were dense and closed (thus all of $l^2(\mathbb{N})$), since $\mathbf{ker}(R - zI)$ is zero, $R - zI$ would be invertible with bounded inverse, while we know that $z \in \sigma(R)$.) Since $\mathfrak{R}(R - zI)$ is not closed, $\partial\mathbb{D}_1 \subset \sigma_{\mathrm{ess},1}(R)$. On the other hand, for any $z \in \mathbb{D}_1$, the operator $R - zI$ has closed range, zero kernel, and one-dimensional cokernel, thus $\mathbb{D}_1 \not\subset \sigma_{\mathrm{ess},k}(R)$ for $1 \le k \le 4$. At the same time, $\sigma_{\mathrm{ess},5}(R)$ contains $\sigma_{\mathrm{ess},1}(R)$ and also $\mathbb{D}_1$, which is a connected component of $\mathbb{C} \setminus \sigma_{\mathrm{ess},1}(R)$ not intersecting the resolvent set of R; thus, $\sigma_{\mathrm{ess},5}(R) = \overline{\mathbb{D}_1}$.

Problem III.120 Compare the decomposition of the spectrum (III.13) to different definitions of the essential spectrum in Definition III.112.

Problem III.121 ([ELZ83, Theorem 1.5], [EE87, Theorem IX.1.6]) Let $\mathbf{H}$ be a Hilbert space and assume that $A \in \mathscr{C}(\mathbf{H})$ is self-adjoint. Then $\sigma_{\mathrm{ess},k}(A)$ for $1 \le k \le 5$ coincide.
Hint: Since for a self-adjoint operator A one has $\sigma(A) \subset \mathbb{R}$, any connected component of $\mathbb{C} \setminus \sigma_{\mathrm{ess},1}(A)$ has a nonempty intersection with the resolvent set of A.

Example III.122 On $l^2(\mathbb{N})$, consider the operator

$$T : \boldsymbol{e}_j \mapsto \frac{1}{j+1}\boldsymbol{e}_{j+1}.$$

Then, for each $n \in \mathbb{N}$, $X_n = \left\{ \sum_{j \ge n} c_j \boldsymbol{e}_j \in l^2(\mathbb{N}) \right\}$ is an invariant subspace of T, and $\|T|_{X_n}\| \le \frac{1}{n+1}$, hence for $1 \le k \le 4$, $\sigma_{\mathrm{ess},k}(T) \subset \mathbb{D}_\epsilon$ for any $\epsilon > 0$, thus $\sigma_{\mathrm{ess},k}(T) = \{0\}$. At the same time, since $\sigma(T) \subsetneq \mathbb{C}$, one also has $\sigma_{\mathrm{ess},5}(T) = \{0\}$, while $\sigma_{\mathrm{p}}(T) = \emptyset$.

Problem III.123 Construct an example of an operator A with an isolated point $\lambda \in \sigma(A)$ in its spectrum such that A restricted to the corresponding spectral subspace, $\mathfrak{R}(P_\lambda)$, has an infinite size Jordan block.
Hint: Describe T^*, where T is from Example III.122.

Theorem III.124 ([EE87, Theorem IX.1.1]) *Let $A : \mathbf{X} \to \mathbf{X}$ be a closed linear operator in the Banach space $\mathbf{X}$ with a dense domain $\mathfrak{D}(A)$. For $1 \le k \le 5$,*

$$\lambda \in \sigma_{\mathrm{ess},k}(A) \quad \textit{if and only if} \quad \bar\lambda \in \sigma_{\mathrm{ess},k}(A^*).$$

Proof. For $1 \le k \le 4$, it is enough to mention that for a closed operator with a dense domain, $\mathfrak{R}(A^*)$ is closed if and only if $\mathfrak{R}(A)$ is closed and then to use the observation that

$$\lambda \in \Phi_A^{\pm} \quad \text{if and only if} \quad \bar\lambda \in \Phi_{A^*}^{\mp}$$

and $\operatorname{ind} A = -\operatorname{ind} A^*$. All these statements follow from Theorem III.73.

To prove the equality $\sigma_{\mathrm{ess},5}(A^*) = \sigma_{\mathrm{ess},5}(A)$, it is enough to notice that λ belongs to a connected component of $\Phi_{A^*}^{+} \cup \Phi_{A^*}^{-}$ which intersects the resolvent set of A^* if and only if $\bar\lambda$ belongs to a connected component of $\Phi_A^{-} \cup \Phi_A^{+}$ which intersects the resolvent set of A. □

Lemma III.125 *Let $A \in \mathscr{C}(\mathbf{X})$, with $\mathfrak{D}(A)$ being dense. The spectrum of A is a disjoint union of its* normal eigenvalues *(Definition* III.87*) and $\sigma_{\mathrm{ess},5}(A)$:*

$$\sigma(A) = \left\{ \textit{normal eigenvalues of } A \right\} \sqcup \sigma_{\mathrm{ess},5}(A). \tag{III.31}$$

Proof. Assume that λ is a normal eigenvalue: thus, $\lambda \in \sigma(A)$ is an isolated point of the spectrum, with $A - \lambda I$ a Fredholm operator. By Theorem III.88 *(3)*, $\operatorname{ind}(A - \lambda I) = 0$. It follows that $\lambda \notin \sigma_{\mathrm{ess},k}(A)$ for $1 \le k \le 4$. Moreover, $\mathbb{D}_\epsilon(\lambda) \in \mathbb{C} \setminus \sigma_{\mathrm{ess},1}(A)$, while $\mathbb{D}_\epsilon(\lambda)$ intersects the resolvent set of A; by Definition III.112, $\lambda \notin \sigma_{\mathrm{ess},5}(A)$.

On the contrary, assume that $\lambda \in \sigma(A)$ is not a normal eigenvalue. Thus, either it is an isolated spectral point but $A - \lambda I$ is not Fredholm, or it is not an isolated point of the spectrum. In the first case, $\lambda \in \sigma_{\mathrm{ess},3}(A)$. In the second case, if $\lambda \in \sigma_{\mathrm{ess},1}(A)$, we are done, so we assume that λ belongs to one of the components of $\mathbb{C} \setminus \sigma_{\mathrm{ess},1}(A)$. We claim that $\lambda \in \sigma_{\mathrm{ess},5}(A)$. Indeed, if λ were in a connected component of $\Phi_A^- \cup \Phi_A^+$ which had a nonempty intersection with the resolvent set, on which $A - zI$ had a bounded inverse except at isolated points, $\lambda \in \sigma(A)$ would imply that $A - zI$ were invertible with bounded inverse for $0 < |z - \lambda| < \epsilon$ for some $\epsilon > 0$, contradicting the assumption that λ is not an isolated point of the spectrum. □

Example III.119 and Lemma III.125 suggest that we use the following definition of the essential spectrum.

Definition III.126 (Essential spectrum) Let $A : \mathbf{X} \to \mathbf{X}$ be a closed linear operator in the Banach space $\mathbf{X}$. The *essential spectrum* $\sigma_{\mathrm{ess}}(A)$ is defined by

$$\sigma_{\mathrm{ess}}(A) = \sigma_{\mathrm{ess},5}(A). \tag{III.32}$$

According to the definition of the discrete spectrum as the set of *normal eigenvalues* (Definition III.95) and the definition of the essential spectrum as $\sigma_{\mathrm{ess},5}$ (Definition III.126), the relation (III.31) from Lemma III.125 shows that the spectrum is the disjoint union of the discrete spectrum and the essential spectrum:

$$\sigma(A) = \sigma_{\mathrm{d}}(A) \sqcup \sigma_{\mathrm{ess}}(A).$$

Problem III.127 Which of the sets

$$\sigma_{\mathrm{p}}(A), \quad \sigma_{\mathrm{cont}}(A), \quad \sigma_{\mathrm{res}}(A) \quad \sigma_{\mathrm{ap}}(A), \quad \sigma_{\mathrm{d}}(A), \quad \text{and} \quad \sigma_{\mathrm{ess},k}(A) \text{ for } 1 \le k \le 5$$

do not have to be closed?

III.7.2 Weyl sequences. There is an alternative characterization of the essential spectrum based on Weyl sequences, which we present for completeness.

Definition III.128 (Weyl sequences) We say that a sequence $u_j \in \mathbf{X}$, $j \in \mathbb{N}$, is a *Weyl sequence* of the operator A corresponding to $\lambda \in \mathbb{C}$ if

$$x_j \in \mathfrak{D}(A), \qquad \|x_j\| = 1 \quad \forall j \in \mathbb{N}, \qquad (A - \lambda I)x_j \to 0,$$

and moreover $\left(x_j\right)_{j \in \mathbb{N}}$ does not contain a convergent subsequence.

A Weyl sequence is also known as a *singular sequence* or a *singular Weyl sequence.*

Problem III.129 Prove that for the left shift L on $l^2(\mathbb{N})$, when $0 \in \sigma_{\mathrm{ess},4}(L)$ but $0 \notin \sigma_{\mathrm{ess},3}(L)$ (see Problem III.115), there is no Weyl sequence corresponding to $\lambda = 0$.

Remark III.130 If the Banach space $\mathbf{X}$ is reflexive, then there exists a Weyl sequence corresponding to $\lambda \in \mathbb{C}$ if and only if there is a sequence $\left(x_j\right)_{j \in \mathbb{N}}$ such that

$$x_j \in \mathfrak{D}(A), \qquad \|x_j\| = 1 \quad \forall j \in \mathbb{N}, \quad x_j \to 0 \text{ weakly in } \mathbf{X}, \quad (A - \lambda I)x_j \to 0.$$

Theorem III.131 (Weyl's criterion [EE87, Theorem IX.1.3]) *Let $A \in \mathscr{C}(\mathbf{X})$, with its domain $\mathfrak{D}(A)$ dense in $\mathbf{X}$. Then $\lambda \in \sigma_{\mathrm{ess},2}(A)$ if and only if A has a* Weyl sequence *corresponding to λ.*

Problem III.132 Prove Theorem III.131.
Hint: For the only if *part, use the Cauchy criterion to prove that if* $\mathfrak{R}(A-\lambda I)$ *is not closed, then there is a* Weyl sequence. *For the* if *part, use Lemma* III.72 *to prove that if* $\mathfrak{R}(A-\lambda I)$ *is closed and there is a Weyl sequence, then* $\mathbf{ker}(A-\lambda I)$ *is infinite-dimensional.*

For more details on the Weyl sequences, see [**Wol59**, **EE87**].

Example III.133 By Problem III.129, the left shift operator L on $l^2(\mathbb{N})$ has no Weyl sequence corresponding to $\lambda = 0$, thus, by Theorem III.131, $0 \notin \sigma_{\mathrm{ess},2}(L)$. This agrees with a stronger conclusion $0 \notin \sigma_{\mathrm{ess},3}(L)$ in Problem III.115.

III.7.3 The Weyl theorem on the stability of the essential spectrum. The stability of the essential spectrum for the self-adjoint operators under addition of symmetric compact operators was proved by Hermann Weyl [**Wey09**]. Informally, the Weyl theorem states that *the essential spectrum does not change under relatively compact perturbations*, although in the case of nonselfadjoint operators one has to be careful as to which of the definitions of the essential spectrum is used. Different definitions (Definition III.112) are important in view of the fact that only $\sigma_{\mathrm{ess},k}(A)$ with $1 \le k \le 4$, which are defined purely in terms of Fredholm properties, are invariant under relatively compact perturbations. At the same time, the spectrum $\sigma_{\mathrm{ess},5}(A)$ is not invariant under compact perturbations, as one can see from the following example.

Example III.134 ([EE87, Example IX.2.2**])** Consider the operator

$$T : l^2(\mathbb{Z}) \to l^2(\mathbb{Z}), \qquad \boldsymbol{e}_0 \mapsto 0, \qquad \boldsymbol{e}_j \mapsto \boldsymbol{e}_{j-1} \text{ for } j \in \mathbb{Z}\setminus\{0\}.$$

One can see that $\sigma_{\mathrm{p}}(T) = \mathbb{D}_1$ (with an eigenvector corresponding to $0 < |\lambda| < 1$ given by $\sum_{j\in\mathbb{N}_0} \lambda^j \boldsymbol{e}_j$), thus $\mathbb{D}_1 \subset \sigma_{\mathrm{ess},5}(T)$. Define a compact operator (which is, moreover, rank one operator) $K : \boldsymbol{e}_0 \mapsto \boldsymbol{e}_{-1}$, $\boldsymbol{e}_j \mapsto 0$ for $j \in \mathbb{Z}\setminus\{0\}$. Then $T + K = L$ is the left shift operator on $\mathbb{Z}$, with $\sigma(L) = \partial\mathbb{D}_1$. This shows that $\sigma_{\mathrm{ess},5}$ is not invariant under compact perturbations.

Now we formulate and prove the Weyl theorem on the invariance of the essential spectra $\sigma_{\mathrm{ess},k}$, $1 \le k \le 4$ from Definition III.112.

Theorem III.135 (Weyl's theorem on the essential spectrum) *Let* $\mathbf{X}$ *be a Banach space and let* A *be a closed densely defined linear operator. Let* $B : \mathbf{X} \to \mathbf{X}$ *be a linear* A*-compact operator. Then*

$$\Phi^{\pm}_{A+B} = \Phi^{\pm}_{A}; \qquad \sigma_{\mathrm{ess},k}(A+B) = \sigma_{\mathrm{ess},k}(A), \qquad 1 \le k \le 4. \tag{III.33}$$

Similar results for nonselfadjoint operators in Banach spaces are proved in [**GK57**], [**Sch66**, Theorem 2.1], and [**EE87**, Theorem IX.2.1]. The approach to the nonselfadjoint operators in the Hilbert space is given in [**Wol59**, §1], with the proofs based on the concept of a Weyl sequence. In [**RS78**, Theorem XIII.14, Corollary 2], the proof is given under assumption that A is a self-adjoint operator in the Hilbert space.

Proof. By Theorem III.79, a relatively compact perturbation of a closed densely defined operator preserves the sets $\Phi^{\pm}_{A}$ (see Definition III.67), so that $\Phi^{\pm}_{A+B} = \Phi^{\pm}_{A}$. Consequently, the Fredholm set $\Phi = \Phi^{+} \cap \Phi^{-}$ and semi-Fredholm set $\Psi = \Phi^{+} \cup \Phi^{-}$ are also preserved:

$$\Phi_{A+B} = \Phi_{A}, \qquad \Psi_{A+B} = \Psi_{A},$$

and also for each $\lambda \in \Psi_{A+B} = \Psi_A$ one has $\operatorname{ind}(A+B-\lambda I) = \operatorname{ind}(A-\lambda I)$ (including the cases when $\operatorname{ind} A = \pm\infty$). It follows that

$$\sigma_{\mathrm{ess},k}(A+B) = \sigma_{\mathrm{ess},k}(A), \qquad 1 \le k \le 4. \qquad \square$$

The following theorem confirms that $\sigma_{\mathrm{ess},4}$ is the largest subset of the spectrum which is preserved by compact perturbations.

Theorem III.136 (**[EE87**, Theorem IX.1.4]) *Let $A \in \mathscr{C}(\mathbf{X})$, with $\mathfrak{D}(A)$ being dense. Then*

$$\sigma_{\mathrm{ess},4}(A) = \bigcap_{B \in \mathscr{B}_0(\mathbf{X})} \sigma(A+B).$$

Proof. Let $\lambda \notin \cap_{K \in \mathscr{B}_0} \sigma(A+K)$. This means that there is $K_1 \in \mathscr{B}_0(\mathbf{X})$ such that $\lambda \in \rho(A+K_1)$, hence $\lambda \in \Phi_{A+K_1}$ and $\operatorname{ind}(A+K_1-\lambda I) = 0$; then, by Theorem III.79, $\lambda \in \Phi_A$ and $\operatorname{ind}(A-\lambda I) = 0$, so $\lambda \notin \sigma_{\mathrm{ess},4}(A)$.

Conversely, assume that $\lambda \notin \sigma_{\mathrm{ess},4}(A)$; that is, $\lambda \in \Phi_A$, $\operatorname{ind}(A-\lambda I) = 0$. We need to show that $\lambda \notin \sigma(A+K_1)$ for some $K_1 \in \mathscr{B}_0(\mathbf{X})$. Let x_i, $1 \le i \le n = \dim \mathbf{ker}(A-\lambda I)$, be the basis in $\mathbf{ker}(A-\lambda I)$, with $\xi_i \in \mathbf{X}^*$ a dual basis (that is, such that $\langle \xi_i, x_j \rangle = \delta_{ij}$ for all $1 \le i, j \le n$). Let $\eta_i \in \mathbf{X}^*$, $1 \le i \le n$, be the basis in $\mathfrak{R}(A-\lambda I)^\perp \subset \mathbf{X}^*$ (we are taking into account that $\mathbf{ker}(A-\lambda I)$ and $\mathbf{coker}(A-\lambda I)$ have the same dimension n since $\operatorname{ind}(A-\lambda I) = 0$); let $y_i \in \mathbf{X}$ be a dual basis. We define $K_1 \in \mathscr{B}_0(\mathbf{X})$ by

$$K_1 = \sum_{i=1}^{n} y_i \otimes \xi_i, \qquad K_1 : \mathbf{X} \to \mathbf{X}.$$

Then, by construction, $A + K_1 - \lambda I$ is closed, has zero kernel, and is onto, so $(A+K_1-\lambda I)^{-1}$ exists, with its domain the whole space $\mathbf{X}$. Since this operator is also closed (see Problem III.13), it is bounded by the closed graph theorem (Theorem III.15); then $\lambda \in \rho(A+K_1)$, finishing the proof. $\square$

Problem III.137 Show that σ_{cont} is not invariant under compact perturbations.

III.8 The Schur complement

We will often use the concept of the Schur complement, which is sometimes called the Feshbach map; its history is given in [**PS05**]. Let us give the formulation in the operator form that we will need (see also [**BFS98**, Theorem IV.1] and [**JN01**, Lemma 2.3]).

Lemma III.138 *Let $\mathbf{X}_1$ and $\mathbf{X}_2$ be Banach spaces and let A be a closed linear operator on $\mathbf{X}_1 \oplus \mathbf{X}_2$ with dense domain $\mathfrak{D}(A) \subset \mathbf{X}_1 \oplus \mathbf{X}_2$, such that*

$$A = \begin{bmatrix} A_{11} & A_{12} \\ A_{21} & A_{22} \end{bmatrix},$$

where $A_{ji} : \mathbf{X}_i \to \mathbf{X}_j$, $1 \le i, j \le 2$ are closed linear operators with domains

$$\mathfrak{D}(A_{j1}) = \{x_1 \in \mathbf{X}_1 :\ (x_1, 0) \in \mathfrak{D}(A)\}, \quad \mathfrak{D}(A_{j2}) = \{x_2 \in \mathbf{X}_2 :\ (0, x_2) \in \mathfrak{D}(A)\}$$

which are dense in $\mathbf{X}_1$ and $\mathbf{X}_2$, respectively. Assume that there are Banach spaces

$$\mathbf{W}_i \subset \mathbf{X}_i \subset \mathbf{Y}_i, \qquad 1 \le i \le 2,$$

with continuous dense embeddings, such that A_{ji} *extends to a bounded operator from* $\mathbf{W}_i$ *to* $\mathbf{Y}_j$*, with* $1 \le i, j \le 2$*. Assume further that* $A_{11} : \mathbf{W}_1 \to \mathbf{Y}_1$ *has a bounded inverse. Let* S *be the Schur complement of* A_{11} *defined by*

$$S = A_{22} - A_{21} A_{11}^{-1} A_{12} : \mathbf{W}_2 \to \mathbf{Y}_2. \tag{III.34}$$

Then A *(considered as an operator from* $\mathfrak{D}(A)$ *to* $\mathbf{X}_1 \oplus \mathbf{X}_2$*) has a bounded inverse if and only if* $S : \mathbf{W}_2 \to \mathbf{Y}_2$ *is invertible (not necessarily with the bounded inverse) and the restrictions* $S^{-1}|_{\mathbf{X}_2}$*,* $A_{11}^{-1} A_{12} S^{-1}|_{\mathbf{X}_2}$*,* $S^{-1} A_{21} A_{11}^{-1}|_{\mathbf{X}_1}$*, and* $A_{11}^{-1} A_{12} S^{-1} A_{21} A_{11}^{-1}|_{\mathbf{X}_1}$ *define the bounded operators*

$$\begin{aligned} S^{-1} : \mathbf{X}_2 &\to \mathbf{X}_2, \\ A_{11}^{-1} A_{12} S^{-1} : \mathbf{X}_2 &\to \mathbf{X}_1, \\ S^{-1} A_{21} A_{11}^{-1} : \mathbf{X}_1 &\to \mathbf{X}_2, \\ A_{11}^{-1} A_{12} S^{-1} A_{21} A_{11}^{-1} : \mathbf{X}_1 &\to \mathbf{X}_1. \end{aligned}$$

PROOF. We consider A as a bounded operator from $\mathbf{W}_1 \oplus \mathbf{W}_2$ to $\mathbf{Y}_1 \oplus \mathbf{Y}_2$. Since $A_{11} : \mathbf{W}_1 \to \mathbf{Y}_1$ has a bounded inverse, the operators

$$\begin{bmatrix} I & A_{11}^{-1} A_{12} \\ 0 & I \end{bmatrix} \qquad \text{and} \qquad \begin{bmatrix} I & 0 \\ A_{21} A_{11}^{-1} & I \end{bmatrix}$$

are bounded as linear operators in $\mathbf{W}_1 \oplus \mathbf{W}_2$ and in $\mathbf{Y}_1 \oplus \mathbf{Y}_2$, respectively, with bounded inverses given by

$$\begin{bmatrix} I & -A_{11}^{-1} A_{12} \\ 0 & I \end{bmatrix} \qquad \text{and} \qquad \begin{bmatrix} I & 0 \\ -A_{21} A_{11}^{-1} & I \end{bmatrix}.$$

The following identities show that S is invertible as a map from $\mathbf{W}_2$ to $\mathbf{Y}_2$ if and only if A is invertible as a map from $\mathbf{W}_1 \oplus \mathbf{W}_2$ to $\mathbf{Y}_1 \oplus \mathbf{Y}_2$:

$$A = \begin{bmatrix} A_{11} & A_{12} \\ A_{21} & A_{22} \end{bmatrix} = \begin{bmatrix} I & 0 \\ A_{21} A_{11}^{-1} & I \end{bmatrix} \begin{bmatrix} A_{11} & 0 \\ 0 & S \end{bmatrix} \begin{bmatrix} I & A_{11}^{-1} A_{12} \\ 0 & I \end{bmatrix}, \tag{III.35}$$

$$\begin{bmatrix} I & 0 \\ -A_{21} A_{11}^{-1} & I \end{bmatrix} \begin{bmatrix} A_{11} & A_{12} \\ A_{21} & A_{22} \end{bmatrix} \begin{bmatrix} I & -A_{11}^{-1} A_{12} \\ 0 & I \end{bmatrix} = \begin{bmatrix} A_{11} & 0 \\ 0 & S \end{bmatrix}. \tag{III.36}$$

Explicitly, the inverse of A in terms of the inverse of S is given by

$$A^{-1} = \begin{bmatrix} A_{11}^{-1} + A_{11}^{-1} A_{12} S^{-1} A_{21} A_{11}^{-1} & -A_{11}^{-1} A_{12} S^{-1} \\ -S^{-1} A_{21} A_{11}^{-1} & S^{-1} \end{bmatrix}; \tag{III.37}$$

this expression is known as the Banachiewicz inversion formula. One can see that A (considered as a map from $\mathfrak{D}(A)$ to $\mathbf{X}_1 \oplus \mathbf{X}_2$) has a bounded inverse if and only if the mappings S^{-1}, $A_{11}^{-1} A_{12} S^{-1}$, $S^{-1} A_{21} A_{11}^{-1}$, and $A_{11}^{-1} A_{12} S^{-1} A_{21} A_{11}^{-1}$ are bounded in appropriate spaces. □

Remark III.139 Swapping the indices in Lemma III.138, we have the following result. Assume that $A_{22} : \mathbf{W}_2 \to \mathbf{Y}_2$ has a bounded inverse. Let T be the Schur complement of A_{22} defined by

$$T = A_{11} - A_{12} A_{22}^{-1} A_{21} : \mathbf{W}_1 \to \mathbf{Y}_1. \tag{III.38}$$

Then A (considered as an operator from $\mathfrak{D}(A)$ to $\mathbf{X}_1 \oplus \mathbf{X}_2$) has a bounded inverse if and only if $T : \mathbf{W}_1 \to \mathbf{Y}_1$ is invertible and the restrictions $T^{-1}|_{\mathbf{X}_1}$, $A_{22}^{-1} A_{21} T^{-1}|_{\mathbf{X}_1}$,

$T^{-1}A_{12}A_{22}^{-1}|_{\mathbf{X}_2}$, and $A_{22}^{-1}A_{21}T^{-1}A_{12}A_{22}^{-1}|_{\mathbf{X}_2}$ define the bounded operators

$$\begin{aligned} T^{-1} : \mathbf{X}_1 &\to \mathbf{X}_1, \\ A_{22}^{-1}A_{21}T^{-1} : \mathbf{X}_1 &\to \mathbf{X}_2, \\ T^{-1}A_{12}A_{22}^{-1} : \mathbf{X}_2 &\to \mathbf{X}_1, \\ A_{22}^{-1}A_{21}T^{-1}A_{12}A_{22}^{-1} : \mathbf{X}_2 &\to \mathbf{X}_2. \end{aligned}$$

Similarly to (III.37), the inverse of A in terms of the inverse of T is given in the explicit form by

$$A^{-1} = \begin{bmatrix} T^{-1} & -T^{-1}A_{12}A_{22}^{-1} \\ -A_{22}^{-1}A_{21}T^{-1} & A_{22}^{-1} + A_{22}^{-1}A_{21}T^{-1}A_{12}A_{22}^{-1} \end{bmatrix}. \tag{III.39}$$

III.9 The Keldysh theory of characteristic roots

We will need the theory of characteristic roots formulated by M. Keldysh [**Kel51**, **Kel71**]; see also [**MS70**, **GS71**]. Informally, characteristic roots can be viewed as "nonlinear eigenvalues". Let $\mathbf{A} : \mathbb{C} \to \operatorname{End}(\mathbb{C}^N)$, $N \in \mathbb{N}$, be a matrix-valued analytic function. We say that $z_0 \in \mathbb{C}$ is a *characteristic root* of $\mathbf{A}(z)$ if $\det \mathbf{A}(z_0) = 0$. For example, let us consider

$$\mathbf{A}(z) = \begin{bmatrix} z^3 & 1 \\ 0 & z^4 \end{bmatrix}, \qquad z \in \mathbb{C}. \tag{III.40}$$

Clearly, $z_0 = 0$ is a characteristic root. How many characteristic roots can bifurcate from $\lambda = 0 \in \sigma(\mathbf{A}(z_0))$ under perturbations? Let us assume that there is a perturbation $\mathbf{A}(\epsilon, z) \in \operatorname{End}(\mathbb{C}^2)$, continuous in $\epsilon \in \mathbb{R}$ and analytic in $z \in \mathbb{C}$, with $\mathbf{A}(0, z) = \mathbf{A}(z)$. Since the order of vanishing of $\det \mathbf{A}(z)$ at $z_0 = 0$ is $\alpha = 7$, given a small open neighborhood U of $z_0 = 0$, by the Rouché theorem, there could be up to $\alpha = 7$ zeros $Z_j(\epsilon)$, $1 \le j \le \alpha$, of $\det \mathbf{A}(\epsilon, z)$ inside U as long as the parameter ϵ is from a sufficiently small open neighborhood of the origin (so that additional zeros do not enter U from the outside).

Example III.140 For $\mathbf{A}(z)$ defined in (III.40), let us consider its perturbation given by

$$\mathbf{A}(\epsilon, z) = \begin{bmatrix} z^3 & 1 \\ \epsilon & z^4 \end{bmatrix};$$

the relation

$$0 = \det \mathbf{A}(\epsilon, z) = z^7 - \epsilon$$

gives seven families of characteristic roots $Z_j(\epsilon)$, $1 \le j \le 7$, bifurcating from the characteristic root $z_0 = 0$. On the other hand, the perturbation $\mathbf{A}(\epsilon, z) = \begin{bmatrix} (z-\epsilon)^3 & 1 \\ 0 & (z+\epsilon)^4 \end{bmatrix}$ leads to two families of characteristic roots, $Z_\pm(\epsilon) = \pm\epsilon$, bifurcating from $z_0 = 0$.

Now let us consider the analytic family of unbounded operators. Let $\mathbf{X}$ be a separable Hilbert space and $\Omega \subset \mathbb{C}$ an open set. We assume that the family of closed operators $A(z) : \mathbf{X} \to \mathbf{X}$, $z \in \Omega$, is a *holomorphic family of type (A)* [**Kat76**, Section VII.2]: that is, we assume that operators have the same domain $\mathfrak{D}(A(z)) = \mathfrak{D}$ for all $z \in \Omega$, with $\mathfrak{D}$ dense in $\mathbf{X}$, and that for any $x \in \mathfrak{D}$ the vector-function $A(z)x$ is holomorphic in Ω. It follows that for any $x, y \in \mathfrak{D}$, each

$z_0 \in \Omega$, and each η in the resolvent set of $A(z_0)$, the function $\langle x, (A(z) - \eta)^{-1} y \rangle$ is analytic in z in an open neighborhood of z_0.

Definition III.141 (Characteristic roots) If $z_0 \in \Omega$ is such that the equation

$$A(z_0)\varphi = 0$$

has a non-trivial solution $\varphi_0 \in \mathfrak{D}$, then we call z_0 a *characteristic root* of A and φ_0 an eigenvector of A corresponding to the characteristic root z_0. The characteristic root z_0 of A is said to be *normal* if $A(z_0)$ is a Fredholm operator and if there is $R > 0$ such that for all $z \in \Omega$ satisfying $0 < |z - z_0| < R$ the operator $A(z)$ has a bounded inverse.

Assumption III.142 We assume that $z_0 \in \Omega$ is an isolated characteristic root of $A(z)$ and that $0 \in \sigma(A(z_0))$ is a normal eigenvalue (in the sense of Definition III.87).

In particular, this means that $\lambda = 0$ is an isolated point of the spectrum of $A(z_0)$; we fix $\delta > 0$ such that

$$\overline{\mathbb{D}_\delta} \cap \sigma(A(z_0)) = \{0\}. \tag{III.41}$$

We note that $A(z)$ is resolvent-continuous in z in the uniform operator topology, in the following sense:

Definition III.143 The operator family $A(z)$, $z \in \Omega$, is *resolvent-continuous* in z at $z_0 \in \Omega$ in the weak (or strong, or uniform) operator topology if for each $\eta \in \rho(A(z_0))$ there is an open subset $U \ni z_0$ of Ω such that for each $z \in U$ one has $\eta \in \mathbb{C} \setminus \sigma(A(z))$ and the resolvent $(A(z) - \eta)^{-1}$ as a function of z is continuous at z_0 in this topology.

Due to the resolvent continuity of A in z in the uniform operator topology, there is an open neighborhood $U \subset \Omega$ of z_0 such that $\Gamma := \partial\mathbb{D}_\delta \subset \rho(A(z))$ for all $z \in U$. For a particular $z \in U$, we consider the *Riesz projector* (see Definition III.83)

$$P_\Gamma(z) = -\frac{1}{2\pi \mathrm{i}} \oint_\Gamma (A(z) - \eta)^{-1}\, d\eta, \qquad z \in U. \tag{III.42}$$

Since $\lambda = 0$ is assumed to be a normal eigenvalue of $A(z_0)$,

$$\nu := \operatorname{rank} P_\Gamma(z_0) = \operatorname{rank} P_\Gamma(z) < \infty.$$

We choose the basis $\{\psi_i\}_{1 \le i \le \nu}$ in $\mathfrak{R}(P_\Gamma(z_0))$ and define

$$\psi_i(z) := P_\Gamma(z)\psi_i \in \mathfrak{R}(P_\Gamma(z)), \qquad 1 \le i \le \nu, \qquad z \in U.$$

We take U smaller if necessary so that for each $z \in U$ the vectors $\psi_i(z)$, $1 \le i \le \nu$, are linearly independent. For $z \in U$, let $\mathbf{A}(z) \in \operatorname{End}(\mathbb{C}^\nu)$ be the matrix representation of $A(z)|_{\mathfrak{R}(P_\Gamma(z))}$ in the basis $\{\psi_i(z)\}_{1 \le i \le \nu}$.

Definition III.144 The *multiplicity* $\alpha \in \mathbb{N}$ of the characteristic root z_0 of $A(z)$ is the order of vanishing of $\det \mathbf{A}(z)$ at z_0.

We note that the order of vanishing of $\det \mathbf{A}(z)$ at z_0 does not depend on the choice of basis $\{\psi_i\}_{1 \le i \le \nu}$ in $\mathfrak{R}(P_\Gamma(z_0))$.

Lemma III.145 *Let z_0 be a normal characteristic root of $A(z)$ of multiplicity $\alpha \in \mathbb{N}$ (see Definition* III.141*). The geometric multiplicity $g := \dim \mathbf{ker}(A(z_0))$ of $0 \in \sigma(A(z_0))$ satisfies $g \le \alpha$.*

PROOF. Denote $\nu = \dim \mathfrak{R}(P_\Gamma(z_0)) < \infty$. We choose the basis $\{\psi_i\}_{1\le i\le\nu}$ in $\mathfrak{R}(P_\Gamma(z_0))$ so that $\psi_i \in \mathbf{ker}(A(z_0))$ for $1 \le i \le g$. Let $\mathbf{A}(z)$ be the matrix representation of $A(z)|_{\mathfrak{R}(P_\Gamma(z))}$ in the basis $\{P_\Gamma(z)\psi_i\}_{1\le i\le\nu}$. Then the first g columns of $\mathbf{A}(z)$ vanish at $z = z_0$, hence $\det \mathbf{A}(z) = O((z-z_0)^g)$. □

Remark III.146 There is no relation between multiplicity of a characteristic root and the algebraic multiplicity of zero eigenvalue. For example, the characteristic root $z_0 = 0$ will be of arbitrary multiplicity $\alpha \in \mathbb{N}$ if one sets $M(z) = \begin{bmatrix} 0 & 1 \\ z^\alpha & 0 \end{bmatrix}$, while the algebraic multiplicity of the zero eigenvalue will not change.

According to the operator version of the Rouché theorem [**GS71**, Theorem 2.1], the sum of multiplicities of characteristic roots is stable under perturbations:

Theorem III.147 *Let $A(\epsilon, z) : \mathbf{X} \to \mathbf{X}$, $\epsilon \in \mathbb{R}$, $z \in \Omega$, be a continuous family of operators with domains $\mathfrak{D}(A(\epsilon,z)) = \mathfrak{D}$, $\forall(\epsilon,z) \in \mathbb{R}\times\Omega$, which is resolvent-continuous in ϵ and z and analytic in $z \in \Omega$. Let z_0 be a normal characteristic root of $A(0,z)$ of multiplicity $\alpha \in \mathbb{N}$. There is an open neighborhood $\Omega \subset \mathbb{C}$ of z_0 such that if $\epsilon \ge 0$ is sufficiently small, then the sum of multiplicities of all characteristic values of $A(\epsilon,z)$ inside Ω equals α.*

PROOF. Fix $\delta > 0$ sufficiently small so that $\sigma(A(0,z_0)) \cap \overline{\mathbb{D}_\delta} = \{0\}$. Due to the resolvent continuity, there is $\epsilon_1 > 0$ and an open neighborhood $\Omega \subset \mathbb{C}$ of z_0 such that for all $\epsilon \in [0,\epsilon_1]$ and all $z \in \Omega$ one has $\sigma(A(\epsilon,z)) \cap \overline{\mathbb{D}_\delta} = \{0\}$. Let $\Gamma = \partial\mathbb{D}_\delta$ and denote

$$P_\Gamma(\epsilon,z) = -\frac{1}{2\pi \mathrm{i}} \oint_\Gamma (A(\epsilon,z) - \eta I)^{-1}\, d\eta, \qquad \epsilon \in [0,\epsilon_1], \quad z \in \Omega.$$

Due to the resolvent continuity of $A(\epsilon,z)$ in ϵ and z, one has

$$\operatorname{rank} P_\Gamma(\epsilon,z) = \operatorname{rank} P_\Gamma(0,z_0) =: \nu, \qquad \forall \epsilon \in [0,1], \quad \forall z \in \Omega.$$

Let $(\psi_i)_{1\le i\le\nu}$ be the basis in $\mathfrak{R}(P_\Gamma(0,z_0))$. Denote

$$\psi_i(\epsilon,z) = P_\Gamma(\epsilon,z)\psi_i, \qquad 1 \le i \le \nu, \quad \epsilon \in [0,1], \quad z \in \Omega;$$

it is a basis in $\mathfrak{R}(P_\Gamma(\epsilon,z))$ (we take Ω and $\epsilon_1 > 0$ smaller if necessary). Let $\mathbf{M}(\epsilon,z)$ be the matrix representation of $A(\epsilon,z)$, restricted onto $\mathfrak{R}(P_\Gamma(\epsilon,z))$, in this basis. If $\epsilon \in [0,\epsilon_1]$ and $z_j \in \Omega$ is a characteristic root of $A(\epsilon,z)$, then its multiplicity equals the order of vanishing of $\det \mathbf{M}(\epsilon,z)$ at z_j. Now the proof follows from Cauchy's argument principle applied to $\det \mathbf{M}(\epsilon,z)$ in Ω when we take Ω and $\epsilon_1 > 0$ small enough so that zeros of $\det \mathbf{M}(\epsilon,z)$ do not intersect $\partial\Omega$. □

III.10 Quantum Mechanics examples

Let us consider a couple of problems from Quantum Mechanics.

Problem III.148 Find the energy levels of a particle trapped in a potential well of infinite height:

$$E\psi = H\psi := -\frac{\hbar^2}{2m}\Delta\psi + V\psi, \qquad x \in \mathbb{R}, \qquad V(x) = \begin{cases} 0, & 0 \le x \le L; \\ +\infty & \text{otherwise.} \end{cases}$$

That is, one needs to find the eigenvalues of the Sturm–Liouville problem

$$E\psi = -\frac{\hbar^2}{2m}\Delta\psi, \qquad 0 \le x \le L; \qquad \psi\big|_{x=0} = \psi\big|_{x=L} = 0.$$

The complex-valued function ψ is called the *wave function* of a particle; $|\psi(x)|^2$ is interpreted as the probability density (the chance to find the electron near the point x).

Problem III.149 A particle described by the Schrödinger equation

$$\text{(III.43)} \qquad \mathrm{i}\hbar\dot{\psi}(t,x) = -\frac{\hbar^2}{2m}\Delta\psi(t,x), \qquad \psi(t,x) \in \mathbb{C}, \qquad x \in \mathbb{R}^3,$$

is contained in the interior Ω of a torus of length L and radius of cross-section $R \ll L$, so that $\psi(t,x)$ in (III.43) is considered as a function in the region Ω, with the Dirichlet boundary condition $\psi|_{\partial\Omega} = 0$. Assume that the particle is initially in the groundstate (the smallest energy state). Estimate the energy needed to cut the torus.
Hint: Apply the Rayleigh quotient

$$E_{\text{groundstate}} = \inf\left\{\frac{\langle u, Hu\rangle}{\langle u, u\rangle}\colon \ u \in C^\infty(\Omega),\ u \ne 0,\ u|_{\partial\Omega} = 0\right\}, \qquad H = -\frac{\hbar^2}{2m}\Delta,$$

with Ω as in the problem and with $\Omega' = [0, L] \times \mathbb{B}^2_R$, to estimate the corresponding groundstate energies. Their difference is the energy required to cut the torus. (When estimating the energies, substitute $\mathbb{B}^2_R$ by a more convenient rectangular cross-section.)

Problem III.150 A particle of mass $m = 1/2$ sits in a well of depth Λ and size $2a$:

$$E\psi = H\psi := -\psi'' + V\psi, \qquad V(x) = \begin{cases} -\Lambda, & x \in (-a, a); \\ 0 & \text{otherwise.} \end{cases}$$

Find approximate values $E < 0$ corresponding to the *even* eigenstates.
Hint: For $x > a$, $\psi(x) = Ae^{-(-E)^{1/2}x}$; for $x \in (-a, a)$, $\psi(x) = B\cos((\Lambda + E)^{1/2}x)$ (since we are interested in even eigenstates). Request the continuity of ψ and ψ' at $x = a$.

In other words, when solving Problem III.150, we are to determine whether for a particular value $E \in (-\Lambda, 0)$ the functions

$$\psi_{\text{inside}}(x, E) := \cos((\Lambda + E)^{1/2}x), \qquad x \in [-a, a];$$

$$\psi_{\text{outside}}(x, E) := e^{-(-E)^{1/2}|x|}, \qquad |x| \ge a$$

correspond to the same initial data at $x = a$ (and hence at $x = -a$).

Definition III.151 The *Evans function* corresponding to even solutions is defined as the Wronskian of $\psi_{\text{inside}}(x)$ and $\psi_{\text{outside}}(x)$:

$$\mathrm{E}_{\text{even}}(z) = \big(\psi_{\text{outside}}(x, z)\partial_x\psi_{\text{inside}}(x, z) - \psi_{\text{inside}}(x, z)\partial_x\psi_{\text{outside}}(x, z)\big)\big|_{x=a}, \quad z \in \mathbb{C}.$$

Zeros of this Evans function correspond to eigenvalues of the Schrödinger operator restricted onto the space of even functions. The Evans function corresponding to odd functions is defined similarly, with sin instead of cos. We note that, as long as V is even, H is invariant in the spaces of even and odd functions; eigenfunctions of H are obtained by restricting it on each of these two subspaces.

Remark III.152 The solutions with purely exponential asymptotics as $x \to +\infty$ or $x \to -\infty$ (such as $\psi_{\text{outside}}(x, \lambda)$) are called *Jost solutions*.

In Problem III.150, the interval $0 \le E < \infty$ is the *essential spectrum* of the equation. To physicists, for any $E \ge 0$, there are solutions $\sim e^{\pm ix\sqrt{E}}$ for $|x|$ large, called *plane waves* (these are particular examples of the Jost solutions). In the mathematical sense, the values $E \ge 0$ are not *eigenvalues* since there are no nontrivial L^2 solutions to $(-\partial_x^2 + V - E)\psi = 0$; yet, they belong to the *essential spectrum* since the operator $-\partial_x^2 + V - E$ has no bounded inverse (on L^2). To show this, take the functions

$$u_j(x) = \rho(x-a)\rho(j-x)\cos(x\sqrt{E}), \qquad j \in \mathbb{N},$$

with $\rho \in C^\infty(\mathbb{R})$ such that $\rho|_{x\le 0} \equiv 0$ and $\rho|_{x\ge 1} \equiv 1$; one can see that $\|u_j\|_{L^2} \to \infty$ as $j \to \infty$, while $\|(-\partial_x^2 + V - E)u_j\|_{L^2}$ remains bounded. This family of functions (or, rather, the rescaled family $u_j/\|u_j\|_{L^2}$) is the *Weyl sequence* from Definition III.128.

Problem III.153 Estimate the groundstate energy of the electron in the Hydrogen atom, described by the stationary Schrödinger equation

$$E\psi = H\psi := -\frac{\hbar^2}{2m}\Delta\psi(x) + \Big(-\frac{e^2}{|x|}\Big)\psi(x), \qquad x \in \mathbb{R}^3. \tag{III.44}$$

Above, $V(x) = -\frac{e^2}{|x|}$ is the potential energy of the electron of negative charge $-e$ at the point $x \in \mathbb{R}^3$ in the Coulomb potential of the nucleus with charge $+e$ and located at the origin.
Hint: Assuming that the eigenfunction ψ that corresponds to the lowest energy bound state, or groundstate, "1s", is spherically symmetric, estimate E using the Rayleigh quotient. As a sample function, take $\psi(x) = e^{-\beta|x|}$ in the Rayleigh quotient and find $\beta > 0$ which gives the best (smallest) value for E.

Solution.

$$\begin{aligned} E_0 &\le \frac{\int_{\mathbb{R}^3}\big(-\frac{\hbar^2}{2m}\psi^*\Delta\psi - \frac{e^2}{|x|}|\psi|^2\big)\,d^3x}{\int_{\mathbb{R}^3}|\psi|^2\,d^3x} = \frac{\int_{\mathbb{R}^3}\big(\frac{\hbar^2}{2m}|\nabla\psi|^2 - \frac{e^2}{|x|}|\psi|^2\big)\,d^3x}{\int_{\mathbb{R}^3}|\psi|^2\,d^3x} \\ &= \frac{\int_{\mathbb{R}_+}\big(\frac{\hbar^2}{2m}|\partial_r\psi|^2 - \frac{e^2}{r}|\psi|^2\big)\,r^2\,dr}{\int_{\mathbb{R}_+}|\psi|^2\,r^2\,dr} = \frac{\int_{\mathbb{R}_+}\big(\frac{\beta^2\hbar^2}{2m}e^{-2\beta r} - \frac{e^2}{r}e^{-2\beta r}\big)\,r^2\,dr}{\int_{\mathbb{R}_+}e^{-2\beta r}\,r^2\,dr} \\ &= \frac{\beta^2\hbar^2}{2m} - e^2\beta. \end{aligned} \tag{III.45}$$

The minimal value is achieved when $\frac{\beta\hbar^2}{m} - e^2 = 0$; we conclude that $\beta = -\frac{me^2}{\hbar^2}$.

The quantity $a = 1/\beta = \frac{\hbar^2}{me^2} \approx 0.5 \cdot 10^{-10}\,\text{m}$ is interpreted as the radius of the Bohr orbit. The second term in the right-hand side of (III.45) is the potential energy of the electron at the distance a from the nucleus. The first term has the meaning of the kinetic energy of a particle moving with the momentum $|p| = \hbar/a$ and with the speed $|p|/m = e^2/\hbar = \alpha c$, where $\alpha = e^2/(\hbar c) \approx 1/137$ is the fine structure constant. Note that

$$\oint_{|q|=a} p\cdot dq = \beta\hbar\cdot 2\pi a = 2\pi\hbar,$$

matching the Bohr–Sommerfeld quantization condition.

Problem III.154 Is $\psi(x) = e^{-\beta|x|}$ an eigenfunction of (III.44)?

Problem III.155 Should the groundstate in the Coulomb potential be smooth?

There are infinitely many solutions to equation (III.44). The solutions from $L^2(\mathbb{R}^3)$ are called eigenstates. Normally, they correspond to negative eigenvalues E, and called *bound states*: *the electron can not escape the nucleus; some energy is needed to pull the electron away to infinity, where its energy would become zero.* The number of bound states of the Schrödinger operator could be either finite or infinite; see [**RS78**, Section XIII.3].

The solutions with $E \geq 0$ normally have infinite L^2-norm. At the same time, there exist examples of potentials $V(x)$, originally constructed by von Neumann and Wigner [**vNW29**], with L^2-eigenstates corresponding to certain positive values of E, which are called *embedded eigenvalues*. Examples of such potentials exist even in one dimension and are characterized by oscillatory behaviour and slow decay at infinity; for more details, see [**RS78**, Section XIII.13].

III.11 Spectrum of the Dirac operator

The standard reference to the Dirac operator and the Dirac equation is Bernd Thaller's book [**Tha92**] (in spatial dimension $n = 3$).

For $N \in \mathbb{N}$, we introduce the L^2 space of vector-valued functions by

$$L^2(\Omega, \mathbb{C}^N) = L^2(\Omega) \otimes_{\mathbb{C}} \mathbb{C}^N.$$

Similarly for other spaces.

Definition III.156 (Dirac operator in $\mathbb{R}^n$) Let $m \geq 0$, n, $N \in \mathbb{N}$. We define the free Dirac operator

$$D_m : L^2(\mathbb{R}^n, \mathbb{C}^N) \to L^2(\mathbb{R}^n, \mathbb{C}^N), \qquad \mathfrak{D}(D_m) = H^1(\mathbb{R}^n, \mathbb{C}^N)$$

by

(III.46) $$D_m : \psi \mapsto D_m\psi = \big(D_0 + \beta m\big)\psi = \big(- \mathrm{i}\boldsymbol{\alpha} \cdot \nabla + \beta m\big)\psi,$$

with

$$D_0 = -\mathrm{i}\boldsymbol{\alpha} \cdot \nabla = -\mathrm{i}\alpha^i \frac{\partial}{\partial x^i},$$

where the summation in $1 \leq i \leq n$ is assumed (we always assume that there is a summation with respect to repeated upper-lower indices, unless specified otherwise). Above, α^i, $1 \leq i \leq n$, and β are the Dirac matrices of size N: that is, they are self-adjoint anticommuting roots of the identity,

$$\{\alpha^i, \alpha^j\} = 2\delta_{ij} I_N, \qquad \beta^2 = I_N, \qquad \{\alpha^i, \beta\} = 0, \qquad 1 \leq i, j \leq n,$$

so that

$$D_m^2 = (-\Delta + m^2) I_N,$$

where I_N is the $N \times N$ identity matrix.

The above definition imposes the restriction $N = 2^d N_0$, with $d = [(n+1)/2]$ and $N_0 \in \mathbb{N}$ (if $m = 0$, one does not need the matrix β and can take $N = 2^d N_0$ with $d = [n/2]$ and $N_0 \in \mathbb{N}$). For the construction and properties of the Dirac matrices in higher dimensions, see Chapter VIII.

Assume that $J \in \mathrm{End}(\mathbb{C}^N)$ satisfies the following properties:

(III.47) $$J^2 = -I_N, \qquad [J, D_m] = 0, \qquad \sigma(J) = \{\pm \mathrm{i}\}.$$

Lemma III.157 *For any $m \geq 0$ and any $\omega \in \mathbb{R}$, one has:*

$$\sigma(D_m) = \sigma_{\mathrm{ess}}(D_m) = \mathbb{R} \setminus (-m, m),$$

$$\sigma(J(D_m - \omega)) = \sigma_{\mathrm{ess}}(J(D_m - \omega)) = \mathrm{i}\big(-\infty, |\omega| - m\big] \cup \mathrm{i}\big[m - |\omega|, +\infty\big).$$

PROOF. Under the Fourier transform, D_m corresponds to the operator $\hat{D}_m$ of multiplication by $\hat{D}_m(\xi) = \boldsymbol{\alpha} \cdot \xi + \beta m$ acting on functions of $\xi \in \mathbb{R}^n$:

$$\hat{D}_m : L^2(\mathbb{R}^n, \mathbb{C}^N) \to L^2(\mathbb{R}^n, \mathbb{C}^N), \qquad u(\xi) \mapsto (\boldsymbol{\alpha} \cdot \xi + \beta m) u(\xi), \tag{III.48}$$

with $\boldsymbol{\alpha} \cdot \xi = \sum_{i=1}^n \alpha^i \xi_i$ the multiplication operator, with the domain

$$\mathfrak{D}(\hat{D}_m) = \{u \in L^2(\mathbb{R}^n, \mathbb{C}^N) \colon \quad \hat{D}_m u \in L^2(\mathbb{R}^n, \mathbb{C}^N)\}. \tag{III.49}$$

One can see that $\hat{D}_m - zI_N$ has a bounded inverse $(\xi^2 + m^2 - z^2)^{-1}(\hat{D}_m + zI_N)$ in $L^2(\mathbb{R}^n, \mathbb{C}^N)$ if and only if $z \notin \mathbb{R} \setminus (-m, m)$. Thus, $\sigma(D_m) = (-\infty, -m] \cup [m, +\infty)$.

Since $[J, D_m] = 0$, $J(D_m - \omega)$ preserves the decomposition into the direct sum

$$L^2(\mathbb{R}^n, \mathfrak{R}(\pi^+)) \oplus L^2(\mathbb{R}^n, \mathfrak{R}(\pi^-)) = L^2(\mathbb{R}^n, \mathbb{C}^N),$$

with $\pi^\pm = (I_N \mp \mathrm{i}J)/2 \in \mathrm{End}(\mathbb{C}^N)$ projectors onto eigenspaces corresponding to eigenvalues $\pm\mathrm{i} \in \sigma(J)$, one has:

$$\sigma\big(J(D_m - \omega)\big) = \bigcup_{\pm} \sigma\big(J(D_m - \omega)|_{\mathfrak{R}(\pi^\pm)}\big) = \{\pm\mathrm{i}z \colon \quad z \in \sigma(D_m - \omega)\},$$

giving the desired expression for $\sigma\big(J(D_m - \omega)\big)$.

Since $\sigma(D_m)$ does not contain isolated points, $\sigma_{\mathrm{d}}(D_m) = \emptyset$, and by Lemma III.125 $\sigma(D_m) = \sigma_{\mathrm{ess},5}(D_m)$, which we chose as our definition of σ_{ess} (Definition III.126). Arguing as in the hint after Problem III.121, we show that all the essential spectra $\sigma_{\mathrm{ess},k}(D_m)$, $1 \leq k \leq 5$, coincide, and so do $\sigma_{\mathrm{ess},k}(J(D_m - \omega))$, $1 \leq k \leq 5$. □

Lemma III.158 *The Dirac operator $D_m : L^2(\mathbb{R}^n, \mathbb{C}^N) \to L^2(\mathbb{R}^n, \mathbb{C}^N)$ with domain $\mathfrak{D}(D_m) = H^1(\mathbb{R}^n, \mathbb{C}^N)$ is self-adjoint.*

PROOF. The proof is by the Fourier transform. One writes

$$D_m = \mathcal{F}^{-1} \circ \hat{D}_m \circ \mathcal{F},$$

with the multiplication operator $\hat{D}_m$ from (III.48) with domain (III.49), which is clearly self-adjoint. It is now enough to mention that $\mathcal{F}^{-1} = (2\pi)^{-n}\mathcal{F}^*$. More details are in [**Kat76**, Chapter V]. □

III.11.1 Stability of the essential spectrum.

To be able to study the spectrum of the linearization at a solitary wave, we need to work with operators of the form L and JL, with

$$L = D_m - \omega + V, \qquad \omega \in [-m, m], \tag{III.50}$$

where $J \in \mathrm{End}(\mathbb{C}^N)$ is a skew-adjoint operator satisfying (III.47) and V is a measurable function of $x \in \mathbb{R}^n$ taking values in $\mathrm{End}(\mathbb{C}^N)$. The domain of L is $\mathfrak{D}(L) = H^1(\mathbb{R}^n, \mathbb{C}^N)$. The operator L is closed.

Problem III.159 ([RS78, Chapter XIII, Problem 41]) Let $V \in L^q(\mathbb{R}^n) + L^\infty_\varepsilon(\mathbb{R}^n)$, $n \geq 1$, with $q = \max\{n/2, 2\}$ if $n \neq 4$ and $q > 2$ if $n = 4$. (That is, for any $\varepsilon > 0$, there is a decomposition $V = V_1 + V_2$ with $V_1 \in L^q$ and $\|V_2\|_{L^\infty} < \varepsilon$.) Prove that V is a relatively compact perturbation of $-\Delta$.

By Weyl's theorem on the essential spectrum (Theorem III.135), $\sigma_{\rm ess}(-\Delta + V) = [0, +\infty)$.

The following result is an adaptation of Problem III.159 to the Dirac operator.

Lemma III.160 *Let $n \ge 1$ and assume that*

$$V \in L^q\big(\mathbb{R}^n, \mathrm{End}(\mathbb{C}^N)\big), \qquad \begin{cases} 2 \le q < \infty, & n = 1; \\ 2 < q < \infty, & n = 2; \\ n \le q < \infty, & n \ge 3. \end{cases} \tag{III.51}$$

Then for any $m \ge 0$ and $\omega \in \mathbb{R}$ the potential V is a relatively compact perturbation of D_m and $J(D_m - \omega)$, so that

$$\sigma_{\rm ess}(D_m + V) = \sigma_{\rm ess}(D_m), \qquad \sigma_{\rm ess}(J(D_m - \omega + V)) = \sigma_{\rm ess}(J(D_m - \omega)).$$

PROOF. We will give the proof for $n \ge 3$; the cases $n \le 2$ are considered similarly. Let A be either D_m or $J(D_m - \omega)$. We need to prove that

$$V(A - z)^{-1} : L^2 \to L^2, \qquad \forall z \in \mathbb{C} \setminus \sigma(A),$$

is compact. Let $2 < p < \infty$ be such that

$$\frac{1}{2} = \frac{1}{p} + \frac{1}{q}; \tag{III.52}$$

note that $p \le \frac{2n}{n-2}$ since $q \ge n$, $n \ge 3$. Let $\mathbb{1}_{\mathbb{B}^n_1}$ be the characteristic function of the unit ball, and set $\chi_j(x) = \mathbb{1}_{\mathbb{B}^n_1}(x/j)$, $x \in \mathbb{R}^n$, $j \in \mathbb{N}$. Then, for $z \in \mathbb{C} \setminus \sigma(A)$,

$$\|(1 - \chi_j)V(A - z)^{-1}\|_{L^2 \to L^2} \le \|(1 - \chi_j)V\|_{L^p \to L^2}\|(A - z)^{-1}\|_{L^2 \to L^p}$$

is bounded (the second factor in the right-hand side is bounded due to the Sobolev embedding $H^1(\mathbb{R}^n) \subset L^p(\mathbb{R}^n)$), and moreover the norm of the operator of multiplication by $(1 - \chi_j)V$ goes to zero as $j \to \infty$:

$$\|(1 - \chi_j)V\|_{L^p \to L^2} \le \|(1 - \chi_j)V\|_{L^q} \underset{j \to \infty}{\longrightarrow} 0.$$

Above, the inequality is due to (III.51) and (III.52). On the other hand, the mapping

$$\chi_j V(A - z)^{-1} : L^2(\mathbb{R}^n, \mathbb{C}^N) \to L^2(\mathbb{R}^n, \mathbb{C}^N), \qquad z \in \mathbb{C} \setminus \sigma(A),$$

is compact, since

$$\chi_j(A - z)^{-1} : L^2(\mathbb{R}^n, \mathbb{C}^N) \to L^p(\mathbb{R}^n, \mathbb{C}^N) \tag{III.53}$$

is compact by the Rellich–Kondrachov compactness theorem (Theorem II.11), if $1 \le p < 2^* = 2n/(n-2)$, that is $q > n$, while the operator of multiplication by V acts continuously from L^p to L^2. Therefore, the operator

$$V(A - z)^{-1} : L^2(\mathbb{R}^n, \mathbb{C}^N) \to L^2(\mathbb{R}^n, \mathbb{C}^N)$$

is a limit of the sequence of compact operators $\chi_j V(A-z)^{-1}$ in the uniform operator norm if $q > n$, and hence is also compact.

To conclude in the case $q = n$, if $V \in L^n\big(\mathbb{R}^n, \mathrm{End}(\mathbb{C}^N)\big)$ then

$$V_j = V\chi_j \circ V = \begin{cases} V, & |V| < j \\ 0, & |V| \ge j \end{cases}$$

is in $L^q(\mathbb{R}^n, \operatorname{End}(\mathbb{C}^N))$ for any $q \in (n, \infty]$ and $V - V_j$ goes to 0 in $L^n(\mathbb{R}^n, \operatorname{End}(\mathbb{C}^N))$ as $j \to \infty$ by dominated convergence. This implies that $V(A-z)^{-1} - V_j(A-z)^{-1}$ goes to 0 in the uniform operator norms as $j \to \infty$. Therefore, the operator

$$V(A-z)^{-1} : L^2(\mathbb{R}^n, \mathbb{C}^N) \to L^2(\mathbb{R}^n, \mathbb{C}^N)$$

is compact as the limit of a sequence of compact operators. □

Problem III.161 Give the proof of Lemma III.160 for $n \leq 2$.

Extend Lemma III.160 to the case

$$V \in L^q(\mathbb{R}^n, \operatorname{End}(\mathbb{C}^N)) + L^\infty_\varepsilon(\mathbb{R}^n, \operatorname{End}(\mathbb{C}^N)), \qquad \begin{cases} 2 \leq q < \infty, & n = 1; \\ 2 < q < \infty, & n = 2; \\ n \leq q < \infty, & n \geq 3. \end{cases}$$

CHAPTER IV

Linear stability of NLS solitary waves

IV.1 Derrick's instability theorem vs. linear stability analysis

To motivate the linear stability approach to solitary waves, we start with the following instability result, known as Derrick's theorem: the instability of stationary solutions to a nonlinear wave equation

$$(\text{IV.1}) \qquad -\partial_t^2 u = -\Delta u + g(u), \qquad u = u(t,x) \in \mathbb{R}, \qquad x \in \mathbb{R}^n, \qquad n \ge 1.$$

Let the nonlinearity $g(s)$ be smooth and $g(0) = 0$. Equation (IV.1) can be written as a Hamiltonian system $\partial_t v = -\delta_u E$, $\partial_t u = \delta_v E$ (here δ_u and δ_v are variational derivatives) with the Hamiltonian

$$E(u,v) = \int_{\mathbb{R}^n} \Big(\frac{v^2}{2} + \frac{|\nabla u|^2}{2} + G(u) \Big)\, dx,$$

where $G(u) = \int_0^u g(s)\, ds$, $u \in \mathbb{R}$.

There is a well-known result by G.H. Derrick **[Der64]** about non-existence of stable localized stationary solutions in dimension $n \ge 3$, known as *Derrick's theorem*, which we briefly recall. Denote

$$T(u) = \frac{1}{2} \int_{\mathbb{R}^n} |\nabla u(x)|^2\, dx, \qquad V(u) = \int_{\mathbb{R}^n} G(u(x))\, dx.$$

We assume that $u(t,x) = \theta(x)$, with some $\theta \in H^1(\mathbb{R}^n) \cap C^2(\mathbb{R}^n)$, is a localized stationary solution, so that

$$0 = \partial_t u = \frac{\delta E}{\delta v}(\theta, 0), \qquad 0 = \partial_t v = -\frac{\delta E}{\delta u}(\theta, 0) = -\Delta\theta + g(\theta).$$

Then for the family of functions $\theta_\lambda(x) = \theta(x/\lambda)$, using the identities $T(\theta_\lambda) = \lambda^{n-2} T(\theta)$, $V(\theta_\lambda) = \lambda^n V(\theta)$, one has:

$$(\text{IV.2}) \qquad 0 = \Big\langle \frac{\delta E}{\delta u}(\theta, 0), \frac{\partial \theta_\lambda}{\partial \lambda}\big|_{\lambda=1} \Big\rangle = \partial_\lambda\big|_{\lambda=1} E(\theta_\lambda, 0) = (n-2)T(\theta) + nV(\theta);$$

this relation is known as the Pokhozhaev identity **[Pok65]** or the virial theorem (for more details, see Section V.1 below). It follows that

$$\partial_\lambda^2\big|_{\lambda=1} E(\theta_\lambda) = (n-2)(n-3)T(\theta) + n(n-1)V(\theta) = -2(n-2)T(\theta),$$

which is strictly negative as long as θ is not identically zero and $n \ge 3$. That is, $\delta^2 E < 0$ for a variation corresponding to the uniform stretching, and the solution $\theta(x)$, *from the physical point of view*, is expected to be unstable (in a certain sense).

Of course, the fact that in Derrick's argument the quantity $\partial_\lambda^2 E(\theta_\lambda)|_{\lambda=1}$ was not negative for $n = 1$ and 2 does not prove that in these dimensions the localized stationary solutions minimize the energy; it just means that a particular family of perturbations failed to catch the "negative" direction. We mention that a similar result – non-minimization of the energy of localized stationary states – was obtained by R.H. Hobart in [**Hob63**]. His approach applies in spatial dimensions $n \geq 2$.

Here is an elementary argument showing the linear instability of localized stationary solutions in any dimension.

Definition IV.1 We say that the solution is linearly unstable if the linearized equation satisfied by its perturbation has the form

$$\partial_t \rho = A\rho,$$

with

$$\sigma_{\rm d}(A) \cap \{z \in \mathbb{C}: \quad \operatorname{Re} z > 0\} \neq \emptyset.$$

Remark IV.2 Often one calls a particular solitary wave linearly stable if the spectrum of the equation linearized at this wave does not contain points with positive real part; we add the requirement about the size of the Jordan block since such blocks seem to lead to instability. A related definition of linear instability is in [**BC12b, Cuc14**].

Lemma IV.3 (Linear instability of stationary states for $n \in \mathbb{N}$) *For any $n \geq 1$, any stationary solution $\theta \in H^1(\mathbb{R}^n) \cap C^2(\mathbb{R}^n)$ to the nonlinear wave equation* (IV.1) *is linearly unstable, in the sense that the spectrum of the linearization operator contains eigenvalues with positive real part.*

PROOF. We recall that g is smooth. Since $\theta(x)$ satisfies $-\Delta\theta + g(\theta) = 0$, $x \in \mathbb{R}^n$, by elliptic regularity (see e.g. [**GT83**, Theorem 6.17]), θ is smooth and we also have

$$-\Delta\partial_{x_1}\theta + g'(\theta)\partial_{x_1}\theta = 0.$$

Due to $\lim_{|x|\to\infty} \theta(x) = 0$, $\partial_{x_1}\theta$ vanishes somewhere. Therefore, by e.g. [**GT83**, Theorem 8.38], there is a nowhere vanishing smooth function $\chi \in H^2(\mathbb{R}^n)$ (due to Δ being elliptic) which corresponds to some smaller (hence negative) eigenvalue of $L = -\Delta + g'(\theta)$:

$$L\chi = -c^2\chi, \qquad x \in \mathbb{R}^n,$$

with some $c > 0$. Taking $u(t,x) = \theta(x) + r(t,x)$, we obtain the linearization at θ, $-\partial_t^2 r = -Lr$, which we rewrite as $\partial_t \begin{bmatrix} r \\ s \end{bmatrix} = \begin{bmatrix} 0 & 1 \\ -L & 0 \end{bmatrix} \begin{bmatrix} r \\ s \end{bmatrix}$, with $s(t,x) := \partial_t r(t,x)$. The matrix in the right-hand side has eigenvectors $\begin{bmatrix} \chi \\ \pm c\chi \end{bmatrix}$, corresponding to the eigenvalues $\pm c \in \mathbb{R}$; thus, the solution θ is linearly unstable. □

Remark IV.4 One has $\partial_\tau^2|_{\tau=0} E(\theta+\tau\chi) < 0$, showing that $\delta^2 E$ at θ is not positive-definite in any spatial dimension $n \geq 1$.

Remark IV.5 Let us mention that, roughly speaking, "the positivity of $\delta^2 E$" is not necessarily required for stability. A straightforward example is to consider the Hamiltonian system $\dot{q} = J\nabla h(q)$, $q(t) \in \mathbb{R}^2$, $h \in \mathrm{C}^1(\mathbb{R}^2)$, $J = \begin{bmatrix} 0 & 1 \\ -1 & 0 \end{bmatrix}$. Trajectories of this system are level curves of the function h, hence both maxima and minima of h are orbitally stable. A similar example is in [**Rañ83a**, Section V]: One considers the Lagrangian which is linear in the time derivatives,

$$\mathscr{L}(q,\dot{q}) = A_j(q)\dot{q}^j - V(q),$$

with the coordinates $q = (q^1, \dots, q^n) \in \mathbb{R}^n$; the summation over repeated indices assumed. The Lagrange equations

$$-\frac{d}{dt}\Big(\frac{\partial \mathscr{L}}{\partial \dot{q}^j}\Big) + \frac{\partial \mathscr{L}}{\partial q^j} = 0, \qquad t \in \mathbb{R}, \qquad 1 \le j \le n$$

result in

$$-\frac{d}{dt}A_j(q) + \partial_{q^j} A_k(q)\dot{q}^k - \partial_{q^j} V(q) = \big(\partial_{q^j} A_k - \partial_{q^k} A_j\big)\dot{q}^k - \partial_{q^j} V(q) = 0, \quad 1 \le j \le n.$$

If the matrix $\eta_{jk} = \partial_{q^j} A_k - \partial_{q^k} A_j$ is invertible (which requires that n is even), then we get

$$\dot{q}^j = \eta^{jk}\partial_{q^k} V(q);$$

this equation itself could be considered as a Hamiltonian system, implying that $V(q)$ is the integral of motion. Again, the critical points of $V(q)$, whether maxima or minima, are stable.

The above considerations indicate that the stability conclusions based on the energy minimization are not as accurate as the linear stability analysis.

Remark IV.6 For a more general relation between instability and linear instability, see **[SS00]**. A more general result on (nonlinear) instability of localized stationary solutions to (IV.1), in any spatial dimension and under rather mild assumptions, is proved in **[KS07]**. The relation between the linear instability in the context of the nonlinear Schrödinger equation, see **[GO12]**.

IV.2 The Kolokolov stability criterion of NLS groundstates

By Derrick's theorem **[Der64]** (see Section IV.1), *any stationary localized solution* is unstable (cf. Lemma IV.3). To get a hold of stable localized solutions, Derrick suggests: *elementary particles might correspond to stable, localized solutions which are periodic in time, rather than time-independent.* The linear stability analysis of solitary wave solutions of the form

$$\psi(t,x) = u(x)e^{-\mathrm{i}\omega t}, \qquad \omega \in \mathbb{R}, \qquad u \in H^1(\mathbb{R}^n), \tag{IV.3}$$

to the nonlinear Schrödinger and Klein–Gordon equations was already started in **[Zak67, AD70, And71]** and culminated in the fundamental paper **[Kol73]**. Let us give the essence of the analytic approach to the linear stability analysis on the example of the nonlinear Schrödinger equation in one dimension:

$$\mathrm{i}\partial_t \psi = -\partial_x^2 \psi - f(|\psi|^2)\psi, \qquad \psi(t,x) \in \mathbb{C}, \qquad x \in \mathbb{R}, \tag{IV.4}$$

where the nonlinearity satisfies

$$f \in C^1(\mathbb{R}_+) \cap C(\overline{\mathbb{R}_+}), \qquad f(0) = 0, \qquad |f'(s)| \le Cs^{\kappa-1} \ \ \forall s \in (0,1), \tag{IV.5}$$

for some $\kappa > 0$.

The amplitude $\varphi_\omega(x)$ of a solitary wave (IV.3) is to satisfy the stationary nonlinear Schrödinger equation

$$\omega u = -\partial_x^2 u - f(|u|^2)u, \qquad u(x) \in \mathbb{C}, \qquad x \in \mathbb{R}. \tag{IV.6}$$

Existence of solitary waves. While more details on the existence and regularity properties of NLS solitary waves are in Section V.2, let us give here a simple construction of solitary waves in the one-dimensional case. We rewrite (IV.6) in the following form:

$$-\partial_x^2 u = g(u) := \omega u + f(u^2)u, \tag{IV.7}$$

where we assume that f satisfies (IV.5) and that u is real-valued. We interpret (IV.7) as a particle in the "effective potential"

$$G(u) := \int_0^u g(s)\,ds = \frac{\omega u^2}{2} + \int_0^u f(s^2)s\,ds, \qquad u \in \mathbb{R},$$

so that x is "the time" and u is "the position" of the particle. The "mechanical" energy corresponding to this system is $\mathcal{E}(u) = |u'|^2/2 + G(u)$. For a particular solution $u(x)$ to (IV.7), $\mathcal{E}(u)$ is constant (it does not depend on the "time" x). We are interested in soliton-like solutions, such that $u \to 0$ and $u' \to 0$ as $|x| \to \infty$, and hence $\mathcal{E}(u(x)) = 0$ for all $x \in \mathbb{R}$. If there is a "turning point" $\zeta > 0$ such that $G(u) < 0$ for $u \in (0, \zeta)$, $G(\zeta) = 0$, and $G'(\zeta) > 0$, then there exists a set of solutions with zero "mechanical" energy $\mathcal{E}$, $u(x + C)$, $C \in \mathbb{R}$, where $u(x)$ satisfies $\lim_{x\to\pm\infty} u(x) = 0$. (This implies that in (IV.7) one has $\omega \le 0$.) Such a soliton is defined up to a shift along x. We fix u by requiring that it assumes its maximum value at the origin: $u(0) = \zeta$. Then u is symmetric and is obtained by integration from $du/dx = -\sqrt{G(u)}$ for $x > 0$, with $u(0) = \zeta > 0$. One can see from (IV.7) that the assumption (IV.5) on f implies that $u \in C^3(\mathbb{R})$.

The above reasoning can be summarized as follows.

Theorem IV.7 ([BL83a, Theorem 5]) *Let $g \in C(\mathbb{R})$ be Lipschitz with $g(0) = 0$. A necessary and sufficient condition for the existence of a solution of problem*

$$-\partial_x^2 u = g(u), \qquad u = u(x), \qquad x \in \mathbb{R} \tag{IV.8}$$

is that

$$\zeta = \inf\{s > 0;\ \ G(s) = 0\} \quad \textit{satisfies} \quad \zeta > 0, \qquad g(\zeta) > 0. \tag{IV.9}$$

Furthermore, if (IV.9) *is satisfied, then* (IV.8) *has a unique solution up to translations of the origin, and this solution satisfies (after a suitable rotation of the origin):*

(1) $u(x) = u(-x) \quad \forall x \in \mathbb{R}$;
(2) $u(x) > 0 \quad \forall x \in \mathbb{R}$;
(3) $u(0) = \zeta$;
(4) $u'(x) < 0 \quad \forall x > 0$.

Linear stability of solitary waves. Let $\varphi_\omega(x)e^{-\mathrm{i}\omega t}$, $x \in \mathbb{R}$, $\omega \in \mathcal{O} \subset \mathbb{R}_-$, with $\mathcal{O}$ an open interval, be a family of solitary wave solutions to the nonlinear Schrödinger equation (IV.4). We assume that this family corresponds to the *ground states*, in the sense that $\varphi_\omega(x) > 0$ for all $x \in \mathbb{R}$, $\omega \in \mathcal{O}$, and that the map

$$\omega \mapsto \varphi_\omega$$

is C^1 as a map from $\mathcal{O}$ to $H^1(\mathbb{R}) \cap C^2(\mathbb{R})$. We also assume that $\varphi_\omega(x)$ are spherically symmetric and monotonically decreasing to zero as $|x| \to \infty$ (see Chapter V for the construction of such solitary waves). To study the linear stability of such solitary waves, we consider the Ansatz $\psi(t, x) = (\varphi_\omega(x) + \rho(t, x))e^{-\mathrm{i}\omega t}$, with $\rho(t, x) \in \mathbb{C}$

which we consider to be a small perturbation. The linearized equation on ρ is called the linearization at a solitary wave:

$$\partial_t \rho = \frac{1}{\mathrm{i}}\big(- \partial_x^2 \rho - \omega\rho - f(\varphi_\omega^2)\rho - 2f'(\varphi_\omega^2)\varphi_\omega^2 \operatorname{Re}\rho\big). \tag{IV.10}$$

Remark IV.8 Because of the term with $\operatorname{Re}\rho$, the operator in the right-hand side is $\mathbb{R}$-linear but not $\mathbb{C}$-linear.

To study the spectrum of the operator in the right-hand side of (IV.10), we first write it in the $\mathbb{C}$-linear form, considering its action onto $\boldsymbol{\rho}(t,x) = \begin{bmatrix}\operatorname{Re}\rho(t,x)\\ \operatorname{Im}\rho(t,x)\end{bmatrix}$:

$$\partial_t \boldsymbol{\rho} = A(\omega)\boldsymbol{\rho}, \qquad \boldsymbol{\rho}(t,x) = \begin{bmatrix}\operatorname{Re}\rho(t,x)\\ \operatorname{Im}\rho(t,x)\end{bmatrix}, \tag{IV.11}$$

where the linearization operator is given by

$$A(\omega) = \begin{bmatrix} 0 & \mathfrak{l}_-(\omega) \\ -\mathfrak{l}_+(\omega) & 0 \end{bmatrix}, \tag{IV.12}$$

$$\mathfrak{l}_-(\omega) = -\Delta - \omega - f(\varphi_\omega^2), \qquad \mathfrak{l}_+(\omega) = \mathfrak{l}_-(\omega) - 2\varphi_\omega^2 f'(\varphi_\omega^2), \tag{IV.13}$$

$$\mathfrak{D}(\mathfrak{l}_\pm) = H^2(\mathbb{R}^1).$$

Since the amplitude φ_ω satisfies the stationary equation $\omega\varphi_\omega = -\Delta\varphi_\omega - f(\varphi_\omega^2)\varphi_\omega$, there are the relations

$$\mathfrak{l}_-(\omega)\varphi_\omega = 0, \qquad \mathfrak{l}_+(\omega)\partial_\omega\varphi_\omega = \varphi_\omega; \tag{IV.14}$$

$$\mathfrak{l}_+(\omega)\partial_x\varphi_\omega = 0, \qquad \mathfrak{l}_-(\omega)(x\varphi_\omega) = 2\partial_x\varphi_\omega. \tag{IV.15}$$

Since $\lim_{|x|\to\infty} u(x) = 0$, $\mathfrak{l}_\pm$ are relatively compact perturbations of the Laplace operator (see Problem III.159 below), and then Weyl's theorem on the essential spectrum (Theorem III.135) yields:

$$\sigma_{\mathrm{ess}}(\mathfrak{l}_-(\omega)) = \sigma_{\mathrm{ess}}(\mathfrak{l}_+(\omega)) = [-\omega, +\infty) = [|\omega|, +\infty).$$

For brevity, usually we will not indicate explicitly that $\mathfrak{l}_\pm$ depend on ω.

We would like to know at which ω the eigenvalues of $A(\omega)$ with $\operatorname{Re}\lambda > 0$ may appear. Note that $A(\omega)$ is resolvent continuous in the uniform operator topology as a function of $\omega \in \mathcal{O}$ due to the assumption that the mapping $\omega \mapsto \varphi_\omega$ is C^1. Since $\lambda^2 \in \mathbb{R}$, real eigenvalues can only bifurcate from $\lambda = 0$. The analysis of the rank of the Riesz projector corresponding to $\lambda = 0$ (see Definition III.83) shows that, due to the resolvent continuity of $A(\omega)$, the bifurcation of eigenvalues from $\lambda = 0$ follows the jump in the dimension of the generalized null space of $A(\omega)$. Such a jump happens at a particular value of ω if one can solve the equation $A(\omega)\boldsymbol{\alpha} = \begin{bmatrix}\partial_\omega\varphi_\omega\\ 0\end{bmatrix}$. This leads to the condition that $\begin{bmatrix}\partial_\omega\varphi_\omega\\ 0\end{bmatrix}$ is orthogonal to the null space of the adjoint to $A(\omega)$,

$$A(\omega)^* = \begin{bmatrix} 0 & -\mathfrak{l}_+(\omega) \\ \mathfrak{l}_-(\omega) & 0 \end{bmatrix},$$

which contains the vector $\begin{bmatrix}\varphi_\omega\\ 0\end{bmatrix}$; this results in

$$\langle \varphi_\omega, \partial_\omega\varphi_\omega\rangle = \partial_\omega \|\varphi_\omega\|_{L^2}^2/2 = 0. \tag{IV.16}$$

A more careful analysis (see Lemma IV.9 below) shows that there are two real eigenvalues $\pm\lambda \in \mathbb{R}$ if ω is such that $\partial_\omega \|\varphi_\omega\|^2_{L^2}$ is positive, leading to a linear instability of the corresponding solitary wave. The opposite condition,

$$\partial_\omega \|\varphi_\omega\|^2_{L^2} < 0, \tag{IV.17}$$

is the *Kolokolov stability criterion* which guarantees the absence of nonzero real eigenvalues for the nonlinear Schrödinger equation. This criterion appeared in [**Kol73**, **VK73**, **CL82**, **Sha83**, **Wei86**, **GSS87**, **BP92a**] in relation to linear and orbital stability of solitary waves.

Theorem IV.9 (Kolokolov stability criterion [Kol73]) *Let there be a family of solitary wave solutions $\varphi_\omega(x)e^{-i\omega t}$, $\omega \in \mathcal{O}$, with $\mathcal{O} \subset \mathbb{R}_-$ an interval. Assume that $\varphi_\omega \in H^1(\mathbb{R}) \cap C^2(\mathbb{R})$ are strictly positive and that $\omega \mapsto \varphi_\omega$ is C^1 as a map from $\mathcal{O}$ to $H^1(\mathbb{R}) \cap C^2(\mathbb{R})$. There is $\lambda \in \sigma_p(A)$, $\lambda > 0$, where A is the linearization (IV.11) at the solitary wave $\varphi_\omega(x)e^{-i\omega t}$, if and only if $\frac{d}{d\omega}\|\varphi_\omega\|^2_{L^2} > 0$ at this value of ω.*

Proof. We follow [**Kol73**]. To find out whether there is linear instability, we need to check whether there is $\lambda \in \sigma_d(A)$ with $\operatorname{Re}\lambda > 0$.

The relation $(A-\lambda)\Xi = 0$ with $\operatorname{Re}\lambda > 0$ and $\Xi(x) = \begin{bmatrix} r(x) \\ s(x) \end{bmatrix} \in \mathbb{C}^2$ not identically zero implies that

$$\lambda^2 r = -\mathfrak{l}_-\mathfrak{l}_+ r, \qquad x \in \mathbb{R}, \tag{IV.18}$$

with $r \not\equiv 0$. It follows that r is orthogonal to the kernel of the selfadjoint operator $\mathfrak{l}_-$ (which is spanned by φ_ω):

$$\langle \varphi, u\rangle = -\lambda^{-2}\langle \varphi, -\mathfrak{l}_-\mathfrak{l}_+ r\rangle = -\lambda^{-2}\langle \mathfrak{l}_-\varphi, -\mathfrak{l}_+ r\rangle = 0; \tag{IV.19}$$

geometrically, this means that Ξ is tangent to the same-charge hypersurface:

$$\langle \partial_\omega Q(\varphi_\omega), \Xi\rangle = \operatorname{Re}\int_{\mathbb{R}} \left\langle \begin{bmatrix} \varphi_\omega \\ 0 \end{bmatrix}, \begin{bmatrix} r \\ s \end{bmatrix} \right\rangle_{\mathbb{C}^2} dx = \operatorname{Re}\int_{\mathbb{R}} \varphi_\omega(x) r(x)\, dx = 0.$$

Due to (IV.19), there is $\eta := \mathfrak{l}_-^{-1} r \in L^2(\mathbb{R},\mathbb{C}) \cap \phi_\omega^\perp$. Pairing (IV.18) with η, we have:

$$\lambda^2\langle \eta, \mathfrak{l}_-\eta\rangle = \lambda^2\langle \eta, r\rangle = \langle \eta, -\mathfrak{l}_-\mathfrak{l}_+ r\rangle = -\langle r, \mathfrak{l}_+ r\rangle.$$

Since the operators $\mathfrak{l}_\pm$ are self-adjoint, we conclude that $\lambda^2 \in \mathbb{R}$, and then our assumption $\operatorname{Re}\lambda > 0$ is equivalent to $\lambda > 0$.

Since $\mathfrak{l}_-$ is positive-definite and $\eta \notin \mathbf{ker}(\mathfrak{l}_-)$, one has $\langle \eta, \mathfrak{l}_-\eta\rangle > 0$. Therefore, if $\lambda > 0$, then one concludes that $\langle u, \mathfrak{l}_+ u\rangle < 0$, and

$$\mu := \inf\{\ \langle \psi, \mathfrak{l}_+\psi\rangle\colon\ \ \langle \psi, \psi\rangle = 1,\ \ \langle \psi, \varphi_\omega\rangle = 0\ \} \tag{IV.20}$$

is negative. According to Lagrange's principle, the function ψ corresponding to the minimum of $\langle \psi, \mathfrak{l}_+\psi\rangle$ under conditions

$$\langle \psi, \psi\rangle = 1, \qquad \langle \psi, \varphi_\omega\rangle = 0 \tag{IV.21}$$

satisfies

$$\mathfrak{l}_+\psi = \mu\psi + \alpha\varphi_\omega, \qquad \mu,\ \alpha \in \mathbb{R}; \tag{IV.22}$$

pairing with ψ shows that μ in (IV.22) is the same as in (IV.20). By (IV.15), one has $z_1 := 0 \in \sigma_p(\mathfrak{l}_+)$. Due to $\partial_x\varphi_\omega$ vanishing at one point ($x = 0$), there is exactly one negative eigenvalue of $\mathfrak{l}_+$ (see e.g. [**LS75**, Theorem 4.2.1]), which we denote by $z_0 \in \sigma_{\rm p}(\mathfrak{l}_+) \cap \mathbb{R}_-$. (For this eigenvalue, one can choose a strictly positive eigenfunction.) Note that $\alpha \neq 0$, or else μ would be equal to z_0, with ψ the corresponding eigenfunction of $\mathfrak{l}_+$, but then ψ, having to be nonzero (we can choose it to be strictly positive), could not be orthogonal to $\varphi_\omega > 0$, contradicting the assumptions (IV.21).

Denote $z_2 = \inf(\sigma(\mathfrak{l}_+) \cap \mathbb{R}_+) > 0$. Let us consider the function

$$h(z) := \langle \varphi_\omega, (\mathfrak{l}_+ - z)^{-1} \varphi_\omega \rangle, \qquad z \in \rho(\mathfrak{l}_+), \tag{IV.23}$$

which in particular is defined for $z \in (z_0, z_2)$. Note that $h(z)$ is defined at $z = z_1 = 0$ since the kernel of $\mathfrak{l}_+$ is spanned by an odd function $\partial_x \varphi_\omega$; hence $\mathfrak{l}_+$ is invertible with bounded inverse on the subspace of even functions (which in particular contains φ_ω). By (IV.22),

$$h(\mu) = \langle \varphi_\omega, (\mathfrak{l}_+ - \mu)^{-1} \varphi_\omega \rangle = \frac{1}{\alpha} \langle \varphi_\omega, \psi \rangle = 0,$$

and since $h'(z) > 0$ for $z \in (z_0, z_2)$, $\mu < 0$ if and only if $h(0) > 0$. At the same time,

$$h(0) = \langle \varphi_\omega, \mathfrak{l}_+^{-1} \varphi_\omega \rangle = \langle \varphi_\omega, \partial_\omega \varphi_\omega \rangle = \frac{1}{2} \frac{d}{d\omega} \int_{\mathbb{R}} |\varphi_\omega(x)|^2 \, dx;$$

so, the linear instability leads to $\mu < 0$, which is equivalent to $\frac{d}{d\omega} \int_{\mathbb{R}} |\varphi_\omega(x)|^2 \, dx > 0$. □

Problem IV.10 Find the explicit form of solitary wave solutions $\psi(t, x) = \varphi_\omega(x) e^{-\mathrm{i}\omega t}$, $\omega < 0$, to the cubic Schrödinger equation

$$\mathrm{i} \partial_t \psi = -\partial_x^2 \psi - |\psi|^2 \psi, \qquad \psi(t, x) \in \mathbb{C}, \qquad x \in \mathbb{R}.$$

CHAPTER V

Solitary waves of nonlinear Schrödinger equation

We consider the nonlinear Schrödinger equation,

$$\text{(V.1)} \qquad \mathrm{i}\partial_t\psi = -\Delta\psi - f(|\psi|^2)\psi, \qquad \psi(t,x) \in \mathbb{C}, \qquad x \in \mathbb{R}^n,$$

and the nonlinear Klein–Gordon equation,

$$\text{(V.2)} \qquad -\partial_t^2\psi = -\Delta\psi - f(|\psi|^2)\psi, \qquad \psi(t,x) \in \mathbb{C}, \qquad x \in \mathbb{R}^n,$$

where $n \in \mathbb{N}$ and the nonlinearity is represented by a real-valued function

$$f : \overline{\mathbb{R}_+} \to \mathbb{R}.$$

Both equations are $\mathbf{U}(1)$-invariant, in the sense that if $\psi(t,x)$ is a solution, so is $e^{\mathrm{i}s}\psi(t,x)$, for any $s \in \mathbb{R}$.

Definition V.1 (Solitary waves) Solitary wave solutions (or, simply, *solitary waves*) to the nonlinear Schrödinger equation (V.1) or the Klein–Gordon equation (V.2) are solutions of the form

$$\psi(t,x) = u(x)e^{-\mathrm{i}\omega t}, \qquad \omega \in \mathbb{R}, \qquad u \in H^1(\mathbb{R}^n).$$

The problems of existence of solitary wave solutions to (V.1) or (V.2) lead to the question of existence of solutions u to the stationary nonlinear Schrödinger equation or the stationary Klein–Gordon equation,

$$\omega u = -\Delta u - f(|u|^2)u, \qquad \omega^2 u = -\Delta u - f(|u|^2)u, \qquad x \in \mathbb{R}^n,$$

with some $\omega \in \mathbb{R}$, which we write in the form

$$\text{(V.3)} \qquad -\Delta u = g(u), \qquad x \in \mathbb{R}^n; \qquad g(0) = 0.$$

The original approach to proving the existence of $H^1(\mathbb{R}^n)$-solutions to (V.3) was developed by Walter Strauss [**Str77a**]. Later results are in [**CGM78**], [**BL83a**], [**BGK83**] (two-dimensional case), and [**Sha85**]. The uniqueness of a symmetric solution $u > 0$ is proved in [**Kwo89, McL93**]. The inclusion $u \in H^1(\mathbb{R}^n) \cap C^2(\mathbb{R}^n)$, monotonicity, and the exponential decay of u follows from [**BL83a**] for $n \geq 3$ and $n = 1$. For $n = 2$, the inclusion $u \in H^1(\mathbb{R}^2) \cap C^2(\mathbb{R}^2)$ is proved in [**BGK83**]. We are going to remind the most essential parts of the arguments; for more details, see the monograph [**Caz03**].

V.1 The Pokhozhaev identity

We derive the Pokhozhaev identity [**Pok65**] following the approach of [**BL83a**]; we give it in the form suitable for all dimensions $n \in \mathbb{N}$ and for vector-valued functions, so that it also applies to the complex-valued case.

Lemma V.2 ([BL83a, Proposition 1]) *Let $N \in \mathbb{N}$. Let $G \in C^1(\mathbb{R}^N, \mathbb{R})$ satisfy*

$$G(0) = 0, \qquad \nabla G(0) = 0,$$

and denote

$$g = \nabla_u G = \Big(\frac{\partial G}{\partial u_1}, \dots, \frac{\partial G}{\partial u_N}\Big) : \mathbb{R}^N \to \mathbb{R}^N, \qquad u \in \mathbb{R}^N.$$

Let $n \in \mathbb{N}$ and let $u \in L^\infty_{\mathrm{loc}}(\mathbb{R}^n, \mathbb{R}^N)$ satisfy

$$-\Delta u = g(u) \quad \text{in} \quad \mathscr{D}'(\mathbb{R}^n, \mathbb{R}^N). \tag{V.4}$$

Assume furthermore that

$$\nabla u \in L^2(\mathbb{R}^n, \mathbb{R}^{n\times N}), \qquad G(u) \in L^1(\mathbb{R}^n).$$

Then u satisfies the virial identity

$$\frac{n-2}{2n}\int_{\mathbb{R}^n} |\nabla u(x)|^2\,dx = \int_{\mathbb{R}^n} G(u(x))\,dx. \tag{V.5}$$

Above, $|\nabla u|^2 = \sum_{a=1}^N |\nabla u_a|^2 = \sum_{a=1}^N \sum_{i=1}^n |\partial_{x^i} u_a|^2$. The following proof applies to $u \in \mathbb{R}^N$ although we will not indicate the indices $1 \le a \le N$ explicitly.

Proof. In Derrick's argument, when deriving the identity (IV.2), there are difficulties which arise from the need to argue that the functions $\partial_\lambda u(x/\lambda)|_{\lambda=1} = x^i \partial_{x^i} u(x)$ belong to an appropriate functional space. A more cautious approach in **[BL83a]** is to derive the identity (V.5) in a ball of radius $R \ge 1$, to estimate the boundary term, and to send $R \to \infty$. Since g is continuous and $u \in L^\infty_{\mathrm{loc}}(\mathbb{R}^n, \mathbb{R}^N)$, $g(u) \in L^\infty_{\mathrm{loc}}(\mathbb{R}^n, \mathbb{R}^N)$ and hence so is Δu. We multiply the relation $0 = g(u) + \Delta u$ by $x^i \partial_{x^i} u$ (more accurately, by the transpose $x^i \partial_{x^i} u^T$, which is in $L^2_{\mathrm{loc}}(\mathbb{R}^n, \mathbb{R}^N)$, where the summation in $1 \le i \le n$ is implied) and use Green's formula, where we need some care to estimate the boundary term. Let $\rho \in C^1(\mathbb{R}^n)$ be such that $\rho(t) = 1$ for $t \le 1$, $\rho(t) = 0$ for $t \ge 2$, and $|\rho'(t)| \le 2$ for all $t \in \mathbb{R}$. For $R \ge 1$, define $\rho_R(x) = \rho(|x|/R)$, $x \in \mathbb{R}^n$. In view of (V.4), for any $R \ge 1$, we have:

$$0 = \int_{\mathbb{R}^n} \rho_R(x)\big(x\cdot\nabla_x G(u) + (x\cdot\nabla u)\Delta u\big)\,dx. \tag{V.6}$$

In the above integral, we will integrate by parts both terms separately. The first term can be rewritten as

$$\int_{\mathbb{R}^n} \rho_R\, x\cdot\nabla G(u(x))\,dx = -\int_{\mathbb{R}^n} \big(x\cdot\nabla\rho_R(x)\,G(u(x)) + n\rho_R(x) G(u(x))\big)\,dx.$$

For the second term, we obtain:

$$\begin{aligned}
\int_{\mathbb{R}^n} (x\cdot\nabla u)\rho_R \Delta u\,dx &= -\int_{\mathbb{R}^n} \big((x\cdot\nabla u)\,(\nabla\rho_R\cdot\nabla u) + \rho_R \nabla(x\cdot\nabla u)\cdot\nabla u\big)\,dx \\
&= -\int_{\mathbb{R}^n} \Big((x\cdot\nabla u)\,(\nabla\rho_R\cdot\nabla u) + \rho_R|\nabla u|^2 + \rho_R\, x\cdot\nabla\frac{|\nabla u|^2}{2}\Big)\,dx
\end{aligned}$$

$$= -\int_{\mathbb{R}^n} \Big((x\cdot\nabla u)\,(\nabla\rho_R\cdot\nabla u) + \rho_R|\nabla u|^2 - \rho_R\, n\frac{|\nabla u|^2}{2} - \frac{|\nabla u|^2}{2} x\cdot\nabla\rho_R\Big)\,dx. \tag{V.7}$$

Let us justify the above derivation, which so far makes sense for $u \in C^\infty(\mathbb{R}^n)$. If $u \in H^1(\mathbb{R}^n)$, we pick a function

$$\eta \in C^\infty_{\mathrm{comp}}(\mathbb{R}^n), \qquad \eta(x) \ge 0 \quad \forall x \in \mathbb{R}^n, \qquad \int_{\mathbb{R}^n} \eta(x)\,dx = 1,$$

and define the mollified sequence $u_j \in C^\infty(\mathbb{R}^n)$ by

$$u_j = \eta_j * u \quad \text{with} \quad \eta_j(x) = \frac{\eta(jx)}{j^n}, \qquad x \in \mathbb{R}^n, \qquad j \in \mathbb{N},$$

so that $u_j \to u$ in $H^1(\mathbb{R}^n)$ and $\Delta u_j \to \Delta u$ in $L^2_{\mathrm{loc}}(\mathbb{R}^n)$ (that is, the convergence takes place on any fixed compact set in $\mathbb{R}^n$).

Problem V.3 Prove the above stated convergences.

As a result, both sides of (V.7) converge in $L^1(\mathbb{R}^n)$.

Substituting the above expressions into (V.6), we arrive at

$$\text{(V.8)} \qquad 0 = \int_{\mathbb{B}^n_R} \Big(- nG(u(x))\,dx + \frac{n-2}{2}|\nabla u(x)|^2 \Big)\,dx \; + \; \text{error term},$$

with the error term contains terms supported in the spherical shell $R \le |x| \le 2R$. Taking into account that $|\nabla \rho_R(x)| \le 2/R$ for all $x \in \mathbb{R}^n$, one may bound the error term by

$$\text{(V.9)} \qquad C \int_{R \le |x| \le 2R} \big(|\nabla u|^2 + |G(u)|\big)\,dx.$$

Since $|\nabla u| + |G(u)|$ is integrable on $\mathbb{R}^n$ due to the assumptions of the lemma, (V.9) tends to zero as $R \to \infty$. It suffices to add that

$$\int_{\mathbb{B}^n_R} |\nabla u|^2\,dx \to \int_{\mathbb{R}^n} |\nabla u|^2\,dx, \qquad \int_{\mathbb{B}^n_R} G(u)\,dx \to \int_{\mathbb{R}^n} G(u)\,dx$$

as $R \to \infty$. This concludes the proof. □

Lemma V.4 *Let $n \ge 3$, $N \in \mathbb{N}$, and let $q > (n+2)/(n-2)$. Then there is no $u \in L^\infty_{\mathrm{loc}}(\mathbb{R}^n, \mathbb{R}^N) \cap H^1(\mathbb{R}^n, \mathbb{R}^N) \cap L^{q+1}(\mathbb{R}^n, \mathbb{R}^N)$ such that*

$$\text{(V.10)} \qquad -\Delta u = |u|^{q-1} u + \omega u \quad \text{in } \mathscr{D}'(\mathbb{R}^n, \mathbb{R}^N)$$

with $\omega \le 0$.

We note that the above lemma also applies to complex-valued solutions.

Proof. This is a consequence of the Pokhozhaev identity. Indeed, multiplying (V.10) by u^T and integrating by parts (like in the proof of Lemma V.2), one has

$$\text{(V.11)} \qquad \int_{\mathbb{R}^n} \frac{|\nabla u|^2}{2}\,dx - \int_{\mathbb{R}^n} \frac{|u|^{q+1}}{2}\,dx - \omega \int_{\mathbb{R}^n} \frac{|u|^2}{2}\,dx = 0.$$

Combining the above with the Pokhozhaev identity (Lemma V.2), which takes the form

$$\text{(V.12)} \qquad \frac{n-2}{n} \int_{\mathbb{R}^n} \frac{|\nabla u|^2}{2}\,dx - \int_{\mathbb{R}^n} \frac{|u|^{q+1}}{q+1}\,dx - \omega \int_{\mathbb{R}^n} \frac{|u|^2}{2}\,dx = 0,$$

we exclude $\int_{\mathbb{R}^n} |u|^{q+1}\,dx$, arriving at

$$\Big(1 - (q+1)\frac{n-2}{2n}\Big) \int_{\mathbb{R}^n} \frac{|\nabla u|^2}{2}\,dx + \frac{q+1}{2}\omega \int_{\mathbb{R}^n} \frac{|u|^2}{2}\,dx = 0.$$

Thus, $\omega \le 0$ is incompatible with $q+1 > 2n/(n-2)$. □

Remark V.5 Subtracting (V.12) from (V.11) gives the relation

$$\frac{2}{n}\int_{\mathbb{R}^n} \frac{|\nabla u|^2}{2}\,dx - (q-1)\int_{\mathbb{R}^n} \frac{|u|^{q+1}}{2q+2}\,dx = 0. \tag{V.13}$$

It follows that the energy of the solitary wave solution $u(x)e^{-i\omega t}$ to (V.1) with $f(\tau) = |\tau|^\kappa$ with $\kappa = (q-1)/2$ is given by

$$E = \int_{\mathbb{R}^n} \frac{|\nabla u|^2}{2}\,dx - \int_{\mathbb{R}^n} \frac{|u|^{q+1}}{q+1}\,dx = \Big(1 - \frac{4}{(q-1)n}\Big)\int_{\mathbb{R}^n} \frac{|\nabla u|^2}{2}\,dx,$$

which is negative for $\kappa = (q-1)/2 < 2/n$ and vanishes in the "charge-critical case" $\kappa = 2/n$.

Remark V.6 The absence of solitary waves with $\omega > 0$ in the case when $V(x) = |u(x)|^{q-1}$ is spherically symmetric, $-\Delta + V$ is self-adjoint, $\int_a^\infty |V(r)|\,dr < \infty$ for some $a > 0$, and $V \in L^2_{\mathrm{loc}}(\mathbb{R}^n \backslash \{0\})$ corresponds to the absence of "embedded" eigenstates which in turn follows from [**RS78**, Theorem XIII.56]. See [**RS78**, Section XIII.13] for more details on the absence of embedded eigenvalues.

V.2 Existence of groundstates

Here we sketch the construction of solitary waves in the case $n \ge 3$. We rewrite the equation (V.3):

$$-\Delta u = g(u), \qquad x \in \mathbb{R}^n, \tag{V.14}$$

where

$$g \in C\big(\overline{\mathbb{R}_+}, \mathbb{R}\big), \qquad g(0) = 0.$$

We denote

$$G(z) = \int_0^{|z|} g(s)\,ds, \qquad z \in \mathbb{C}.$$

Theorem V.7 ([BL83a, Theorem 1]) *Let $n \ge 3$. Assume that*

$$-\infty < \liminf_{s\to 0+} \frac{g(s)}{s} \le \limsup_{s\to 0+} \frac{g(s)}{s} < 0; \tag{V.15}$$

$$-\infty \le \limsup_{s\to +\infty} \frac{g(s)}{s^l} \le 0, \qquad l = \frac{n+2}{n-2}; \tag{V.16}$$

$$\exists \zeta > 0 : \qquad G(\zeta) = \int_0^\zeta g(s)\,ds > 0. \tag{V.17}$$

Then there exists a solution u to (V.14) such that

(1) $u > 0$ on $\mathbb{R}^n$;
(2) $u \in H^1_r(\mathbb{R}^n)$, and $u(x)$ is a decreasing function of $r = |x|$;
(3) $u \in C^2(\mathbb{R}^n)$;
(4) u and its derivatives up to order 2 have exponential decay at infinity.

The existence of a positive solution $u \in H^1_r(\mathbb{R}^n)$ (*Parts (1) (2)* of the theorem) is proved in the remainder of this section. The exponential decay of $u(x)$ and its regularity (*Parts (3) (4)* of the theorem) will be proved in Section V.3, in a slightly stronger form; see Lemma V.19 and Lemma V.20 for the exact formulation.

We notice that ζ in (V.17) can be chosen such that $g(\zeta) \geq 0$ (taking $\zeta > 0$ smaller if necessary). Following **[BL83a]**, if for some $s_0 > \zeta$ one has $g(s_0) < 0$, then instead of (V.14) we consider the equation

$$
\text{(V.18)} \qquad -\Delta u = \tilde{g}(u), \quad x \in \mathbb{R}^n; \qquad \tilde{g}(s) = \begin{cases} g(s), & 0 \leq s \leq s_0; \\ g(s_0), & s > s_0. \end{cases}
$$

Let us show that if $u \in C^2(\mathbb{R}^n)$ is a solution to (V.18), then it is also a solution to (V.14).

Lemma V.8 *If $u(x)$ is a solution to* (V.18) *which satisfies all the conclusions of Theorem* V.7, *then* $\sup_{x \in \mathbb{R}^n} u(x) < s_0$.

PROOF. Since $u(x) > 0$ is C^2 and tends to zero at infinity, there is a point $x_0 \in \mathbb{R}^n$ such that $u(x_0) = \sup_{x \in \mathbb{R}^n} u(x)$. If $u(x_0) \geq s_0$, then by (V.18) $\Delta u|_{x_0} > 0$, contradicting the assumption that it is the point of the global maximum. □

Thus, a solution to (V.18) with properties as stated in Theorem V.7 will automatically be a solution to (V.14). Without loss of generality, we may substitute g by $\tilde{g}$ if necessary, and now (V.16) takes the form

$$
\text{(V.19)} \qquad \lim_{s \to +\infty} \frac{\tilde{g}(s)}{s^l} = 0.
$$

We note that, by (V.15) and (V.19), there are $C_1 > c_1 > 0$ and $c_2 > 0$ such that

$$
\text{(V.20)} \qquad -C_1|z|^2 - c_2|z|^{l+1} \leq G(z) \leq -c_1|z|^2 + c_2|z|^{l+1}, \qquad \forall z \in \mathbb{C}.
$$

This implies that $G(u(x)) \in L^1(\mathbb{R}^n)$ as a function of $x \in \mathbb{R}^n$ as long as $u \in H^1(\mathbb{R}^n)$:

$$
\int_{\mathbb{R}^n} |G(u(x))|\,dx \leq \int_{\mathbb{R}^n} \left(C_1|u(x)|^2 + c_2|u(x)|^{l+1}\right) dx \leq C_1\|u\|_{L^2}^2 + c_2\|u\|_{L^{l+1}}^{l+1} < \infty
$$

due to the Sobolev embedding $H^1(\mathbb{R}^n) \subset L^{l+1}(\mathbb{R}^n)$, with $n \geq 3$, $l = (n+2)/(n-2)$.

The following minimization problem was considered by Keller **[Kel83]** and then by Shatah **[Sha85]**:

$$
\text{(V.21)} \qquad \text{minimize} \left\{ \int_{\mathbb{R}^n} |\nabla u|^2\,dx\colon \quad K(u) = 0, \quad u \in H^1(\mathbb{R}^n), \quad u \not\equiv 0 \right\},
$$

where

$$
\text{(V.22)} \qquad K(u) = \frac{n-2}{2} \int_{\mathbb{R}^n} |\nabla u|^2\,dx - n \int_{\mathbb{R}^n} G(u)\,dx.
$$

The constraint $K(u) = 0$ comes from the Pokhozhaev identity; see Section V.1.

Remark V.9 We consider the above minimization problem on the space H^1 of complex-valued functions, although we will see that a function which delivers its solution could be chosen real and positive.

Problem V.10 Show that K is a C^1 map from $H^1(\mathbb{R}^n)$ to $\mathbb{R}$.
Hint: Expand the expression $K(u+h)$ to the first order and use the dominated convergence to analyze the remainder.

Lemma V.11 *Let $n \geq 3$.*

$$\mathscr{M} = \{u \in H^1(\mathbb{R}^n): \quad K(u) \leq 0, \quad u \not\equiv 0\} \tag{V.23}$$

is a closed nonempty set (in particular, its closure in the H^1-norm does not contain $u = 0$).

PROOF. To see that $\mathscr{M}$ is nonempty, we use $\zeta > 0$ from (V.17) to define the function $u_R \in H^1(\mathbb{R}^n)$ by

$$u_R(x) = \begin{cases} \zeta, & |x| \leq R; \\ \zeta + R - |x|, & R < |x| \leq R + \zeta; \\ 0, & |x| > R + \zeta. \end{cases}$$

Then $K(u_R) \leq cR^{n-1} - c' R^n G(\zeta)$, with some $c,\, c' > 0$, which becomes negative for R sufficiently large.

Since the functional K defined in (V.22) is continuous on H^1, the condition $K \leq 0$ defines a closed set, and we only need to check that $u = 0$ is not in the closure of $\mathscr{M}$. Using (V.20),

$$\begin{aligned} K(u) \quad &\geq \quad \frac{n-2}{2} \int_{\mathbb{R}^n} |\nabla u|^2 \, dx - n \int_{\mathbb{R}^n} \big(- c_1 |u|^2 + c_2 |u|^{l+1} \big) \, dx \\ &\geq \quad \frac{n-2}{2} \int_{\mathbb{R}^n} |\nabla u|^2 \, dx + n c_1 \int_{\mathbb{R}^n} |u|^2 \, dx - C \|u\|_{H^1}^{l+1}, \end{aligned}$$

with some $C > 0$. In the last inequality, we used the Sobolev embedding $H^1(\mathbb{R}^n) \subset L^{l+1}(\mathbb{R}^n)$. We see that the right-hand side becomes strictly positive for $0 < \|u\|_{H^1} < \varepsilon$, with some $\varepsilon > 0$ small enough. □

Lemma V.12 *Let $n \geq 3$. $K'(u)$ does not vanish on*

$$\mathscr{M}_0 = \{u \in H^1(\mathbb{R}^n): \quad K(u) = 0, \quad u \not\equiv 0\}. \tag{V.24}$$

PROOF. Let $u_0 \in H^1(\mathbb{R}^n)$ be such that $K'(u_0) = 0$. This means that u_0 satisfies

$$-(n-2)\Delta u_0 - n g(u_0) = 0, \qquad x \in \mathbb{R}^n,$$

with the Laplacian of H^1-functions understood in the distributional sense. Proceeding as in Lemma V.2, we get the relation

$$\frac{(n-2)^2}{2} \int_{\mathbb{R}^n} |\nabla u_0|^2 \, dx - n^2 \int_{\mathbb{R}^n} G(u_0) \, dx = 0.$$

Together with (V.5) this gives $u_0 \equiv 0$. □

Lemma V.13 *Let $n \geq 3$. The minimum of the minimization problem*

$$\inf \Big\{ \int_{\mathbb{R}^n} |\nabla u|^2 \, dx, \quad u \in \mathscr{M} \Big\} \tag{V.25}$$

is equal to the minimum of

$$\inf \Big\{ \int_{\mathbb{R}^n} |\nabla u|^2 \, dx, \quad u \in \mathscr{M}_0 \Big\}, \tag{V.26}$$

and is attained at

$$u_e \in \mathscr{M}_0 \cap H^1_r(\mathbb{R}^n), \qquad u_e^* = u_e.$$

Above, $\mathscr{M}$ and $\mathscr{M}_0$ are defined in (V.23) and (V.24) and u_e^* denotes the Schwarz symmetrization of u_e (Definition II.27).

PROOF. Let $u_j \in \mathscr{M} \subset H^1(\mathbb{R}^n)$ be a minimizing sequence. One can show that it suffices to search for the minimizer among functions which do not change under the Schwarz symmetrization. Indeed, the Schwarz symmetrization u^* satisfies the Pólya–Szegő inequality (see Lemma II.28):

$$\|\nabla u^*\|_{L^2} \le \|\nabla u\|_{L^2}, \qquad \forall u \in H^1(\mathbb{R}^n). \tag{V.27}$$

This inequality implies that $K(u^*) \le K(u)$; therefore, the minimizing sequence (u_j) can be substituted by the sequence of the Schwarz symmetrizations, (u_j^*).

So, let (u_j), $u_j \in \mathscr{M} \cap H^1_r(\mathbb{R}^n)$, be a minimizing sequence which satisfies $u_j^* = u_j$, $j \in \mathbb{N}$. Then $\int_{\mathbb{R}^n} |\nabla u_j|^2\,dx$ are bounded. Therefore, by (V.20),

$$\begin{aligned} 0 \ge K(u_j) &= \frac{n-2}{2}\int_{\mathbb{R}^n} |\nabla u_j|^2\,dx - n\int_{\mathbb{R}^n} G(u_j)\,dx \\ &\ge \frac{n-2}{2}\int_{\mathbb{R}^n} |\nabla u_j|^2\,dx + nc_1\int_{\mathbb{R}^n} |u_j|^2\,dx - nc_2\int_{\mathbb{R}^n} |u_j|^{l+1}\,dx. \end{aligned} \tag{V.28}$$

By Gagliardo–Nirenberg–Sobolev inequality (Lemma II.13), since $l+1 = 2n/(n-2) = 2^*$, there is $C > 0$ such that

$$\int_{\mathbb{R}^n} |u_j(x)|^{l+1}\,dx \le C\|\nabla u_j\|_{L^2}^{l+1};$$

then (V.28) yields

$$0 \ge \frac{n-2}{2}\int_{\mathbb{R}^n} |\nabla u_j|^2\,dx + nc_1\int_{\mathbb{R}^n} |u_j|^2\,dx - C\|\nabla u_j\|_{L^2}^{l+1}. \tag{V.29}$$

Due to the uniform boundedness of $\|\nabla u_j\|_{L^2}$, we conclude from (V.29) that $\|u_j\|_{L^2}$ are also uniformly bounded, hence so are $\|u_j\|_{H^1}$, $j \in \mathbb{N}$. Therefore, there is a subsequence, also denoted u_j, such that u_j converges weakly to $u_e \in H^1_r(\mathbb{R}^n)$. By the compactness of the embedding $H^1_r(\mathbb{R}^n) \subset L^q(\mathbb{R}^n)$ (see Section II.3), this convergence is strong in $L^q(\mathbb{R}^n)$ for $q \in [2, 2^*)$. It follows that u_e is also spherically symmetric and monotonically decreasing, so that $u_e^* = u_e$.

By the lower semicontinuity of norms with respect to weak limits, one has

$$\int_{\mathbb{R}^n} |\nabla u_e|^2\,dx \le \liminf_{j\to\infty} \int_{\mathbb{R}^n} |\nabla u_j|^2\,dx,$$

hence $K(u_e) \le \liminf_{j\to\infty} K(u_j) \le 0$. By the choice of u_j, the above inequalities are equalities, hence the convergence $u_j \to u_e$ is in fact strong in $H^1_r(\mathbb{R}^n)$. Note that $u_e \not\equiv 0$ since $u_j \in \mathscr{M}$ while $\mathscr{M} \cap \mathbb{B}_\varepsilon(H^1) = \emptyset$ with some $\varepsilon > 0$ (Lemma V.11).

Let us now argue that we can choose $u_e \in \mathscr{M}_0 \cap H^1_r(\mathbb{R}^n)$. Assume that $v_1 := u_e \in H^1_r(\mathbb{R}^n)$ satisfies $K(v_1) < 0$; let $v_\beta = v_1(x/\beta)$. Then

$$K(v_\beta) = \frac{\beta^{n-2}(n-2)}{2}\int_{\mathbb{R}^n} |\nabla v_1|^2\,dx - \beta^n n\int_{\mathbb{R}^n} G(v_1)\,dx.$$

Since $K(v_1) < 0$ and $K(v_\beta) > 0$ for β close to zero (due to $n \ge 3$), there is $\beta_0 \in (0,1)$ such that $K(v_{\beta_0}) = 0$, while

$$\int_{\mathbb{R}^n} |\nabla v_{\beta_0}|^2\,dx = \beta_0^{n-2}\int_{\mathbb{R}^n} |\nabla v_1|^2\,dx \le \int_{\mathbb{R}^n} |\nabla v_1|^2\,dx.$$

At the same time, since u_e is the minimizer of (V.25), the above inequality has to be equality; thus, one could take $u_e = v_{\beta_0} \in \mathscr{M}_0 \cap H^1_r(\mathbb{R}^n)$. □

Lemma V.14 *The function u_e satisfies $-\Delta u_e = g(u_e)$ in $\mathscr{D}'(\mathbb{R}^n)$.*

PROOF. According to the minimization problem we were considering, u_e satisfies

$$-\Delta u_e + \eta K'(u_e) = -\Delta u_e + \eta\big(-(n-2)\Delta u_e - n g(u_e)\big) = 0 \tag{V.30}$$

in $\mathscr{D}'(\mathbb{R}^n)$, where η is the Lagrange multiplier; for more details, see [**Zei84**, Chapter 43]. We note that $K'(u_e) \neq 0$ by Lemma V.12. As in Lemma V.2, it follows that

$$\begin{aligned} &\frac{n-2}{2}\int_{\mathbb{R}^n} |\nabla u_e|^2 + \eta\left(\frac{(n-2)^2}{2}\int_{\mathbb{R}^n} |\nabla u_e|^2 - n^2 \int_{\mathbb{R}^n} G(u_e)\right) \\ &\qquad = \frac{n-2}{2}T + \eta\left(\frac{(n-2)^2}{2}T - n^2 V\right) = 0, \end{aligned}$$

where

$$T := \int_{\mathbb{R}^n} |\nabla u_e|^2\,dx \qquad\text{and}\qquad V := \int_{\mathbb{R}^n} G(u_e)\,dx.$$

On the other hand, $\frac{n-2}{2}T - nV = 0$, $V = \frac{n-2}{2n}T$. This allows us to find η:

$$0 = \frac{n-2}{2} + \eta\left(\frac{(n-2)^2}{2} - \frac{n(n-2)}{2}\right) = \frac{n-2}{2} - \eta\frac{n-2}{2}2,$$

hence $\eta = 1/2$, and (V.30) takes the form $-\Delta u_e = g(u_e)$. □

V.3 Decay and regularity of solitary waves

Now let us prove that the solitary wave u_e is bounded in the origin; we will prove a slightly more general result.

Lemma V.15 (Continuity at the origin) *Let $n \in \mathbb{N}$. Let $g : \mathbb{R} \to \mathbb{R}$ be measurable, and assume that there is $q > 0$ and $c_0 > 0$ such that*

$$|g(u)| \le c_0 \langle u\rangle^q \quad \textit{almost everywhere in } \mathbb{R}. \tag{V.31}$$

If $n \ge 3$, additionally assume that $q < (n+2)/(n-2)$. Let $\delta > 0$. If $u \in H^1_r(\mathbb{B}^n_\delta, \mathbb{R})$ satisfies

$$-\Delta u = g(u), \qquad x \in \mathbb{B}^n_\delta,$$

then there is a finite limit $\lim_{r\to 0+} u(r)$.

PROOF. In the case $n = 1$, there is nothing to prove since $H^1((-\delta,\delta))$ consists of continuous functions. Let $n \ge 2$. Without loss of generality, we can assume that $\delta = 1$. Since u is spherically symmetric, we have:

$$\partial_r^2 u + \frac{n-1}{r}\partial_r u = \Delta u = -g(u), \qquad r > 0. \tag{V.32}$$

The above relation is considered in the sense of distributions on $\mathbb{R}_+$, dual to the space of smooth spherically symmetric functions with compact support in $\mathbb{R}^n \setminus \{0\}$. Writing the above equation in the form

$$\partial_r(r^{n-1}\partial_r u) = -r^{n-1}g(u), \qquad r > 0, \tag{V.33}$$

and integrating it over the interval $[r_0, r] \subset (0,1)$ we have:

$$r^{n-1}\partial_r u(r) - r_0^{n-1}\partial_r u(r_0) = -\int_{r_0}^{r} g(u(s))s^{n-1}\,ds. \tag{V.34}$$

Let us show that the integral in the right-hand side of (V.34) is absolutely convergent over $(0,1)$. We have:

$$\int_0^1 |g(u(s))| s^{n-1}\,ds \le c_0 \int_0^1 \langle u(s)\rangle^q s^{n-1}\,ds \le C\big(1+\|u\|^q_{L^q(\mathbb{B}^n_1)}\big),$$

which is finite since $u \in H^1(\mathbb{R}^n) \subset L^q(\mathbb{R}^n)$ by the Sobolev embedding theorem (note that if $n \ge 3$, one has $q < (n+2)/(n-2) < 2n/(n-2) = 2^*$). Once the integral in the right-hand side of (V.34) is absolutely convergent on $(0,1)$, we fix $r \in (0,1)$ and send $r_0 \to 0+$, concluding that there exists a finite limit $C = \lim_{r_0\to 0+} r_0^{n-1}|\partial_r u(r_0)|$. This limit equals zero, since otherwise in a small disc around the origin one would have $|\partial_r u(r)| \ge Cr^{1-n}/2$, in contradiction to $\nabla u \in L^2(\mathbb{B}^n_1, \mathbb{R}^n)$. Therefore, sending $r_0 \to 0$ in (V.34), we arrive at

$$r^{n-1}\partial_r u(r) = -\int_0^r g(u(s))s^{n-1}\,ds. \tag{V.35}$$

We claim that

$$u \in L^Q(\mathbb{B}^n_1), \tag{V.36}$$

with arbitrarily large $q < Q < \infty$. By the Sobolev embedding theorem, if $n = 2$, then we already have the inclusion (V.36) with arbitrarily large $Q < \infty$. Let now $n \ge 3$. Then in (V.36) we have $u \in L^Q(\mathbb{B}^n_1)$ with $Q = 2^* := 2n/(n-2) > q$. We will use the bootstrapping to show that Q could be chosen arbitrarily large. From (V.35), using the estimate (V.31) and the inclusion (V.36), one derives:

$$\begin{aligned} |r^{n-1}\partial_r u(r)| &\le c_0 \int_0^r \langle u(s)\rangle^q s^{n-1}\,ds \le C\big(r^n + \||u|^q\|_{L^{Q/q}(\mathbb{B}^n_r)}\|1\|_{L^{(Q/q)'}(\mathbb{B}^n_r)}\big) \\ &\le C r^{\frac{n}{(Q/q)'}} = C r^{n-\frac{qn}{Q}}, \qquad \forall r \in (0,1), \end{aligned} \tag{V.37}$$

where $1/(Q/q) + 1/(Q/q)' = 1$. By (V.37), we have

$$|\partial_r u(r)| \le C r^{1-qn/Q}, \qquad r \in (0,1),$$

and then $|u(r)| \le C(1 + r^{2-qn/Q}\langle \ln r\rangle)$, with some $C > 0$ (with the logarithmic term only appearing when $qn/Q = 2$). Therefore, $u \in L^{\tilde{Q}}(\mathbb{B}^n_1)$, with any $1 \le \tilde{Q} < n/(qn/Q - 2)$. This leads to the restriction $1/\tilde{Q} > q/Q - 2/n$, which we write as

$$\frac{1}{Q} - \frac{1}{\tilde{Q}} < \frac{2}{n} - \frac{q-1}{Q}.$$

We note that the right-hand side is strictly positive as long as $Q \ge 2^* := 2n/(n-2)$:

$$\frac{2}{n} - \frac{q-1}{Q} \ge \delta := \frac{2}{n} - \frac{q-1}{2^*} > \frac{2}{n} - \frac{(n+2)/(n-2) - 1}{2n/(n-2)} = 0.$$

Hence, $\tilde{Q} \ge 1$ could be chosen so that $1/Q - 1/\tilde{Q} < \delta$. By induction (repeating the above argument with $Q = \tilde{Q}$ and so on), we conclude that Q in (V.36) could be chosen arbitrarily large. Using the inequality (V.37) with $Q < \infty$ sufficiently large, we see that $\partial_r u(r) = o(1)$ for $r \in (0,1)$; it follows that there is a finite limit of $u(r)$ as $r \to 0$. □

Problem V.16 Let $u \in C^2(\mathbb{R})$, $|u(x)| \le 1$, $|u''(x)| \le 1$, $\forall x \in \mathbb{R}$. Find a bound on $|u'|$.

Lemma V.17 (Boundedness in C^2) *Let $n \in \mathbb{N}$, $g \in C(\mathbb{R}, \mathbb{R})$. If $u \in H^1_r(\mathbb{R}^n, \mathbb{R}) \cap C(\mathbb{R}^n, \mathbb{R})$ satisfies $-\Delta u = g(u)$, then*

$$u \in C^2_{\mathrm{b}}(\mathbb{R}^n, \mathbb{R}) := \Big\{u \in C^2(\mathbb{R}^n, \mathbb{R}): \quad \max_{\alpha \in \mathbb{N}^n_0,\, |\alpha| \le 2} \ \sup_{x \in \mathbb{R}^n} |\partial^\alpha_x u(x)| < \infty\Big\}.$$

Note that we do not need any additional assumptions on g.

PROOF. In the case $n = 1$, the inclusion $u \in H^1(\mathbb{R})$ shows that both u and then $u'' = -g(u)$ are bounded uniformly on $\mathbb{R}$. Then so is u' (see Problem V.16).

Let $n \ge 2$. Since $u \in C(\mathbb{R}^n)$, by (V.33), we have $u \in C(\mathbb{R}^n) \cap C^2(\mathbb{R}^n \setminus \{0\})$. Since there is a finite limit of $u(r)$ as $r \to 0$, the right-hand side of (V.34) has a finite limit as $r_0 \to 0$, hence so does $r_0^{n-1}\partial_r u(r_0)$, which moreover has to be zero (or else, by the argument in the proof of Lemma V.15, there would be a contradiction to $u \in H^1_r(\mathbb{B}^n_1)$); we arrive at (V.35), and then, due to the continuity of u at the origin, there is a finite limit

$$\frac{\partial_r u}{r} = -\frac{1}{r^n}\int_0^r g(u(s))s^{n-1}\,ds \to -\frac{1}{n}g(u(0)) \qquad \text{as} \quad r \to 0,$$

thus

$$\partial_r^2 u\big|_{r=0} = \lim_{r\to 0}\frac{\partial_r u(r) - 0}{r} = -\frac{1}{n}g(u(0)).$$

By (V.32), this agrees with the limit of $\partial_r^2 u$ as $r \to 0$; hence $u \in C^2(\mathbb{R}^n)$.

By Lemma II.15, any $u \in H^1_r(\mathbb{R}^n)$ satisfies $u \in C_{\mathrm{b}}(\mathbb{R}^n \setminus \mathbb{B}^n_1)$; it follows that

$$\text{(V.38)} \qquad \partial_r(r^{n-1}\partial_r u) = -r^{n-1}g(u(r)) = O(r^{-(n-1)/2}), \qquad r \ge 1,$$

therefore $r^{n-1}\partial_r u(r) = O(r^n)$, $r \ge 1$, and then from (V.32) one concludes that $\partial_r^2 u(r)$ is uniformly bounded. By Problem V.16, so is $\partial_r u(r)$. Thus, $u \in C^2_{\mathrm{b}}(\mathbb{R}^n)$. □

Before proceeding to the exponential decay and improved regularity of spherically symmetric solitary waves, let us prove the exponential decay of solutions to the Schrödinger equation in one dimension.

Lemma V.18 (Exponential decay of Schrödinger eigenstates in 1D) *Assume that*

$$\varphi \in C_{\mathrm{b}}(\mathbb{R}_+, \mathbb{R}) := \big\{u \in C(\mathbb{R}_+, \mathbb{R}): \quad \sup_{x>0}|u(x)| < \infty\big\},$$

$\varphi \not\equiv 0$, is a solution which satisfies

$$\text{(V.39)} \qquad -\varphi = -\varphi'' + V(x)\varphi, \qquad x > 0.$$

(1) If $V \in C(\mathbb{R}_+, \mathbb{R})$ is such that

$$\text{(V.40)} \qquad \nu := \inf_{x \in \mathbb{R}_+} V > -1,$$

then $|\varphi(x)|$ is strictly monotonically decreasing: either $\varphi(x) > 0$ and $\varphi'(x) < 0$ for all $x > 0$, or $\varphi(x) < 0$ and $\varphi'(x) > 0$ for all $x > 0$. Moreover,

$$\lim_{x\to+\infty}\varphi(x) = \lim_{x\to+\infty}\varphi'(x) = 0.$$

(2) *Moreover, assume that there are* $G \in L^1(\mathbb{R}_+, \mathbb{R})$, $c_0 > 0$, *and* $\kappa > 0$ *such that* V *and* φ *satisfy the following relation:*

$$|V(x)| \le G(x) + c_0|\varphi(x)|^{2\kappa}, \qquad \forall x > 0. \tag{V.41}$$

Then there are constants $C > c > 0$ *such that*

$$ce^{-x} \le |\varphi(x)| \le Ce^{-x}, \qquad \forall x > 0.$$

PROOF. Let us argue that $\varphi(x)$, $x > 0$, is strictly nonzero and strictly monotonically decaying to zero. Pick $x_0 > 0$. If $\varphi(x_0) = 0$ and $\varphi'(x_0) = 0$, then, by the uniqueness of solutions to ODEs, $\varphi \equiv 0$ for all $x > 0$, contradicting our assumptions. If $\varphi(x_0) > 0$ and $\varphi'(x_0) \ge 0$, or if $\varphi(x_0) = 0$ and $\varphi'(x_0) > 0$ then, from $\varphi'' = (1+V)\varphi$, φ'' is of the same sign as φ, both φ and φ' would be strictly monotonically increasing for $x \ge x_0$, and φ would be unbounded, again contradicting our assumptions. The cases $\varphi(x_0) < 0$, $\varphi'(x_0) \le 0$ and $\varphi(x_0) = 0$, $\varphi'(x_0) < 0$ are considered by changing the sign of φ. Thus, either $\varphi(x_0) > 0$ and $\varphi'(x_0) < 0$, or $\varphi(x_0) < 0$ and $\varphi'(x_0) > 0$, and by continuity these inequalities hold for all $x > x_0$. Changing the sign of φ if necessary, we may assume that

$$\varphi(x) > 0, \qquad \varphi'(x) < 0, \qquad \forall x > 0. \tag{V.42}$$

Therefore, there is a finite limit $a = \lim_{x\to+\infty} \varphi(x) \ge 0$. This limit has to be zero, or else, taking into account (V.40),

$$\varphi'(x) = \varphi'(x_0) + \int_{x_0}^{x} \big(\varphi(y) + V(y)\varphi(y)\big)\,dy \ge \varphi'(x_0) + (1+\nu)\int_{x_0}^{x} \varphi(y)\,dy$$

would become positive for x sufficiently large, and then both φ, φ' would start growing, hence $\varphi|_{\mathbb{R}_+}$ would be unbounded. This proves the first part of the lemma.

Let us prove the exponential decay. By (V.42), we assume that

$$\varphi(x) > 0, \qquad \varphi'(x) < 0, \qquad \forall x > 0.$$

Multiplying (V.39) by φ' and integrating from $x > 0$ to $+\infty$ and taking into account that $\lim_{x\to+\infty} \varphi(x) = \lim_{x\to+\infty} \varphi'(x) = 0$, we have:

$$\varphi(x)^2 - \varphi'(x)^2 = 2\int_x^\infty V(y)\varphi(y)\varphi'(y)\,dy, \qquad x > 0, \tag{V.43}$$

where the integral in the right-hand side is estimated by

$$\int_x^\infty |V(y)\varphi(y)\varphi'(y)|\,dy \le \varphi(x)|\varphi'(x)|\int_x^\infty |G(y)|\,dy + c_0\varphi(x)^{2\kappa+2}, \quad \forall x > 0.$$

We took into account the bound (V.41) and monotonicity of φ and φ'. Then (V.43) leads to

$$\left|1 - \left|\frac{\varphi'(x)}{\varphi(x)}\right|^2\right| \le 2\frac{|\varphi'(x)|}{\varphi(x)}\int_x^\infty |G(y)|\,dy + c_0\varphi(x)^{2\kappa}, \qquad \forall x > 0. \tag{V.44}$$

Since both $\int_x^\infty |G(y)|\,dy$ and $\varphi(x)$ go to zero as $x \to +\infty$, we deduce from the inequality (V.44) that there is $x_0 > 0$ such that

$$\frac{1}{2} \le \frac{|\varphi'(x)|}{\varphi(x)} \le \frac{3}{2}, \qquad \forall x \ge x_0. \tag{V.45}$$

The relation (V.43) yields the following inequality:

$$1 - \frac{2}{\varphi(x)^2}\int_x^\infty |V\varphi\varphi'|\,dy \le \frac{|\varphi'(x)|^2}{\varphi(x)^2} \le 1 + \frac{2}{\varphi(x)^2}\int_x^\infty |V\varphi\varphi'|\,dy, \qquad x > 0.$$

Taking into account that the inequality $1-a \le X^2 \le 1+b$, where $a,\, b \ge 0$, implies that $1-a \le |X| \le 1+b/2$, we have:

$$(V.46)\qquad 1-\frac{2}{\varphi(x)^2}\int_x^\infty |V\varphi\varphi'|\,dy \le \frac{|\varphi'(x)|}{\varphi(x)} \le 1+\frac{1}{\varphi(x)^2}\int_x^\infty |V\varphi\varphi'|\,dy, \quad x>0.$$

As long as

$$K := \int_{x_0}^\infty \Big[\int_x^\infty |V\varphi\varphi'|\,dy\Big]\frac{dx}{\varphi(x)^2} < \infty,$$

we would integrate (V.46) from some $x_0 > 0$ to $x > x_0$, arriving at

$$x - x_0 - 2K < \ln|\varphi(x_0)| - \ln|\varphi(x)| < x - x_0 + K,$$

which is the desired conclusion; we are left to show that $K < \infty$. Taking into account (V.45), we estimate K via the integration by parts, by

$$\int_{x_0}^\infty \Big[\int_x^\infty |V\varphi\varphi'|\,dy\Big]\frac{|\varphi'|\,dx}{\varphi^3} \le \frac{\int_{x_0}^\infty |V\varphi\varphi'|\,dy}{2\varphi(x_0)^2} + \int_{x_0}^\infty \frac{|V\varphi'|}{2\varphi}\,dx \le \int_{x_0}^\infty \frac{|V\varphi'|}{\varphi}\,dx;$$

we used the monotonic decay of $\varphi(x)$. Taking into account the relation (V.41), we have:

$$K \le \int_{x_0}^\infty \frac{|V(x)\varphi'(x)|}{\varphi(x)}dx \le \int_{x_0}^\infty G(x)\frac{|\varphi'(x)|}{\varphi(x)}\,dx + c_0\int_0^{\varphi(x_0)} s^{2\kappa-1}\,ds < \infty.$$

To estimate the first integral in the right-hand side, we took into account the inequality (V.45). □

Lemma V.19 (Exponential decay of NLS solitary waves) *Let $F \in C(\mathbb{R})$. Assume that there is $q > 1$ and $c_0 > 0$ such that*

$$(V.47)\qquad |F(u)| \le c_0|u|^q, \qquad \forall u \in \mathbb{R}.$$

Let $n \in \mathbb{N}$ and assume that $u \in H^1_r(\mathbb{R}^n) \cap C(\mathbb{R}^n)$, $u \not\equiv 0$, is a solution to

$$(V.48)\qquad -u = -\Delta u - F(u), \qquad x \in \mathbb{R}^n.$$

Then there is $R \ge 1$ such that the function u is strictly monotonically decreasing as a function of $r = |x|$ for $r \ge R$, and there are constants $C > c > 0$ such that

$$(V.49)\qquad c\langle x\rangle^{-(n-1)/2}e^{-|x|} \le |u(x)| \le C\langle x\rangle^{-(n-1)/2}e^{-|x|}, \qquad x \in \mathbb{R}^n \setminus \mathbb{B}^n_R.$$

Similar results are proved in [**GNNN81**, Theorem 2] and in [**BL83a**, Section 4.2, Lemma 1].

Proof. In the case $n = 1$, the result immediately follows from Lemma V.18, where we take $\varphi = u$, $V(u) = -F(u)/u$ (with $V(0) = 0$), and use the bound $|V(u)| \le c_0|u|^{q-1}$, $\forall u \in \mathbb{R}$.

Now let $n \ge 2$. By Lemma V.17, $u \in C^2_{\mathrm{b}}(\mathbb{R}^n)$. Denote

$$(V.50)\qquad \varphi(r) = r^{-(n-1)/2}u(r), \qquad r > 0,$$

where u is considered as a function of $r = |x|$. By (V.48), $\varphi(r)$ satisfies

$$(V.51)\qquad \partial_r^2\varphi = \varphi + \frac{b_n}{r^2}\varphi - \frac{F(u)}{u}\varphi, \qquad \text{where} \quad b_n := \frac{(n-1)(n-3)}{4}.$$

Since u decays to zero at infinity (Lemma II.15), there is $R \ge 1$ such that

$$(V.52)\qquad V(r) := \frac{b_n}{r^2} - \frac{F(u(r))}{u(r)} \ge \frac{b_n}{r^2} - c_0|u(r)|^{q-1}, \qquad r > 0$$

is bounded from below by $-1/2$ for all $r \geq R$. Therefore, we may apply Lemma V.18 to equation (V.51) for $r \geq R$, deducing that (changing the sign of φ if necessary) $\varphi(r) > 0$ and $\varphi'(r) < 0$ for all $r \geq R$, with $\lim_{r\to+\infty}\varphi(r) = \lim_{r\to+\infty}\varphi'(r) = 0$. Moreover, since V from (V.52) satisfies the relation (V.41) for $r \geq R \geq 1$, by Lemma V.18 there are constants $C > c > 0$ such that $ce^{-r} \leq \varphi(r) \leq Ce^{-r}$ for all $r \geq R$. It remains to use the relation (V.50). $\square$

Lemma V.20 (Regularity of NLS solitary waves) *Let $M \in \mathbb{N}$,*

$$F \in C(\mathbb{R}) \cap C^M(\mathbb{R}\setminus\{0\}),$$

and let there be $q > 1$ and $c_j > 0$, $1 \leq j \leq M$, such that

$$|\nabla_\tau^j F(\tau)| \leq c_j|\tau|^{q-j} \quad \text{for all} \quad \tau \in [-1,1]\setminus\{0\}, \quad 0 \leq j \leq M. \tag{V.53}$$

Let $n \in \mathbb{N}$ and assume that there is a spherically symmetric function $u \in H_r^1(\mathbb{R}^n) \cap C(\mathbb{R}^n)$ which is strictly positive and satisfies

$$-u = -\Delta u - F(u), \qquad x \in \mathbb{R}^n. \tag{V.54}$$

- *If $n = 1$, then there are $C_j > 0$, $1 \leq j \leq M+2$, such that*

$$|\partial_x^j u(x)| \leq C_j u(x), \qquad \forall x > 0, \qquad 1 \leq j \leq M+2, \tag{V.55}$$

  and one has $u \in H^{M+2}(\mathbb{R})$.

- *If $n \geq 2$, then there are $C_j > 0$, $1 \leq j \leq M+2$, such that*

$$|\partial_r^j u(x)| \leq C_j \left(\frac{\langle x\rangle}{|x|}\right)^{j-2} u(x), \qquad \forall x \in \mathbb{R}^n\setminus\{0\}, \qquad 1 \leq j \leq M+2. \tag{V.56}$$

  For any $s \leq M+2$ such that $s < \frac{n}{2}+2$, one has $u \in H^s(\mathbb{R}^n)$.

Proof. We start with the case $n \geq 2$, postponing the case $n = 1$ until the very end of the proof. Since u is spherically symmetric, the relation (V.54) takes the form

$$\partial_r^2 u + \frac{n-1}{r}\partial_r u = u - F(u), \qquad r > 0. \tag{V.57}$$

By Lemma V.19, there is $R \geq 1$ and constants $C > c > 0$ such that

$$c\langle r\rangle^{-(n-1)/2}e^{-r} \leq u(r) \leq C\langle r\rangle^{-(n-1)/2}e^{-r}, \qquad r \geq R,$$

where u is considered as a function of $r = |x|$, and such that $u(r)$ is strictly monotonically decreasing for $r \geq R$; then

$$w(r) := -\frac{\partial_r u(r)}{u(r)}, \quad r > 0 \qquad \text{satisfies} \quad w(r) \geq 0 \quad \forall r \geq R. \tag{V.58}$$

Let us prove (V.56) for $j = 1$, showing that there is $C_1 > 0$ such that

$$|w(r)| \leq C_1 r/\langle r\rangle \quad \forall r > 0. \tag{V.59}$$

Using (V.57), we arrive at

$$\partial_r w(r) = -\frac{\partial_r^2 u}{u} + w^2 = -1 + \frac{F(u)}{u} - \frac{n-1}{r}w + w^2, \qquad \forall r > 0. \tag{V.60}$$

Let us show the assumption that $w(r)$ is unbounded from above for $r \geq R$ would lead to a contradiction. Indeed, denote

$$a := \sup_{r>0}|1 - F(u(r))/u(r)| < \infty.$$

Assuming that $\limsup_{r\to+\infty} w(r) = +\infty$, there is $r_1 \ge R$ such that $w(r_1) > 0$ is large enough so that

$$-a - (n-1)w(r_1) + w(r_1)^2 > 0.$$

Let $W(r)$ is a solution to

$$\partial_r W = -a - (n-1)W + W^2, \qquad r \ge r_1, \tag{V.61}$$

with the initial data $W(r_1) = w(r_1) > 0$. Comparing the right-hand sides of (V.60) and (V.61), we see that $w(r) \ge W(r)$ for $r \ge r_1$ (as long as both solutions are defined), while $W(r)$ for $r \ge r_1$ blows up on a finite interval, in contradiction to $u \in C^2(\mathbb{R}^n)$. We conclude that $w(r)$ remains uniformly bounded from above for $r \ge R$. We also conclude from the inclusion $u \in C^2(\mathbb{R}^n)$, with u spherically symmetric, that $|\partial_r u(r)| = O(r)$ in the ball $\mathbb{B}^n_1$; due to $u(0) > 0$, the bound (V.59) follows.

For $j = 2$, the bound $|\partial_r^2 u(r)| \le C_2 u(r)$ with some $C_2 > 0$ follows from the relation (V.57) and the bound (V.59).

For $j \ge 3$, the proof is by induction. Assume that (V.56) is proved for $j \le k$, with some $k \in \mathbb{N}$, $k \ge 2$. The relation (V.57) allows us to express $u^{(k+1)}$ in terms of lower order derivatives:

$$\partial_r^{k+1} u = \partial_r^{k-1} u - \partial_r^{k-1}\Big(F(u) + (n-1)\frac{\partial_r u}{r}\Big), \qquad r > 0. \tag{V.62}$$

We notice that each of $k-1$ derivatives of the expression in the brackets, when acting on $F(u)$ or its derivative of certain order, due to the assumptions (V.53), contributes a factor of $\partial_r u/u$ (which is uniformly bounded); or else it changes one of the factors $u^{(i)}$ to $u^{(i+1)}$ with $i < k$ (worsening the bound by $\langle r\rangle/r$ by the induction assumptions); or else it acts on $1/r$, contributing another $1/r$; therefore, after each differentiation, the resulting estimate deteriorates by the factor $C\langle r\rangle/r$, with some $C > 0$. This allows us to bound (V.62) by $(\langle r\rangle/r)^{k-1}u(r)$ (times a constant factor), concluding the induction argument.

The inequality (V.56) and the interpolation arguments show that $u \in H^s(\mathbb{R}^n)$ as long as $|x|^{2-s}$ is L^2 locally near the origin; this imposes the restriction $s-2 < n/2$. We leave the details as an exercise:

Problem V.21 Show that $u \in H^s(\mathbb{R}^n)$ for any $s \le M+2$ as long as $s < n/2+2$.

Solution. Define $v(x) = u(x) - u(0)$, $x \in \mathbb{R}^n$, so that, by (V.56),

$$|\partial_r^j v(x)| \le C_j \left(\frac{\langle x\rangle}{|x|}\right)^{j-2} u(x), \qquad 0 \le j \le M+2, \qquad 0 < |x| \le 2, \tag{V.63}$$

with the same C_j, $j \in \mathbb{N}$, as in (V.56), and with some $C_0 > 0$. That is, now we extended (V.56) to $j = 0$, although x has to be inside a bounded region. Let $\beta_0 \in C^\infty_{\mathrm{comp}}([-2,2])$ be such that

$$\beta_0(t) \equiv 1 \ \ \forall t \le 1, \qquad \beta_0(t) \equiv 0 \ \ \forall |t| \ge 2, \qquad 0 \le \beta_0(t) \le 1 \ \ \forall t \in \mathbb{R}.$$

Then $\beta(t) = \beta_0(t) - \beta_0(2t)$ for $t \ge 0$, $\beta(t) = 0$ for $t < 0$ is such that

$$\beta \in C^\infty_{\mathrm{comp}}([1/2,2]), \qquad \beta(t) \ge 0, \qquad \sum_{k\in\mathbb{N}_0} \beta(2^k t) = \beta_0(t) \quad \forall t \in \mathbb{R}_+. \tag{V.64}$$

We set

$$v_k(x) = \beta(2^k|x|)v(x), \quad k \in \mathbb{N}_0, \quad x \in \mathbb{R}^n,$$

so that

$$u(x) = (1 - \beta_0(|x|))u(x) + u(0)\beta_0(|x|) + \sum_{k\in\mathbb{N}_0} v_k(x), \qquad x \in \mathbb{R}^n;$$

we note that for each $x \in \mathbb{R}^n$ the summation contains finitely many terms. By (V.56), the first term in the right-hand side is bounded in $H^s(\mathbb{R}^n)$ for any $s \le M+2$. Using (V.63), one can see that for each $j \in \mathbb{N}_0$ there is $c_j > 0$ such that

$$\|v_k\|_{H^j(\mathbb{R}^n)} \le c_j(1 + (2^k)^{j-2})2^{-kn/2}, \qquad \forall k \in \mathbb{N}_0, \qquad \forall j \in \mathbb{N}_0, \quad j \le M+2.$$

The complex interpolation shows that for each $s \in [0, M+2]$ there is $C_s > 0$ such that

$$\text{(V.65)} \qquad \|v_k\|_{H^s(\mathbb{R}^n)} \le c_s(1 + (2^k)^{s-2})2^{-kn/2}, \qquad \forall k \in \mathbb{N}_0, \quad \forall s \in [0, M+2].$$

There is the inequality

$$\|u\|_{H^s} \le \|(1 - \beta_0)u\|_{H^s} + u(0)\|\beta_0\|_{H^s} + \sum_{k\in\mathbb{N}_0} \|v_k\|_{H^s};$$

using the bound from (V.65), one concludes that the summation converges as long as one has $s - 2 - n/2 < 0$.

In the case $n = 1$, we claim that (V.56) improves to (V.55); the inclusion $u \in H^{M+2}(\mathbb{R})$ follows. Again, for $j = 1$ and $j = 2$, the statement follows from (V.59) and (V.57). For larger values of j, the proof is by the same induction as above; the improvement over (V.56) (absence of $(\langle r\rangle/r)^{j-2}$ factor in the right-hand side) is due to absence of the term $(n-1)\partial_r u/r$ in the right-hand side of (V.62). □

Lemma V.22 (Strict monotonicity) *Assume that $f \in C(\mathbb{R}_+, \mathbb{R})$ is Lipschitz. If $n \ge 1$ and $u \in C^2(\mathbb{R}_+, \mathbb{R})$ is a nonincreasing solution to*

$$\text{(V.66)} \qquad -u = -\partial_r^2 u - \frac{n-1}{r}\partial_r u - f(u^2)u, \qquad r > 0,$$

then either u is strictly monotonically decreasing,

$$u(r) > u(s) \qquad \text{as long as} \quad 0 < r < s,$$

or $u(r)$ is constant for all $r > 0$.

If, moreover,

$$\text{(V.67)} \qquad u(r) > 0 \quad \forall r > 0 \qquad \text{and} \qquad \lim_{r\to\infty} \partial_r u(r) = 0,$$

and $f(\tau)$ is strictly monotonically increasing for $\tau > 0$, then

$$\partial_r u(r) < 0 \quad \forall r > 0.$$

Proof. The strict monotonicity of u follows from the ODE theory. Indeed, let us assume that there are $r_1 > r_0 > 0$ such that $u(r_1) = u(r_0)$. Then, by monotonicity, $u(r) = c := u(r_0)$ for all $r \in [r_0, r_1]$ and then $u(r) = c$ for $r > 0$ would be the unique solution to the Cauchy problem

$$\begin{cases} -u = -\partial_r^2 u - \frac{n-1}{r}\partial_r u - f(u^2)u, & r > 0; \\ u(r_0) = c, & \partial_r u(r_0) = 0. \end{cases}$$

Thus, $u(r)$ is either strictly monotonically decreasing or constant.

Now we assume that $u(r)$ satisfies (V.67) and that $f \in C(\mathbb{R}_+, \mathbb{R})$ is not only Lipschitz but also strictly monotonically increasing. Let us assume that, contrary to the statement of the Lemma, there is $r_0 > 0$ such that $\partial_r u(r_0) = 0$. Since

$\partial_r u(r) \le 0$ for all $r > 0$, one has $\partial_r^2 u(r_0) = 0$; then, by (V.66), $f(u(r_0)^2) = 1$; we took into account that $u(r)$ is positive. Since $u(r)^2$ is strictly monotonically decreasing and so is $f(u(r)^2)$, we have:

$$u(r)\big(1 - f(u(r)^2)\big) > 0 \qquad \forall r > r_0. \tag{V.68}$$

From $\partial_r u(r_0) = 0$, $\partial_r u(r) \le 0$, and $\partial_r u(r) \to 0$ as $r \to \infty$ we conclude that there is $r_1 > r_0$ such that $\partial_r^2 u(r_1) \le 0$. Together with $\partial_r u(r_1) \le 0$ and the relation (V.68), this leads to a contradiction with (V.66) at $r = r_1$. Thus, $\partial_r u(r) < 0$ for all $r > 0$. □

V.4 Regularity and linear stability in pure power case

For the applications to the nonrelativistic limit of the nonlinear Dirac equation, we need to consider the linearization of the nonlinear Schrödinger equation with the pure power nonlinearity $f(u) = |u|^{2\kappa} u$, $u \in \mathbb{C}$, with $\kappa > 0$:

$$\mathrm{i}\partial_t \psi = -\Delta\psi - |\psi|^{2\kappa}\psi, \qquad \psi(t,x) \in \mathbb{C}, \qquad x \in \mathbb{R}^n. \tag{V.69}$$

If $n \ge 3$, additionally assume that $\kappa < 2/(n-2)$, or else (V.69) has no solitary wave solutions (see Lemma V.4). We need properties of a solitary wave $u(x)e^{-\mathrm{i}\omega t}$ with $\omega = -1$; in the pure power case, solitary waves for all other values $\omega < 0$ are obtained by rescaling.

Problem V.23 Find the relation between κ and n so that the Kolokolov stability condition (IV.17) holds for solitary wave solutions to (V.69).
Hint: Show that if $u_e(x)$ is a profile of a solitary wave corresponding to $\omega = -1$, then for any $\omega < 0$ there is a solitary wave with the profile

$$\varphi_\omega(x) = |\omega|^{\frac{1}{2\kappa}} u_e(|\omega|^{\frac{1}{2}} x), \qquad x \in \mathbb{R}^n.$$

Lemma V.24 (Solitary waves in the pure power NLS) *Let $n \in \mathbb{N}$, $\kappa > 0$. Moreover, assume that*

$$\kappa < \frac{2}{n-2} \qquad \text{if} \quad n \ge 3. \tag{V.70}$$

(1) There is a function $u_e \in H^1_r(\mathbb{R}^n)$ which satisfies

$$-u_e = -\Delta u_e - |u_e|^{2\kappa} u_e, \qquad x \in \mathbb{R}^n, \tag{V.71}$$

which is strictly positive and monotonically decreasing as a function of $r = |x|$;

(2) u_e is strictly monotonically decreasing as a function of $r = |x|$ with

$$\partial_r u_e(x) < 0 \qquad \forall x \in \mathbb{R}^n \setminus \{0\},$$

belongs to $H^s(\mathbb{R}^n)$ for any $s < n/2 + 2$, and satisfies the following inequalities:

$$c\langle x\rangle^{-\frac{n-1}{2}} e^{-|x|} \le u_e(x) \le C\langle x\rangle^{-\frac{n-1}{2}} e^{-|x|}, \qquad \forall x \in \mathbb{R}^n,$$

with some constants $C > c > 0$ which depend on n and κ;

$$|\partial_r^j u_e(x)| \le C_j \Big(\frac{\langle x\rangle}{|x|}\Big)^{j-2} u_e(x), \qquad \forall x \in \mathbb{R}^n \setminus \{0\}, \quad \forall j \in \mathbb{N}, \tag{V.72}$$

with some $C_j > 0$, $j \in \mathbb{N}$ which depend on n and κ;

(3) *The operator of multiplication by* u_e^κ *is a bounded linear operator in* $L^2(\mathbb{R}^n)$ *and in* $H^1(\mathbb{R}^n)$.

PROOF. The existence of a solution $u_e \in H_r^1(\mathbb{R}^n)$ for $n \geq 3$ follows from Theorem V.7. The case $n = 1$ is covered in Theorem IV.7. For the case $n = 2$, see **[BGK83, Caz03]** (or see an alternative argument below which works for the pure power case). Then, by Lemma V.15, due to the bound $1+2\kappa < (n+2)/(n-2)$ if $n \geq 3$, one has $u_e \in H_r^1(\mathbb{R}^n) \cap C(\mathbb{R}^n)$, and then, by Lemma V.17, $u_e \in H_r^1(\mathbb{R}^n) \cap C_b^2(\mathbb{R}^n)$.

For the pure power case, we will need an alternative proof based on finding optimal constants in the Gagliardo–Nirenberg inequalities **[Wei83]** (we will use it later for the linear stability results). Let $n \in \mathbb{N}$ and $\kappa > 0$ ($\kappa < 2/(n-2)$ if $n \geq 3$). One defines

$$J^{\kappa,n}(u) = \frac{\|\nabla u\|_{L^2}^{\kappa n} \|u\|_{L^2}^{2+\kappa(2-n)}}{\|u\|_{L^{2\kappa+2}}^{2\kappa+2}} \tag{V.73}$$

and considers the extremum problem

$$\varkappa = \inf_{u \in H^1(\mathbb{R}^n),\, u \neq 0} J^{\kappa,n}(u). \tag{V.74}$$

We note that the infimum $\varkappa$ in (V.74) is strictly positive due to the Gagliardo–Nirenberg inequality from Lemma II.13 *(2)* with $r = 2$, $q = 2$, $\alpha = \kappa n/(2\kappa+2)$, and $p_\alpha = 2\kappa + 2$; then one has:

$$\|u\|_{L^{2\kappa+2}} \leq (\varkappa)^{1/(2\kappa+2)} \|\nabla u\|^\alpha \|u\|_{L^q}^{1-\alpha}, \qquad \forall u \in H^1(\mathbb{R}^n).$$

Given $u \in H^1(\mathbb{R}^n)$, one defines $u^{\lambda,\mu} \in H^1(\mathbb{R}^n)$, λ, $\mu > 0$, by

$$u^{\lambda,\mu}(x) = \lambda u(\mu x), \qquad x \in \mathbb{R}^n,$$

and notices that $J^{\kappa,n}(u^{\lambda,\mu}) = J^{\kappa,n}(u)$ for all λ, $\mu > 0$. Therefore, if $u_j \in H^1(\mathbb{R}^n)$, $u_j \neq 0$, is a minimizing subsequence for (V.74) (we can assume that this sequence is Schwarz-symmetrized, so that $u_j = u_j^*$; see the proof of Theorem V.7), we can rescale each of the functions u_j, $j \in \mathbb{N}$, using appropriate λ_j, $\mu_j > 0$, so that $\psi_j(x) = \lambda_j u_j(\mu_j x)$ satisfies

$$\|\psi_j\|_{L^2} = 1, \qquad \|\nabla \psi_j\|_{L^2} = 1, \qquad \forall j \in \mathbb{N}; \qquad \lim_{j\to\infty} J^{\kappa,n}(\psi_j) = \varkappa > 0.$$

Like in the proof of Theorem V.7, since ψ_j are radial, we may assume that there is a subsequence (again denoted ψ_j) which converges strongly in $L^{2\kappa+2}(\mathbb{R}^n)$ to $\psi^* \in H^1(\mathbb{R}^n)$. Due to the weak convergence of ψ_j to ψ^* in $H^1(\mathbb{R}^n)$,

$$\|\psi^*\| \leq 1, \qquad \|\nabla \psi^*\| \leq 1.$$

Therefore,

$$\varkappa \leq J^{\kappa,n}(\psi^*) \leq \frac{1}{\|\psi^*\|_{L^{2\kappa+2}}^{2\kappa+2}} = \lim_{j\to\infty} \frac{1}{\|\psi_j\|_{L^{2\kappa+2}}^{2\kappa+2}} = \lim_{j\to\infty} J^{\kappa,n}(\psi_j) = \varkappa,$$

leading to

$$\|\psi^*\|_{L^2} = 1, \qquad \|\nabla \psi^*\|_{L^2} = 1,$$

thus $\psi_j \to \psi^*$ strongly in H^1.

Define $u_e \in H^1(\mathbb{R}^n)$ by $u_e(x) = \lambda\psi^*(\mu x)$, with some λ, $\mu > 0$ to be chosen later. For any $\eta \in \mathscr{D}(\mathbb{R}^n)$, one has:

$$\frac{\partial_\epsilon J^{\kappa,n}(u_e+\epsilon\eta)}{J^{\kappa,n}(u_e+\epsilon\eta)} = \frac{\kappa n}{2}\frac{\partial_\epsilon\|\nabla(u_e+\epsilon\eta)\|^2}{\|\nabla(u_e+\epsilon\eta)\|^2} + \frac{2+\kappa(2-n)}{2}\frac{\partial_\epsilon\|u_e+\epsilon\eta\|^2}{\|u_e+\epsilon\eta\|^2}$$
$$-(2\kappa+2)\frac{\partial_\epsilon\|u_e+\epsilon\eta\|^{2\kappa+2}_{L^{2\kappa+2}}}{\|u_e+\epsilon\eta\|^{2\kappa+2}_{L^{2\kappa+2}}}. \tag{V.75}$$

Evaluating the above expression at $\epsilon = 0$, we arrive at

$$0 = \left\langle \frac{-\kappa n\Delta u_e}{\|\nabla u_e\|^2_{L^2}} + \frac{(2+\kappa(2-n))u_e}{\|u_e\|^2_{L^2}} - \frac{(2\kappa+2)u_e^{2\kappa+1}}{\|u_e\|^{2\kappa+2}_{L^{2\kappa+2}}}, \eta\right\rangle. \tag{V.76}$$

Choosing $\lambda > 0$ and $\mu > 0$ so that

$$\frac{\|\nabla u_e\|^2_{L^2}}{\kappa n} = \frac{\|\nabla u_e\|^2_{L^2}}{2+\kappa(2-n)} = \frac{\|\nabla u_e\|^{2\kappa+2}_{L^{2\kappa+2}}}{2\kappa+2} =: \Lambda, \tag{V.77}$$

we see from (V.76) that u_e satisfies (V.71). This finishes *Part (1)*.

For *Part (2)*, the exponential decay follows from Lemma V.19, while the estimates on $\partial_r u_e$ and the improved Sobolev regularity follow from Lemma V.20. The strict monotonicity of $u_e(x)$ as a function of $r = |x|$ and the relation $\partial_r u_e(x) < 0$ for all $x \neq 0$ follow from Lemma V.22.

For *Part (3)*, it is enough to notice that, by *Part (2)*, both u_e^κ and $u_e^{\kappa-1}\partial_r u_e$ are from $L^\infty(\mathbb{R}^n)$. □

We need the detailed knowledge of the spectrum of the linearization at the solitary wave $u_e(x)e^{-\mathrm{i}\omega t}$, $\omega = -1$; we closely follow the exposition in Section IV.2. By (IV.12), the linearization at this solitary wave is represented by the operator

$$A = \begin{bmatrix} 0 & \mathfrak{l}_- \\ -\mathfrak{l}_+ & 0 \end{bmatrix},$$

$$\mathfrak{l}_- = 1 - \Delta - u_e^{2\kappa}, \qquad \mathfrak{l}_+ = 1 - \Delta - (1+2\kappa)u_e^{2\kappa},$$

with domains $\mathfrak{D}(A) = H^2(\mathbb{R}^n, \mathbb{C}^2)$, $\mathfrak{D}(\mathfrak{l}_\pm) = H^2(\mathbb{R}^n)$. We recall that $u_e^{2\kappa}$ is a multiplier in $L^2(\mathbb{R}^n)$ by Lemma V.24 *(3)*.

Taking into account that u_e decays exponentially, we conclude that $\mathfrak{l}_\pm$ are relatively compact perturbations of the Laplace operator (see Problem III.159 below), and then, by Weyl's theorem on the essential spectrum (Theorem III.135), one has

$$\sigma_{\mathrm{ess}}(\mathfrak{l}_-) = \sigma_{\mathrm{ess}}(\mathfrak{l}_+) = [1, +\infty).$$

Since the amplitude u_e satisfies the stationary equation (V.71), there are the relations (cf. (IV.14), (IV.15))

$$\mathfrak{l}_- u_e = 0, \qquad \mathfrak{l}_+\theta = u_e; \tag{V.78}$$

$$\mathfrak{l}_+\partial_{x^i}u_e = 0, \qquad \mathfrak{l}_-(x^i u_e) = 2\partial_{x^i}u_e, \qquad 1 \le i \le n. \tag{V.79}$$

Above, just like in (IV.14), $\theta = \partial_\omega|_{\omega=-1}\varphi_\omega(x)$ with $\varphi_\omega(x) = |\omega|^{1/(2\kappa)}u_e(|\omega|^{1/2}x)$ the profile of the solitary wave $\varphi_\omega(x)e^{-\mathrm{i}\omega t}$ (see Problem V.23); the explicit expression for θ is

$$\theta(x) = \partial_\omega|_{\omega=-1}\varphi_\omega(x) = \partial_\omega|_{\omega=-1}|\omega|^{\frac{1}{2\kappa}}u_e(|\omega|^{\frac{1}{2}}x) = -\frac{u_e(x)}{2\kappa} - \frac{x\cdot\nabla u_e(x)}{2}. \tag{V.80}$$

Lemma V.25 *Let $n \in \mathbb{N}$ and let $\kappa > 0$ satisfy* (V.70). *One has:*

$$\mathfrak{l}_- \geq 0; \qquad \mathfrak{l}_-\big|_{\{u_e\}^\perp} > 0. \tag{V.81}$$

The statement of the lemma follows from noticing that $u_e(x) > 0$ satisfies $\mathfrak{l}_- u_e = 0$ and thus u_e is a groundstate, so the smallest nondegenerate eigenvalue of $\mathfrak{l}_-$ is $\lambda = 0$; this smallest eigenvalue is simple. For more details, see e.g. [**GT83**, Theorem 8.38].

The following result is proved in [**Wei85, Kwo89**]; here we present the argument from [**CGNT08**, Lemma 2.1].

Lemma V.26 *Let $n \in \mathbb{N}$ and let $\kappa > 0$ satisfy* (V.70). *Then:*

(1) $\mathfrak{l}_+$ has exactly one negative eigenvalue;
(2) $\mathbf{ker}(\mathfrak{l}_+) = \mathrm{Span}\{\partial_{x^i} u_e \colon \ 1 \leq i \leq n\}$.

PROOF. Since u_e is spherically symmetric, we may decompose $L^2(\mathbb{R}^n)$ into invariant subspaces corresponding to spherical harmonics. In each of these invariant subspaces, $\mathfrak{l}_+$ is represented by the operator

$$\mathfrak{l}_+^{(k)} = 1 - \Delta_r + \frac{k(k+n-2)}{r^2} - (2\kappa+1)u_e^{2\kappa}, \qquad \Delta_r := \partial_r^2 + \frac{n-1}{r}\partial_r, \quad k \in \mathbb{N}_0.$$

By (V.81), there is at least one negative eigenvalue of $\mathfrak{l}_+$; let us argue that there is exactly one. To do so, following [**CGNT08**, Lemma 2.2], let us prove that

$$\mathfrak{l}_+\big|_{\{u_e^{2\kappa+1}\}^\perp} \geq 0, \tag{V.82}$$

where orthogonality is with respect to the scalar product in $L^2(\mathbb{R}^n)$. Taking into account that $J^{\kappa,n}(u_e) > 0$, $\partial_\epsilon\big|_{\epsilon=0} J^{\kappa,n}(u_e+\epsilon\eta) = 0$, $\partial_\epsilon^2\big|_{\epsilon=0} J^{\kappa,n}(u_e+\epsilon\eta) \geq 0$, we differentiate (V.75) with respect to ϵ at $\epsilon = 0$ and use (V.77) to derive:

$$\begin{aligned} 0 \ \leq \ & \frac{\partial_\epsilon^2\big|_{\epsilon=0} J^{\kappa,n}(u_e+\epsilon\eta)}{J^{\kappa,n}(u_e+\epsilon\eta)} = \partial_\epsilon\big|_{\epsilon=0} \frac{\partial_\epsilon J^{\kappa,n}(u_e+\epsilon\eta)}{J^{\kappa,n}(u_e+\epsilon\eta)} \\ = \ & \frac{\langle \eta, \mathfrak{l}_+\eta\rangle}{\Lambda} - \frac{\kappa n}{2}\Big(\frac{\partial_\epsilon\big|_{\epsilon=0}\|\nabla(u_e+\epsilon\eta)\|^2}{\|\nabla u_e\|^2}\Big)^2 - \frac{2+\kappa(2-n)}{2}\Big(\frac{\partial_\epsilon\big|_{\epsilon=0}\|u_e+\epsilon\eta\|^2}{\|u_e\|^2}\Big)^2 \\ & +(2\kappa+2)\Big(\frac{\partial_\epsilon\big|_{\epsilon=0}\|u_e+\epsilon\eta\|_{L^{2\kappa+2}}^{2\kappa+2}}{\|u_e\|_{L^{2\kappa+2}}^{2\kappa+2}}\Big)^2, \end{aligned}$$

with $\Lambda > 0$ from (V.77). By (V.70), $2+\kappa(2-n) > 0$. We conclude that

$$\langle \eta, \mathfrak{l}_+\eta\rangle \geq 0 \qquad \text{as long as} \qquad \partial_\epsilon\big|_{\epsilon=0}\|u_e+\epsilon\eta\|_{L^{2\kappa+2}}^{2\kappa+2} = 0, \tag{V.83}$$

that is, as long as η is orthogonal to $u_e^{2\kappa+1}$. This proves (V.82).

If $u_e^{2\kappa+1}$ were an eigenfunction corresponding to some negative eigenvalue, then, by (V.83), there would be no other negative eigenvalues of $\mathfrak{l}_+$, finishing the proof. Now let us consider the case when $u_e^{2\kappa+1}$ is not an eigenfunction corresponding to a negative eigenvalue. If there were two different negative eigenvalues of $\mathfrak{l}_+$, there would be a nontrivial linear combination of the corresponding eigenfunctions which is orthogonal to $u_e^{2\kappa+1}$, and then (V.83) would lead to a contradiction.

Let us show that $\mathfrak{l}_+^{(0)}$ does not have a zero eigenvalue; we give an argument from [**CGNT08**, Lemma 2.2]. Let us assume that, on the contrary, there is a spherically symmetric function $v_0 \in L^2(\mathbb{R}^n)$ such that $\mathfrak{l}_+^{(0)} v_0 = 0$. By (V.78),

$$\mathfrak{l}_+^{(0)} u_e = \mathfrak{l}_+ u_e = -2\kappa u_e^{2\kappa+1}, \qquad \mathfrak{l}_+^{(0)}\theta = \mathfrak{l}_+\theta = u_e;$$

then we would have

$$\begin{aligned}\langle u_e^{2\kappa+1}, v_0\rangle &= -\tfrac{1}{2\kappa}\langle \mathfrak{l}_+^{(0)} u_e, v_0\rangle = -\tfrac{1}{2\kappa}\langle u_e, \mathfrak{l}_+^{(0)} v_0\rangle = 0,\\ \langle u_e, v_0\rangle &= \langle \mathfrak{l}_+^{(0)}\theta, v_0\rangle = \langle \theta, \mathfrak{l}_+^{(0)} v_0\rangle = 0.\end{aligned} \tag{V.84}$$

Now let us consider u_e and v_0 as functions of $r = |x|$. With $v_0(r)$ corresponding to the second lowest eigenvalue $\lambda = 0$ of $\mathfrak{l}_+^{(0)}$ on $(0,\infty)$ (with the Neumann boundary condition at $r = 0$), by the Sturm–Liouville theory (see e.g. [**Zet05**, Theorem 2.6.3]), $v_0(r)$ would have exactly one positive zero, $r_0 > 0$. Changing the sign of v_0 if necessary, we may assume that $v_0(r) < 0$ for $r \in (0, r_0)$ and $v_0(r) > 0$ for $r > r_0$. Now consider the function

$$w(r) = u_e(r)^{2\kappa+1} - u_e(r_0)^{2\kappa} u_e(r) = \big(u_e(r)^{2\kappa} - u_e(r_0)^{2\kappa}\big) u_e(r), \qquad r > 0,$$

which is positive for $r \in (0, r_0)$ and negative for $r > r_0$ (since $u_e(r)$ is strictly monotonically decreasing to zero). Thus, on one hand, $v_0(r)w(r) > 0$ for all $r > 0$, $r \neq r_0$; on the other hand, by (V.84),

$$\langle w, v_0\rangle = \big\langle u_e^{2\kappa+1} - u_e(r_0)^{2\kappa} u_e,\, v_0\big\rangle = 0,$$

with the scalar product with respect to $L^2(\mathbb{R}^n)$. This contradiction proves that the second lowest eigenvalue of $\mathfrak{l}_+^{(0)}$ is positive.

Since $\mathfrak{l}_+^{(1)}\partial_r u_e = 0$ and $\partial_r u_e < 0$ for $r > 0$ (see Lemma V.24), $\partial_r u_e$ is the ground state of $\mathfrak{l}_+$ on $\mathbb{R}_+$ corresponding to the Dirichlet boundary conditions at $r = 0$ and corresponds to a nondegenerate eigenvalue $\lambda = 0$. The multiplicity n of this eigenvalue comes from the dimension of the spherical harmonics $\operatorname{Span}\big\{\frac{x^i}{r} : \ 1 \le i \le n\big\}$ which coincides with the dimension of the space of degree one harmonic polynomials.

Since $\mathfrak{l}_+^{(k)} - \mathfrak{l}_+^{(1)} > 0$ for $k \ge 2$ while $\mathfrak{l}_+^{(1)}$ has a (nondegenerate) groundstate with $\lambda = 0$, the operators $\mathfrak{l}_+^{(k)}$ for $k \ge 2$ have trivial kernel. □

The following result is well-known; for the sake of completeness, we sketch it here.

Lemma V.27 (Null space of linearization of pure power NLS) *Let $n \in \mathbb{N}$, $\kappa > 0$. If $n \ge 3$, additionally assume that $\kappa < 2/(n-2)$. The dimensions of the null space and the generalized null space of $A = \begin{bmatrix} 0 & \mathfrak{l}_- \\ -\mathfrak{l}_+ & 0\end{bmatrix}$ are given by*

$$\mathbf{ker}(A) = n+1, \qquad\qquad \mathfrak{L}(A) = \begin{cases} 2n+2, & \kappa \neq 2/n;\\ 2n+4, & \kappa = 2/n.\end{cases}$$

We remind that $\mathfrak{L}(A)$ is the generalized null space (or, more generally, a root lineal; see Section III.1).

Proof. By Lemma V.25 and Lemma V.26, one has

$$\mathbf{ker}(A) = \operatorname{Span}\left\{\begin{bmatrix}0\\u_e\end{bmatrix}, \begin{bmatrix}\partial_{x^i} u_e\\ 0\end{bmatrix}, \ 1 \le i \le n\right\}; \qquad \dim \mathbf{ker}(A) = n+1.$$

Now let us study the generalized eigenvectors. By (V.79),

$$\begin{bmatrix} 0 & \mathfrak{l}_- \\ -\mathfrak{l}_+ & 0\end{bmatrix}\begin{bmatrix}\partial_{x^i} u_e\\ 0\end{bmatrix} = 0, \qquad \begin{bmatrix} 0 & \mathfrak{l}_- \\ -\mathfrak{l}_+ & 0\end{bmatrix}\begin{bmatrix}0\\ x^i u_e\end{bmatrix} = -2\begin{bmatrix}\partial_{x^i} u_e\\ 0\end{bmatrix}, \tag{V.85}$$

for all $1 \le i \le n$. This Jordan block structure can not be extended, in the sense that there is no v such that

$$\begin{bmatrix} 0 & \mathfrak{l}_- \\ -\mathfrak{l}_+ & 0 \end{bmatrix} \begin{bmatrix} v \\ 0 \end{bmatrix} = \begin{bmatrix} 0 \\ x^i u_e \end{bmatrix}, \tag{V.86}$$

since $x^i u_e$ is not orthogonal to the kernel of $\mathfrak{l}_+$. Indeed, as follows from the identity

$$\langle x^i u_e, \partial_{x^i} u_e \rangle = \langle (-u_e - x^i \partial_{x^i} u_e), u_e \rangle, \tag{V.87}$$

with no summation in i, one has

$$\langle x^i u_e, \partial_{x^i} u_e \rangle = -\langle u_e, u_e \rangle / 2 < 0. \tag{V.88}$$

The relations (V.78) can be written as

$$\begin{bmatrix} 0 & \mathfrak{l}_- \\ -\mathfrak{l}_+ & 0 \end{bmatrix} \begin{bmatrix} 0 \\ u_e \end{bmatrix} = 0, \qquad \begin{bmatrix} 0 & \mathfrak{l}_- \\ -\mathfrak{l}_+ & 0 \end{bmatrix} \begin{bmatrix} -\theta \\ 0 \end{bmatrix} = \begin{bmatrix} 0 \\ u_e \end{bmatrix}. \tag{V.89}$$

By (V.85) and (V.89), $\dim \mathfrak{L}(A) \ge 2n + 2$. The dimension jumps above $2n + 2$ in the case when one can find $\alpha \in L^2(\mathbb{R}^n)$ such that

$$\mathfrak{l}_- \alpha = \theta, \quad \text{hence} \quad \begin{bmatrix} 0 & \mathfrak{l}_- \\ -\mathfrak{l}_+ & 0 \end{bmatrix} \begin{bmatrix} 0 \\ \alpha \end{bmatrix} = \begin{bmatrix} \theta \\ 0 \end{bmatrix}. \tag{V.90}$$

This happens when θ is orthogonal to $\mathbf{ker}(\mathfrak{l}_-) = \mathrm{Span}\{u_e\}$. Using the identity (V.87), we see that

$$2\langle \theta, u_e \rangle = -\Big\langle \frac{1}{\kappa} u_e + x \cdot \nabla u_e, u_e \Big\rangle = -\Big(\frac{1}{\kappa} \langle u_e, u_e \rangle - \frac{n}{2} \langle u_e, u_e \rangle \Big). \tag{V.91}$$

One can see that (V.91) vanishes when $\kappa = 2/n$ (that is, when the nonlinear Schrödinger equation is charge-critical).

So, we assume in the rest of the argument that $\kappa = 2/n$. Since $\theta(x)$ is real-valued and spherically symmetric while $\mathfrak{l}_\pm$ are invariant in $L^2_r(\mathbb{R}^n, \mathbb{R})$, α can also be chosen real-valued and spherically symmetric; moreover, it then follows that $\alpha \in H^2(\mathbb{R}^n)$. By (V.80),

$$\mathfrak{l}_- x^2 u_e = -2n u_e - 4x \cdot \nabla u_e = -8\theta,$$

hence one can choose $\alpha(x) = -4x^2 u_e(x)$. Then α is orthogonal to $\mathbf{ker}(\mathfrak{l}_+)$ which is spanned by $\partial_{x^i} u_e$, $1 \le i \le n$ (see Lemma V.26); since $z = 0$ is a point of the discrete spectrum of $\mathfrak{l}_+$, this operator is invertible on $\mathbf{ker}(\mathfrak{l}_+))^\perp$, and there is $\beta \in L^2(\mathbb{R}^n)$ such that

$$\mathfrak{l}_+ \beta = \alpha. \tag{V.92}$$

Arguing as above, β can be chosen spherically symmetric and real-valued, and moreover $\beta \in H^2(\mathbb{R}^n)$. Thus,

$$\begin{bmatrix} 0 & \mathfrak{l}_- \\ -\mathfrak{l}_+ & 0 \end{bmatrix} \begin{bmatrix} -\beta \\ 0 \end{bmatrix} = \begin{bmatrix} 0 \\ \alpha \end{bmatrix}.$$

This Jordan block structure can not be extended: there is no $\gamma \in L^2(\mathbb{R}^n)$ such that $\mathfrak{l}_- \gamma = \beta$ since β is never orthogonal to $\mathbf{ker}(\mathfrak{l}_-)$; indeed, due to semi-positivity of $\mathfrak{l}_-$ (see Lemma V.25), one has

$$\langle \beta, u_e \rangle = \langle \beta, \mathfrak{l}_+ \theta \rangle = \langle \beta, \mathfrak{l}_+ \mathfrak{l}_- \alpha \rangle = \langle \alpha, \mathfrak{l}_- \alpha \rangle > 0.$$

This finishes the proof. □

Later we will also need the following technical result.

Lemma V.28 *For $z \in \rho(\mathfrak{l}_-)$, the operator $(\mathfrak{l}_- - z)^{-1} : L^2(\mathbb{R}^n) \to H^2(\mathbb{R}^n)$ extends to a continuous mapping*

$$(\mathfrak{l}_- - z)^{-1} : H^{-1}(\mathbb{R}^n) \to H^1(\mathbb{R}^n).$$

PROOF. By means of the Fourier transform, the operator

$$(1 - \Delta - z)^{-1} : L^2(\mathbb{R}^n) \to H^2(\mathbb{R}^n), \qquad z \in [1, +\infty),$$

extends to a continuous mapping

$$(1 - \Delta - z)^{-1} : H^{-1}(\mathbb{R}^n) \to H^1(\mathbb{R}^n), \qquad z \in [1, +\infty).$$

Now the statement of the lemma follows from the resolvent identity

$$(\mathfrak{l}_- - z)^{-1} = R(z) + R(z)u_e^{2\kappa}R(z) - R(z)u_e^{2\kappa}(\mathfrak{l}_- - z)^{-1}u_e^{2\kappa}R(z),$$

with $R(z) = (-\Delta + 1 - z)^{-1}$, valid for all $z \in \rho(\mathfrak{l}_-)$. We note that

$$\sigma(1 - \Delta) = [1, +\infty) \subset \sigma(\mathfrak{l}_-).$$

This completes the proof. □

CHAPTER VI

Limiting absorption principle

VI.1 Agmon's Appendix A

We start with reproducing – almost verbatim – several results from Agmon's article [**Agm75**, Appendix A].

Lemma VI.1 ([Agm75], Lemma A.1) *Let $u \in H^1(\mathbb{R})$, $\lambda \in \mathbb{C}$, $s > 1/2$. The following inequality holds:*

$$\|u\|_{L^2_{-s}} \le c_s \Big\| \Big(\frac{d}{dx} - \lambda\Big) u \Big\|_{L^2_s}, \qquad c_s = \int_{\mathbb{R}} (1+x^2)^{-s}\, dx. \tag{VI.1}$$

PROOF. We set

$$f(x) = \Big(\frac{d}{dx} - \lambda\Big) u(x), \qquad u \in H^1(\mathbb{R}). \tag{VI.2}$$

We may assume without loss of generality that $f \in L^1(\mathbb{R})$ (or else $\|f\|_{L^2_s} = \infty$ and (VI.1) holds trivially), and that $\operatorname{Re}\lambda \le 0$ (or else replace $u(x)$ by $u(-x)$). Solving (VI.2) for u we get

$$u(x) = \int_{-\infty}^{x} f(y) e^{\lambda(x-y)}\, dy. \tag{VI.3}$$

From (VI.3) it follows that
(VI.4)

$$|u(x)|^2 \le \Big(\int_{-\infty}^{x} |f(y)|\, dy\Big)^2 \le \Big(\int_{\mathbb{R}} (1+y^2)^{-s}\, dy\Big)\Big(\int_{\mathbb{R}} (1+y^2)^{s} |f(y)|^2\, dy\Big).$$

We multiply (VI.4) by $(1+x^2)^{-s}$ and integrate:

$$\int_{\mathbb{R}} (1+x^2)^{-s} |u(x)|^2\, dx \le \Big(\int_{\mathbb{R}} (1+y^2)^{-s}\, dy\Big)^2 \int_{\mathbb{R}} (1+y^2)^{s} |f(y)|^2\, dy. \qquad \square$$

Lemma VI.2 ([Agm75], Lemma A.2) *Let $P(D) = P(D_1, \dots, D_n)$, $D_j = -\mathrm{i}\frac{\partial}{\partial x^j}$, be a partial differential operator with constant coefficients of order m. Then for any $u \in H^m(\mathbb{R}^n)$ and any given $s > \frac{1}{2}$, the following inequality holds:*

$$\int_{\mathbb{R}^n} (1+(x^j)^2)^{-s} |P^{(j)}(D)u|^2\, dx \le m^2 c_s^2 \int_{\mathbb{R}^n} (1+(x^j)^2)^{s} |P(D)u|^2\, dx, \tag{VI.5}$$

for $j = 1, \dots, m$, where c_s is the constant from Lemma VI.1.

Above, $P^{(j)}(\xi) = \partial_{\xi_j} P(\xi)$, $\xi \in \mathbb{R}^n$.

PROOF. First we assume that $n = 1$. Let $\lambda_1, \dots, \lambda_k$ be the roots of P repeated according to their multiplicities. Then

$$P'(\xi) = \sum_{i=1}^{k} Q_i(\xi), \quad \text{with} \quad Q_i(\xi) := \frac{P(\xi)}{\xi - \lambda_i}.$$

From Lemma VI.1, we have:

$$\|Q_i(\partial_x)u\|^2_{L^2_{-s}} \leq c_s^2 \Big\| \Big(\partial_x - \lambda_i\Big) Q_i(\partial_x)u\Big\|^2_{L^2_s} = c_s^2 \Big\| P(\partial_x)u\Big\|^2_{L^2_s}.$$

Summing with respect to i provides the result in dimension $n = 1$.

For higher dimension, we can assume without loss of generality that $j = 1$ and consider $\mathcal{F}_{x'}$ partial Fourier transform in $x' = (x^2, \dots, x^n)$. For any $(\eta_2, \dots, \eta_n) \in \mathbb{R}^{n-1}$, let $\widetilde{P}(\xi) := P(\xi, \eta_2, \dots, \eta_n)$. We thus have

$$\|\widetilde{P}'(\partial_x)\mathcal{F}_{x'}u\|^2_{L^2_{-s}(dx^1)} \leq m^2 c_s^2 \|P(\partial_x)\mathcal{F}_{x'}u\|^2_{L^2_s(dx^1)}.$$

Integrating in $(\eta_2, \dots, \eta_n)$ over $\mathbb{R}^{n-1}$ and using the fact that $\mathcal{F}_{x'}$ is unitary in $L^2(dx')$ we conclude the proof. $\square$

Let $P(D) = P(D_1, \dots, D_n)$ be a partial differential operator with constant coefficients of order m, acting on functions on $\mathbb{R}^n$. We denote its principal part by $P_m(D)$. The operator P is said to be *of principal type* if $\operatorname{grad} P_m(\xi) \neq 0$ $\forall \xi \in \mathbb{R}^n \setminus \{0\}$; P is said to be *elliptic* if $P_m(\xi) \neq 0$ $\forall \xi \in \mathbb{R}^n \setminus \{0\}$.

Problem VI.3 Show that an elliptic operator is an operator of principal type. *Hint: Use the fact that $m \neq 0$ and that P_m is homogeneous.*

We shall say that a number $z \in \mathbb{C}$ is a *critical value* of P if there exists $\xi_0 \in \mathbb{R}^n$ such that $P(\xi_0) = z$, $\operatorname{grad} P(\xi_0) = 0$. We shall denote the set of all critical values of P by $\Lambda_C(P)$.

Lemma VI.4 ([Agm75], Lemma A.3) *Let $P(D) = P(D_1, \dots, D_n)$, with P of principal type, with $D_j = -\mathrm{i}\frac{\partial}{\partial x^j}$, be a partial differential operator of order m. Set $m' = m$ if P is elliptic, $m' = m - 1$ otherwise. Let $\mathcal{K}$ be a compact set in $\mathbb{C} \setminus \Lambda_C(P)$ and let s be a real number. The following estimate holds:*

$$\|u\|_{H^{m'}_s} \leq C_s\Big(\|(P(D) - z)u\|_{L^2_s} + \sum_{j=1}^n \|P^{(j)}(D)u\|_{L^2}\Big) \tag{VI.6}$$

for any $u \in H^m_s(\mathbb{R}^n)$ and any $z \in \mathcal{K}$, where C_s is a constant not depending on z or u.

Proof. Since P is of principal type, $P_m(\xi) \neq 0$ for $\xi \in \mathbb{R}^n \setminus \{0\}$, and hence, for some $R_0 > 0$ and $C > 0$,

$$(1 + |\xi|^2)^{m-1} \leq C \sum_{j=1}^n |P^{(j)}(\xi)|^2, \qquad \forall \xi \in \mathbb{R}^n, \quad |\xi| > R_0. \tag{VI.7}$$

For any $z \notin \Lambda_C(P)$,

$$\sum_{j=1}^n |P^{(j)}(\xi)|^2 + |P(\xi) - z|^2 \neq 0, \qquad \forall \xi \in \mathbb{R}^n,$$

and hence

$$\min_{z \in \mathcal{K},\, |\xi| \leq R_0} \Big(|P(\xi) - z|^2 + \sum_{j=1}^n |P^{(j)}(\xi)|^2\Big) \geq c_0 > 0. \tag{VI.8}$$

Thus, combining (VI.7) and (VI.8), we conclude that there is $C_0 > 0$ such that

$$(1+|\xi|^2)^{m-1} \leq C_0\Big(|P(\xi)-z|^2 + \sum_{j=1}^{n} |P^{(j)}(\xi)|^2\Big), \qquad \forall \xi \in \mathbb{R}^n. \tag{VI.9}$$

This leads to (VI.6) in Fourier variables with $s = 0$.

Problem VI.5 Assuming that P is elliptic, revise the proof to extend (VI.9) from $m' = m-1$ to $m' = m$.

Problem VI.6 Finish the proof in the case $s \neq 0$.
Hint: Consider the weight $\rho_1(x) = \langle x\rangle$ *by* $\rho_\varepsilon(x) = \rho_1(\varepsilon x)$ *with* $\varepsilon > 0$. *When applying* (VI.6) *with* $s = 0$ *to* $\rho_\varepsilon^s u$, *choose* $\varepsilon > 0$ *so that unwanted terms coming from commutators are controlled by those appearing in* (VI.6). □

Remark VI.7 The constant C_0 in the proof depends on P, n, m, and a priori also on $\mathcal{K}$. More accurately, in (VI.8) the dependence of c_0 on $\mathcal{K}$ can be specified as follows:

$$\min_{z\in\mathbb{C},\, \operatorname{dist}(z,\Lambda_C(P))>\delta,\, |\xi|\leq R_0} \Big(|P(\xi)-z|^2 + \sum_{j=1}^{n} |P^{(j)}(\xi)|^2\Big) \geq c_\delta > 0$$

for any $\delta > 0$ and some $c_\delta > 0$. Hence, in the above theorem, one can replace $\mathcal{K}$ by $\{z \in \mathbb{C}, \operatorname{dist}(z, \Lambda_C(P)) > \delta\}$; then C_0 and thus C_s depend on $\mathcal{K}$ via $\operatorname{dist}(\mathcal{K}, \Lambda_C(P))$.

Theorem VI.8 ([Agm75], Theorem A.1) *Let $P(D)$ be a differential operator with constant coefficients of order m and of principal type. Set $m' = m$ if P is elliptic, $m' = m-1$ otherwise. Let $\mathcal{K}$ be a compact set in $\mathbb{C}\setminus\Lambda_C(P)$ and let $s > \frac{1}{2}$. The following estimate holds:*

$$\|u\|_{H^{m'}_{-s}} \leq C\|(P(D)-z)u\|_{L^2_s} \tag{VI.10}$$

for any $u \in H^m(\mathbb{R}^n)$, where C is some constant not depending on z or u.

Problem VI.9 Prove Theorem VI.8 using Lemmata VI.2–VI.4.

Problem VI.10 Show that for any $\delta > 0$ there is $C > 0$ such that for any measurable u,

$$\sup_{z\in\mathbb{C},\operatorname{dist}(z,\Lambda_C(P))>\delta} \sum_{k=0}^{m'} \langle z\rangle^{\frac{m-1-k}{m}} \|u\|_{k,-s} \leq C\|(P(D)-z)u\|. \tag{VI.11}$$

Hint: From Remark VI.7 *and* (VI.7), *deduce the following relation:*

$$(1+|\xi|^2)^{m'} \leq C_\delta \min_{|z|=1,\operatorname{dist}(z,\Lambda_C(P))>\delta} \Big(|P_m(\xi)-z|^2 + \sum_{j=1}^{n} |P_m^{(j)}(\xi)|^2\Big), \quad \forall \xi \in \mathbb{R}^n,$$

with P_m the principal part of P. By rescaling $\xi \to |z|^{-1/m}\xi$, conclude that for all $z \in \mathbb{C}$ with $z \neq 0$ and $\operatorname{dist}(z, \Lambda_C(P)) > \delta)$ *and for all ξ in $\mathbb{R}^n$, one has:*

$$|z|^{2-2m'/m}(|z|^{2/m}+|\xi|^2)^{m'} \leq \Big(|P_m(\xi)-z|^2 + \sum_{j=1}^{n} |z|^{2-2(m-1)/m}|P_m^{(j)}(\xi)|^2\Big). \tag{VI.12}$$

Use (VI.12) *to revise the proof of Theorem* VI.8, *arriving at* (VI.11).

We also give the following form of the limiting absorption principle for the free resolvent [**Agm75**, Remark 2 in Appendix A] and [**JK79**, Theorem 8.1].

Lemma VI.11 (Limiting absorption principle for the Laplace operator) *Let $n \ge 1$. For any $k \in \mathbb{N}_0$, $\nu \le 2+2k$, $s > 1/2+k$, and $\delta > 0$, there is $C = C(n,s,k,\nu,\delta) > 0$ such that*

$$\|\partial_z^k(-\Delta - z)^{-1}\|_{L^2_s(\mathbb{R}^n)\to H^\nu_{-s}(\mathbb{R}^n)} \le C|z|^{-(k+1-\nu)/2}, \qquad z \in \mathbb{C}\setminus(\mathbb{D}_\delta \cup \mathbb{R}_+).$$

Proof. For $\nu = 0$, the lemma rephrases [**JK79**, Theorem 8.1] and follows from the generalization of Theorem VI.8 in Problem VI.10. That the recurrence based on the identities

$$-\Delta(-\Delta-z)^{-1} = 1+z(-\Delta-z)^{-1} \quad \text{and} \quad \partial_z^k(-\Delta-z)^{-1} = k!(-\Delta-z)^{-k-1}, \quad k \ge 0,$$

provides all the cases starting from $\nu = k = 0$. □

VI.2 Improvement at the continuous spectrum

For λ at the continuous spectrum, in [**Agm75**, Appendix B], Agmon gives an improvement of Theorem VI.8, which we illustrate in the following lemma.

Lemma VI.12 *For any $s > 1/2$ there is $C_s > 0$ such that for any $u \in H^1(\mathbb{R})$ and $\Lambda \in \mathbb{R}$ the following inequality holds:*

$$\|u\|_{L^2_{s-1}} \le C_s\left\|\left(\frac{d}{dx} - \mathrm{i}\Lambda\right)u\right\|_{L^2_s}. \tag{VI.13}$$

Proof. Considering $v(x) = e^{-\mathrm{i}\Lambda x}u(x)$, $v \in H^1(\mathbb{R})$, we see that it is enough to prove (VI.13) for $\Lambda = 0$. Let $u \in H^1(\mathbb{R})$,

$$\partial_x u(x) = f(x) \in L^2_s(\mathbb{R}), \qquad s > 1/2.$$

We need to prove that

$$\|u\|_{L^2_{s-1}} \le C_s\|u'\|_{L^2_s}, \qquad u \in H^1(\mathbb{R}), \quad u' \in L^2_s(\mathbb{R}),$$

for some $C_s > 0$ which only depends on $s > 1/2$. We notice that since $u \in H^1(\mathbb{R})$, $\hat{u} \in L^1(\mathbb{R})$, one has

$$\lim_{|x|\to\infty} |u(x)| = 0. \tag{VI.14}$$

Since $u \in H^1(\mathbb{R})$, $\lim_{x\to\pm\infty} u(x) = 0$. Therefore, since we assume that $\partial_x u \in L^2_s(\mathbb{R})$, for $x \ge 0$, we have:

$$|u(x)| \le \int_x^{+\infty} |u'(y)|\,dy \le \left(\int_x^{+\infty} (1+y^2)^s|u'(y)|^2\,dy \int_x^{+\infty} (1+y^2)^{-s}\,dy\right)^{1/2},$$

$$|u(x)| \le C_s\left(\int_x^{+\infty} (1+y^2)^s|u'(y)|^2\,dy\right)^{1/2} \langle x\rangle^{-s+1/2}, \qquad x \ge 0,$$

where $C_s > 0$ depends on $s > 1/2$. There is a similar inequality for $x \le 0$. Therefore,

$$\lim_{x\to\pm\infty} \langle x\rangle^{s-1/2}u(x) = 0. \tag{VI.15}$$

Using the integration by parts, we obtain the following identity for $u \in H^1(\mathbb{R})$:

$$(2s-1)\int_x^X y^{2s-2}u(y)^2\,dy \le y^{2s-1}u(y)^2|_x^X - 2\int_x^X y^{2s-1}u(y)u'(y)\,dy.$$

Sending $x \to -\infty$, $X \to +\infty$, and using (VI.15) to remove the boundary terms, we arrive at

$$(2s-1)\int_{\mathbb{R}} y^{2s-2}u(y)^2\,dy \le 2\int_{\mathbb{R}} y^{2s-1}|u(y)u'(y)|\,dy.$$

We notice that

$$\left|\int_{\mathbb{R}} y^{2s-1}u(y)u'(y)\,dy\right| \le \left|\int_{\mathbb{R}} y^{2s-2}u(y)^2\,dy\right|^{1/2}\left|\int_{\mathbb{R}} y^{2s}u'(y)^2\,dy\right|^{1/2},$$

hence

$$(2s-1)\left(\int_{\mathbb{R}} y^{2s-2}u(y)^2\,dy\right)^{1/2} \le 2\left(\int_{\mathbb{R}} y^{2s}|u'(y)|^2\,dy\right)^{1/2}.$$

Instead of the Hardy inequality, we can base the argument on Cotlar–Stein's almost orthogonality lemma (Lemma II.18). By (VI.14), we have:

$$u(x) = \begin{cases} -\int_x^{+\infty} f(y)\,dy, & x \ge 0;\\ \int_{-\infty}^x f(y)\,dy, & x \le 0.\end{cases}$$

Without loss of generality, we may consider f with support $\operatorname{supp} f \subset \mathbb{R}_+$ and to study the norm of $\mathbb{1}_{\mathbb{R}_+}u$. For $x \ge 0$, we use the dyadic decomposition

$$u(x) = -\int_x^{+\infty} f(y)\,dy = -\sum_{j\in\mathbb{Z}}\int_{\mathbb{R}_+}\mathbb{1}_{(x,+\infty)}(y)\beta(y/2^j)f(y)\,dy,$$

with $\beta = \mathbb{1}_{[1,2]}$. To prove the boundedness of the mapping $f \mapsto u$, $L^2_s(\mathbb{R}_+) \to L^2_{s-1}(\mathbb{R}_+)$, we need to show that the operators

$$T_jf(x) = \int_{\mathbb{R}_+}\mathbb{1}_{(x,+\infty)}(y)\langle x\rangle^{s-1}\beta(y/2^j)\langle y\rangle^{-s}f(y)\,dy, \qquad j\in\mathbb{Z},$$

are bounded in L^2 and are almost orthogonal.

Let us consider the integral kernel of $T_j^*T_k$:

$$|K(T_j^*T_k)(y,w)| = \int_{\mathbb{R}_+}\mathbb{1}_{(x,+\infty)}(y)\mathbb{1}_{(x,+\infty)}(w)\beta(y/2^j)\beta(w/2^k)\langle y\rangle^{-s}\langle x\rangle^{2s-2}\langle w\rangle^{-s}\,dx.$$

Computing

$$\int_{\mathbb{R}_+}|K(T_j^*T_k)(y,w)|\,dy \le CY\langle Y\rangle^{-s}\langle W\rangle^{-s}\langle\min(Y,W)\rangle^{2s-1},$$

with $Y = 2^j$ and $W = 2^k$, and applying the Schur test, we have:

$$\text{(VI.16)}\qquad \|T_j^*T_k\| \le C\sqrt{YW}\langle Y\rangle^{-s}\langle W\rangle^{-s}\langle\min(Y,W)\rangle^{2s-1},\quad Y = 2^j,\quad W = 2^k.$$

Let us consider the integral kernel of $T_jT_k^*$:

$$|K(T_jT_k^*)(x,z)| = \int_{\mathbb{R}_+}\mathbb{1}_{(x,+\infty)}(y)\mathbb{1}_{(z,+\infty)}(y)\langle x\rangle^{s-1}\beta(y/2^j)\beta(y/2^k)\langle y\rangle^{-2s}\langle z\rangle^{s-1}\,dy.$$

Because of the support properties of β, the above is zero unless $j = k$; in this last case, the bound on $\|T_jT_j^*\| = \|T_j^*T_j\|$ follows from (VI.16).

It follows that there is $C > 0$ such that for each fixed $j\in\mathbb{Z}$,

$$\sum_{k\in\mathbb{N}}\|T_j^*T_k\|^{1/2} + \sum_{k\in\mathbb{N}}\|T_jT_k^*\|^{1/2} \le C.$$

By the Cotlar–Stein almost orthogonality lemma (Lemma II.18), $\sum_{j\in\mathbb{Z}} T_j$ converges (in the strong operator norm) to an operator bounded in L^2. □

In higher dimensional setting, this leads to the following result:

Lemma VI.13 ([Agm75, Lemma B.2 Bis]) *Let $g(x')\in C^\infty_{\mathrm{comp}}(\mathbb{R}^{n-1})$. Set*

$$\varGamma=\{(x^1,x')\in\mathbb{R}\times\mathbb{R}^{n-1}:\quad x^1=g(x')\}.$$

Let $u\in H^s(\mathbb{R}^n)$ with $s>1/2$ such that $u|_{\varGamma}=0$. Then

$$\frac{u(x)}{x^1-g(x')}\in H^{s-1}(\mathbb{R}^n)\cap L^1_{\mathrm{loc}}(\mathbb{R}^n)$$

and

$$\left\|\frac{u}{x^1-g(x')}\right\|_{H^{s-1}}\le\gamma_s\|u\|_{H^s},$$

where γ_s is a constant depending only on s and g.

The proof is obtained by first considering the one-dimensional case with x^1 (shifted by $g(x')$) and applying Lemma VI.12 (in the Fourier variables) and then integrating in x'.

The conclusion that $u/(x^1-g(x'))\in L^1_{\mathrm{loc}}(\mathbb{R}^n)$ comes from the Sobolev embedding in x^1 (since $s>1/2$) and from the Hölder inequality in x'.

VI.3 Limiting absorption principle for the Laplacian near the threshold

In this section, we consider the limiting absorption principle for the resolvent of the Laplace operator, $(-\Delta-z)^{-1}$, $z\in\mathbb{C}\setminus\overline{\mathbb{R}_+}$, near the threshold $z=0$ in dimensions $n\ge 3$. For $n=3$, the uniform boundedness of

$$(-\Delta-z)^{-1}:\ L^2_s(\mathbb{R}^n)\to L^2_{-s'}(\mathbb{R}^n),\qquad z\in\mathbb{C}\setminus\overline{\mathbb{R}_+},$$

for $s,\ s'>1/2$, $s+s'>2$ is proved in [**JK79**, Lemma 2.1] by noticing that the operator with the integral kernel of $(-\Delta-z)^{-1}$ at $z=0$, $R_0(x,y,0)=(4\pi|x-y|)^{-1}$, is of Hilbert–Schmidt class:

$$\int_{\mathbb{R}^3}\int_{\mathbb{R}^3}\langle x\rangle^{-2s}|x-y|^{-2}\langle y\rangle^{-2s'}\,dx\,dy<\infty,\qquad\forall s,\ s'>1/2,\quad s+s'>2,$$

while the integral kernel $R_0(x,y,z)$ of $(-\Delta-z)^{-1}$, $z\in\mathbb{C}\setminus\overline{\mathbb{R}_+}$ satisfies

$$|R_0(x,y,z)|\le R_0(x,y,0),\qquad\forall x,\ y\in\mathbb{R}^3.$$

Below, it will be convenient to consider the resolvent of the fractional powers of the Laplace operator in all dimensions.

Let $0<\alpha<n$. For $z=0$, with $I_\alpha:=R_0^{(\alpha)}(0)=(-\Delta)^{-\alpha/2}$ (understood as a Fourier multiplier) the Riesz potentials, the sharp version ($s+s'\ge\alpha$) follows from [**NW73**, Lemma 2.1] (see also [**Ily61**]) and [**Jen80**, Lemma 2.3]. We recall that for $n\in\mathbb{N}$ and $0<\alpha<n$, the Riesz potentials I_α could also be defined by

$$(I_\alpha f)(x)=\frac{1}{c_\alpha}\int_{\mathbb{R}^n}\frac{f(y)}{|x-y|^{n-\alpha}}\,dy,\qquad c_\alpha=\pi^{n/2}2^\alpha\frac{\Gamma(\alpha/2)}{\Gamma((n-\alpha)/2)},\tag{VI.17}$$

where $f\in L^p(\mathbb{R}^n)$, with $1\le p<n/\alpha$.

Problem VI.14 Show that $|\xi|^{-\alpha}$, $0<\alpha<n$, is a tempered distribution, and that its Fourier transform equals $C|x|^{\alpha-n}$, with some $C>0$. Find the value of C. *Hint: Let $g_\alpha:\xi\mapsto|\xi|^{-\alpha}$.*

(1) *Show that* $\widehat{g_\alpha}$ *is a radial and homogeneous distribution of order* $n-\alpha$.
(2) *Show that* $x\cdot\nabla\widehat{g_\alpha}=(n-\alpha)\widehat{g_\alpha}$.
(3) *Find the solution* $u\in\mathscr{S}'(\mathbb{R}^n)$ *to the equation* $x\cdot\nabla u-(n-\alpha)u=0$ *and prove that* $\widehat{g_\alpha}=C|x|^{\alpha-n}$.
(4) *Use Gaussian functions to compute* C.

Let us remind the classical results.

Lemma VI.15 (**[NW73**, Lemma 2.1]) *For real numbers* a *and* b *whose sum is positive, consider the kernel*

$$K(x,y)=\frac{1}{|x|^a|x-y|^{n-a-b}|y|^b},\qquad \text{for } x\neq y \text{ in } \mathbb{R}^n.$$

The integral operator $Ku(x)=\int_{\mathbb{R}^n}K(x,y)u(y)\,dy$ *is a bounded operator in* $L^p(\mathbb{R}^n)$ *if and only if* $a<n/p$ *and* $b<n/p'$.

Lemma VI.16 (**[Jen80**, Lemma 2.3])

(1) *Assume* $0<\alpha<n/2$, $s\geq 0$, $s'\geq 0$.
If $s+s'\geq\alpha$, *then* $I_\alpha\in\mathscr{B}\big(L^2_s(\mathbb{R}^n),L^2_{-s'}(\mathbb{R}^n)\big)$.
If $s+s'>\alpha$, *then* $I_\alpha\in\mathscr{B}_0\big(L^2_s(\mathbb{R}^n),L^2_{-s'}(\mathbb{R}^n)\big)$.
(2) *Assume* $n/2\leq\alpha<n$, $s>\alpha-n/2$, $s'>\alpha-n/2$.
If $s+s'\geq\alpha$, *then* $I_\alpha\in\mathscr{B}\big(L^2_s(\mathbb{R}^n),L^2_{-s'}(\mathbb{R}^n)\big)$.
If $s+s'>\alpha$, *then* $I_\alpha\in\mathscr{B}_0\big(L^2_s(\mathbb{R}^n),L^2_{-s'}(\mathbb{R}^n)\big)$.

The uniform boundedness of $R_0(z)$ in $z\in\mathbb{C}\backslash\overline{\mathbb{R}_+}$ for $n\geq 3$, $s+s'>2$, s, $s'>1/2$ is proved by Ginibre and Moulin [**GM74**, Proposition 2.4]; as the matter of fact, they prove a stronger result that

$$(-\Delta+z)^{-1}:\ L^2_s(\mathbb{R}^3)\otimes L^2(\mathbb{R}^{n-3})\to L^2_{-s'}(\mathbb{R}^3)\otimes L^2(\mathbb{R}^{n-3})$$

is bounded uniformly for $z\in\mathbb{C}\setminus\overline{\mathbb{R}_+}$. For $n\geq 3$ and under a stronger requirement $s+s'>(n+1)/2$, s, $s'>1/2$, the proof is in [**Jen80**, Lemma 3.1]. See also the monograph by Kuroda [**Kur78**, Appendix to Chapter IV, Theorem 2].

We give an elementary proof of this result arriving at a slightly stronger statement than in [**GM74**, Proposition 2.4]: we reduce the assumption on the weights from $s+s'>2$ to $s+s'\geq 2$. The main technical part of the proof is the following lemma.

Lemma VI.17 *Let* $n\geq 1$ *and assume that* $0<s<n/2$. *Then* $|x|^{-s}\circ\mathcal{F}^{-1}\circ|\xi|^{-s}$, *with* $\mathcal{F}$ *the Fourier transform, extends to a continuous map in* $L^2(\mathbb{R}^n)$.

Let us mention the following result proved in [**BKS91**, Proposition 2.8]: if a, $b\in L^{r,w}(\mathbb{R}^n)$ with $r>2$, then $a\circ\mathcal{F}\circ b$, with $\mathcal{F}$ the Fourier transform (or a certain more general operator), is bounded. More general Cwikel-type estimates are obtained in [**Slo09, Slo13**].

PROOF. Let $\beta\in C^\infty_{\rm comp}(\mathbb{R})$ be chosen so that $\beta\geq 0$, $\operatorname{supp}\beta\subset[1/2,2]$, and $\sum_{j\in\mathbb{Z}}\beta(t/2^j)=1$ for each $t\in\mathbb{R}_+$. For R, $\Lambda>0$ (they will be integer powers of 2), define $T_{R,\Lambda}:\ \mathscr{D}(\mathbb{R}^n)\to\mathscr{D}'(\mathbb{R}^n)$ by

$$T_{R,\Lambda}u(x)=\int_{\mathbb{R}^n}|x|^{-s}\beta(|x|/R)e^{\mathrm{i}x\cdot\xi}|\xi|^{-s}\beta(|\xi|/\Lambda)u(\xi)\,d\xi.$$

Then, integrating by parts $N = n + 1$ times in ξ in the integral kernel $K(x, y)$ of $T_{R,\Lambda}T^*_{R,\Lambda}$,

$$\text{(VI.18)} \quad K(x,y) = \int_{\mathbb{R}^n} |x|^{-s}\beta(|x|/R)e^{\mathrm{i}x\cdot\xi}|\xi|^{-2s}\beta(|\xi|/\Lambda)^2 e^{-\mathrm{i}y\cdot\xi}|y|^{-s}\beta(|y|/R)\,d\xi,$$

with the aid of the identity $Le^{\mathrm{i}(x-y)\cdot\xi} = e^{\mathrm{i}(x-y)\cdot\xi}$, where $L := \frac{(x-y)\cdot\nabla_\xi}{\mathrm{i}|x-y|^2}$, we get a factor $C(\Lambda|x-y|)^{-N}$, which benefits us if $\Lambda|x-y| > 1$; this leads to the following bound on $|K(x,y)|$:

$$\frac{C}{R^{2s}\Lambda^{2s}} \int_{\mathbb{R}^n} \frac{\beta(|x|/R)\mathbb{1}_{[1/2,2]}(|\xi|/\Lambda)\beta(|y|/R)}{1 + (\Lambda|x-y|)^N}\,d\xi \le \frac{C\Lambda^n}{R^{2s}\Lambda^{2s}} \frac{\beta(|x|/R)\beta(|y|/R)}{1 + (\Lambda|x-y|)^N}.$$

Applying the Schur test to $T_{R,\Lambda}T^*_{R,\Lambda}$, we have:

$$\text{(VI.19)} \qquad \|T_{R,\Lambda}\|^2 \le \frac{C\Lambda^n}{R^{2s}\Lambda^{2s}} \min(R^n, \Lambda^{-n}) = C \min\big((\Lambda R)^{n-2s}, (\Lambda R)^{-2s}\big).$$

Let $m \in \mathbb{Z}$. Because of the support properties, the operators $T_{R,\Lambda}$ with $R = 2^j$, $\Lambda = 2^k$, $j, k \in \mathbb{Z}$ such that $\Lambda R = 2^m$ are *almost orthogonal*:

$$T_{R,\Lambda}T^*_{R',\Lambda'} = 0$$

whenever $\Lambda/\Lambda' > 2$ or $\Lambda/\Lambda' < 1/2$, with $\Lambda R = \Lambda' R' = 2^m$ (so that $R/R' < 1/2$ or $R/R' > 2$, respectively). Therefore, by the Cotlar–Stein almost orthogonality lemma (Lemma II.18), the operator

$$T_m := \sum_{R=2^j,\, \Lambda=2^k,\, j,\, k\in\mathbb{Z},\, \Lambda R = 2^m} T_{R,\Lambda}$$

satisfies the same estimate (up to a factor of 3) as each of $T_{R,\Lambda}$ with $\Lambda R = 2^m$:

$$\|T_m\| \le 3C \min((2^m)^{n/2-s}, (2^m)^{-s});$$

cf. (VI.19). Therefore, as long as $0 < s < n/2$, one has

$$\|T\| = \|\sum_{m\in\mathbb{Z}} T_m\| \le \sum_{m\in\mathbb{Z}} \|T_m\| < \infty. \qquad \square$$

Lemma VI.18 (Free resolvent of the Laplacian at $z = 0$) *Let $n \ge 3$. The convolution with $|x|^{2-n}$ extends to a continuous map $L^2_s \to L^2_{-s'}$ with s, $s' > 2 - n/2$, s, $s' \ge 0$, $s + s' \ge 2$.*

PROOF. As long as s, $s' > 2 - n/2$, s, $s' \ge 0$, $s + s' = 2$, we also have s, $s' < n/2$. If $s > 0$ and $s' > 0$, it remains to decompose

$$|x|^{-s} \circ \mathcal{F}^{-1} \circ |\xi|^{-2} \circ \mathcal{F} \circ |x|^{-s'} = \big(|x|^{-s} \circ \mathcal{F}^{-1} \circ |\xi|^{-s}\big)\big(|\xi|^{-s'} \circ \mathcal{F} \circ |x|^{-s'}\big),$$

with $\mathcal{F}$ the Fourier transform, and to apply Lemma VI.17 to each factor.

When $n \ge 5$, we need to consider the case $s = 2$, $s' = 0$ (and a similar case $s = 0$, $s' = 2$). In this case, we may apply Lemma VI.17 to $|x|^{-2} \circ \mathcal{F}^{-1} \circ |\xi|^{-2}$.

Finally, one may reduce the condition $s + s' = 2$ to the condition $s + s' \ge 2$ when $|x|$ is substituted by $\langle x \rangle$. $\square$

Lemma VI.19 (Free resolvent of the Laplacian at $z \notin \overline{\mathbb{R}^+}$) *Let $n \ge 3$. Then, for any $z \in \mathbb{C} \setminus \overline{\mathbb{R}_+}$,*

$$R_0(z) : L^2_\sigma(\mathbb{R}^n) \to L^2_{-\sigma'}(\mathbb{R}^n),$$

with $\sigma + \sigma' \geq 2$, $\sigma, \sigma' > 1/2$, *with the norm uniformly bounded in* z. *As* $z \to 0$, $R_0(z)$ *has a limit in the strong operator topology.*

PROOF. In the case $n = 3$, the proof follows from the inequality

$$|e^{\mathrm{i}|x|\sqrt{z}}/(4\pi|x|)| \leq (4\pi|x|)^{-1}, \qquad \forall x \in \mathbb{R}^3 \setminus \{0\}, \quad \forall z \in \mathbb{C} \quad \text{with} \quad \operatorname{Im}\sqrt{z} \geq 0.$$

The strong continuity follows from the dominated convergence theorem.

For the case $n \geq 4$, we reproduce the argument by Ginibre and Moulin [**GM74**, Proposition (2.4)], decomposing $x = (x^1, x^2) \in \mathbb{R}^3 \times \mathbb{R}^{n-3}$ and taking the partial Fourier transform with respect to x^2, $\varphi(x^1, x^2) \mapsto \tilde{\varphi}(x^1, p_2)$. Then, for each $p_2 \in \mathbb{R}^{n-3}$,

$$((-\Delta - z)^{-1}\tilde{\varphi})(x^1, p_2) = \int_{\mathbb{R}^3} \frac{e^{\mathrm{i}|x^1-y^1|\sqrt{z-p_2^2}}}{4\pi|x^1 - y^1|}\tilde{\varphi}(y^1, p_2)\, dy^1, \qquad \operatorname{Im}\sqrt{z - p_2^2} \geq 0.$$

Using the result from dimension $n = 3$ which is already available to us, we have:

$$\|((-\Delta - z)^{-1}\tilde{\varphi})(\cdot, p_2)\|^2_{L^2_{-\sigma'}(\mathbb{R}^3)} \leq \|\tilde{\varphi}(\cdot, p_2)\|^2_{L^2_\sigma(\mathbb{R}^3)}, \qquad \forall \sigma, \sigma' > 1/2, \quad \sigma + \sigma' \geq 2.$$

Integrating over p_2 proves the continuity of the mapping

$$L^2_\sigma(\mathbb{R}^3) \otimes L^2(\mathbb{R}^{n-3}) \to L^2_{-\sigma'}(\mathbb{R}^3) \otimes L^2(\mathbb{R}^{n-3})$$

and hence $L^2_\sigma(\mathbb{R}^n) \to L^2_{-\sigma'}(\mathbb{R}^n)$. The continuity in the strong operator topology follows from the dominated convergence theorem. □

Problem VI.20 Show that one can not bound $\|u\|_{L^2(\mathbb{R}^3)}$ by $\|\Delta u\|_{L^2_s(\mathbb{R}^3)}$ with arbitrarily large $s \geq 0$.

VI.4 Limiting absorption principle for the Dirac operator

The limiting absorption principle for the Dirac operator was studied in [**Yam73**, **BH92**, **BdMBMP93**, **IM99**, **BG10**].

Theorem VI.21 ([Yam73], Theorem 3.1) *Let* $\mathcal{K}$ *be a compact set in* $\mathbb{C} \setminus \{\pm m\}$. *For any* $s > \frac{1}{2}$ *there exists* $C = C(s, \mathcal{K})$ *such that*

$$\|u\|_{H^1_{-s}} \leq C\|(D_m - z)u\|_{L^2_s}$$

for all $u \in H^1_s(\mathbb{R}^n, \mathbb{C}^N)$ *and* $z \in \mathcal{K} \setminus \mathbb{R}$.

PROOF. We show that $(D_m - z)^{-1} : L^2_s(\mathbb{R}^n, \mathbb{C}^N) \to H^1_{-s}(\mathbb{R}^n, \mathbb{C}^N)$ is uniformly bounded for $z \in \mathcal{K} \setminus \mathbb{R}$. Indeed, by Theorem VI.8,

$$(-\Delta + m^2 - z^2)^{-1} : L^2_s(\mathbb{R}^n, \mathbb{C}^N) \to H^2_{-s}(\mathbb{R}^n, \mathbb{C}^N)$$

is uniformly bounded for $z \in \mathcal{K} \setminus \mathbb{R}$, and then

$$D_m + z : H^2_{-s}(\mathbb{R}^n, \mathbb{C}^N) \to H^1_{-s}(\mathbb{R}^n, \mathbb{C}^N)$$

is uniformly bounded for $z \in \mathcal{K} \setminus \mathbb{R}$. It follows that

$$\|(D_m + z)(-\Delta + m^2 - z^2)^{-1}f\|_{H^1_{-s}} \leq C\|f\|_{L^2_s}.$$

Denote $u := (D_m + z)(-\Delta + m^2 - z^2)^{-1}f$; then

$$(D_m - z)u = (D_m - z)(D_m + z)(-\Delta + m^2 - z^2)^{-1}f = f.$$

We conclude that $\|u\|_{H^1_{-s}} \leq C\|(D_m - z)u\|_{L^2_s}$. □

Problem VI.22 Show that the constant C can be chosen in such a way that it only depends on the distance from $\mathcal{K}$ to $\pm m$.

Problem VI.23 Prove that $D_m : H^2_s(\mathbb{R}^n, \mathbb{C}^N) \to H^1_s(\mathbb{R}^n, \mathbb{C}^N)$ is bounded for any $s \in \mathbb{R}$.

Let us reformulate this result in the form valid for a spectral parameter from a non-compact set:

Lemma VI.24 (Limiting absorption principle for the Dirac operator) *Let $s > 1/2$, $\delta > 0$, and $m \geq 0$. There exists $C_0 = C_0(s, \delta, m) > 0$ (locally bounded in s, δ, and m) such that for all $z \in \mathbb{C} \setminus ((-\infty, -m] \cup [m, +\infty))$ with $|z^2 - m^2| \geq \delta$ one has*

$$\text{(VI.20)} \qquad \|u\|_{L^2_{-s}} \leq C_0(s, \delta, m) \|(D_m - z)u\|_{L^2_s}, \qquad \forall u \in L^2(\mathbb{R}^n, \mathbb{C}^N).$$

Let $s > 1/2$, $m \geq 0$, and $z \in \mathbb{C} \setminus ((-\infty, -m] \cup [m, +\infty))$. There exists $C_1 = C_1(s, z, m) > 0$ (locally bounded in s, z, and m) such that

$$\text{(VI.21)} \qquad \|u\|_{H^1_{-s}} \leq C_1(s, z, m) \|(D_m - z)u\|_{L^2_s}, \qquad \forall u \in L^2(\mathbb{R}^n, \mathbb{C}^N).$$

Proof. By [**Agm75**, Remark 2 in Appendix A] (see Lemma VI.11), for any $s > 1/2$ and $\delta > 0$, there is $C_{s,\delta} > 0$ such that for all $v \in H^2(\mathbb{R}^n)$ and $\zeta \in \mathbb{C}$, $|\zeta| \geq \delta$, one has

$$\text{(VI.22)} \qquad (|\zeta| + 1)^{\frac{1-\nu}{2}} \|v\|_{H^\nu_{-s}} \leq C_{s,\delta} \|(-\Delta - \zeta)v\|_{L^2_s}, \qquad 0 \leq \nu \leq 2.$$

We will apply this inequality to vector-valued functions $v \in H^2(\mathbb{R}^n, \mathbb{C}^N)$.

Let $u \in L^2(\mathbb{R}^n, \mathbb{C}^N)$ and $z \in \mathbb{C} \setminus ((-\infty, -m] \cup [m, \infty))$. Without loss of generality, we can assume that $(D_m - z)u \in L^2_s(\mathbb{R}^n, \mathbb{C}^N)$ (or else there is nothing to prove); it then follows that $u \in H^1(\mathbb{R}^n, \mathbb{C}^N)$ and $v := (D_m + z)^{-1} u \in H^2(\mathbb{R}^n, \mathbb{C}^N)$. One has:

$$\text{(VI.23)} \qquad \|u\|_{H^{\nu-1}_{-s}} = \|(D_m + z)v\|_{H^{\nu-1}_{-s}} \leq C(s) \|v\|_{H^\nu_{-s}} + (m + |z|) \|v\|_{H^{\nu-1}_{-s}},$$

where $C(s) > 0$ depends on s only. Applying (VI.22) with $\zeta = z^2 - m^2$, $|\zeta| \geq \delta > 0$, to the right-hand side of (VI.23), we have:

$$\|u\|_{H^{\nu-1}_{-s}} \leq \left(C(s)(|\zeta| + 1)^{\frac{\nu-1}{2}} + (m + |z|)(|\zeta| + 1)^{\frac{\nu-2}{2}} \right) C_{s,\delta} \|(-\Delta - \zeta)u\|_{L^2_s},$$

for $1 \leq \nu \leq 2$. Taking $\nu = 1$ and $\nu = 2$ and using the identity $(-\Delta - \zeta)v = (D_m - z)u$, we arrive at the inequalities (VI.20) and (VI.21). □

We also need the following Hardy-type inequality, along the lines of [**Agm75**, Appendix B].

Lemma VI.25 *For any $s > -1/2$ and $z \in \mathbb{R} \setminus [-m, m]$, there is $C_2 = C_2(s, z, m) > 0$ (locally bounded in s and z) such that*

$$\text{(VI.24)} \qquad \|u\|_{H^1_s} \leq C_2 \|(D_m - z)u\|_{L^2_{s+1}}, \qquad \forall u \in L^2(\mathbb{R}^n, \mathbb{C}^N) \cap H^1_{\mathrm{loc}}(\mathbb{R}^n, \mathbb{C}^N).$$

Proof. While the result [**BG87**, Theorem 2] is stated in the three-dimensional case, a careful look at the proof shows that it is independent of the dimension, being based on [**Agm75**, Appendix B], which treats any dimension.

Let us provide the proof of a weaker result, for $s > 0$, based on the Carleman estimates; see Theorem VII.5 below. We apply that theorem with $\varphi(r) = s \log\langle r \rangle$;

this gives some $R_s > 0$ and $C(s,z) > 0$ such that for any $R \geq R_s$ and for any $v \in H^1(\mathbb{R}^n, \mathbb{C}^N)$, $\operatorname{supp} v \subset \Omega_R^n$, which satisfies

$$\langle r\rangle^{s+1}(D_m - z)v = \langle r\rangle e^{\varphi}(D_m - z)v \in L^2(\mathbb{R}^n, \mathbb{C}^N)$$

one has $\langle r\rangle^s v = e^{\varphi}v \in L^2(\mathbb{R}^n, \mathbb{C}^N)$ and moreover

$$\|\langle r\rangle^s v\| \leq C(s,z)\|\langle r\rangle^{s+1}(D_m - z)v\|. \tag{VI.25}$$

Let $\eta \in C^\infty(\mathbb{R}^n)$ be such that $\operatorname{supp}\eta \subset \Omega_R^n$ and $\eta|_{\Omega_{R+1}^n} = 1$ for $R = R_s + 1$. Let $u \in L^2(\mathbb{R}^n, \mathbb{C}^N) \cap H^1_{\mathrm{loc}}(\mathbb{R}^n, \mathbb{C}^N)$. Without loss of generality, we may assume that $\|(D_m - z)u\|_{L^2_{s+1}}$ is finite (or else there is nothing to prove); then we conclude that $u \in H^1(\mathbb{R}^n, \mathbb{C}^N)$. Applying (VI.25) to $\eta u \in H^1(\mathbb{R}^n, \mathbb{C}^N)$, $\operatorname{supp}\eta u \subset \Omega_R^n$, one has:

$$\|\eta u\|_{H^1_s} \leq C(s,z)\|(D_m - z)\eta u\|_{L^2_{s+1}}. \tag{VI.26}$$

At the same time, since $\operatorname{supp}(1-\eta)u \subset \mathbb{B}_{R+1}^n$, we deduce that

$$\begin{aligned}
\|(1-\eta)u\|_{H^1_s} &\leq \langle R+1\rangle^{s+1}\|(1-\eta)u\|_{H^1_{-1}} \\
&\leq \langle R+1\rangle^{s+1}C_1(1,z,m)\|(D_m - z)(1-\eta)u\|_{L^2_1};
\end{aligned} \tag{VI.27}$$

in the second inequality, we applied (VI.21) (where we take $s = 1$). Using (VI.26) and (VI.27), we have:

$$\begin{aligned}
\|u\|_{H^1_s} &\leq \|\eta u\|_{H^1_s} + \|(1-\eta)u\|_{H^1_s} \\
&\leq C(s,z)\|(D_m - z)\eta u\|_{L^2_{s+1}} + \langle R+1\rangle^{s+1}C_1\|(D_m - z)(1-\eta)u\|_{L^2_1} \\
&\leq (C(s,z) + \langle R+1\rangle^{s+1}C_1)\Big(\|(D_m - z)u\|_{L^2_{s+1}} + \|(\boldsymbol{\alpha}\cdot\nabla\eta)u\|_{L^2_{s+1}}\Big).
\end{aligned}$$

Due to the compact support of $\nabla\eta$, the inequality (VI.21) shows that the second term in the brackets in the right-hand side is dominated by the first term, which concludes the proof. □

Remark VI.26 In the above theorem, one can prove the bound (VI.24) with $s = 0$ using in the proof $\varphi(r) = \log\log\langle r\rangle$.

We now consider an extension of this result for values outside the real line, using a simple commutator estimate.

Lemma VI.27 *Assume that* $\lambda \in \mathbb{C} \setminus ((-\infty, -m] \cup [m, +\infty))$. *If* $s \in \mathbb{R}$ *satisfies* $|s| < \operatorname{dist}(\lambda, \sigma(D_m))$, *then*

$$\|u\|_{L^2_s} \leq \frac{1}{\operatorname{dist}(\lambda, \sigma(D_m)) - |s|}\|(D_m - \lambda)u\|_{L^2_s}, \qquad \forall u \in L^2_{s-1}(\mathbb{R}^n, \mathbb{C}^N).$$

Proof. First, we notice that for any $u \in C^\infty_{\mathrm{comp}}(\mathbb{R}^n, \mathbb{C}^N)$, one has

$$\|[\langle r\rangle^s, D_m]u\| = \|(D_0\langle r\rangle^s)u\| = \left\|\frac{\boldsymbol{\alpha}\cdot x}{r}s\langle r\rangle^{s-1}u\right\| \leq |s|\|u\|_{L^2_{s-1}}; \tag{VI.28}$$

note that $\|\boldsymbol{\alpha}\cdot x\|_{\mathrm{End}(\mathbb{C}^N)} = \|(\boldsymbol{\alpha}\cdot x)(\boldsymbol{\alpha}\cdot x)\|^{1/2}_{\mathrm{End}(\mathbb{C}^N)} = \|x^2\|^{1/2}_{\mathrm{End}(\mathbb{C}^N)} = r$. Using (VI.28), we compute for such u:

$$\begin{aligned}
\|\langle r\rangle^s(D_m - \lambda)u\| &\geq \|(D_m - \lambda)(\langle r\rangle^s u)\| - \|[D_m, \langle r\rangle^s]u\| \\
&\geq \|(D_m - \lambda)(\langle r\rangle^s u)\| - |s|\|u\|_{L^2_{s-1}}.
\end{aligned}$$

The above inequality shows that if $u \in L^2_{s-1}(\mathbb{R}^n, \mathbb{C}^N)$ and $(D_m - \lambda)u \in L^2_s(\mathbb{R}^n, \mathbb{C}^N)$ (if the latter inclusion were not satisfied then there would be nothing to prove), then $(D_m - \lambda)(\langle r\rangle^s u) \in L^2(\mathbb{R}^n, \mathbb{C}^N)$. Since D_m is self-adjoint, one has

$$\|(D_m - \lambda)^{-1}\| = \frac{1}{\operatorname{dist}(\lambda, \sigma(D_m))};$$

therefore, $\langle r\rangle^s u \in H^1(\mathbb{R}^n, \mathbb{C}^N)$, and

$$\begin{aligned} \|(D_m - \lambda)u\|_{L^2_s} &\geq \operatorname{dist}(\lambda, \sigma(D_m))\|\langle r\rangle^s u\| - |s|\|u\|_{L^2_{s-1}} \\ &\geq \big(\operatorname{dist}(\lambda, \sigma(D_m)) - |s|\big)\|u\|_{L^2_s}. \end{aligned}$$

This concludes the proof. □

VI.5 Analytic continuation of the free resolvent

Let $E_\mu : L^2(\mathbb{R}^n) \to L^2(\mathbb{R}^n)$ denote the operator of multiplication by $e^{-\mu\langle r\rangle}$, $\mu \in \mathbb{R}$. Following [**Rau78**], not to confuse the regularized resolvent

$$R^0_\mu(\zeta^2) := E_\mu R^0(\zeta^2) E_\mu = E_\mu(-\Delta - \zeta^2)^{-1} E_\mu, \qquad \forall \zeta \in \mathbb{C}, \quad \operatorname{Im}\zeta > 0,$$

with its analytic continuation through the line $\operatorname{Im}\zeta = 0$, we will denote the latter by $F^0_\mu(\zeta)$.

Proposition VI.28 (Analytic continuation of the resolvent) *Let* $n \geq 1$, $\mu > 0$.

(1) *There is an analytic function* $F^0_\mu(\zeta)$,

$$F^0_\mu : \{\operatorname{Im}\zeta > -\mu\} \setminus (-\mathrm{i}\overline{\mathbb{R}_+}) \longrightarrow \mathscr{B}(L^2(\mathbb{R}^n), L^2(\mathbb{R}^n)),$$

such that $F^0_\mu(\zeta) = R^0_\mu(\zeta^2)$ *for* $\operatorname{Im}\zeta > 0$, *and for any* $\nu \leq 2$, $\delta > 0$, *there is* $C = C(n, \nu, \mu, \delta) > 0$ *such that*

$$\|F^0_\mu(\zeta)\|_{L^2 \to H^\nu} \leq \frac{C}{(1 + |\zeta|)^{1-\nu}}, \tag{VI.29}$$

where $\zeta \in \mathbb{C}$ *satisfies* $\operatorname{Im}\zeta \geq -\mu + \delta$ *and* $\operatorname{dist}(\zeta, -\mathrm{i}\mathbb{R}_+) > \delta$.

(2) *If* n *is odd and satisfies* $n \geq 3$, *then* (VI.29) *holds for all* $\zeta \in \mathbb{C}$, $\operatorname{Im}\zeta \geq -\mu + \delta$.

Remark VI.29 This result in dimension $n = 3$ was stated and proved in [**Rau78**, Proposition 3], as a consequence of the explicit expression for the integral kernel of $R^0_\mu(\zeta^2)$,

$$-\frac{e^{-\mu\langle y\rangle} e^{\mathrm{i}\zeta|y-x|} e^{-\mu\langle x\rangle}}{4\pi|y-x|}, \qquad \operatorname{Im}\zeta > 0, \qquad x,\, y \in \mathbb{R}^3,$$

which could be extended analytically to the region $\operatorname{Im}\zeta > -\mu$ as a holomorphic function of ζ with values in $L^2(\mathbb{R}^3 \times \mathbb{R}^3)$. In [**Rau78**], the restriction on ζ was stronger: $\operatorname{Im}\zeta > -\mu/2 + \delta$ (with any $\delta > 0$); this was a pay-off for using an elegant argument based on the Huygens principle. (We note that our signs and inequalities are often the opposite to those of [**Rau78**] since we consider the resolvent of $-\Delta$ instead of Δ.)

PROOF. Let us define the analytic continuation of $F^0_\mu(\zeta)$. For $u,\ v \in L^2(\mathbb{R}^n)$ we define $u_\mu,\ v_\mu \in L^{2,\mu}(\mathbb{R}^n)$ by $u_\mu(x) = e^{-\mu\langle x\rangle}u(x)$, $v_\mu(x) = e^{-\mu\langle x\rangle}v(x)$ and consider

$$I(\zeta) = \langle v, F^0_\mu(\zeta)u\rangle = \int_{\mathbb{R}^n} \overline{\widehat{v_\mu}(\xi)}\frac{1}{\xi^2-\zeta^2}\widehat{u_\mu}(\xi)\,\frac{d^n\xi}{(2\pi)^n}, \tag{VI.30}$$

which is an analytic function in $\zeta \in \mathbb{C}^+ := \{z \in \mathbb{C} : \ \operatorname{Im} z > 0\}$.

Let us prove analyticity in ζ for $\operatorname{Im}\zeta > -\mu$, $\operatorname{Re}\zeta > 0$ (the case $\operatorname{Re}\zeta < 0$ is considered similarly). It is enough to prove that for any $a > 0$ and any $\delta > 0$, $\delta \le a/3$, $I(\zeta)$ extends analytically into the rectangular neighborhood

$$\mathcal{K}^\delta_a = \{\zeta \in \mathbb{C}: \ \ a-\delta \le \operatorname{Re}\zeta \le a+\delta, \ -\mu+\delta \le \operatorname{Im}\zeta \le 0\} \tag{VI.31}$$

(see Figure VI.1), satisfying there the bounds (VI.29) with constants c_j independent of a.

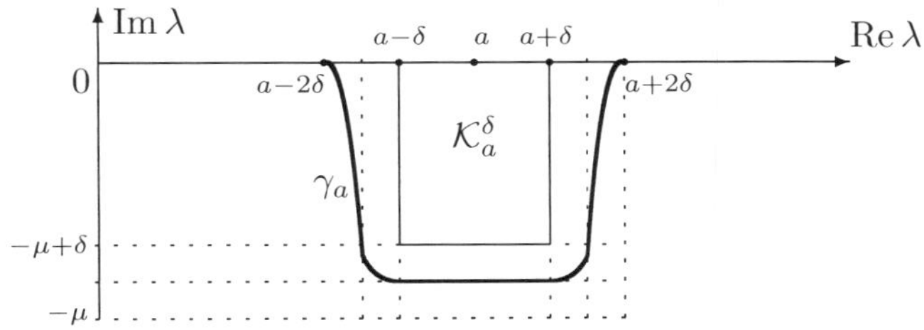

FIGURE VI.1. The rectangular semi-neighborhood $\mathcal{K}^\delta_a$ around a surrounded by the contour $\gamma_a = \{\lambda : \ \operatorname{Im}\lambda = g_a(\operatorname{Re}\lambda),\ a-2\delta \le \operatorname{Re}\lambda \le a+2\delta\}$; $\operatorname{dist}(\mathcal{K}^\delta_a, \gamma_a) \ge \delta/2$.

We pick $a > 0$ and $\delta > 0$, with $a \ge 3\delta$, and break the integral (VI.30) into two:

$$I(\zeta) = I^{(\delta)}_1(\zeta) + I^{(\delta)}_2(\zeta) = \int_{||\xi|-a|>2\delta} + \int_{||\xi|-a|<2\delta}. \tag{VI.32}$$

The first integral in (VI.32) is finite, being bounded by

$$|I^{(\delta)}_1(\zeta)| \le \int\limits_{||\xi|-a|>2\delta} |\hat v_\mu(\xi)||\hat u_\mu(\xi)|\frac{1}{2|\zeta|}\left|\frac{1}{|\xi|-\zeta} - \frac{1}{|\xi|+\zeta}\right|\frac{d^n\xi}{(2\pi)^n}$$

$$\le \int\limits_{\mathbb{R}^n} \frac{|\hat v_\mu(\xi)||\hat u_\mu(\xi)|}{2|\zeta|}\frac{2}{\delta}\frac{d^n\xi}{(2\pi)^n} \le \frac{\|u_\mu\|\|v_\mu\|}{|\zeta|\delta},$$

and therefore is analytic in ζ and is bounded by $C/|\zeta|$. Above, to estimate the denominators, we took into account that for $\zeta \in \mathcal{K}^\delta_a$ and $||\xi|-a| > 2\delta$,

$$||\xi| \pm \zeta| \ge |(|\xi|-a) + (a \pm \operatorname{Re}\zeta)| \ge ||\xi|-a| - |a \pm \operatorname{Re}\zeta| > 2\delta - \delta = \delta.$$

To analyze the second integral in (VI.32), we will deform the contour of integration in ξ. Let $g_0 \in C^\infty_{\mathrm{comp}}(\mathbb{R})$ be even, $g_0 \le 0$, $\operatorname{supp} g_0 \in [-2\delta, 2\delta]$, with $g_0(0) = -\mu+\delta/2$ and monotonically increasing to zero as $|t| \to 2\delta$. Moreover, we may assume that $|g_0'| < 4\mu/\delta$ and that $\operatorname{dist}(\gamma_0, \mathcal{K}^\delta_0) \ge \delta/2$, where $\mathcal{K}^\delta_a$ is defined in (VI.31) and $\gamma_0 = \{(\lambda, g_0(\lambda)) : \ |\lambda| \le 2\delta\}$; see Figure VI.1. Define $g_a(t) = g_0(t-a)$.

Lemma VI.30 *Let $u \in L^{2,\mu}(\mathbb{R}^n)$, so that $\|u\|_{L^{2,\mu}(\mathbb{R}^n)} := \|e^{\mu\langle r\rangle}u\|_{L^2(\mathbb{R}^n)} < \infty$. Then its Fourier transform, $\hat{u}(\xi)$, can be extended analytically into the μ-neighborhood of $\mathbb{R}^n \subset \mathbb{C}^n$, which we denote by*

$$\Omega_\mu(\mathbb{R}^n) = \{\xi \in \mathbb{C}^n: \quad |\operatorname{Im}\xi| < \mu\} \subset \mathbb{C}^n,$$

and there is $C_\mu > 0$ such that

$$\|\hat{u}\|_{L^2(\Omega_\mu(\mathbb{R}^n))} \le C_\mu \|u\|_{L^{2,\mu}(\mathbb{R}^n)}, \tag{VI.33}$$

where $\Omega_\mu(\mathbb{R}^n)$ is interpreted as a region in $\mathbb{R}^{2n} \cong \mathbb{C}^n$.

Problem VI.31 Prove this lemma.
Hint: Define $\hat{u}$ on Ω_μ as the Fourier–Laplace transform of u. Then compute the L^2-norm for fixed $\operatorname{Im}\zeta$.

By Lemma VI.30, the functions $U(\xi) = \widehat{u_\mu}(\xi)$ and $V(\xi) = \overline{\widehat{v_\mu}(\xi)}$ could be extended analytically in $\xi \in \mathbb{R}^n$ into the strip $\xi \in \mathbb{C}^n$, $|\operatorname{Im}\xi| < \mu$. We rewrite the second integral in (VI.32) in polar coordinates, denoting $\lambda = |\xi| \in [a-2\delta, a+2\delta]$, $\boldsymbol{\theta} = \xi/|\xi| \in \mathbb{S}^{n-1}$, and then deform the contour of integration in λ, arriving at

$$I_2^{(\delta)}(\zeta) = \int_{\gamma_a \times \mathbb{S}^{n-1}} \frac{V(\boldsymbol{\theta}\lambda)U(\boldsymbol{\theta}\lambda)}{\lambda^2 - \zeta^2} \lambda^{n-1}\, d\lambda \frac{d\Omega_{\boldsymbol{\theta}}}{(2\pi)^n}, \tag{VI.34}$$

with γ_a as on Figure VI.1. Clearly, (VI.34) is analytic for $\operatorname{Re}\zeta > 0$ and $\operatorname{Im}\zeta > 0$ (since $\operatorname{Im}\lambda^2 \le 0$ while $\operatorname{Im}\zeta^2 > 0$).

Let us argue that (VI.34) can also be extended analytically into the box $\mathcal{K}_a^\delta$. For $\lambda \in \gamma_a$ and $\zeta \in \mathcal{K}_a^\delta$, taking into account that

$$|\lambda - \zeta| \ge \delta/2, \qquad |\lambda + \zeta| \ge \operatorname{Re}\lambda + \operatorname{Re}\zeta \ge (a-2\delta) + (a-\delta) = 2a - 3\delta \ge a$$

(recall that $\delta \le a/3$), we see that (VI.34) defines an analytic function which is bounded by

$$\begin{aligned} &|I_2^{(\delta)}(\zeta)| \\ &\le \frac{2}{a\delta}\left[\int_{\gamma_a \times \mathbb{S}^{n-1}} |V(\boldsymbol{\theta}\lambda)|^2 |\lambda|^{n-1}\, |d\lambda| \frac{d\Omega_{\boldsymbol{\theta}}}{(2\pi)^n} \int_{\gamma_a \times \mathbb{S}^{n-1}} |U(\boldsymbol{\theta}\lambda)|^2 |\lambda|^{n-1}\, |d\lambda| \frac{d\Omega_{\boldsymbol{\theta}}}{(2\pi)^n}\right]^{\frac{1}{2}}, \end{aligned} \tag{VI.35}$$

where the integration in λ over the contour γ_a could be parametrized by

$$\lambda(t) = t + \mathrm{i}g_a(t), \quad t \in (a-2\delta, a+2\delta), \quad d\lambda = (1 + \mathrm{i}g_a'(t))\, dt, \tag{VI.36}$$

so that $|d\lambda|$ is understood as $(1+(g_a'(t))^2)^{1/2}\, dt$. Our assumption that $a \ge 3\delta$ allows us to bound the first factor in (VI.35) by $\frac{2}{a\delta} \le \frac{2}{3\delta^2}$. Moreover, if $|\zeta| \ge 2(\mu+\delta)$, the first factor in (VI.35) is also bounded by

$$\frac{2}{a\delta} < \frac{2}{(|\operatorname{Re}\zeta| - \delta)\delta} < \frac{2}{(|\zeta| - \mu - \delta)\delta} < \frac{4}{|\zeta|\delta}, \qquad \forall \zeta \in \mathcal{K}_a^\delta \setminus \mathbb{D}_{2(\mu+\delta)}.$$

Therefore, that factor is bounded by $c/(1+|\zeta|)$ with certain $c = c(\mu,\delta) > 0$. To study the integrals in (VI.35), we define $\boldsymbol{G} : \mathbb{R}^n \to \mathbb{R}^n$ by

$$\boldsymbol{G}(\eta) = \frac{\eta}{|\eta|} g_a(|\eta|)\rho(|\eta|/\delta), \qquad \eta \in \mathbb{R}^n, \tag{VI.37}$$

where $\rho \in C^\infty(\mathbb{R})$ satisfies $\rho(t) \equiv 1$ for $|t| \geq 1$, $\rho(t) \equiv 0$ for $|t| \leq 1/2$, and parametrize ξ as follows:

$$\xi = \eta + \mathrm{i}\boldsymbol{G}(\eta) \in \mathbb{C}^n, \quad \eta \in \mathbb{R}^n; \qquad ||\eta| - a| \leq 2\delta.$$

We have:

$$\int\limits_{\gamma_a \times \mathbb{S}^{n-1}} |U(\boldsymbol{\theta}\lambda)|^2 \, |\lambda|^{n-1} \, |d\lambda| \frac{d\Omega_{\boldsymbol{\theta}}}{(2\pi)^n} \leq \Big(1 + \Big(\frac{4\mu}{\delta}\Big)^2\Big)^{\frac{n}{2}} \int\limits_{||\eta|-a|<2\delta} |U(\eta + \mathrm{i}\boldsymbol{G}(\eta))|^2 \, d^n\eta,$$

where we took into account that both $|\lambda/\operatorname{Re}\lambda|$ and $|\dot\lambda/\operatorname{Re}\dot\lambda|$, which are equal to $(t+g(t)^2)^{1/2}/t$ and $(1+(g_a'(t))^2)^{1/2}$ (see the parametrization (VI.36)), are bounded by

$$\sqrt{1+(g_0')^2} \leq \sqrt{1+(4\mu/\delta)^2}.$$

One has:

$$U(\eta + \mathrm{i}\boldsymbol{G}(\eta)) = A_g u(\eta),$$

where

$$A_g u(\eta) := \int_{\mathbb{R}^n} e^{-\mathrm{i}x\cdot\eta} e^{x\cdot\boldsymbol{G}(\eta)} e^{-\mu\langle x\rangle} u(x) \, d^n x$$

is an oscillatory integral operator with the non-degenerate phase function $\phi(x,\eta) = x\cdot\eta$ and a bounded smooth symbol $a(x,\eta) = e^{x\cdot\boldsymbol{G}(\eta)-\mu\langle x\rangle}$. Note that, by (VI.37),

$$x\cdot\boldsymbol{G}(\eta) - \mu\langle x\rangle \leq |x||g_a(|\eta|)| - \mu\langle x\rangle < 0, \qquad \forall (x,\eta) \in \mathbb{R}^n \times \mathbb{R}^n.$$

One can show that A_g is continuous in $L^2(\mathbb{R}^n)$ (see Problems II.24, II.25, and II.26). Then it follows that there is $c = c(\mu,\delta) > 0$ such that

$$\int\limits_{\gamma_a \times \mathbb{S}^{n-1}} |U(\boldsymbol{\theta}\lambda)|^2 |\lambda|^{n-1} \, |d\lambda| \frac{d\Omega_{\boldsymbol{\theta}}}{(2\pi)^n} \leq c(\mu,\delta)\|u\|^2.$$

There is a similar bound for V. Thus, there is $C = C(\mu,\delta) > 0$ such that $|I_2^{(\delta)}(\zeta)| \leq \frac{C(\mu,\delta)}{|\zeta|\delta}\|v\|\|u\|$, which is the desired bound.

The estimates on $\partial_\zeta^j F_\mu^0(\zeta)$, $j \in \mathbb{N}$, are proved similarly, writing out the derivatives of $(\xi^2-\zeta^2)^{-1}$ and proceeding with the same decomposition as in (VI.32); the only difference is the contribution from higher powers of $\xi^2-\zeta^2$ in the denominator.

This settles the first part of Proposition VI.28.

Before we prove the second part of Proposition VI.28, we need the following technical lemma.

Lemma VI.32 *Let $\rho > 0$ and let $N \in \mathbb{N}$ be odd and satisfy $N \geq 3$. The analytic function*

$$F_{N,\rho}(\zeta) = \int_0^\rho \frac{\lambda^{N-1}\,d\lambda}{\lambda^2-\zeta^2}, \qquad \zeta \in \mathbb{C}, \qquad \operatorname{Im}\zeta > 0,$$

extends analytically into an open disc $\mathbb{D}_\rho$. Moreover, one has

$$\text{(VI.38)} \quad |F_{N,\rho}(\zeta)| \leq \frac{\rho^{N-2}}{2}\left(2 + \ln N + \pi + \ln\frac{\rho+|\zeta|}{\rho-|\zeta|}\right), \qquad \zeta \in \mathbb{D}_\rho.$$

PROOF. Using the identity $\frac{\lambda^2}{\lambda^2-\zeta^2} = 1 + \frac{\zeta^2}{\lambda^2-\zeta^2}$ (note that the denominator is nonzero since $\lambda \geq 0$ and $\operatorname{Im}\zeta > 0$) and remembering that N is odd, we have:

$$F_{N,\rho}(\zeta) = \int_0^\rho \frac{\lambda^{N-1}\,d\lambda}{\lambda^2-\zeta^2} = \int_0^\rho \left(\lambda^{N-3} + \zeta^2\lambda^{N-5} + \cdots + \zeta^{N-3} + \frac{\zeta^{N-1}}{\lambda^2-\zeta^2}\right) d\lambda$$

$$= \frac{\rho^{N-2}}{N-2} + \frac{\zeta^2\rho^{N-4}}{N-4} + \cdots + \zeta^{N-3}\rho + \frac{\zeta^{N-2}}{2}\left[\operatorname{Ln}\left(\frac{\rho-\zeta}{\rho+\zeta}\right) + \pi\mathrm{i}\right]. \tag{VI.39}$$

Above, Ln denotes the analytic branch of the natural logarithm on $\mathbb{C}\setminus\overline{\mathbb{R}_-}$ specified by $\operatorname{Ln}(1) = 0$; we also took into account that, since $\operatorname{Im}\zeta > 0$,

$$\lim_{\lambda\to 0+} \operatorname{Ln}\frac{\lambda-\zeta}{\lambda+\zeta} = \lim_{\lambda\to 0+} \operatorname{Ln}\left(-1 + \frac{2\lambda}{\zeta}\right) = \operatorname{Ln}(-1-0\mathrm{i}) = -\pi\mathrm{i}.$$

Due to the assumption $N \geq 3$, the right-hand side of (VI.39) extends to an analytic function of ζ as long as $\zeta \in \mathbb{D}_\rho$. The bound (VI.38) immediately follows from the inequalities

$$|\zeta| < \rho, \qquad 1 + \frac{1}{3} + \frac{1}{5} + \cdots + \frac{1}{N-2} \leq 1 + \frac{1}{2}\sum_{j=2}^{N-2}\frac{1}{j} \leq 1 + \frac{1}{2}\ln(N-2)$$

(with the understanding that the summation gives no contribution when $N = 3$), and the bound

$$\left|\operatorname{Ln}\left(\frac{\rho-\zeta}{\rho+\zeta}\right) + \pi\mathrm{i}\right| \leq \pi + \ln\frac{\rho+|\zeta|}{\rho-|\zeta|}$$

valid for $\zeta \in \mathbb{D}_\rho \cap \mathbb{C}_+$ (when $\arg\frac{\rho-\zeta}{\rho+\zeta} \in (-\pi, 0)$). □

Remark VI.33 Note that the conclusion of the lemma would not hold if N were even: in that case, one arrives at functions which have a branching point at $\zeta = 0$; e.g.

$$\int_0^\rho \frac{\lambda\,d\lambda}{\lambda^2-\zeta^2} = \frac{1}{2}\ln\left(1 - \frac{\rho^2}{\zeta^2}\right),$$

$$\int_0^\rho \frac{\lambda^3\,d\lambda}{\lambda^2-\zeta^2} = \int_0^\rho \left(\lambda + \frac{\zeta^2\lambda}{\lambda^2-\zeta^2}\right) d\lambda = \frac{\rho^2}{2} + \frac{\zeta^2}{2}\ln\left(1 - \frac{\rho^2}{\zeta^2}\right),$$

which behave like $\ln\left(-\frac{\rho}{\zeta}\right)$ and $\zeta^2\ln\left(-\frac{\rho}{\zeta}\right)$ when $|\zeta| \ll \rho$ (hence have a branching point at $\zeta = 0$). □

Now let us prove the second part of Proposition VI.28; from now on, we assume that n is odd and satisfies $n \geq 3$. It is enough to prove that the function $I(\zeta)$ defined in (VI.30) is analytic inside the disc $\mathbb{D}_\mu \subset \mathbb{C}$.

We pick $\rho \in (0, \mu)$ and break the integral (VI.30) into two parts:

$$I(\zeta) = \int_{\mathbb{R}^n} \frac{V(\xi)U(\xi)}{\xi^2-\zeta^2}\,d^n\xi = I_1^{(\rho)}(\zeta) + I_2^{(\rho)}(\zeta), \tag{VI.40}$$

where

$$I_1^{(\rho)}(\zeta) := \int_{|\xi|\leq\rho} \frac{V(\xi)U(\xi)}{\xi^2-\zeta^2}\,d^n\xi, \tag{VI.41}$$

$$I_2^{(\rho)}(\zeta) := \int_{|\xi|>\rho} \frac{V(\xi)U(\xi)}{\xi^2-\zeta^2}\,d^n\xi. \tag{VI.42}$$

The function $I_2^{(\rho)}(\zeta)$ in (VI.42) is analytic in the disc $\zeta \in \mathbb{D}_\rho$, and moreover for any $r \in (0,\rho)$ one has

$$\sup_{\zeta\in\mathbb{D}_r} |I_2^{(\rho)}(\zeta)| \le \sup_{\zeta\in\mathbb{D}_r} \left| \int_{|\xi|>\rho} \frac{V(\xi)U(\xi)}{\xi^2-\zeta^2}\, d^n\xi \right| \le \frac{1}{\rho^2-r^2}\|v\|_{L^2}\|u\|_{L^2}.$$

Let us consider $I_1^{(\rho)}(\zeta)$ defined in (VI.41). Since both $V(\xi)$ and $U(\xi)$ are analytic for $\xi \in \mathbb{C}^n$, $|\xi| < \mu$, we have the power series expansions

$$V(\xi) = \sum_{\alpha\in\mathbb{N}_0^n} V_\alpha \xi^\alpha, \qquad U(\xi) = \sum_{\alpha\in\mathbb{N}_0^n} U_\alpha \xi^\alpha, \qquad V(\xi)U(\xi) = \sum_{\alpha\in\mathbb{N}_0^n} C_\alpha \xi^\alpha,$$

which are absolutely convergent for $|\xi| < \mu$. Denote $\lambda = |\xi|$, $\boldsymbol{\theta} = \xi/|\xi| \in \mathbb{S}^{n-1}$. Then

$$\text{(VI.43)} \qquad I_1^{(\rho)}(\zeta) = \int\limits_{|\xi|\le\rho} \frac{V(\xi)U(\xi)}{\xi^2-\zeta^2}\, d^n\xi = \sum_{\alpha\in\mathbb{N}_0^n} C_\alpha \int\limits_{\mathbb{S}^{n-1}} \boldsymbol{\theta}^\alpha\, d\Omega_{\boldsymbol{\theta}} \int_0^\rho \frac{\lambda^{|\alpha|+n-1}\, d\lambda}{\lambda^2-\zeta^2},$$

where $\boldsymbol{\theta}^\alpha = \theta_1^{\alpha_1}\dots\theta_n^{\alpha_n}$, $\boldsymbol{\theta} = (\theta_1,\dots,\theta_n) \in \mathbb{S}^{n-1}$. We note that, by parity considerations, the terms corresponding to at least one α_j being odd are equal to zero, hence the summation in the right-hand side is only over $\alpha \in (2\mathbb{N}_0)^n$. We claim that the series (VI.43) defines an analytic function in $\mathbb{D}_\rho$, and moreover for each $r \in (0,\rho)$ there is $C > 0$ such that

$$\sup_{\zeta\in\mathbb{D}_r} |I_1^{(\rho)}(\zeta)| \le C\|v\|_{L^2(\mathbb{R}^n)}\|u\|_{L^2(\mathbb{R}^n)}.$$

We have:

$$\begin{aligned} I_1^{(\rho)}(\zeta) &= \sum_{\alpha\in(2\mathbb{N}_0)^n} C_\alpha \int_{\mathbb{S}^{n-1}} \boldsymbol{\theta}^\alpha\, d\Omega_{\boldsymbol{\theta}} F_{|\alpha|+n,\rho}(\zeta) \\ \text{(VI.44)} \qquad &= \sum_{\alpha\in(2\mathbb{N}_0)^n} C_\alpha \int_{\mathbb{S}^{n-1}} \boldsymbol{\theta}^\alpha\, d\Omega_{\boldsymbol{\theta}} R^{|\alpha|} \frac{F_{|\alpha|+n,\rho}(\zeta)}{R^{|\alpha|}}, \end{aligned}$$

where $R \in (\rho,\mu)$ and the analytic function $F_{N,\rho}(\zeta)$ was defined in Lemma VI.32. By that lemma,

$$\left| \frac{(|\alpha|+n)F_{|\alpha|+n,\rho}(\zeta)}{R^{|\alpha|}} \right| \le (|\alpha|+n)\frac{\rho^{n+|\alpha|-2}}{2R^{|\alpha|}} \left(2 + \ln(|\alpha|+n) + \pi + \ln\frac{\rho+|\zeta|}{\rho-|\zeta|} \right)$$

are analytic functions of $\zeta \in \mathbb{D}_r$, $r \in (0,\rho)$, which are bounded uniformly in $\alpha \in \mathbb{N}_0^n$ and $\zeta \in \mathbb{D}_r$, by some $c_{r,\rho,R} > 0$, $0 < r < \rho < R < \mu$. Using this bound in (VI.44), one has:

$$\text{(VI.45)} \qquad |I_1^{(\rho)}(\zeta)| \le c_{r,\rho,R} \sum_{\alpha\in(2\mathbb{N}_0)^n} \int_{\mathbb{S}^{n-1}} |C_\alpha \boldsymbol{\theta}^\alpha R^{|\alpha|}|\, d\Omega_{\boldsymbol{\theta}}, \qquad \zeta \in \mathbb{D}_r.$$

Now we can argue that the series (VI.44) is absolutely convergent. To bound the right-hand side in (VI.45), we use the following lemma with $\xi = R\boldsymbol{\theta}$.

Lemma VI.34 *For any $0 < R < \mu$ there is $C_{R,\mu} > 0$ such that for any analytic function $a(\xi) = \sum_{\alpha\in\mathbb{N}_0^n} a_\alpha \xi^\alpha$, $\xi \in \mathbb{D}_\mu^n \subset \mathbb{C}^n$, which has finite norm in $L^1(\mathbb{B}_\mu^{2n})$, where $\mathbb{B}_\mu^{2n} \subset \mathbb{R}^{2n}$ is identified with $\mathbb{D}_\mu^n \subset \mathbb{C}^n$, one has*

$$\sup_{\xi\in\mathbb{D}_R^n} \sum_{\alpha\in\mathbb{N}_0^n} |a_\alpha \xi^\alpha| \le C_{R,\mu}\|a\|_{L^1(\mathbb{B}_\mu^{2n})}.$$

Problem VI.35 Prove Lemma VI.34 using the Cauchy estimates

$$|a_\alpha| \le \int\limits_{\mathbb{S}^1_M \times \cdots \times \mathbb{S}^1_M} \frac{|a(Me^{\mathrm{i}\varphi_1}, \dots, Me^{\mathrm{i}\varphi_n})|}{M^{|\alpha|}} \frac{d\varphi_1}{2\pi} \cdots \frac{d\varphi_n}{2\pi},$$

with $M \in (R, \mu)$, and changing each of the integrations from φ_j over $\mathbb{S}^1_M$ to the integration over a thin closed annulus included in the region $R < |z_j| < \mu$.

This lemma, together with the estimate (VI.33) from Lemma VI.30, shows that, for $\zeta \in \mathbb{D}_r$, (VI.45) is bounded by

$$\begin{aligned} |I_1^{(\rho)}(\zeta)| &\le c_{r,\rho,R} \mathrm{vol}\,(\mathbb{S}^{n-1}) \sup_{\boldsymbol{\theta} \in \mathbb{S}^{n-1}} \sum_{\alpha \in (2\mathbb{N}_0)^n} |C_\alpha \boldsymbol{\theta}^\alpha R^{|\alpha|}| \\ &\le c_{r,\rho,R} C_{R,\mu} \mathrm{vol}\,(\mathbb{S}^{n-1}) \|VU\|_{L^1(\mathbb{B}^{2n}_\mu)} \\ &\le c_{r,\rho,R} C_{R,\mu} \mathrm{vol}\,(\mathbb{S}^{n-1}) \|V\|_{L^2(\mathbb{B}^{2n}_\mu)} \|U\|_{L^2(\mathbb{B}^{2n}_\mu)} \\ &\le c_{r,\rho,R} C_{R,\mu} C_\mu^2 \mathrm{vol}\,(\mathbb{S}^{n-1}) \|v\|_{L^{2,\mu}(\mathbb{R}^n)} \|u\|_{L^{2,\mu}(\mathbb{R}^n)}, \end{aligned}$$

where $V(\xi)$ and $U(\xi)$, $\xi \in \Omega_\mu(\mathbb{R}^n) \subset \mathbb{C}^n$, denote the analytic continuations of $\hat{v}(\xi)$ and $\hat{u}(\xi)$, $\xi \in \mathbb{R}^n$, into the μ-neighborhood of $\mathbb{R}^n$ in $\mathbb{C}^n$. We conclude that the series (VI.43) is absolutely convergent and therefore defines an analytic function.

Thus, $I_1^{(\rho)}(\zeta)$ (and hence $I(\zeta)$ in (VI.40)) has an analytic continuation into the disc $\mathbb{D}_\rho$ for arbitrary $\rho \in (0, \mu)$, and for any $r \in (0, \rho)$ the function $I_1^{(\rho)}(\zeta)$ (and hence $I(\zeta)$) is bounded by $C(r)\|v\|\|u\|$ as long as $\zeta \in \mathbb{D}_r$. This concludes the proof of Proposition VI.28.

CHAPTER VII

Carleman–Berthier–Georgescu estimates

One of the main tools in the next chapters is the Carleman–Berthier–Georgescu estimates derived in [**BG87**, Theorem 5]. We generalize these estimates to any dimension (Section VII.2) and adapt them to the context of linearization at solitary waves (Section VII.3).

VII.1 Heuristics

We will illustrate the technique of the Carleman estimates on the following easy result:

Proposition VII.1 *Let $L = \mathrm{i}\partial_x + V(x)$, $\mathfrak{D}(L) = H^1(\mathbb{R})$, with $V \in C(\mathbb{R}, \mathbb{C})$ such that $|V(x)| \le Ce^{-\epsilon|x|}$ with some $\epsilon > 0$ and $C > 0$. Then there are no embedded eigenvalues in $\sigma_{\mathrm{ess}}(L) = \mathbb{R}$.*

Above, V is not assumed to be real-valued.

The change of variable $\phi = e^{\mathrm{i}\int_0^x (V-\lambda)}\psi$ makes the problem trivial, but this approach does not extend easily to higher dimensions and PDE problems. Below we consider another approach. The argument is as follows:

1. *First prove that $|\psi| \le C_N e^{-N|x|}$, $\forall N > 0$;*
2. *Then deduce that ψ has compact support;*
3. *Eventually conclude that $\psi \equiv 0$.*

Lemma VII.2 *If $\psi \in H^1(\mathbb{R})$ satisfies $(\mathrm{i}\partial_x + V(x))\psi = \lambda\psi$ with $\lambda \in \mathbb{R}$, then for any $N > 0$ there is $C_N > 0$ such that*

$$|\psi(x)| \le C_N e^{-N|x|}, \qquad x \in \mathbb{R}.$$

This follows using the ideas of Problem II.30.

Lemma VII.3 *Let $\varphi \in C^1(\mathbb{R})$, $\varphi' > 0$. Then one has*

$$\|\sqrt{\varphi'}e^{\varphi}u\| \le \|\frac{1}{\sqrt{\varphi'}}e^{\varphi}u'\| \qquad \textit{for any } u \in H^1(\mathbb{R}) \textit{ with compact support.}$$

Proof. It is enough to consider $u \in C^1_{\mathrm{comp}}(\mathbb{R}, \mathbb{R})$. We integrating by parts and apply the Cauchy–Schwarz inequality:

$$\Big|\int_{\mathbb{R}} \varphi' e^{2\varphi} u^2\,dx\Big| = \Big|\int_{\mathbb{R}} e^{2\varphi} u u'\,dx\Big| \le \Big(\int_{\mathbb{R}} \varphi' e^{2\varphi} u^2\,dx\Big)^{1/2} \Big(\int_{\mathbb{R}} \frac{1}{\varphi'} e^{2\varphi} (u')^2\,dx\Big)^{1/2}. \quad \square$$

Let $\varphi(x) = \tau x$, $\tau \ge 1$, $\operatorname{supp} u \subset (0, \infty)$. Then Lemma VII.3 yields

$$\text{(VII.1)} \qquad \tau\|e^{x\tau}u\| \le \|e^{x\tau}u'\|,$$

where $\tau \ge 1$ could be arbitrarily large; such estimates are called the *Carleman estimates.*

Given $u \in H^1(\mathbb{R})$ with compact support, we substitute $e^{\lambda x}u(x)$ with $\lambda \in \mathbb{R}$ into (VII.1) in place of $u(x)$, arriving at

$$(\tau - |\lambda|)\|e^{\tau x}u\| \le \|e^{\tau x}(u' - \lambda u)\|, \tag{VII.2}$$

with $\tau \ge 1$ arbitrarily large (we choose $\tau > |\lambda|$ so that the left-hand side is positive).

Now we prove Proposition VII.1. Let $\rho \in C^\infty(\mathbb{R})$ be such that

$$\rho\big|_{(-\infty,0]} = 0, \qquad \rho\big|_{[1,+\infty)} = 1, \qquad \sup|\rho'| \le 2.$$

Let $a \ge 0$, $b \ge a + 2$. Denote

$$\rho_a(x) = \rho(x-a), \qquad \operatorname{supp}\rho_a \subset [a, +\infty); \qquad \sup_{x\in\mathbb{R}}|\partial_x\rho_a(x)| \le 2;$$

$$\rho_{a,b}(x) = \rho(x-a)\rho(b-x), \qquad \operatorname{supp}\rho_{a,b} \subset [a, b]; \qquad \sup_{x\in\mathbb{R}}|\partial_x\rho_{a,b}(x)| \le 2.$$

Let $\lambda \in \mathbb{R}$ and assume that $\psi \in H^1(\mathbb{R})$ satisfies

$$(\mathrm{i}\partial_x + V)\psi = \lambda\psi.$$

Then the product $u = \rho_{a,b}\psi \in C^1_{\mathrm{comp}}(\mathbb{R})$ satisfies

$$(\mathrm{i}\partial_x - \lambda)\rho_{a,b}(x)\psi(x) = -V(x)\rho_{a,b}(x)\psi(x) + \mathrm{i}\psi(x)\partial_x\rho_{a,b}(x).$$

Therefore, by (VII.2),

$$(\tau - |\lambda|)\|e^{\tau x}\rho_{a,b}\psi\| \le \|e^{\tau x}(-V\rho_{a,b}\psi + \mathrm{i}\psi\partial_x\rho_{a,b})\|.$$

Let $a > 0$ be so large that $|V(x)| \le 1$ for $x \ge a$; then

$$(\tau - |\lambda| - 1)\|e^{\tau x}\rho_{a,b}\psi\| \le \|\psi\partial_x\rho_{a,b}\|.$$

Since $\operatorname{supp}\partial_x\rho_{a,b} \subset [a, a+1] \cup [b-1, b]$, we have:

$$(\tau - |\lambda| - 1)\|e^{\tau x}\rho_{a,b}\psi\| \le \|e^{\tau x}\psi\partial_x\rho_{a,b}\| \le 2\|e^{\tau x}\psi\|_{L^2(a,a+1)} + 2\|e^{\tau x}\psi\|_{L^2(b-1,b)}.$$

Fix $\tau \ge |\lambda| + 2$, so that the coefficient on the left is not smaller than 1. Sending $b \to \infty$ and noticing that $\|e^{\tau x}\psi\|_{L^2(b-1,b)} \to 0$ due to Lemma VII.2, we conclude by the monotone convergence theorem that $\rho_a\psi$ satisfies $\|e^{\tau x}\rho_a\psi\| \le 2\|e^{\tau x}\psi\|_{L^2(a,a+1)}$, and moreover we have

$$e^{(a+2)\tau}\|\rho_a\psi\|_{L^2(a+2,\infty)} \le \|e^{\tau x}\rho_a\psi\| \le 2\|e^{\tau x}\psi\|_{L^2(a,a+1)} \le 2e^{(a+1)\tau}\|\psi\|_{L^2(a,a+1)}.$$

Since τ could be arbitrarily large, $\|\rho_a\psi\|_{L^2(a+2,\infty)} = 0$, and then $\psi(x) = 0$ for $x \ge a + 2$.

Finally, one needs the *unique continuation property*; this is a property of certain equations that once a solution to such an equation vanishes on an open interval then it is identically zero. For the sake of this introduction in one dimension, we could say that this is immediate due to the local well-posedness: Since the solution to $\mathrm{i}\psi' + V\psi = \lambda\psi$ vanishes identically on an open interval (in the 1D case, cancellation at a single point is enough), one has $\psi(x) \equiv 0$.

VII.2 Carleman–Berthier–Georgescu estimates for $D_m + V$

Following the proof of [**BG87**, Theorem 3], we define the following set of admissible functions.

Definition VII.4 Let $\lambda \in \mathbb{R} \setminus [-m, m]$. Let $M \geq 1$, $\mathcal{N} \geq 1$, $\rho \geq 1$ and $\nu > 0$. We define the subset

$$\mathscr{C}_\lambda(M, \mathcal{N}, \rho, \nu) \subset C^2(\mathbb{R}_+)$$

to be the set of functions which satisfy the following properties:

(1) $0 < \varphi'(r) \leq \mathcal{N} r, \ \forall r \geq \rho$;
(2) $r|\varphi''(r)| \leq M\varphi'(r), \ \forall r \geq \rho$;
(3) $\lambda^2 - m^2 + \varphi'^2 + 2r\varphi'\varphi'' \geq \nu, \ \forall r \geq \rho$;
(4) if φ' is unbounded then $\varphi''(r) \geq 0$ for $r \geq \rho$ and $\lim_{r\to\infty} \varphi'(r) = +\infty$.

We introduce the weight functions

$$\mu(r) = 2\left(n + 16\lambda^2 r^2 + 8r\varphi'(r)\right)^{1/2}, \qquad r \geq \rho; \tag{VII.3}$$

$$\gamma(r) = \left(\lambda^2 - m^2 + \varphi'(r)^2 + 2r\varphi'(r)\varphi''(r)\right)^{1/2}, \qquad r \geq \rho. \tag{VII.4}$$

Below, we will use the notation

$$\Omega^n_R = \mathbb{R}^n \setminus \overline{\mathbb{B}^n_R} = \{x \in \mathbb{R}^n : \ |x| > R\}, \qquad R > 0, \qquad n \in \mathbb{N}.$$

The following result generalizes the Carleman estimates for the Dirac operator in $\mathbb{R}^3$ due to Berthier and Georgescu [**BG87**, Theorem 5]. to any dimension.

Theorem VII.5 (Carleman–Berthier–Georgescu estimates for $D_m + V$) *Let $n \geq 1$, $m > 0$, $\lambda \in \mathbb{R} \setminus [-m, m]$.*

(1) Assume that $\varphi \in \mathscr{C}_\lambda(M, \mathcal{N}, \rho, \nu)$ with some $M \geq 1$, $\mathcal{N} \geq 1$, $\rho \geq 1$, and $\nu > 0$. Let $V \in \mathscr{B}\left(H^1_{\mathrm{comp}}(\Omega^n_\rho, \mathbb{C}^N), L^2(\Omega^n_\rho, \mathbb{C}^N)\right)$ be a multiplication operator (an operator of multiplication by a matrix-valued function), and assume that there are $\kappa \in (0, 1)$ and $R_1 = R_1(\varphi, V) \geq \rho$ such that

$$\|\mu V v\| \leq \kappa \left(\|\nabla v\|^2 + \|\gamma v\|^2\right)^{1/2}, \qquad \forall v \in H^1_0(\Omega^n_{R_1}, \mathbb{C}^N). \tag{VII.5}$$

Then there is $R = R(\varphi, V) \geq R_1$ such that for any $u \in H^1(\mathbb{R}^n, \mathbb{C}^N)$ with $\operatorname{supp} u \subset \Omega^n_R$ and

$$\mu e^{\varphi}(D_m + V - \lambda)u \in L^2(\mathbb{R}^n, \mathbb{C}^N),$$

the functions $\gamma e^{\varphi} u$, $\nabla(e^{\varphi} u)$, $(r\varphi')^{1/2}\partial_r(e^{\varphi} u)$ are in $L^2(\mathbb{R}^n, \mathbb{C}^N)$, and moreover

$$\begin{aligned} &\|\nabla(e^{\varphi}u)\|^2 + 2\|(r\varphi')^{1/2}\partial_r(e^{\varphi}u)\|^2 + \|\gamma e^{\varphi}u\|^2 \\ &\qquad \leq \frac{1}{(1-\kappa)^2}\|\mu e^{\varphi}(D_m + V - \lambda)u\|^2. \end{aligned} \tag{VII.6}$$

We remind that the weights $\mu(r)$ and $\gamma(r)$ are defined in (VII.3) *and* (VII.4).

(2) Assume that there are $\varkappa \in (0, 1)$ and $R_2(V) > 0$ such that

$$\|V(x)\|_{\mathrm{End}(\mathbb{C}^N)} \leq \varkappa \frac{\sqrt{\lambda^2 - m^2}}{4|\lambda||x|}, \qquad \forall x \in \Omega^n_{R_2(V)}. \tag{VII.7}$$

Let $M \geq 1$, $\mathcal{N} \geq 1$, $\rho \geq 1$, $\nu > 0$, and $\kappa \in (\varkappa, 1)$. Then there is

$$R = R(M, \rho, m, n, \lambda, V) \geq \max(\rho, R_2(V))$$

such that for any $\varphi \in \mathscr{C}_\lambda(M, \mathcal{N}, \rho, \nu)$ *with* $\varphi''(r) \ge 0$ $\forall r \ge \rho$ *and for any*

$$u \in H^1_0(\Omega^n_R, \mathbb{C}^N)$$

which satisfies

$$\mu e^{\varphi}(D_m + V - \lambda)u \in L^2(\mathbb{R}^n, \mathbb{C}^N),$$

the functions $\gamma e^{\varphi} u$, $\nabla(e^{\varphi}u)$, $(r\varphi')^{1/2}\partial_r(e^{\varphi}u)$ *are in* $L^2(\mathbb{R}^n, \mathbb{C}^N)$, *and moreover*

$$\text{(VII.8)} \qquad \begin{aligned} &\|\nabla(e^{\varphi}u)\|^2 + 2\|(r\varphi')^{1/2}\partial_r(e^{\varphi}u)\|^2 + \|\gamma e^{\varphi}u\|^2 \\ &\le \frac{1}{(1-\kappa)^2}\|\mu e^{\varphi}(D_m + V - \lambda)u\|^2. \end{aligned}$$

Above, φ, μ, *and* γ *(see* (VII.3) *and* (VII.4)*) are considered as functions of* $r = |x|$.

Let us rewrite (VII.7) in the form more convenient for the use in the estimates.

Lemma VII.6 *Let* $n \ge 1$. *Assume that there is* $R_2(V) > 0$ *and* $\varkappa \in (0,1)$ *such that*

$$\text{(VII.9)} \qquad \|V(x)\|_{\mathrm{End}(\mathbb{C}^N)} \le \varkappa \frac{\sqrt{\lambda^2 - m^2}}{4|\lambda||x|}, \qquad \forall x \in \Omega^n_{R_2(V)}.$$

Then for any $\kappa \in (\varkappa, 1)$ *we can take* $R_2(V)$ *larger if necessary so that*

$$(n + 16\lambda^2 r^2 + 8r\theta)\|V(x)\|^2_{\mathrm{End}(\mathbb{C}^N)} \le \kappa^2\left(\lambda^2 - m^2 + \theta^2\right), \quad \forall x \in \Omega^n_{R_2(V)}, \quad \forall \theta \ge 0.$$

In particular,

$$\text{(VII.10)} \qquad \|\left(n + 16\lambda^2 r^2 + 8r\theta\right)^{\frac{1}{2}} V v\| \le \kappa \|\left(\lambda^2 - m^2 + \theta^2\right)^{\frac{1}{2}} v\|, \qquad \forall \theta \ge 0,$$

valid for all $v \in H^1_0(\Omega^n_{R_2(V)}, \mathbb{C}^N)$.

Proof. Let $\kappa \in (\varkappa, 1)$. We need to have

$$\text{(VII.11)} \qquad \|V(x)\|^2_{\mathrm{End}(\mathbb{C}^N)} \le \kappa^2 \frac{\lambda^2 - m^2 + \theta^2}{n + 16\lambda^2|x|^2 + 8|x|\theta}, \qquad \forall \theta \ge 0,$$

for $|x|$ sufficiently large. For $0 \le \theta \le 1$, one has

$$\text{(VII.12)} \qquad \kappa^2 \frac{\lambda^2 - m^2 + \theta^2}{n + 16\lambda^2 r^2 + 8r\theta} \ge \kappa^2 \frac{\lambda^2 - m^2}{n + 16\lambda^2 r^2 + 8r} \ge \varkappa^2 \frac{\lambda^2 - m^2}{16\lambda^2 r^2},$$

with the last inequality satisfied as long as $r \ge R$ with R sufficiently large. The inequality (VII.12) together with (VII.9) yields (VII.11), and hence the desired inequality (VII.10) in the case $\theta \in [0,1]$. At the same time, taking the derivative of the right-hand side of (VII.11) in θ, we arrive at

$$\frac{2\theta(n + 16\lambda^2 r^2 + 8r\theta) - (\lambda^2 - m^2 + \theta^2)8r}{(n + 16\lambda^2 r^2 + 8r\theta)^2},$$

which is strictly positive for all $\theta \ge 1$ and for all $r \ge R$, with R sufficiently large. We see that the right-hand side of (VII.11), considered as a function of θ, is strictly monotonically increasing for $\theta \ge 1$ as long as $|x| \ge R$. This completes the proof. □

Proof of Theorem VII.5. We give a detailed proof which closely follows the argument in **[BG87]**. We start with several lemmata. Let $D_m = -\mathrm{i}\boldsymbol{\alpha} \cdot \nabla + \beta m$, $\varphi \in C^2(\mathbb{R}_+)$, and denote

$$\text{(VII.13)} \qquad D^{\varphi}_m = e^{\varphi} \circ D_m \circ e^{-\varphi} = D_m + \mathrm{i}\boldsymbol{\alpha} \cdot \nabla\varphi.$$

The starting point of the analysis is the following lemma.

Lemma VII.7 (Lemma 3, [BG87]) *Let $\Omega \subset \mathbb{R}^n$ be an open set and $\varphi : \Omega \to \mathbb{R}$ a C^1 map. For $v \in H^1_{\mathrm{comp}}(\Omega, \mathbb{C}^N)$,*

$$\mathrm{Re}\langle (D_m - \mathrm{i}\boldsymbol{\alpha}\cdot\nabla\varphi + \lambda)v, (D^\varphi_m - \lambda)v\rangle = \|\nabla v\|^2 + \langle v, [m^2 - \lambda^2 - (\nabla\varphi)^2]v\rangle. \tag{VII.14}$$

PROOF. Taking into account that $\nabla\varphi$ is continuous, due to density of smooth functions with compact support in $H^1_{\mathrm{comp}}(\Omega, \mathbb{C}^N)$, it is enough to give the proof assuming that $v \in H^\infty_{\mathrm{comp}}(\Omega, \mathbb{C}^N)$.

The statement of the lemma is a consequence of the following computation performed in **[BG87]**:

$$\begin{aligned}(D^\varphi_m + \lambda)(D^\varphi_m - \lambda) &= (D^\varphi_m)^2 - \lambda^2 = e^\varphi \circ (D^2_m - \lambda^2) \circ e^{-\varphi} \\ &= e^\varphi \circ (-\Delta + m^2 - \lambda^2) \circ e^{-\varphi} \\ &= -\Delta + m^2 - \lambda^2 - (\nabla\varphi)^2 + \nabla\varphi\cdot\nabla + \nabla\cdot\nabla\varphi, \end{aligned} \tag{VII.15}$$

where the last term is understood as the multiplication by $\nabla\varphi$ and then taking the divergence. In the real part of the corresponding quadratic form the last two terms from the right-hand side of (VII.15) cancel, while the real part of the left-hand side turns into that of (VII.14). □

Lemma VII.7 helps us to establish the exponential decay of eigenfunctions associated to eigenvalues in the gap; see Section VII.5.

For brevity, we adopt the following notations from **[BG87]**:

$$\hat{X} = x\cdot\nabla, \qquad \mathscr{D} = \frac{1}{2}\{x, -\mathrm{i}\nabla\} = -\mathrm{i}\hat{X} - \frac{\mathrm{i}n}{2}.$$

Notice that $\mathscr{D}$ is the generator of dilations and thus in the sense of quadratic forms on $H^1_{\mathrm{comp}}(\Omega, \mathbb{C}^N)$, with $\Omega \subset \mathbb{R}^n$ an open set,

$$[\mathscr{D}, D_m] = [\mathscr{D}, D_0] = [-\mathrm{i}x\cdot\nabla, -\mathrm{i}\boldsymbol{\alpha}\cdot\nabla] = [\boldsymbol{\alpha}\cdot\nabla, x\cdot\nabla] = \boldsymbol{\alpha}\cdot\nabla = \mathrm{i}D_0. \tag{VII.16}$$

In order to analyze the eigenvectors associated to embedded eigenvalues, it will be convenient to subtract the identity (VII.15) from another one (which also involves $(D^\varphi_m - \lambda)$ in the left hand side) that controls the $\dot{H}^1$-norm. Starting with $D_0(D^\varphi_m - \lambda)$ and trying to eliminate inconvenient terms (with a factor λ for instance), Berthier and Georgescu **[BG87]** have the following lemma (Lemma 4, **[BG87]**), which we rewrite for arbitrary dimension $n \geq 1$.

Lemma VII.8 *Let $\Omega \subset \mathbb{R}^n$ be an open set and let $\varphi \in C^2(\Omega)$. Then for any $u \in H^1_{\mathrm{comp}}(\Omega, \mathbb{C}^N)$,*

(VII.17)

$$\begin{aligned}&2\,\mathrm{Re}\langle (D_0 + 2\mathrm{i}\lambda\mathscr{D} + \{\mathscr{D}, \boldsymbol{\alpha}\cdot\nabla\varphi\})v, (D^\varphi_m - \lambda)v\rangle \\ &\quad = 2\|\nabla v\|^2 + 4\,\mathrm{Re}\langle \hat{X}v, (\nabla\varphi)\cdot\nabla v\rangle + 2\,\mathrm{Re}\langle \hat{X}v, \Delta\varphi\, v\rangle + \langle v, \big(\hat{X}(\nabla\varphi)^2\big)v\rangle.\end{aligned}$$

PROOF. We present the proof from **[BG87]** stripped of the external fields. Again, since $\nabla\varphi$ and $\Delta\varphi$ are continuous, and since smooth functions with compact support are dense in $H^1_{\mathrm{comp}}(\Omega, \mathbb{C}^N)$, the computations below, made in the former case, will provide the proof for the latter. First, using (VII.16), we get

$$\begin{aligned}4\,\mathrm{Im}\langle \mathscr{D}v, (D^\varphi_m - \lambda)v\rangle &= \frac{2}{\mathrm{i}}\langle v, [\mathscr{D}, D_0]v\rangle + 4\,\mathrm{Im}\langle \mathscr{D}v, (\beta m + \mathrm{i}\boldsymbol{\alpha}\cdot\boldsymbol{F} - \lambda)v\rangle \\ &= 2\langle v, D_0 v\rangle + 4\,\mathrm{Re}\langle \mathscr{D}v, \boldsymbol{\alpha}\cdot\boldsymbol{F}v\rangle,\end{aligned} \tag{VII.18}$$

where $\boldsymbol{F} = \nabla\varphi$. Using (VII.13) and the identity

$$(\boldsymbol{\alpha}\cdot\boldsymbol{S})(\boldsymbol{\alpha}\cdot\boldsymbol{T}) = \boldsymbol{S}\cdot\boldsymbol{T} + \mathrm{i}\Sigma(\boldsymbol{S},\boldsymbol{T}), \qquad \boldsymbol{S},\ \boldsymbol{T}\in\mathbb{C}^n, \tag{VII.19}$$

where $\Sigma(\boldsymbol{S},\boldsymbol{T}) = \Sigma_{ij}S_iT_j$, with the matrices $\Sigma_{ij} = \frac{1}{2\mathrm{i}}[\alpha^i,\alpha^j]$ which are hermitian for each $1\le i,\ j\le n$, we have

$$D_0(D_m^\varphi - \lambda) = -\Delta + D_0\beta m + (\boldsymbol{\alpha}\cdot\nabla)\circ(\boldsymbol{\alpha}\cdot\boldsymbol{F}) - \lambda D_0$$

$$= -\Delta + mD_0\beta + \boldsymbol{F}\cdot\nabla + \Delta\varphi + \mathrm{i}\Sigma(\nabla,\boldsymbol{F}) - \lambda D_0.$$

But $\operatorname{Re}\langle v, D_0\beta v\rangle = 0$ (due to $\{D_0,\beta\}=0$), $2\operatorname{Re}\langle v,\boldsymbol{F}\cdot\nabla v\rangle = -\langle v,\Delta\varphi\, v\rangle$, and

$$\Sigma(\nabla,\boldsymbol{F}) = \Sigma_{ij}\partial_i\circ\varphi_j = \Sigma_{ij}\varphi_{ij} + \Sigma_{ij}\varphi_j\partial_i = -\Sigma_{ji}\varphi_j\partial_i = -\Sigma(\boldsymbol{F},\nabla),$$

for all $1\le i,\ j\le n$, hence

$$2\operatorname{Re}\langle D_0v,(D_m^\varphi-\lambda)v\rangle = 2\|\nabla v\|^2 + \langle v,\Delta\varphi\, v\rangle - 2\mathrm{i}\langle v,\Sigma(\boldsymbol{F},\nabla)v\rangle - 2\lambda\langle v,D_0v\rangle. \tag{VII.20}$$

The last term is inconvenient to us, since, due to the factor λ, it cannot be controlled uniformly in λ. Adding (VII.18) (multiplied by λ) to (VII.20), to get rid of $2\lambda\langle v, D_0v\rangle$, we obtain:

$$\begin{aligned}&2\operatorname{Re}\langle D_0v,(D_m^\varphi-\lambda)v\rangle + 4\lambda\operatorname{Im}\langle\mathscr{D}v,(D_m^\varphi-\lambda)v\rangle\\ &= 2\|\nabla v\|^2 + \langle v,\big(\Delta\varphi - 2\mathrm{i}\Sigma(\boldsymbol{F},\nabla)\big)v\rangle + 4\lambda\operatorname{Re}\langle\mathscr{D}v,\boldsymbol{\alpha}\cdot\boldsymbol{F}v\rangle;\end{aligned} \tag{VII.21}$$

see (VII.16). Now we eliminate the inconvenient term $4\lambda\operatorname{Re}\langle\mathscr{D}v,\boldsymbol{\alpha}\cdot\boldsymbol{F}v\rangle$. Recalling that $[\mathscr{D},D_0] = \mathrm{i}D_0$, we derive the identity

$$\begin{aligned}&\{\mathscr{D},\boldsymbol{\alpha}\cdot\boldsymbol{F}\}D_0 = \{\mathscr{D},\boldsymbol{\alpha}\cdot\boldsymbol{F}D_0\} + \boldsymbol{\alpha}\cdot\boldsymbol{F}[\mathscr{D},D_0]\\ &= \{\mathscr{D},\boldsymbol{\alpha}\cdot\boldsymbol{F}D_0\} + \mathrm{i}\boldsymbol{\alpha}\cdot\boldsymbol{F}D_0 = \{\mathscr{D}+\frac{\mathrm{i}}{2},\boldsymbol{\alpha}\cdot\boldsymbol{F}D_0\}\\ &= \{\mathscr{D}+\frac{\mathrm{i}}{2},(\boldsymbol{\alpha}\cdot\boldsymbol{F})(-\mathrm{i}\boldsymbol{\alpha}\cdot\nabla)\} = -\mathrm{i}\{\mathscr{D}+\frac{\mathrm{i}}{2},\boldsymbol{F}\cdot\nabla\} + \{\mathscr{D}+\frac{\mathrm{i}}{2},\Sigma(\boldsymbol{F},\nabla)\},\end{aligned}$$

where in the last line we used (VII.19). The above relation leads to

$$\begin{aligned}2\operatorname{Re}\{\mathscr{D},\boldsymbol{\alpha}\cdot\boldsymbol{F}\}D_0 &= \operatorname{Im}\{2\mathscr{D}+\mathrm{i},\boldsymbol{F}\cdot\nabla\} + \operatorname{Re}\{2\mathscr{D}+\mathrm{i},\Sigma(\boldsymbol{F},\nabla)\}\\ &= \operatorname{Im}\{2\mathscr{D}+\mathrm{i},\boldsymbol{F}\cdot\nabla\} + 2\mathrm{i}\Sigma(\boldsymbol{F},\nabla),\end{aligned} \tag{VII.22}$$

where we used the identity $\operatorname{Re}\{\mathscr{D},\Sigma(\boldsymbol{F},\nabla)\}=0$. The first term in the right-hand side of (VII.22) can be written as

$$\begin{aligned}&\operatorname{Im}\{2\mathscr{D}+\mathrm{i},\boldsymbol{F}\cdot\nabla\}\\ &= -\mathrm{i}\big(\mathscr{D}\boldsymbol{F}\cdot\nabla + \boldsymbol{F}\circ\nabla\mathscr{D} + \nabla\circ\boldsymbol{F}\mathscr{D} + \mathscr{D}\boldsymbol{F}\cdot\nabla + \mathrm{i}\boldsymbol{F}\cdot\nabla - \mathrm{i}\nabla\circ\boldsymbol{F}\big)\\ &= -\mathrm{i}\{\mathscr{D},\{\boldsymbol{F},\nabla\}\} - \Delta\varphi = 2\operatorname{Im}(\mathscr{D}\{\boldsymbol{F},\nabla\}) - \Delta\varphi\\ &= -2\operatorname{Re}(\nabla\circ x\{\boldsymbol{F},\nabla\}) - \Delta\varphi\\ &= -4\operatorname{Re}(\nabla\circ x\,\boldsymbol{F}\cdot\nabla) - 2\operatorname{Re}(\nabla\circ x\Delta\varphi) - \Delta\varphi.\end{aligned} \tag{VII.23}$$

From (VII.22) and (VII.23) we obtain

$$2\operatorname{Re}\{\mathscr{D},\boldsymbol{\alpha}\nabla\varphi\}D_0 = -4\operatorname{Re}(\nabla\circ x\,\boldsymbol{F}\cdot\nabla) - 2\operatorname{Re}(\nabla\circ x\Delta\varphi) - \Delta\varphi + 2\mathrm{i}\Sigma(\boldsymbol{F},\nabla),$$

(VII.24)

$$\begin{aligned}
&2\,\mathrm{Re}\langle\{\mathscr{D},\boldsymbol{\alpha}\cdot\boldsymbol{F}\}v,(D_m^\varphi-\lambda)v\rangle\\
&\quad=2\,\mathrm{Re}\langle\{\mathscr{D},\boldsymbol{\alpha}\cdot\boldsymbol{F}\}v,D_0v\rangle+2\,\mathrm{Re}\langle\{\mathscr{D},\boldsymbol{\alpha}\cdot\boldsymbol{F}\}v,\mathrm{i}\boldsymbol{\alpha}\cdot\boldsymbol{F}v\rangle-2\lambda\,\mathrm{Re}\langle\{\mathscr{D},\boldsymbol{\alpha}\cdot\boldsymbol{F}\}v,v\rangle\\
&\quad=4\,\mathrm{Re}\langle\hat{X}v,\boldsymbol{F}\cdot\nabla v\rangle+2\,\mathrm{Re}\langle\hat{X}v,\Delta\varphi\,v\rangle-\langle v,(\Delta\varphi-2\mathrm{i}\Sigma(\boldsymbol{F},\nabla))v\rangle\\
&\quad+\langle v,\hat{X}(\boldsymbol{F})^2v\rangle-4\lambda\,\mathrm{Re}\langle\mathscr{D}v,\boldsymbol{\alpha}\cdot\boldsymbol{F}v\rangle.
\end{aligned}$$

Above, we used the identity

$$2\,\mathrm{Re}\langle\{\mathscr{D},\boldsymbol{\alpha}\cdot\boldsymbol{F}\}v,\mathrm{i}\boldsymbol{\alpha}\cdot\boldsymbol{F}v\rangle=\langle v,(\{\mathscr{D},\boldsymbol{\alpha}\cdot\boldsymbol{F}\}\mathrm{i}\boldsymbol{\alpha}\cdot\boldsymbol{F}-\mathrm{i}\boldsymbol{\alpha}\cdot\boldsymbol{F}\{\mathscr{D},\boldsymbol{\alpha}\cdot\boldsymbol{F}\})v\rangle=\langle v,\hat{X}(\boldsymbol{F})^2v\rangle.$$

Adding (VII.21) and (VII.24) yields (VII.17). □

The following lemma parallels [**BG87**, Lemma 6] with explicit constants.

Lemma VII.9 *Let $n\in\mathbb{N}$, $\rho\ge 1$. Let $\varphi\in C^2([\rho,+\infty))$ with $\varphi'>0$, and let us define $Z\in C([\rho,+\infty))$ by*

$$\begin{aligned}
Z(r)\;&=\;2\left(\lambda^2-m^2+\varphi'^2+2r\varphi'\varphi''-(n-1)\varphi''-\frac{(n-1)^2\varphi'}{r}-\frac{r\varphi''^2}{\varphi'}\right)\\
&\quad-4\left(\frac{(2n-1)(|\lambda|+\varphi')+m+2r|\varphi''|}{\mu}\right)^2,\qquad r\ge\rho.
\end{aligned}\tag{VII.25}$$

Then for any $v\in H^1_{\mathrm{comp}}(\Omega^n_\rho,\mathbb{C}^N)$ one has

$$\|\nabla v\|^2+2\|(r\varphi')^{1/2}\partial_r v\|^2+\langle v,Zv\rangle\le\|\mu(D_m^\varphi-\lambda)v\|^2,\tag{VII.26}$$

and for any $u\in H^1_{\mathrm{comp}}(\Omega^n_\rho,\mathbb{C}^N)$ one has

$$\|\nabla(e^\varphi u)\|^2+2\|(r\varphi')^{1/2}\partial_r(e^\varphi u)\|^2+\langle e^\varphi u,Ze^\varphi u\rangle\le\|\mu e^\varphi(D_m-\lambda)u\|^2.\tag{VII.27}$$

Above, φ, μ, and Z (see (VII.3), (VII.25)*) are considered as functions of $r=|x|$.*

Proof. Denote $\hat{\alpha}=r^{-1}\boldsymbol{\alpha}\cdot x$, where $r=|x|$. We subtract (VII.14) from (VII.17) with the aid of the identity

$$\{\mathscr{D},\boldsymbol{\alpha}\cdot\boldsymbol{F}\}=\{-\mathrm{i}\hat{X}-\frac{\mathrm{i}n}{2},\hat{\alpha}\varphi'\}=-2\mathrm{i}\hat{\alpha}\varphi'\hat{X}-\mathrm{i}\hat{\alpha}r\varphi''-\mathrm{i}n\hat{\alpha}\varphi',$$

arriving at

$$\begin{aligned}
&2\,\mathrm{Re}\Big\langle\Big[D_0+2\mathrm{i}\lambda\mathscr{D}+\{\mathscr{D},\boldsymbol{\alpha}\cdot\boldsymbol{F}\}-\frac{1}{2}(D_m-\mathrm{i}\boldsymbol{\alpha}\cdot\boldsymbol{F}+\lambda)\Big]v,\ (D_m^\varphi-\lambda)v\Big\rangle\\
&=2\,\mathrm{Re}\Big\langle\Big[-\frac{\mathrm{i}}{2}\boldsymbol{\alpha}\cdot\nabla+2(\lambda-\mathrm{i}\hat{\alpha}\varphi')\hat{X}+\Big(n-\frac{1}{2}\Big)\lambda-\frac{m}{2}\beta\\
&\qquad\qquad-\mathrm{i}\Big(n-\frac{1}{2}\Big)\hat{\alpha}\varphi'-\mathrm{i}\hat{\alpha}r\varphi''\Big]v,\ (D_m^\varphi-\lambda)v\Big\rangle\\
&=\|\nabla v\|^2+4\,\mathrm{Re}\Big\langle\hat{X}v,\frac{\varphi'}{r}\hat{X}v\Big\rangle+2\,\mathrm{Re}\langle\hat{X}v,\Delta\varphi\,v\rangle\\
&\qquad+\Big\langle v,\Big[\lambda^2-m^2+\varphi'^2+2r\varphi'\varphi''\Big]v\Big\rangle.
\end{aligned}\tag{VII.28}$$

Since

$$\begin{aligned}
&2\,\mathrm{Re}\Big\langle\hat{X}v,\frac{\varphi'}{r}v\Big\rangle=\Big\langle\hat{X}v,\frac{\varphi'}{r}v\Big\rangle+\Big\langle v,\frac{\varphi'}{r}\hat{X}v\Big\rangle\\
&=\Big\langle v,\Big(-\hat{X}\circ\frac{\varphi'}{r}-n\frac{\varphi'}{r}+\frac{\varphi'}{r}\hat{X}\Big)v\Big\rangle=-\Big\langle v,\Big[\varphi''+(n-1)\frac{\varphi'}{r}\Big]v\Big\rangle,
\end{aligned}$$

which is valid for $v \in H^1_{\mathrm{comp}}(\Omega^n_\rho, \mathbb{C}^N)$ with $\rho \geq 1$, we have

$$2\,\mathrm{Re}\langle \hat{X} v, \Delta\varphi\, v\rangle = 2\,\mathrm{Re}\langle \hat{X} v, \varphi'' v\rangle - \Big\langle v, \Big[(n-1)\varphi'' + (n-1)^2 \frac{\varphi'}{r}\Big] v\Big\rangle.$$

We use the above relation to rewrite (VII.28) as

$$\begin{aligned}
\mathrm{Re}\Big\langle \Big[-\mathrm{i}\boldsymbol{\alpha}\cdot\nabla + 4(\lambda - \mathrm{i}\hat{\alpha}\varphi')\hat{X} + (2n-1)(\lambda - \mathrm{i}\hat{\alpha}\varphi') \\
- \beta m - 2\mathrm{i}\hat{\alpha} r\varphi''\Big] v, (D^\varphi_m - \lambda) v\Big\rangle
\end{aligned}$$

$$\begin{aligned}
&= \|\nabla v\|^2 + 4\|(\frac{\varphi'}{r})^{\frac12}\hat{X} v\|^2 + 2\,\mathrm{Re}\langle \hat{X} v, \varphi'' v\rangle \\
&\quad + \Big\langle v, \Big[\lambda^2 - m^2 + \varphi'^2 + 2r\varphi'\varphi'' - (n-1)\varphi'' - (n-1)^2\frac{\varphi'}{r}\Big] v\Big\rangle.
\end{aligned}$$

For any positive continuous function $\mu(r)$, the above relation yields the following inequality:

$$\begin{aligned}
&\frac12 \Big\| \frac{1}{\mu}\big[-\mathrm{i}\boldsymbol{\alpha}\cdot\nabla + 4(\lambda - \mathrm{i}\hat{\alpha}\varphi')\hat{X} + (2n-1)(\lambda - \mathrm{i}\hat{\alpha}\varphi') - \beta m - 2\mathrm{i}\hat{\alpha} r\varphi''\big] v\Big\|^2 \\
&\qquad + \frac{\|\mu(D^\varphi_m - \lambda) v\|^2}{2} \\
&\geq \|\nabla v\|^2 + 3\|(\frac{\varphi'}{r})^{\frac12}\hat{X} v\|^2 - \|(\frac{r}{\varphi'})^{\frac12}\varphi'' v\|^2 \\
&\qquad + \Big\langle v, \Big[\lambda^2 - m^2 + \varphi'^2 + 2r\varphi'\varphi'' - (n-1)\varphi'' - (n-1)^2\frac{\varphi'}{r}\Big] v\Big\rangle.
\end{aligned}$$

Since $\frac12(a+b+c+d)^2 \leq 2(a^2+b^2+c^2+d^2)$, the above inequality leads to

$$\begin{aligned}
&2\Big[\Big\|\frac{\boldsymbol{\alpha}\cdot\nabla u}{\mu}\Big\|^2 + \Big\|\frac{4\lambda\hat{X} u}{\mu}\Big\|^2 + \Big\|\frac{4\hat{\alpha}\varphi'\hat{X} u}{\mu}\Big\|^2 + \Big\|\frac{[(2n-1)(|\lambda| + \varphi') + m + 2r|\varphi''|] v}{\mu}\Big\|^2\Big] \\
&\qquad + \frac{\|\mu(D^\varphi_m - \lambda) v\|^2}{2} \\
&\geq \|\nabla v\|^2 + 3\|(\frac{\varphi'}{r})^{\frac12}\hat{X} v\|^2 - \|(\frac{r}{\varphi'})^{\frac12}\varphi'' v\|^2
\end{aligned}$$

(VII.29)

$$+ \Big\langle v, \Big[\lambda^2 - m^2 + \varphi'^2 + 2r\varphi'\varphi'' - (n-1)\varphi'' - (n-1)^2\frac{\varphi'}{r}\Big] v\Big\rangle.$$

To eliminate the first three terms from the left-hand side of (VII.29) (with the help of the first two terms from the right-hand side), we require that $\mu(r)$ be such that

$$2\left(\Big\|\frac{\boldsymbol{\alpha}\cdot\nabla v}{\mu}\Big\|^2 + \Big\|\frac{4\lambda\hat{X} v}{\mu}\Big\|^2\right) \leq \frac12\|\nabla v\|^2, \tag{VII.30}$$

$$\Big\|\frac{4\hat{\alpha}\varphi'\hat{X} v}{\mu}\Big\|^2 \leq \Big\|\Big(\frac{\varphi'}{r}\Big)^{\frac12}\hat{X} v\Big\|^2, \tag{VII.31}$$

where

$$\|\nabla v\| := \Big(\sum_{i=1}^{n} \|\partial_{x^i} v\|^2\Big)^{\frac{1}{2}}.$$

Since

$$\Big\|\frac{\boldsymbol{\alpha}\cdot\nabla v}{\mu}\Big\|^2 \le \Big(\sum_{i=1}^{n}\Big\|\frac{\partial_{x^i} v}{\mu}\Big\|\Big)^2 \le n\sum_{i=1}^{n}\Big\|\frac{\partial_{x^i} v}{\mu}\Big\|^2 =: n\Big\|\frac{\nabla v}{\mu}\Big\|^2,$$

while $\hat{X}v = r\partial_r v$, resulting in

$$\text{(VII.32)} \qquad \|\hat{X}v\|_{\mathbb{C}^N} \le r\|\nabla v\|_{\mathbb{C}^N}, \qquad \forall v\in C^1(\mathbb{R}^n,\mathbb{C}^N),$$

we see that (VII.30) will hold whenever

$$2\left(\frac{n+(4\lambda r)^2}{\mu^2}\right) \le \frac{1}{2},$$

$$\text{(VII.33)} \qquad \mu(r)^2 \ge 4n + 64\lambda^2 + r^2.$$

To satisfy (VII.31), again in view of (VII.32), it is enough to have

$$\text{(VII.34)} \qquad \frac{32r\varphi'}{\mu(r)^2} \le 1.$$

To comply with both (VII.33) and (VII.34), it is enough to require that

$$\text{(VII.35)} \qquad \mu(r) \ge 2\Big(n + 16\lambda^2 r^2 + 8r\varphi'\Big)^{1/2}.$$

Taking into account (VII.30) and (VII.31), the inequality (VII.29) yields

$$\begin{aligned}
&2\Big\|\frac{1}{\mu}\Big[(2n-1)(|\lambda|+\varphi') + m + 2r|\varphi''|\Big]v\Big\|^2 + \frac{\|\mu(D_m^\varphi-\lambda)v\|^2}{2}\\
&\ge \frac{1}{2}\|\nabla v\|^2 + \|\big(\frac{\varphi'}{r}\big)^{1/2}\hat{X}v\|^2\\
&\quad + \Big\langle v, \Big[\lambda^2 - m^2 + \varphi'^2 + 2r\varphi'\varphi'' - (n-1)\varphi'' - (n-1)^2\frac{\varphi'}{r} - \frac{r\varphi''^2}{\varphi'}\Big]v\Big\rangle,
\end{aligned}$$

hence

$$\begin{aligned}
\|\mu(D_m^\varphi-\lambda)v\|^2 &\ge \|\nabla v\|^2 + 2\|(r\varphi')^{1/2}\partial_r v\|^2\\
&\quad +2\left\langle v, \left[\lambda^2 - m^2 + \varphi'^2 + 2r\varphi'\varphi'' - (n-1)\varphi'' - \frac{(n-1)^2\varphi'}{r} - \frac{r\varphi''^2}{\varphi'}\right]v\right\rangle\\
&\quad -4\left\langle v, \left[\frac{(2n-1)(|\lambda|+\varphi') + m + 2r|\varphi''|}{\mu}\right]^2 v\right\rangle,
\end{aligned}$$

and (VII.26) follows.

For $u\in H^1_{\mathrm{comp}}(\Omega^n_\rho,\mathbb{C}^N)$, substituting $v = e^\varphi u$ into (VII.26) and using the identity $D_m^\varphi(e^\varphi u) = e^\varphi D_m u$ (cf. (VII.13)), we also have

$$\text{(VII.36)} \quad \|\nabla(e^\varphi u)\|^2 + 2\|(r\varphi')^{1/2}\partial_r(e^\varphi u)\|^2 + \langle e^\varphi u, Z e^\varphi u\rangle \le \|\mu e^\varphi(D_m-\lambda)u\|^2$$

for any $u\in H^1_{\mathrm{comp}}(\Omega^n_\rho,\mathbb{C}^N)$, proving (VII.27). □

To use the inequality (VII.27) from Lemma VII.9, we need to bound $Z(r)$ from below. This is the purpose of the following lemma.

Lemma VII.10 *For any $\varphi \in \mathscr{C}_\lambda(M,\mathcal{N},\rho,\nu)$ there is $R_0 = R_0(\varphi) \ge \rho$ such that for any $r \ge R_0$ the following inequality is satisfied:*

$$Z(r) \ge \lambda^2 - m^2 + \varphi'(r)^2 + 2r\varphi'(r)\varphi''(r), \tag{VII.37}$$

with $Z(r)$ defined in (VII.25)*. If additionally*

$$\varphi''(r) \ge -\frac{\varphi'(r)}{4r}, \qquad r \ge \rho, \tag{VII.38}$$

then

$$R_0 = R_0(M,\rho,m,n,\lambda), \tag{VII.39}$$

independent of $\mathcal{N} > 0$ and ν, and it can be chosen uniformly in $\varphi \in \mathscr{C}_\lambda(M,\mathcal{N},\rho,\nu)$.

Proof. As follows from the definition (VII.25), we need to satisfy the following inequality:

$$2(n-1)|\varphi''(r)| + (n-1)^2\tfrac{\varphi'(r)}{r} + \tfrac{r\varphi''(r)^2}{\varphi'(r)} + 4\left[\tfrac{(2n-1)(|\lambda|+\varphi'(r))+m+2r|\varphi''(r)|}{\mu}\right]^2$$
$$\le \lambda^2 - m^2 + \varphi'(r)^2 + 2r\varphi'(r)\varphi''(r). \tag{VII.40}$$

Taking into account the bound $r|\varphi''| \le M\varphi'$ which takes place for all $r \ge \rho$ (see Definition VII.4) and simplifying some coefficients, we see that the inequality (VII.40) will follow from

$$2(M+n)^2\frac{\varphi'}{r} + 4\left[\frac{2\,(n|\lambda| + m + (M+n)\varphi')}{\mu}\right]^2 \le \lambda^2 - m^2 + \varphi'^2 + 2r\varphi'\varphi''.$$

Taking into account the bound $\mu \ge 8|\lambda|r$ which follows from (VII.3), we see that it suffices to satisfy the inequality

$$2(M+n)^2\frac{\varphi'}{r} + \frac{\big(n|\lambda| + m + (M+n)\varphi'\big)^2}{4\lambda^2 r^2} \le \lambda^2 - m^2 + \varphi'^2 + 2r\varphi'\varphi''. \tag{VII.41}$$

First we consider the case when φ' is bounded. Using the bound

$$\lambda^2 - m^2 + \varphi'^2 + 2r\varphi'\varphi'' \ge \nu, \qquad \forall r \ge \rho,$$

we will have (VII.41) satisfied for all $r \ge R$ as long as

$$2\frac{(M+n)^2}{r}\sup_{r\ge R}\varphi'(r) + \frac{\big(n|\lambda| + m + (M+n)\sup_{r\ge R}\varphi'(r)\big)^2}{4\lambda^2 r^2} \le \nu. \tag{VII.42}$$

The above inequality holds for all $r \ge R_0(\varphi)$ with $R_0(\varphi) = C\sup_{r\ge\rho}\varphi'(r)$ as long as the constant $C > 0$ is large enough, and then (VII.40) holds true.

Now we assume that the inequality (VII.38) is satisfied; this case includes the situation when φ' is unbounded (hence $\varphi'' \ge 0$ for all $r \ge \rho$ and $\lim_{r\to\infty}\varphi'(r) = +\infty$; see Definition VII.4). Due to this bound from below on $r\varphi''$, (VII.41) will be satisfied if we provide

$$2(M+n)^2\frac{\varphi'}{r} + \frac{\big(n|\lambda| + m + (M+n)\varphi'\big)^2}{4\lambda^2 r^2} \le \frac{\lambda^2 - m^2 + \varphi'^2}{2}. \tag{VII.43}$$

The above inequality will hold for all $\varphi' \ge 0$ as long as r is large enough to ensure that

$$\frac{\lambda^2 - m^2 + \zeta^2}{2} - 2(M+n)^2\frac{\zeta}{r} - \frac{\big(n|\lambda| + m + (M+n)\zeta\big)^2}{4\lambda^2 r^2} > 0, \qquad \forall\zeta \ge 0.$$

One can see that the required lower bound on r only depends on M, m, n, and $|\lambda|$. Note that one needs $|\lambda| > m$. □

Lemma VII.11 *Let $\varphi \in \mathscr{C}_\lambda(M, \mathcal{N}, \rho, \nu)$ and let $R_0 = R_0(\varphi)$ be as in Lemma* VII.10.

(1) There is the following inequality for any $u \in H^1_{\mathrm{comp}}(\Omega^n_{R_0}, \mathbb{C}^N)$:

$$\|\nabla(e^\varphi u)\|^2 + 2\|(r\varphi')^{1/2}\partial_r(e^\varphi u)\|^2 + \|\gamma e^\varphi u\|^2 \le \|\mu e^\varphi (D_m - \lambda)u\|^2, \tag{VII.44}$$

with μ and γ defined in (VII.3), (VII.4).

(2) Let $V \in \mathscr{B}\left(H^1_{\mathrm{comp}}(\Omega^n_\rho, \mathbb{C}^N), L^2(\Omega^n_\rho, \mathbb{C}^N)\right)$ be a multiplication operator and assume that there are $\kappa \in (0,1)$ and $R_1 = R_1(\varphi, V) > 0$ such that

$$\|\mu V v\| \le \kappa \left(\|\nabla v\|^2 + \|\gamma v\|^2\right)^{1/2}, \qquad \forall v \in H^1_{\mathrm{comp}}(\Omega^n_{R_1}, \mathbb{C}^N). \tag{VII.45}$$

Then

$$\begin{aligned}(1-\kappa)&\left[\|\nabla(e^\varphi u)\|^2 + 2\|(r\varphi')^{1/2}\partial_r(e^\varphi u)\|^2 + \|\gamma e^\varphi u\|^2\right]^{1/2} \\ &\le \|\mu e^\varphi (D_m + V - \lambda)u\|,\end{aligned} \tag{VII.46}$$

for any $u \in H^1_{\mathrm{comp}}(\Omega^n_R, \mathbb{C}^N)$ with $R = \max(R_0(\varphi), R_1(\varphi, V))$.

Proof. The proof of Lemma VII.11 *(1)* follows from Lemmata VII.9 and VII.10.

To prove Lemma VII.11 *(2)*, we apply the assumption (VII.45), where we take

$$v = e^\varphi u \in H^1_{\mathrm{comp}}(\Omega^n_R, \mathbb{C}^N), \qquad R = \max(R_0, R_1),$$

to the inequality (VII.44), obtaining

$$(1-\kappa)\left[\|\nabla(e^\varphi u)\|^2 + 2\|(r\varphi')^{\frac12}\partial_r(e^\varphi u)\|^2 + \|\gamma e^\varphi u\|^2\right]^{\frac12} \le \|\mu e^\varphi (D_m + V - \lambda)u\|.$$

This proves (VII.46). □

Let us extend (VII.27) to functions $u \in H^1_0(\Omega^n_\rho, \mathbb{C}^N)$ which are no longer compactly supported.

Lemma VII.12 *Let $\varphi \in \mathscr{C}_\lambda(M, \mathcal{N}, \rho, \nu)$.*

(1) Let $R_0 = R_0(\varphi) \ge \rho$ be as in Lemma VII.10, *so that* (VII.37) *is satisfied for $r \ge R_0$. Then, for any $u \in H^1_0(\Omega^n_{R_0}, \mathbb{C}^N)$, one has:*

$$\|\nabla(e^\varphi u)\|^2 + 2\|(r\varphi')^{1/2}\partial_r(e^\varphi u)\|^2 + \|\gamma e^\varphi u\|^2 \le \|\mu e^\varphi (D_m - \lambda)u\|^2. \tag{VII.47}$$

(2) Let $V \in L^n_{\mathrm{loc}}(\mathbb{R}^n, \mathrm{End}(\mathbb{C}^N))$ and assume that there are $\kappa \in (0,1)$ and $R_1 = R_1(\varphi, V) \ge \rho$ such that for any $v \in H^1_0(\Omega^n_{R_1}, \mathbb{C}^N)$ one has

$$\|\mu V v\| \le \kappa \left(\|\nabla v\|^2 + \|\gamma v\|^2\right)^{1/2}; \tag{VII.48}$$

if $V = 0$, then we set $R_1(\varphi, 0) = \rho$. Then for any $u \in H^1_0(\Omega^n_R, \mathbb{C}^N)$ with $R = \max(R_0(\varphi), R_1(\varphi, V))$, one has

$$\begin{aligned}(1-\kappa)&\left[\|\nabla(e^\varphi u)\|^2 + 2\|(r\varphi')^{1/2}\partial_r(e^\varphi u)\|^2 + \|\gamma e^\varphi u\|^2\right]^{1/2} \\ &\le \|\mu e^\varphi (D_m + V - \lambda)u\|.\end{aligned} \tag{VII.49}$$

Remark VII.13 It is enough to assume that the inequality (VII.48) takes place for $v \in H^1_{\mathrm{comp}}(\Omega^n_{R_1}, \mathbb{C}^N)$, since the latter implies (VII.48) for $v \in H^1_0(\Omega^n_{R_1}, \mathbb{C}^N)$ using ideas of **Step 0** of the proof below.

PROOF. We will prove Lemma VII.12 *(2)*; then Lemma VII.12 *(1)* will follow as well.

Step 0. First, we consider the case when $\varphi(r)$ is bounded. Let

$$\eta \in C^\infty_{\mathrm{comp}}([-2,2]), \qquad 0 \le \eta \le 1, \qquad \eta|_{[-1,1]} \equiv 1,$$

and let $\eta_j(x) = \eta(x/j)$, $x \in \mathbb{R}$. Let $u \in H^1(\mathbb{R}^n, \mathbb{C}^N)$ with $\operatorname{supp} u \subset \Omega^n_R$, and define

$$u_j = \eta_j u \in H^1_{\mathrm{comp}}(\Omega^n_R, \mathbb{C}^N).$$

Applying Lemma VII.11 *(2)* to u_j, we have

$$(1-\kappa)\Big[\|\nabla(e^\varphi u_j)\|^2 + 2\|(r\varphi')^{1/2}\partial_r(e^\varphi u_j)\|^2 + \|\gamma e^\varphi u_j\|^2\Big]^{1/2} \le \|\mu e^\varphi (D_m + V - \lambda) u_j\|.$$

Using the identities

$$\nabla(e^\varphi u_j) = \eta_j \nabla(e^\varphi u) + e^\varphi u \nabla \eta_j$$

and

$$(D_m + V - \lambda) u_j = \eta_j (D_m + V - \lambda) u - \mathrm{i}(\boldsymbol{\alpha}\cdot\nabla\eta_j) u,$$

one has

$$\begin{aligned}(1-\kappa)\Big[&\big(\|\eta_j\nabla(e^\varphi u)\| - \|e^\varphi u \nabla\eta_j\|\big)^2 \\ &+ 2\Big(\|\eta_j (r\varphi')^{1/2}\partial_r(e^\varphi u)\| - \|(r\varphi')^{1/2} e^\varphi u \partial_r\eta_j\|\Big)^2 + \|\gamma e^\varphi u \eta_j\|^2\Big]^{1/2} \\ &\le \|\mu e^\varphi \eta_j (D_m + V - \lambda) u\| + \|\mu e^\varphi (\boldsymbol{\alpha}\cdot\nabla\eta_j) u\|.\end{aligned} \tag{VII.50}$$

We claim that

$$\lim_{j\to\infty} \|e^\varphi u \nabla\eta_j\| = 0, \qquad \lim_{j\to\infty} \|(r\varphi')^{1/2} e^\varphi u \partial_r \eta_j\| = 0, \tag{VII.51}$$

$$\lim_{j\to\infty} \|\mu e^\varphi (\boldsymbol{\alpha}\cdot\nabla\eta_j) u\| = 0. \tag{VII.52}$$

Let us prove the inequality (VII.52). According to our assumptions, u is in L^2 and φ is bounded, hence $e^\varphi u \in L^2$ and $\lim_{j\to\infty} \|\mathbb{1}_{[j,2j]}(|x|) e^\varphi u\|_{L^2} = 0$, while

$$\|\mu \nabla\eta_j\|_{L^\infty} \le \|\nabla\eta_j\|_{L^\infty} \|\mathbb{1}_{[-2j,2j]}\mu\|_{L^\infty} = \frac{1}{j}\|\nabla\eta\|_{L^\infty} \|\mathbb{1}_{[-2j,2j]}\mu\|_{L^\infty} \tag{VII.53}$$

is bounded since $\mu(r) = O(\langle r\rangle)$ (cf. (VII.3)), due to φ assumed bounded, so that $\varphi' \to 0$ as $r \to \infty$. The proof of the second inequality in (VII.51) is the same since

$$(r\varphi')^{1/2} \le (\mathcal{N} r^2)^{1/2} = O(\langle r\rangle),$$

while the proof of the first inequality is slightly simpler since there is no linearly growing factor. Taking into account the limits (VII.51) and (VII.52) and applying the Fatou lemma to $\|\eta_j\nabla(e^\varphi u)\|$, $\|\gamma e^\varphi u \eta_j\|$, and $\|\mu e^\varphi \eta_j (D_m + V - \lambda) u\|$, we conclude from (VII.50) that, as we stated,

$$(1-\kappa)\Big(\|\nabla(e^\varphi u)\|^2 + 2\|(r\varphi')^{1/2}\partial_r(e^\varphi u)\|^2 + \|\gamma e^\varphi u\|^2\Big)^{1/2} \le \|\mu e^\varphi (D_m + V - \lambda) u\|. \tag{VII.54}$$

Unbounded φ are considered precisely as in [**BG87**, Theorem 3], closely following the approach of [**ABG82**]. We assume that

$$\varphi_0 := \varphi \in \mathscr{C}_\lambda(M, \mathcal{N}, \rho, \nu), \qquad \text{for some } M, \mathcal{N}, \rho \geq 1 \text{ and } \nu > 0.$$

Without loss of generality, we also assume that $\varphi_0(1) = 0$. Below we will consider sequences $(\varphi_\epsilon)_{\epsilon>0}$ converging pointwise to φ_0 from below.

Step 1. Let us assume that $\varphi \in \mathscr{C}_\lambda(M, \mathcal{N}, \rho, \nu)$ is unbounded but that $\varphi' \to 0$ as $r \to \infty$. We approximate $\varphi_0 = \varphi$ by $\varphi_\epsilon \in C^2(\mathbb{R}_+)$, $\epsilon \in (0, 1)$:

$$\varphi_\epsilon(r) = \int_1^r \frac{\varphi_0'(t)}{1 + \epsilon t^2}\, dt, \qquad r \in \mathbb{R}_+.$$

For each $\epsilon \in (0, 1)$, the function $\varphi_\epsilon(r)$, $r \geq \rho$, is monotonically increasing, satisfies $\sup_{r \geq 0} \varphi_\epsilon(r) < \infty$, and

$$\varphi_\epsilon(r) \nearrow \varphi_0(r), \qquad \varphi_\epsilon'(r) \nearrow \varphi_0'(r), \qquad \varphi_\epsilon''(r) \to \varphi_0''(r), \qquad r \geq \rho. \tag{VII.55}$$

To reduce the argument to **Step 0** (based on Lemma VII.11 *(2)*), we need to check that the inequality (VII.37) in Lemma VII.10 is satisfied by φ_ϵ and the corresponding Z_ϵ (given by (VII.25) with φ_ϵ instead of φ) for $r \geq R_0 = R_0(\varphi)$ as long as $\epsilon > 0$ is sufficiently small. Since

$$0 < \varphi_\epsilon'(r) = \frac{\varphi_0'(r)}{1 + \epsilon r^2} < \varphi_0'(r) \leq \mathcal{N} r, \qquad r \geq \rho, \tag{VII.56}$$

one has $\varphi_\epsilon''(r) = \frac{\varphi_0''(r)}{1+\epsilon r^2} - \frac{2\varphi_0'(r)}{r} \frac{\epsilon r^2}{(1+\epsilon r^2)^2}$ and hence

$$\frac{\varphi_\epsilon''(r)}{\varphi_\epsilon'(r)} = \frac{\varphi_0''(r)}{\varphi_0'(r)} - \frac{2}{r} \frac{\epsilon r^2}{1 + \epsilon r^2}, \qquad \left| \frac{r \varphi_\epsilon''(r)}{\varphi_\epsilon'(r)} \right| \leq \left| \frac{r \varphi_0''(r)}{\varphi_0'(r)} \right| + 2 \leq M + 2. \tag{VII.57}$$

We claim that for any $s < 1$ arbitrarily close to 1 there exists $\epsilon_0 = \epsilon_0(s) \in (0, 1)$ such that

$$\varphi_\epsilon \in \mathscr{C}_\lambda(M + 2, \mathcal{N}, R_0, s\nu), \qquad \forall \epsilon \in (0, \epsilon_0); \tag{VII.58}$$

according to Definition VII.4, we are left to verify that

$$\lambda^2 - m^2 + \varphi'^2 + 2r\varphi'\varphi'' \geq s\nu, \qquad r \geq R_0. \tag{VII.59}$$

Define

$$\nu_\epsilon = \inf_{r > \rho} \big(\lambda^2 - m^2 + \varphi_\epsilon'(r)^2 + 2r\varphi_\epsilon'(r)\varphi_\epsilon''(r)\big). \tag{VII.60}$$

Since

$$2r\varphi_\epsilon'\varphi_\epsilon'' = \frac{2r\varphi_0'\varphi_0''(r)}{(1 + \epsilon r^2)^2} - \frac{4\epsilon r^2 \varphi_0'(r)^2}{(1 + \epsilon r^2)^3},$$

one has

$$\begin{aligned} &\varphi_\epsilon'(r)^2 + 2r\varphi_\epsilon'(r)\varphi_\epsilon''(r) - \varphi_0'(r)^2 - 2r\varphi_0'(r)\varphi_0''(r) \\ &= \Big(\frac{1}{(1 + \epsilon r^2)^2} - 1\Big)\big(\varphi_0'(r)^2 + 2r\varphi_0'\varphi_0''(r)\big) - \frac{4\epsilon r^2 \varphi_0'(r)^2}{(1 + \epsilon r^2)^3}. \end{aligned} \tag{VII.61}$$

Due to $2r|\varphi_0''(r)| \leq M\varphi_0'(r)$ and $\varphi_0'(r) \to 0$ as $r \to \infty$, the absolute value of the right-hand side of (VII.61) goes to zero as $r \to \infty$ uniformly in $\epsilon \in (0, 1)$. Therefore, for any fixed $s \in (0, 1)$, we may choose some finite $R^* = R^*(s) \geq R_0$ (with $R_0 = R_0(\varphi)$ from Lemma VII.10) large enough so that the right-hand side of (VII.61) is smaller than $s\nu$ for $r \geq R^*$, $\epsilon \in (0, 1)$, while the left-hand side of (VII.41) (with φ_ϵ instead of φ and with $M+2$ instead of M) is smaller than $(1-s)\nu$. Then for $\epsilon \in (0,1)$ the functions φ_ϵ will satisfy (VII.41) (with $M+2$ instead of M) and hence (VII.37) for $r \geq R^*$.

At the same time, the convergences (VII.55) are uniform on the interval $r \in [R_0, R^*]$. Since φ satisfies the inequality (VII.37) for $r \geq R_0$, there is $\epsilon_0 = \epsilon_0(s) \in (0,1)$ sufficiently small so that the functions φ_ϵ with $\epsilon \in (0, \epsilon_0)$ satisfy

$$Z_\epsilon(r) \geq s\Big(\lambda^2 - m^2 + \varphi_\epsilon'(r)^2 + 2r\varphi_\epsilon'(r)\varphi_\epsilon''(r)\Big), \qquad r \in [R_0, R^*], \tag{VII.62}$$

where Z_ϵ is defined by (VII.25) with φ_ϵ instead of φ, while ν_ϵ defined in (VII.60) will satisfy

$$\nu_\epsilon > s\nu, \qquad \forall \epsilon \in (0, \epsilon_0). \tag{VII.63}$$

Together with (VII.56) and (VII.57), this leads to the desired inclusion (VII.58) and to the inequality (VII.62) satisfied for all $r \geq R_0 = R_0(\varphi)$.

The previous argument shows that there is $\epsilon_0 \in (0,1)$ so that for any $\epsilon \in (0, \epsilon_0)$,

$$\big|\varphi_\epsilon'(r)^2 + 2r\varphi_\epsilon'(r)\varphi_\epsilon''(r) - \varphi_0'(r)^2 - 2r\varphi_0'(r)\varphi_0''(r)\big| \leq (1-s)\nu, \quad r \geq R_0. \tag{VII.64}$$

Then notice that since (VII.48) is satisfied for φ_0, one has

$$\|\mu_\epsilon V v\| \leq \|\mu V v\| \leq \kappa \left(\|\nabla v\|^2 + \|\gamma v\|^2\right)^{1/2}, \qquad \forall v \in H_0^1(\Omega_{R_1}^n, \mathbb{C}^N), \tag{VII.65}$$

where μ_ϵ is the expression μ in (VII.3) for φ_ϵ and $R_1 = R_1(\varphi, V)$.

Due to (VII.65), we deduce that for all $\epsilon \in (0, \epsilon_0)$ and $v \in H_0^1(\Omega_{R_1}^n, \mathbb{C}^N)$, one has

$$\|\mu_\epsilon V v\| \leq \kappa \max\left\{1, \left\|\frac{\gamma}{\gamma_\epsilon}\right\|_{L^\infty(\operatorname{supp} v)}\right\} \left(\|\nabla v\|^2 + \|\gamma_\epsilon v\|^2\right)^{1/2}, \tag{VII.66}$$

where γ_ϵ is defined by the expression (VII.4) with φ_ϵ in place of φ. We notice that, in view of (VII.60), (VII.63), and (VII.64), for $\epsilon \in (0, \epsilon_0)$,

$$\max\left\{1, \left|\frac{\gamma}{\gamma_\epsilon}\right|\right\} \leq \max\left\{1, \frac{\gamma^2}{\gamma_\epsilon^2}\right\} \leq 1 + \frac{|\gamma^2 - \gamma_\epsilon^2|}{\gamma_\epsilon^2} \leq 1 + \frac{(1-s)\nu}{s\nu} = \frac{1}{s}, \qquad r \geq R_0,$$

hence (VII.66) yields

$$\|\mu_\epsilon V v\| \leq \frac{\kappa}{s}\left(\|\nabla v\|^2 + \|\gamma_\epsilon v\|^2\right)^{1/2}, \qquad \forall v \in H_0^1(\Omega_R^n, \mathbb{C}^N),$$

with $R = \max(R_0(\varphi), R_1(\varphi, V))$. Due to this bound, (VII.54) leads to the inequality

$$\begin{aligned}\left(1 - \frac{\kappa}{s}\right)\Big(\|\nabla(e^{\varphi_\epsilon}u)\|^2 + 2\|(r\varphi_\epsilon')^{1/2}\partial_r(e^{\varphi_\epsilon}u)\|^2 + \|\gamma e^{\varphi_\epsilon}u\|^2\Big)^{1/2}& \\ \leq \|\mu e^{\varphi_\epsilon}(D_m + V - \lambda)u\|,&\end{aligned} \tag{VII.67}$$

valid for any $\epsilon \in (0, \epsilon_0)$ and any $u \in H_0^1(\Omega_R^n, \mathbb{C}^N)$.

By the Fatou lemma applied to the left-hand side of (VII.67) (where we use the decomposition $\nabla(e^{\varphi_\epsilon}u) = e^{\varphi_\epsilon}\nabla u + e^{\varphi_\epsilon}(\nabla\varphi_\epsilon)u$) and the dominated convergence theorem applied to the right-hand side, the same inequality (VII.67) holds for $\varphi_0 = \varphi$ instead of φ_ϵ. Since $s \in (0,1)$ could be chosen arbitrarily close to 1, we also have

$$(1-\kappa)\Big(\|\nabla(e^{\varphi}u)\|^2 + 2\|(r\varphi')^{1/2}\partial_r(e^{\varphi}u)\|^2 + \|\gamma e^{\varphi}u\|^2\Big)^{1/2} \le \|\mu e^{\varphi}(D_m + V - \lambda)u\|,$$

for any $u \in H_0^1(\Omega_R^n, \mathbb{C}^N)$.

Step 2. Assume that $\varphi \in \mathscr{C}_\lambda(M, \mathcal{N}, \rho, \nu)$ is such that φ' is bounded at infinity. Let $R_0 = R_0(\varphi) \ge \rho$, $R_0 > 0$ be as in Lemma VII.10, so that φ satisfies the inequality (VII.37) for $r \ge R_0$. (Since φ' is bounded, such R_0 exists by (VII.42).) We approximate $\varphi_0 := \varphi$ by $\varphi_\epsilon \in C^2(\mathbb{R}_+)$, $\epsilon \in (0, 1/4)$, as follows:

$$\varphi_\epsilon(r) = \varphi_0(r^{1-\epsilon}), \qquad r \in \mathbb{R}_+.$$

Since $\varphi_0(r)$ is increasing and $\rho \ge 1$, for each $\epsilon \in (0, 1/4)$ and $r \ge \rho$ we have:

$$\varphi_\epsilon(r) \le \varphi_0(r), \qquad \varphi_\epsilon'(r) > 0, \qquad \lim_{r\to\infty} \varphi_\epsilon'(r) = 0;$$

$$\varphi_\epsilon(r) \nearrow \varphi_0(r), \qquad \varphi_\epsilon'(r) \to \varphi_0'(r) \qquad \text{as} \quad \epsilon \searrow 0.$$

Thus,

$$0 < \varphi_\epsilon'(r) = (1-\epsilon)r^{-\epsilon}\varphi_0'(r^{1-\epsilon}) \le (1-\epsilon)r^{-\epsilon}\mathcal{N}r^{1-\epsilon} \le \mathcal{N}r,$$

$$\varphi_\epsilon''(r) = (1-\epsilon)^2 r^{-2\epsilon}\varphi_0''(r^{1-\epsilon}) - \epsilon(1-\epsilon)r^{-\epsilon-1}\varphi_0'(r^{1-\epsilon}),$$

$$\frac{\varphi_\epsilon''(r)}{\varphi_\epsilon'(r)} = (1-\epsilon)r^{-\epsilon}\frac{\varphi_0''(r^{1-\epsilon})}{\varphi_0'(r^{1-\epsilon})} - \epsilon r^{-1},$$

$$\left|\frac{r\varphi_\epsilon''(r)}{\varphi_\epsilon'(r)}\right| \le (1-\epsilon)\left|\frac{r^{1-\epsilon}\varphi_0''(r^{1-\epsilon})}{\varphi_0'(r^{1-\epsilon})}\right| + \epsilon \le M.$$

In the last inequality, we took into account that $M \ge 1$.

From the identity

$$\begin{aligned}
&\varphi_\epsilon'(r)^2 + 2r\varphi_\epsilon'(r)\varphi_\epsilon''(r)\\
&= (1-\epsilon)^2(1-2\epsilon)r^{-2\epsilon}(\varphi_0'(r^{1-\epsilon}))^2 + 2r(1-\epsilon)^3 r^{-3\epsilon}\varphi_0'(r^{1-\epsilon})\varphi_0''(r^{1-\epsilon})\\
&= (1-\epsilon)^2(1-2\epsilon)r^{-2\epsilon}\left[(\varphi_0'(r^{1-\epsilon}))^2 + 2r\varphi_0'(r^{1-\epsilon})\varphi_0''(r^{1-\epsilon})\right]\\
&\quad + \left[(1-\epsilon)^3 r^{-3\epsilon} - (1-\epsilon)^2(1-2\epsilon)r^{-2\epsilon}\right] 2r\varphi_0'(r^{1-\epsilon})\varphi_0''(r^{1-\epsilon})
\end{aligned}$$

we deduce as in the previous step that for any $s \in (0,1)$ there is $\epsilon_0 = \epsilon_0(s) \in (0, 1/4)$ such that for $\epsilon \in (0, \epsilon_0)$ one has $\varphi_\epsilon \in \mathscr{C}_\lambda(M, \mathcal{N}, R_0(\varphi), s\min\{\nu, \lambda^2 - m^2\})$. We also deduce that γ_ϵ converges uniformly to γ_0.

Now we consider the term

$$\mu_\epsilon(r) = 2\Big(n + 16\lambda^2 r^2 + 8r\varphi_\epsilon'\Big)^{1/2}, \qquad r \ge \rho,$$

rewritten as

$$\mu_\epsilon(r) = 2r\mathsf{p}_\epsilon(r), \qquad r \ge \rho,$$

with

$$\mathsf{p}_\epsilon(r) = \Big(\frac{n}{r^2} + 16\lambda^2 + 8\frac{\varphi_\epsilon'}{r}\Big)^{1/2}, \qquad r \ge \rho.$$

Due to the local uniform convergence of φ'_ϵ to $\varphi'_0 = \varphi$, ρ_ϵ converges locally uniformly to ρ_0. Since φ'_ϵ is bounded at infinity uniformly in ϵ, $\rho_\epsilon - \rho_0$ is small at infinity uniformly in ϵ. Hence $\rho_\epsilon - \rho_0$ is smaller than ρ_0 (which is bounded from below by $4|\lambda|$) multiplied by an arbitrarily small constant, uniformly in ϵ. We thus obtain from (VII.48) that for any $\kappa' \in (0,\kappa)$ there exists $\epsilon_1 = \epsilon_1(s,\kappa') \in (0,\epsilon_0(s))$ such that, for any $\epsilon \in (0,\epsilon_1)$ and $v \in H^1_0(\Omega^n_R, \mathbb{C}^N)$, with $R = \max(R_0(\varphi), R_1(\varphi, V))$, one has

$$\|\mu_\epsilon V v\| \le \kappa \left(\|\nabla v\|^2 + \|\gamma v\|^2\right)^{1/2} \le \kappa' \left(\|\nabla v\|^2 + \|\gamma_\epsilon v\|^2\right)^{1/2}.$$

This allows us to conclude the argument as in **Step 1**, proving that

$$(1-\kappa')\left(\|\nabla(e^{\varphi_\epsilon}u)\|^2 + 2\|(r\varphi'_\epsilon)^{1/2}\partial_r(e^{\varphi_\epsilon}u)\|^2 + \|\gamma e^{\varphi_\epsilon}u\|^2\right)^{1/2} \le \|\mu e^{\varphi_\epsilon}(D_m + V - \lambda)u\|,$$

with $R = \max(R_0(\varphi), R_1(\varphi, V))$ independent of $\epsilon \in (0,\epsilon_1)$. Using the same reasoning as above, we conclude that this inequality is also satisfied by $\varphi_0 = \varphi$. Finally, since $\kappa' \lessapprox \kappa$ could be chosen arbitrarily close to κ, we also have

$$(1-\kappa)\left(\|\nabla(e^{\varphi}u)\|^2 + 2\|(r\varphi')^{1/2}\partial_r(e^{\varphi}u)\|^2 + \|\gamma e^{\varphi}u\|^2\right)^{\frac{1}{2}} \le \|\mu e^{\varphi}(D_m + V - \lambda)u\|.$$

Step 3. Assume that $\varphi \in \mathscr{C}_\lambda(M, \mathcal{N}, \rho, \nu)$ is such that φ' is unbounded at infinity (this implies that $\varphi'' \ge 0$). It follows from Lemma VII.10 that there is $R_0 = R_0(M, \rho, m, n, \lambda)$, $R_0 \ge \rho$, such that φ satisfies the inequality (VII.37) for all $r \ge R_0$.

We approximate $\varphi_0 := \varphi$ by $\varphi_\epsilon \in C^2(\mathbb{R}_+)$, $\epsilon \in (0,1)$:

$$\varphi_\epsilon(r) = \int_1^r \frac{\varphi'_0(t)}{1+\epsilon\varphi'_0(t)}\,dt, \qquad r \in \mathbb{R}_+.$$

Then φ_ϵ are monotonically increasing, satisfy $\sup_{r\ge 0}\varphi'_\epsilon(r) < \infty$, and, for each $r \ge \rho$, $\varphi_\epsilon(r) \nearrow \varphi_0(r)$, $\varphi'_\epsilon(r) \nearrow \varphi'_0(r)$. Moreover, for $r \ge \rho$, the following inequalities hold:

$$0 < \varphi'_\epsilon(r) = \frac{\varphi'_0(r)}{1+\epsilon\varphi'_0(r)} \le \mathcal{N} r,$$

$$\varphi''_\epsilon(r) = \frac{\varphi''_0(r)}{1+\epsilon\varphi'_0(r)} - \frac{\epsilon\varphi'_0(r)\varphi''_0(r)}{(1+\epsilon\varphi'_0(r))^2} = \frac{\varphi''_0(r)}{(1+\epsilon\varphi'_0(r))^2} \ge 0,$$

$$\frac{\varphi''_\epsilon(r)}{\varphi'_\epsilon(r)} = \frac{\varphi''_0(r)}{\varphi'_0(r)}\frac{1}{1+\epsilon\varphi'_0(r)}, \qquad \left|\frac{r\varphi''_\epsilon(r)}{\varphi'_\epsilon(r)}\right| \le M.$$

One has

$$(\varphi'_\epsilon(r))^2 + 2r\varphi'_\epsilon(r)\varphi''_\epsilon(r) = \frac{\varphi'_0(r)^2 + 2r\varphi'_0(r)\varphi''_0(r)}{(1+\epsilon\varphi'_0(r))^3} + \frac{\epsilon\varphi'_0(r)^3}{(1+\epsilon\varphi'_0(r))^3};$$

in the case $\epsilon\varphi' < 2^{1/3} - 1$, one concludes that

$$\lambda^2 - m^2 + (\varphi'_\epsilon(r))^2 + 2r\varphi'_\epsilon(r)\varphi''_\epsilon(r) \ge \lambda^2 - m^2 + \frac{\nu - (\lambda^2 - m^2)}{2} \ge \frac{\lambda^2 - m^2 + \nu}{2},$$

while in the case $\epsilon\varphi' \ge 2^{1/3} - 1$, one deduces

$$\lambda^2 - m^2 + (\varphi_\epsilon'(r))^2 + 2r\varphi_\epsilon'(r)\varphi_\epsilon''(r) \ge \lambda^2 - m^2 - \frac{|\nu - (\lambda^2 - m^2)|}{2} + \frac{(2^{\frac{1}{3}} - 1)^3}{2\epsilon^2},$$

which is also larger than $\frac{\lambda^2 - m^2 + \nu}{2}$ provided that $\epsilon \in (0, \epsilon_0)$, with $\epsilon_0 \in (0,1)$ sufficiently small. One concludes that $\varphi_\epsilon \in \mathscr{C}_\lambda(M, \mathcal{N}, \rho, \frac{\lambda^2 - m^2 + \nu}{2})$ for all $\epsilon \in (0, \epsilon_0)$ and uses the result and the ideas from the previous step to prove the inequality

$$\begin{aligned}(1-\kappa)\left(\|\nabla(e^{\varphi_\epsilon}u)\|^2 + 2\|(r\varphi_\epsilon')^{1/2}\partial_r(e^{\varphi_\epsilon}u)\|^2 + \|\gamma e^{\varphi_\epsilon}u\|^2\right)^{\frac{1}{2}} \\ \le \|\mu e^{\varphi_\epsilon}(D_m + V - \lambda)u\|, \qquad \forall u \in H_0^1(\Omega_R^n, \mathbb{C}^N),\end{aligned}$$

with $R = \max(R_0(\varphi), R_1(\varphi, V))$ independent of $\epsilon \in (0, \epsilon_0)$, and applies the Fatou lemma and the dominated convergence theorem to the above inequality to prove (VII.47). □

Theorem VII.5 *(1)* follows from Lemma VII.12 *(2)*.

Now we proceed to the proof of Theorem VII.5 *(2)*. Let $\varphi \in \mathscr{C}_\lambda(M, \mathcal{N}, \rho, \nu)$, with $\varphi''(r) \ge 0$ for $r \ge \rho$. Then, by Lemma VII.10, the inequality (VII.37) is satisfied for $r \ge R_0 = R_0(M, \rho, m, n, \lambda)$, independent of a particular representative $\varphi \in \mathscr{C}_\lambda(M, \mathcal{N}, \rho, \nu)$. At the same time, using (VII.10) and taking into account that $\varphi'' \ge 0$, we see that the inequality (VII.5) is satisfied for $r \ge R_1 = R_1(\varphi, V)$ if one chooses $R_1 = R_1(\varphi, V) = R_2(V)$, which would not depend on φ. As follows from Lemma VII.12, the inequality (VII.49) is satisfied for $u \in H_0^1(\Omega_R^n, \mathbb{C}^N)$ with

$$R = \max\big(R_0(\varphi), R_1(\varphi, V)\big) = \max\big(R_0(M, \rho, m, n, \lambda), R_2(V)\big)$$

independent of a particular φ. This finishes the proof of Theorem VII.5. □

VII.3 Carleman–Berthier–Georgescu estimates for $J(D_m - \omega + V)$

Similarly to Definition VII.4, we define the following set of admissible functions.

Definition VII.14 Let $\lambda \in \mathbb{C}$ and $\omega \in [-m, m]$ be such that

$$|\lambda| - |\omega| > m.$$

Let $M \ge 1$, $\mathcal{N} \ge 1$, $\rho \ge 1$ and $\nu > 0$. We define the subset

$$\mathscr{C}_{\lambda,\omega}(M, \mathcal{N}, \rho, \nu) \subset C^2(\mathbb{R}_+)$$

to be the set of functions which satisfy the following properties:

(1) $0 < \varphi'(r) \le \mathcal{N} r, \ \forall r \ge \rho$;
(2) $r|\varphi''(r)| \le M\varphi'(r), \ \forall r \ge \rho$;
(3) $(|\lambda| - |\omega|)^2 - m^2 + \varphi'^2 + 2r\varphi'\varphi'' \ge \nu, \ \forall r \ge \rho$;
(4) if φ' is unbounded then $\varphi''(r) \ge 0$ for $r \ge \rho$ and $\lim_{r\to\infty}\varphi'(r) = +\infty$.

We note that

$$\mathscr{C}_\lambda(M, \mathcal{N}, \rho, \nu) := \mathscr{C}_{\lambda,0}(M, \mathcal{N}, \rho, \nu),$$

in agreement with Definition VII.4. We define

$$\Lambda_+ = |\lambda| + |\omega|, \qquad \Lambda_- = |\lambda| - |\omega| > m, \tag{VII.68}$$

and introduce the weight functions

$$\mu_\pm(r) = 2\left(n + 16\Lambda_\pm^2 r^2 + 8r\varphi'(r)\right)^{1/2}, \qquad r \ge \rho; \tag{VII.69}$$

$$\gamma_\pm(r) = \left(\Lambda_\pm^2 - m^2 + \varphi'(r)^2 + 2r\varphi'(r)\varphi''(r)\right)^{1/2}, \qquad r \ge \rho. \tag{VII.70}$$

Theorem VII.15 (Carleman–Berthier–Georgescu estimates for $J(D_m - \omega + V)$) *Let $n \ge 1$, $m > 0$, $\omega \in [-m, m]$. Let $J \in \operatorname{End}(\mathbb{C}^N)$ be skew-adjoint and invertible, such that $J^2 = -I_N$, $[J, D_m] = 0$. Let*

$$\lambda \in \mathrm{i}\mathbb{R}, \qquad |\lambda| > m + |\omega|.$$

(1) *Assume that $\varphi \in \mathscr{C}_{\lambda,\omega}(M, \mathcal{N}, \rho, \nu)$ with some $M \ge 1$, $\mathcal{N} \ge 1$, $\rho \ge 1$, and $\nu > 0$.*

Let $V \in \mathscr{B}\left(H^1_{\mathrm{comp}}(\Omega^n_\rho, \mathbb{C}^N),\, L^2(\Omega^n_\rho, \mathbb{C}^N)\right)$ be a multiplication operator (an operator of multiplication by a matrix-valued function), and assume that there are $\kappa \in (0,1)$ and $R_1 = R_1(\varphi, V) \ge \rho$ such that

$$\|\mu_+ V v\| \le \kappa \left(\|\nabla v\|^2 + \|\gamma_- v\|^2\right)^{1/2}, \qquad \forall v \in H^1_0(\Omega^n_{R_1}, \mathbb{C}^N). \tag{VII.71}$$

Then there is $R = R(\varphi, V) \ge R_1$ such that for any $u \in H^1(\mathbb{R}^n, \mathbb{C}^N)$ with $\operatorname{supp} u \subset \Omega^n_R$ and

$$\mu_+ e^\varphi (J(D_m - \omega + V) - \lambda) u \in L^2(\mathbb{R}^n, \mathbb{C}^N),$$

the functions $\gamma_- e^\varphi u$, $\nabla(e^\varphi u)$, $(r\varphi')^{1/2}\partial_r(e^\varphi u)$ are in $L^2(\mathbb{R}^n, \mathbb{C}^N)$, and moreover

$$\begin{aligned} &\|\nabla(e^\varphi u)\|^2 + 2\|(r\varphi')^{1/2}\partial_r(e^\varphi u)\|^2 + \|\gamma_- e^\varphi u\|^2 \\ &\le \frac{1}{(1-\kappa)^2}\|\mu_+ e^\varphi (J(D_m - \omega + V) - \lambda)u\|^2. \end{aligned} \tag{VII.72}$$

(2) *Assume that there are $\varkappa \in (0,1)$ and $R_2(V) > 0$ such that*

$$\|V(x)\|_{\operatorname{End}(\mathbb{C}^N)} \le \varkappa \frac{\sqrt{\Lambda_-^2 - m^2}}{4\Lambda_+ |x|}, \qquad \forall x \in \Omega^n_{R_2(V)}. \tag{VII.73}$$

Let $M \ge 1$, $\mathcal{N} \ge 1$, $\rho \ge 1$, $\nu > 0$, and $\kappa \in (\varkappa, 1)$. Then there is

$$R = R(M, \rho, m, n, \lambda, \omega, V) \ge \max(\rho, R_2(V))$$

such that for any $\varphi \in \mathscr{C}_{\lambda,\omega}(M, \mathcal{N}, \rho, \nu)$ with $\varphi''(r) \ge 0\ \forall r \ge \rho$ and for any

$$u \in H^1_0(\Omega^n_R, \mathbb{C}^N)$$

which satisfies

$$\mu_+ e^\varphi (J(D_m - J + V) - \lambda) u \in L^2(\mathbb{R}^n, \mathbb{C}^N),$$

the functions $\gamma_- e^\varphi u$, $\nabla(e^\varphi u)$, $(r\varphi')^{1/2}\partial_r(e^\varphi u)$ are in $L^2(\mathbb{R}^n, \mathbb{C}^N)$, and moreover

$$\begin{aligned} &\|\nabla(e^\varphi u)\|^2 + 2\|(r\varphi')^{1/2}\partial_r(e^\varphi u)\|^2 + \|\gamma_- e^\varphi u\|^2 \\ &\le \frac{1}{(1-\kappa)^2}\|\mu_+ e^\varphi (J(D_m - \omega + V) - \lambda)u\|^2. \end{aligned} \tag{VII.74}$$

Remark VII.16 That is, in Theorem VII.15 *(2)* we state that if $\varphi'' \ge 0$, then $R(\varphi, V)$ from Theorem VII.15 *(1)* depends on the class $\mathscr{C}_{\lambda,\omega}(M, \mathcal{N}, \rho, \nu)$, but not on a particular representative $\varphi \in \mathscr{C}_{\lambda,\omega}(M, \mathcal{N}, \rho, \nu)$.

Proof. In the proof, one needs the substitute for the inequality (VII.10), which comes from the following lemma:

Lemma VII.17 *Let $n \geq 1$. Assume that there is $R_2(V) > 0$ and $\varkappa \in (0,1)$ such that*

$$\|V(x)\|_{\mathrm{End}(\mathbb{C}^N)} \leq \varkappa \frac{\sqrt{\Lambda_-^2 - m^2}}{4\Lambda_+|x|}, \qquad \forall x \in \Omega^n_{R_2(V)}, \tag{VII.75}$$

with $\Lambda_\pm = |\lambda| \pm |\omega|$ from (VII.68). *Then for any $\kappa \in (\varkappa, 1)$ we can take $R_2(V)$ larger if necessary so that*

$$(n + 16\Lambda_+^2 r^2 + 8r\theta)\|V(x)\|^2_{\mathrm{End}(\mathbb{C}^N)} \leq \kappa^2 \left(\Lambda_-^2 - m^2 + \theta^2\right), \quad \forall x \in \Omega^n_{R_2(V)}, \quad \forall \theta \geq 0.$$

In particular,

$$\|\left(n + 16\Lambda_+^2 r^2 + 8r\theta\right)^{\frac{1}{2}} V v\| \leq \kappa \|\left(\Lambda_-^2 - m^2 + \theta^2\right)^{\frac{1}{2}} v\|, \qquad \forall \theta \geq 0, \tag{VII.76}$$

valid for all $v \in H_0^1(\Omega^n_{R_2(V)}, \mathbb{C}^N)$.

The proof is an immediate adaptation of that of Lemma VII.6.

We first give the proof for compactly supported functions $u \in H^1_{\mathrm{comp}}(\Omega^n_R, \mathbb{C}^N)$. For the functions $u_\pm = \Pi^\pm u$, with

$$\Pi^\pm = \frac{1}{2}(1 \mp \mathrm{i}J)$$

the projections onto eigenspaces of J corresponding to $\pm\mathrm{i} \in \sigma(J)$, we obtain the following inequalities from Lemma VII.11 *(1)*:

$$\|e^\varphi \nabla u_+\|^2 + 2\|(r\varphi')^{1/2} u_+\| + \|\gamma_+ e^\varphi u_+\|^2 \leq \|\mu_+ e^\varphi((D_m - \omega) + \mathrm{i}\lambda)u_+\|^2,$$
$$\|e^\varphi \nabla u_-\|^2 + 2\|(r\varphi')^{1/2} u_-\| + \|\gamma_- e^\varphi u_-\|^2 \leq \|\mu_- e^\varphi((D_m - \omega) - \mathrm{i}\lambda)u_-\|^2,$$

with $\mu_\pm$ and $\gamma_\pm$ from (VII.69) and (VII.70), respectively. Since the projectors $\Pi^\pm$ are self-adjoint, we may add up the above inequalities, arriving at

$$\|e^\varphi \nabla u\|^2 + 2\|(r\varphi')^{1/2} u\| + \|\min_\pm(\gamma_\pm) e^\varphi u\|^2 \leq \|\max_\pm(\mu_\pm) e^\varphi (J(D_m - \omega) - \lambda)u\|^2.$$

It remains to notice that $\max_\pm(\mu_\pm) = \mu_+$ and that $\min_\pm(\gamma_\pm) = \gamma_-$; we arrive at (VII.72) (for $u \in H^1(\Omega^n_R, \mathbb{C}^N)$ compactly supported and with $V = 0$).

Introducing V into the estimates for $J(D_m - \omega)$ is done as in Lemma VII.11 *(2)* for D_m. The extension to $u \in H_0^1(\Omega^n_\rho, \mathbb{C}^N)$ which are not necessarily compactly supported is done as for $D_m + V$ in Lemma VII.12. □

VII.4 Absence of embedded eigenstates

The immediate consequence of the Carleman–Berthier–Georgescu estimates is the following result on the absence of embedded eigenstates for $L = D_m - \omega + V$ and JL, for rather general potentials V:

Theorem VII.18 *Let $n \geq 1$, $\omega \in [-m, m]$, and*

$$V \in L^p_{\mathrm{loc}}(\mathbb{R}^n, \mathrm{End}(\mathbb{C}^N)), \qquad \textit{with} \quad p = \begin{cases} \frac{3n-2}{2}, & n \geq 3; \\ \frac{7}{2}, & n = 2; \\ 1, & n = 1. \end{cases} \tag{VII.77}$$

(1) *Let* $\lambda \in \mathbb{R} \setminus [-m-\omega, m-\omega]$ *and assume that there are* $\varkappa \in (0,1)$ *and* $R > 0$ *such that*

$$\|V(x)\|_{\mathrm{End}(\mathbb{C}^N)} \le \varkappa \frac{\sqrt{\Lambda_-^2 - m^2}}{4\Lambda_+ |x|}, \qquad \forall x \in \Omega_R^n, \tag{VII.78}$$

where

$$\Lambda_+ = |\lambda| + |\omega|, \qquad \Lambda_- = |\lambda| - |\omega|.$$

Then $\lambda \notin \sigma_{\mathrm{p}}(D_m - \omega + V)$.

(2) *Let* $J \in \mathrm{End}(\mathbb{C}^N)$ *be skew-adjoint and invertible, and assume that it satisfies* $J^2 = -I_N$ *and* $[J, D_m] = 0$. *Let*

$$\lambda \in \mathbb{R} \setminus [-m - |\omega|, m + |\omega|]$$

and assume that there are $\varkappa \in (0,1)$ *and* $R > 0$ *such that*

$$\|V(x)\|_{\mathrm{End}(\mathbb{C}^N)} \le \varkappa \frac{\sqrt{\Lambda_-^2 - m^2}}{4\Lambda_+ |x|}, \qquad \forall x \in \Omega_R^n. \tag{VII.79}$$

Then

$$\pm \mathrm{i}\lambda \notin \sigma_{\mathrm{p}}\left(J(D_m - \omega + V)\right).$$

PROOF. Let us prove Theorem VII.18 *(2)*. Let $L = D_m - \omega + V$ and assume that $\lambda \in \mathrm{i}\mathbb{R}$, $|\lambda| > m + |\omega|$, is an embedded eigenvalue of JL, with $\zeta \in L^2(\mathbb{R}^n, \mathbb{C}^N)$ the corresponding eigenvector.

We are going to use Theorem VII.15 *(2)*, where we take $\varphi(r) = \tau r$ with $\tau \ge 1$. We note that, due to (VII.79), Assumption (VII.73) in Theorem VII.15 *(2)* is satisfied. Let $\kappa \in (0,1)$ and $R = R(1, 1, m, n, \lambda, \omega, V)$ be as in Theorem VII.15 *(2)* (note that this value of R is independent of $\tau \ge 1$). Let $\Theta \in C^\infty(\mathbb{R}^n)$ be a smooth radially symmetric cut-off function with support in the closure of Ω_{R+1}^n and with value 1 in Ω_{R+2}^n. By Theorem VII.15 *(2)*,

$$\begin{aligned} &\|((|\lambda| - |\omega|)^2 - m^2 + \tau^2)^{1/2} e^{\tau r} \Theta \zeta\| \\ &\le \frac{1}{1-\kappa} \|2(n + 16(|\lambda| + |\omega|)^2 r^2 + 8\tau r)^{1/2} e^{\tau r} (JL - \lambda) \Theta \zeta\|. \end{aligned} \tag{VII.80}$$

Since $JL\zeta = \lambda\zeta$, we have $(JL - \lambda)\Theta\zeta = [JL, \Theta]\zeta = J(D_0\Theta)\zeta$. Using (VII.80) and taking into account that $(|\lambda| - |\omega|)^2 - m^2 > 0$, we have:

$$\forall \tau \ge 1, \qquad \|e^{\tau r}\Theta\zeta\| \le \frac{2}{1-\kappa} \left\| \left(\frac{n}{\tau^2} + 16(|\lambda| + |\omega|)^2 \frac{r^2}{\tau^2} + 8\frac{r}{\tau} \right)^{1/2} e^{\tau r} (D_0\Theta)\zeta \right\|.$$

Taking into account that $D_0\Theta = -\mathrm{i}\boldsymbol{\alpha} \cdot \nabla\Theta$ is identically zero outside of the ball $\mathbb{B}_{R+2}^n$, we conclude that

$$\forall \tau \ge 1, \qquad \|e^{\tau r}\zeta\|_{L^2(\Omega_{R+2}^n, \mathbb{C}^N)} \le C e^{(R+2)\tau} \|r\nabla\Theta\|_{L^\infty} \|\zeta\|_{L^2(\mathbb{B}_{R+2}^n, \mathbb{C}^N)},$$

with $C > 0$ independent of $\tau \ge 1$. Since τ could be arbitrarily large, we conclude that $\operatorname{supp}\zeta \cap \Omega_{R+2}^n = \varnothing$.

We now use the unique continuation property for the Dirac operator, which states that if $|D_0\psi(x)| \le |V(x)\psi(x)|$ for almost all x from a connected open neighborhood $\Omega \subset \mathbb{R}^n$ and moreover ψ vanishes identically in an open subset of Ω, then

$\psi = 0$ almost everywhere in Ω, as long as V satisfies certain conditions:

- for $n \geq 3$, by [**Jer86**, Corollary 2], one needs
$$V \in L^p(\mathbb{R}^n, \mathrm{End}(\mathbb{C}^N)), \qquad p = \frac{3n-2}{2};$$
- for $n = 2$, the conclusion follows from the three-dimensional case, by extending in the relation $D_0\psi = V\psi$ the functions V and ψ of $(x, y) \in \mathbb{R}^2$ trivially to functions of $(x, y, z) \in \mathbb{R}^3$ and using the above-mentioned three-dimensional result, which imposes the restriction for V to be in $L^p_{\mathrm{loc}}(\mathbb{R}^2, \mathrm{End}(\mathbb{C}^N))$ with $p = \frac{7}{2}$;
- for $n = 1$, the requirement $V \in L^1_{\mathrm{loc}}(\mathbb{R}, \mathrm{End}(\mathbb{C}^N))$ follows from the ODE theory; see Problem VII.20 below.

Since ζ vanishes identically in Ω^n_{R+2} while V is assumed to be sufficiently regular, the unique continuation property ensures that $\zeta \equiv 0$ everywhere in $\mathbb{R}^n$, contradicting our assumption that there were an embedded eigenvalue $\lambda \in \mathrm{i}\mathbb{R}$, $|\lambda| > m + |\omega|$.

Theorem VII.18 *(1)* is proved similarly, by using Theorem VII.5 *(2)* instead of Theorem VII.15 *(2)*. □

Remark VII.19 The value of p in (VII.77) is likely not optimal; the limitation is due to the absence of the optimal results on the unique continuation property for the Dirac operator.

Problem VII.20 Complete the proof of Theorem VII.18 in the case $n = 1$ by establishing a unique continuation result for $V \in L^1_{\mathrm{loc}}(\mathbb{R}, \mathrm{End}(\mathbb{C}^N))$.
Hint: Use Grönwall's lemma.

VII.5 Exponential decay of eigenstates

Let us study the exponential decay of eigenfunctions of the Dirac operator. We start with the following inequality adapted from [**BG87**, Theorem 1].

Lemma VII.21 *Let $\lambda \in (-m; m)$ and $R_0 > 0$. Let $\varphi : [R_0, +\infty) \to \mathbb{R}$ be a monotonically increasing C^1 function such that*

$$\limsup_{r\to\infty} \varphi'(r) < \sqrt{m^2 - \lambda^2}. \tag{VII.81}$$

Assume that $V : \Omega^n_{R_0} \to \mathrm{End}(\mathbb{C}^N)$ satisfies the following condition:
(VII.82)
$$\forall \varepsilon > 0 \quad \exists R > 0 \quad \textit{such that} \quad \|Vu\| \leq \varepsilon \|u\|_{H^1} \quad \forall u \in H^1_{\mathrm{comp}}(\Omega^n_R, \mathbb{C}^N).$$

Then there are constants c, R such that

$$\|e^\varphi u\|_{H^1} \leq c\|e^\varphi (D_m + V - \lambda)u\| \quad \forall u \in L^2(\Omega^n_R, \mathbb{C}^N) \cap H^1_{0,\mathrm{loc}}(\Omega^n_R, \mathbb{C}^N). \tag{VII.83}$$

Above, the notation $H^k_{0,\mathrm{loc}}(\Omega)$, $k \in \mathbb{N}_0$, stands for the set of functions $u \in H^k(\Omega)$ such that $\eta u \in H^k_0(\Omega)$ for any $\eta \in C^\infty_{\mathrm{comp}}(\mathbb{R}^n)$.

Remark VII.22 Since $\|D_m u\| = \|(-\Delta^2 + m^2)^{1/2} u\| \geq m\|u\|$, the assumption (VII.82) is weaker than $\|V(x)\|_{\mathrm{End}(\mathbb{C}^N)} \to 0$ as $|x| \to \infty$.

Proof. Without loss of generality, we may assume that $R_0 = 1$. The general case follows by the same ideas and can be recovered by rescaling.

We deduce from Lemma VII.7 that for any $v \in H^1_{\mathrm{comp}}(\mathbb{R}^n, \mathbb{C}^N)$,

$$\mathrm{Re}\langle (D_m - \mathrm{i}\boldsymbol{\alpha}\cdot\nabla\varphi + \lambda)v, (D^\varphi_m - \lambda)v\rangle = \|\nabla v\|^2 + \langle v, [m^2 - \lambda^2 - (\nabla\varphi)^2]v\rangle, \tag{VII.84}$$

where $D_m^\varphi = e^\varphi \circ D_m \circ e^{-\varphi} = D_m + \mathrm{i}\boldsymbol{\alpha}\cdot\nabla\varphi$ was introduced in (VII.13). For any $\varepsilon > 0$, we have

$$\begin{aligned} &\operatorname{Re}\langle(D_m - \mathrm{i}\boldsymbol{\alpha}\cdot\nabla\varphi + \lambda)v, (D_m^\varphi - \lambda)v\rangle \\ &\le \frac{\varepsilon}{2}\|(D_m - \mathrm{i}\boldsymbol{\alpha}\cdot\nabla\varphi + \lambda)v\|^2 + \frac{1}{2\varepsilon}\|(D_m^\varphi - \lambda)v\|^2 \\ &\le \frac{3\varepsilon}{2}\|\nabla v\|^2 + \frac{3\varepsilon}{2}\||\nabla\varphi| v\|^2 + \frac{3\varepsilon}{2}\|(\beta m + \lambda)v\|^2 + \frac{1}{2\varepsilon}\|(D_m^\varphi - \lambda)v\|^2, \end{aligned} \tag{VII.85}$$

where we took into account that

$$\begin{aligned} \|(D_m - \mathrm{i}\boldsymbol{\alpha}\cdot\nabla\varphi + \lambda)u\|^2 &\le \big(\|D_0 u\| + \|(\boldsymbol{\alpha}\cdot\nabla\varphi)u\| + \|(\beta m + \lambda)u\|\big)^2 \\ &\le 3\big(\|D_0 u\|^2 + \|(\boldsymbol{\alpha}\cdot\nabla\varphi)u\|^2 + \|(\beta m + \lambda)u\|^2\big) \end{aligned}$$

and $\|\boldsymbol{\alpha}\cdot\nabla\varphi\|_{\operatorname{End}(\mathbb{C}^N)} = |\nabla\varphi|$. We deduce from (VII.84) and (VII.85) that

$$\Big(1-\frac{3\varepsilon}{2}\Big)\|\nabla v\|^2+\Big\langle v, \Big[m^2-\lambda^2-(1+\frac{3\varepsilon}{2})(\nabla\varphi)^2-\frac{3\varepsilon}{2}(\beta m+\lambda)^2\Big]v\Big\rangle \le \frac{1}{2\varepsilon}\|(D_m^\varphi-\lambda)v\|^2.$$

Thus, due to the assumption (VII.81), there exist $c > 0$ and $R > 0$ such that

$$\|v\|_{H^1} \le c\|(D_m^\varphi - \lambda)v\|, \qquad \forall v \in H^1_{\mathrm{comp}}(\Omega_R^n, \mathbb{C}^N). \tag{VII.86}$$

Then, due to the assumption on V, there exist $c > 0$ and $R > 0$ such that

$$\|v\|_{H^1} \le c\|(D_m^\varphi + V - \lambda)v\|, \qquad \forall v \in H^1_{\mathrm{comp}}(\Omega_R^n, \mathbb{C}^N).$$

Substituting $e^\varphi u$ in place of v and using the identity $(D_m^\varphi + V)(e^\varphi u) = e^\varphi(D_m + V)u$, we conclude that there exist $c > 0$ and $R > 0$ such that

$$\|e^\varphi u\|_{H^1} \le c\|e^\varphi(D_m + V - \lambda)u\|, \qquad \forall u \in H^1_{\mathrm{comp}}(\Omega_R^n, \mathbb{C}^N). \tag{VII.87}$$

Let us extend (VII.87) to functions $u \in H^1_0(\Omega_R^n, \mathbb{C}^N)$ which are no longer compactly supported. First, we consider the case when $\varphi(r)$ is bounded. Let

$$\eta \in C^\infty_{\mathrm{comp}}([-2,2]), \qquad 0 \le \eta \le 1, \qquad \eta|_{[-1,1]} \equiv 1,$$

and define $\eta_j(x) = \eta(x/j)$. Let $u \in L^2(\Omega_R^n, \mathbb{C}^N) \cap H^1_{\mathrm{loc}}(\Omega_R^n, \mathbb{C}^N)$ with $\operatorname{supp} u \subset \Omega_R^n$, and define

$$w_j = \eta_j u \in H^1_{\mathrm{comp}}(\Omega_R^n, \mathbb{C}^N).$$

Using the identity $(D_m - \lambda)w_j = \eta_j(D_m - \lambda)u - \mathrm{i}(\alpha\nabla\eta_j)u$, one has

$$\|e^\varphi w_j\| \le c\|e^\varphi \eta_j (D_m + V - \lambda)u\| + c\|e^\varphi(\alpha\nabla\eta_j)u\|. \tag{VII.88}$$

The second term in the right-hand side tends to zero as $j \to \infty$. Indeed, according to our assumptions, u is in L^2 and φ is bounded, hence $e^\varphi u \in L^2$ and

$$\lim_{j\to\infty} \|\mathbb{1}_{[j,2j]}(|x|)e^\varphi u\|_{L^2} = 0,$$

while

$$\|\nabla\eta_j\|_{L^\infty} \le \|\nabla\eta_j\|_{L^\infty}\|\mathbb{1}_{[-2j,2j]}\|_{L^\infty} = \frac{1}{j}\|\nabla\eta\|_{L^\infty}\|\mathbb{1}_{[-2j,2j]}\|_{L^\infty}$$

is bounded. Applying the dominated convergence theorem to the first term in the right-hand side of (VII.88) and the Fatou lemma to the left-hand side, we conclude that

$$\|e^\varphi u\| \le c\|e^\varphi(D_m + V - \lambda)u\|, \qquad \forall u \in H^1_0(\Omega_R^n, \mathbb{C}^N).$$

Unbounded φ are considered precisely as in [**BG87**, Theorem 3], which in turn follows the approach of [**ABG82**]. We already presented the details in the proof of Theorem VII.5. □

Theorem VII.23 (Exponential decay of eigenstates) *Let $n \geq 1$ and assume that $V \in L^\infty(\mathbb{R}^n, \mathrm{End}(\mathbb{C}^N))$.*

(1) Assume that for any $\varepsilon > 0$ there is $R > 0$ such that

$$\|Vv\| \leq \varepsilon \|v\|_{H^1}, \qquad \forall v \in H^1_{\mathrm{comp}}(\Omega^n_R, \mathbb{C}^N). \tag{VII.89}$$

Assume that

$$\lambda \in \sigma_{\mathrm{p}}(D_m + V) \cap (-m, m).$$

Then the corresponding eigenfunctions are exponentially decaying: if ζ is an eigenfunction corresponding to λ, then for any $\mu < \sqrt{m^2 - \lambda^2}$ one has

$$e^{\mu \langle r \rangle} \zeta \in H^1(\mathbb{R}^n, \mathbb{C}^N).$$

(2) Let $J \in \mathrm{End}(\mathbb{C}^N)$ be skew-adjoint and invertible, and assume that it satisfies $J^2 = -I_N$ and $[J, D_m] = 0$. Let $\omega \in [-m, m]$ and assume that

$$\lambda \in \sigma_{\mathrm{p}}(JL(\omega)) \cap \mathrm{i}\mathbb{R}.$$

(a) *If $|\lambda| < m - |\omega|$ and for any $\varepsilon > 0$ there is $R > 0$ such that*

$$\|Vv\| \leq \varepsilon \|v\|_{H^1}, \qquad \forall v \in H^1_{\mathrm{comp}}(\Omega^n_R, \mathbb{C}^N), \tag{VII.90}$$

then the corresponding eigenfunctions are exponentially decaying. More precisely, if ζ is an eigenfunction corresponding to λ, then for any

$$\mu < \sqrt{m^2 - (|\lambda| + |\omega|)^2}$$

one has

$$e^{\mu \langle r \rangle} \zeta \in H^1(\mathbb{R}^n, \mathbb{C}^N).$$

(b) *If $m - |\omega| < |\lambda| < m + |\omega|$ and for any $\varepsilon > 0$ there is $R > 0$ such that*

$$\|\langle r \rangle V v\| \leq \varepsilon \|v\|_{H^1}, \qquad \forall v \in H^1_{\mathrm{comp}}(\Omega^n_R, \mathbb{C}^N), \tag{VII.91}$$

then the corresponding eigenfunctions are exponentially decaying. More precisely, if ζ is an eigenfunction corresponding to λ, then for any

$$\mu < \sqrt{m^2 - (|\lambda| - |\omega|)^2}$$

one has

$$e^{\mu \langle r \rangle} \zeta \in H^1(\mathbb{R}^n, \mathbb{C}^N).$$

Remark VII.24 In the above theorem, the potential V is not necessarily self-adjoint.

PROOF. We will prove Theorem VII.23 *(*2b*)*. The proofs of Theorem VII.23 *(*1*)* and Theorem VII.23 *(*2a*)* are slightly shorter and readily follow along the same lines.

Let $n \geq 1$. Let $V : \mathbb{R}^n \to \mathrm{End}(\mathbb{C}^N)$ be measurable, and assume that for any $\varepsilon > 0$ there exists $R > 0$ such that (VII.91) is satisfied.

Lemma VII.25 *Let $J \in \mathrm{End}(\mathbb{C}^N)$ be skew-adjoint and invertible, such that $J^2 = -I_N$, $[J, D_m] = 0$. Let $\lambda \in \sigma_{\mathrm{p}}(JL(\omega))$ satisfy $\lambda \in \mathrm{i}\mathbb{R}$, $m - |\omega| < |\lambda| < m + |\omega|$. Then, for any $\mu < \sqrt{m^2 - (|\lambda| - |\omega|)^2}$, an eigenfunction ζ corresponding to λ satisfies*

$$e^{\mu \langle r \rangle} \zeta \in H^1(\mathbb{R}^n, \mathbb{C}^N).$$

PROOF. Let $M, \mathcal{N}, \rho \geq 1$, $\nu > 0$. Assume that $\varphi \in C^2(\mathbb{R}_+)$ satisfies (cf. Definition VII.14 and (VII.81))

(1) $0 < \varphi' \leq \sqrt{m^2 - (|\lambda| - |\omega|)^2}$, $\forall r \geq \rho$;
(2) $\limsup_{r\to\infty} \varphi'(r) < \sqrt{m^2 - (|\lambda| - |\omega|)^2}$;
(3) $r|\varphi''| \leq M\varphi'$, $\forall r \geq \rho$;
(4) $\Lambda_-^2 - m^2 + \varphi'^2 + 2r\varphi'\varphi'' \geq \nu$, $\forall r \geq \rho$.

Let

$$\Pi^{\pm} = (1 \mp \mathrm{i}J)/2. \tag{VII.92}$$

Assume that $\mathrm{i}\lambda$ is of the same sign as ω (the other case is treated verbatim by exchanging the treatment of $\zeta^{\pm}$); then $\omega + \mathrm{i}\lambda$ is outside the spectral gap of D_m while $\omega - \mathrm{i}\lambda$ is inside. So, $|\omega + \mathrm{i}\lambda| > m$, and Lemma VII.12 *(1)* yields

$$\|\nabla(e^{\varphi}u)\|^2 + \|\gamma_- e^{\varphi}u\|^2 \leq \|\mu_+ e^{\varphi}(D_m - (\omega + \mathrm{i}\lambda))u\|^2, \tag{VII.93}$$

valid for all $u \in H_0^1(\Omega^n_{R_0}, \mathbb{C}^N)$. Above, $\mu_+(r)$ and $\gamma_-(r)$ are defined in (VII.69) and (VII.70):

$$\begin{aligned}
\mu_+(r) &= 2\left(n + 16\Lambda_+^2 r^2 + 8r\varphi'(r)\right)^{1/2}, \qquad r \geq \rho;\\
\gamma_-(r) &= \left(\Lambda_-^2 - m^2 + \varphi'(r)^2 + 2r\varphi'(r)\varphi''(r)\right)^{1/2} \geq \nu^{1/2} > 0, \qquad r \geq \rho,
\end{aligned}$$

with some $\rho \geq 1$ and $\Lambda_+ = |\lambda| + |\omega|$, $\Lambda_- = |\lambda| - |\omega| > m$.

Since $|\omega - \mathrm{i}\lambda| < m$ and $|\varphi'| < \sqrt{m^2 - (\omega - \mathrm{i}\lambda)^2}$, we apply Lemma VII.21 (with $\omega - \mathrm{i}\lambda$ in place of λ and with $V = 0$):

$$\|e^{\varphi}u\|_{H^1} \leq c\|e^{\varphi}(D_m - (\omega - \mathrm{i}\lambda))u\|, \qquad \forall u \in H^1_{\mathrm{comp}}(\Omega^n_R, \mathbb{C}^N). \tag{VII.94}$$

Summing up the inequality (VII.93) (applied to $\Pi^- u$) and (VII.94) (applied to $\Pi^+ u$), with $\Pi^{\pm}$ from (VII.92)), we have:

$$\begin{aligned}
&\|\nabla(e^{\varphi}\Pi^- u)\|^2 + \|\gamma_- e^{\varphi}\Pi^- u\|^2 + \|e^{\varphi}\Pi^+ u\|^2_{H^1}\\
&\quad\leq \|\mu_+ e^{\varphi}(D_m - \omega - \mathrm{i}\lambda)\Pi^- u\|^2 + c^2\|e^{\varphi}(D_m - \omega + \mathrm{i}\lambda)\Pi^+ u\|^2\\
&\quad\leq \|\mu_+ e^{\varphi}\Pi^-(D_m - \omega - J^{-1}\lambda)u\|^2 + c^2\|e^{\varphi}\Pi^+(D_m - \omega - J^{-1}\lambda)u\|^2,
\end{aligned} \tag{VII.95}$$

valid for all $u \in H^1_{\mathrm{comp}}(\Omega^n_R, \mathbb{C}^N)$; we conclude from (VII.95) that

$$\min\left(1, \nu^{1/2}\right)\|e^{\varphi}u\|_{H^1} \leq \langle c\rangle\|(1 + \mu_+)e^{\varphi}\big((D_m - \omega) - J^{-1}\lambda\big)u\|.$$

Increasing R if necessary, we use (VII.91) to arrive at

$$\|e^{\varphi}u\|_{H^1} \leq C\|(1 + \mu_+)e^{\varphi}\big(J(D_m - \omega + V) - \lambda\big)u\|, \qquad \forall u \in H^1_{\mathrm{comp}}(\Omega^n_R, \mathbb{C}^N),$$

with $C = 2\langle c\rangle \max\big(1, \nu^{-1/2}\big)$. The extension of the above estimate to $u \in H_0^1(\Omega^n_R, \mathbb{C}^N)$ can be done as in the proof of Lemma VII.12 up to **Step 2**; it follows that

$$\|e^{\varphi}u\|_{H^1} \leq C\|(1 + \mu_+)e^{\varphi}\big(J(D_m - \omega + V) - \lambda\big)u\|, \quad \forall u \in H_0^1(\Omega^n_R, \mathbb{C}^N). \tag{VII.96}$$

Now we conclude by applying this estimate to a smooth localization of an eigenfunction to the region Ω_R^n. Let $\psi \in L^2(\mathbb{R}^n, \mathbb{C}^N)$ be an eigenfunction corresponding to λ. Pick $\eta \in C^\infty(\mathbb{R})$, $\operatorname{supp} \eta \subset (1, +\infty)$, $0 \le \eta \le 1$ and $\eta|_{[2,+\infty)} \equiv 1$. Denote $\eta_R(r) = \eta(r/R)$, and use the inequality (VII.96) with $u(x) = \eta_R(|x|)\psi(x)$ and $\varphi(r) = \upmu r$. This yields the relation

$$\|e^{\upmu r}\eta_R\psi\|_{H^1} \le C\|(1+\mu_+)e^{\upmu r}(\boldsymbol{\alpha}\cdot\nabla\eta_R)\psi\|,$$

finishing the proof. □

This completes the proof of Theorem VII.23 *(*2b*)*. □

CHAPTER VIII

The Dirac matrices

This chapter is devoted to the study of elementary properties of the Dirac matrices in all dimensions.

As we mentioned in the introduction, Paul Dirac started from the identity

$$(\sigma_1 p_1 + \sigma_2 p_2 + \sigma_3 p_3)^2 = p_1^2 + p_2^2 + p_3^2, \qquad \forall\, \mathbf{p} = (p_1, p_2, p_3) \in \mathbb{R}^3,$$

with σ_i the Pauli matrices

$$\text{(VIII.1)} \qquad \sigma_1 = \begin{bmatrix} 0 & 1 \\ 1 & 0 \end{bmatrix}, \qquad \sigma_2 = \begin{bmatrix} 0 & -\mathrm{i} \\ \mathrm{i} & 0 \end{bmatrix}, \qquad \sigma_3 = \begin{bmatrix} 1 & 0 \\ 0 & -1 \end{bmatrix},$$

and doubled the matrix size to be able to extract the square root of $E^2 = p_1^2 + p_2^2 + p_3^2 + m^2$, with $m \geq 0$ the mass of the electron, arriving at the first-order relation $E = \boldsymbol{\alpha} \cdot \mathbf{p} + \beta m$, with the matrices

$$\text{(VIII.2)} \qquad \alpha^i = \begin{bmatrix} 0 & \sigma_i \\ \sigma_i & 0 \end{bmatrix}, \qquad 1 \leq i \leq 3; \qquad \beta = \begin{bmatrix} I_2 & 0 \\ 0 & -I_2 \end{bmatrix},$$

now known as the Dirac matrices. Substituting E and $\mathbf{p}$ by the operators $\mathrm{i}\hbar\partial_t$ and $-\mathrm{i}\hbar\nabla$ led to the Dirac equation

$$\mathrm{i}\partial_t \psi = D_m \psi,$$

with the Dirac operator given by

$$D_m = -\mathrm{i}\boldsymbol{\alpha} \cdot \nabla + \beta m : \ L^2(\mathbb{R}^n, \mathbb{C}^N) \to L^2(\mathbb{R}^n, \mathbb{C}^N),$$

with the domain $\mathfrak{D}(D_m) = \{\psi \in L^2(\mathbb{R}^n, \mathbb{C}^N) \colon \ D_m\psi \in L^2(\mathbb{R}^n, \mathbb{C}^N)\}$; this operator was introduced in Section III.11.

In this chapter, we use the following convention:

> *Instead of the Dirac matrices α^i, $1 \leq i \leq n$, and β, where $n \in \mathbb{N}$, we consider $\ell = n + 1$ Dirac matrices*
>
> $$\alpha^i, \quad 1 \leq i \leq \ell, \quad \textit{with} \quad \alpha^\ell = \beta.$$

Definition VIII.1 (Dirac matrices) Let $N \in \mathbb{N}$. We say that $\alpha^i \in \operatorname{End}(\mathbb{C}^N)$, $1 \leq i \leq \ell$, with $\ell \in \mathbb{N}$, are the Dirac matrices if

$$(\alpha^i)^* = \alpha^i, \qquad \alpha^i \alpha^j + \alpha^j \alpha^i = 2\delta_{ij} I_N, \qquad 1 \leq i,\, j \leq \ell.$$

Since $(\alpha^i)^2 = I_N$, one has $\sigma(\alpha^i) \subset \{\pm 1\}$. If $\ell \geq 2$, then from the relation

$$\alpha^i = -\alpha^j \alpha^i \alpha^j, \qquad i \neq j,$$

one concludes that $\operatorname{Tr}\alpha^i = 0$. It follows that if $\ell \geq 2$, then N is even; without loss of generality, we may then assume that

$$\alpha^\ell = \begin{bmatrix} I_{N/2} & 0 \\ 0 & -I_{N/2} \end{bmatrix}.$$

Then the anticommutation relations $\{\alpha^i, \alpha^\ell\} = 0$, $1 \leq i \leq \ell - 1$, show that the matrices $(\alpha^i)_{1 \leq i \leq \ell-1}$ are block-antidiagonal; being self-adjoint, they have to be of the form

$$\alpha^i = \begin{bmatrix} 0 & \sigma_i^* \\ \sigma_i & 0 \end{bmatrix}, \qquad 1 \leq i \leq \ell - 1,$$

where the "generalized Pauli matrices" $(\sigma_i)_{1 \leq i \leq \ell-1}$ satisfy

(VIII.3) $$\sigma_i \; sp * \sigma_j + \sigma_j^* \sigma_i = 2\delta_{ij}, \quad \sigma_i \sigma_j^* + \sigma_j \sigma_i^* = 2\delta_{ij}, \quad 1 \leq i, j \leq \ell - 1.$$

Remark VIII.2 The first relation in (VIII.3) implies the second one (and vice versa). Indeed, it was pointed out by A. Sukhtayev that the identities $\sigma_i^* \sigma_i = \sigma_i \sigma_i^* = I_{N/2}$, $1 \leq i \leq \ell - 1$, allow one to turn the former relation in (VIII.3) into the latter multiplying it by σ_i from the left and by σ_i^* from the right.

Remark VIII.3 In Theorem VIII.7 below, we will see that there are $\ell = 2d + 1$ Dirac matrices of size $N = 2^d$, $d \in \mathbb{N}$; for the future reference, let us give the explicit construction of such matrices. Formally, for $d = 0$, $N = 1$, we can take $\alpha = 1$. For $d = 1$, $N = 2$, we define α^i, $1 \leq i \leq 3$, to be the standard Pauli matrices σ_i. In higher dimensions the Dirac matrices can be constructed recursively. Given the Dirac matrices of size $N = 2^d$, α^i, $1 \leq i \leq \ell = 2d + 1$, one can define the Dirac matrices $\hat{\alpha}^i$, $1 \leq i \leq 2d + 3$, of size $2N$, by the Kronecker products

(VIII.4) $$\begin{aligned} \hat{\alpha}^i &= \sigma_1 \otimes \alpha^i = \begin{bmatrix} 0 & \alpha^i \\ \alpha^i & 0 \end{bmatrix}, \quad 1 \leq i \leq 2d + 1; \\ \hat{\alpha}^{2d+2} &= \sigma_2 \otimes I_N = \begin{bmatrix} 0 & -\mathrm{i} I_N \\ \mathrm{i} I_N & 0 \end{bmatrix}, \\ \hat{\beta} := \hat{\alpha}^{2d+3} &= \sigma_3 \otimes I_N = \begin{bmatrix} I_N & 0 \\ 0 & -I_N \end{bmatrix}. \end{aligned}$$

We note that the constructed matrices are such that in $\mathbb{C}^N$ with $N = 2^d$ there are $d + 1$ Dirac matrices with real entries and d Dirac matrices with imaginary entries; we will analyze the possible numbers of real and imaginary Dirac matrices in Section VIII.2.

Remark VIII.4 More generally, if there are Dirac matrices

$$\alpha^i \in \operatorname{End}(\mathbb{C}^{N_1}), \qquad 1 \leq i \leq \ell_1, \qquad N_1 = 2^{d_1},$$

$$\tilde{\alpha}^i \in \operatorname{End}(\mathbb{C}^{N_2}), \qquad 1 \leq i \leq \ell_2, \qquad N_2 = 2^{d_2},$$

then one can construct $\ell_1 + \ell_2 - 1$ Dirac matrices of size $N_1 N_2$ taking the following Kronecker products:

(VIII.5) $$\alpha^i \otimes I_{N_2}, \quad 1 \leq i \leq \ell_1 - 1; \qquad \alpha^{\ell_1} \otimes \tilde{\alpha}^j, \quad 1 \leq j \leq \ell_2.$$

We remind that the *Kronecker product* of $A = \begin{bmatrix} a_{11} & \dots & a_{1n} \\ \vdots & \ddots & \vdots \\ a_{m1} & \dots & a_{mn} \end{bmatrix} \in \mathbb{C}^{m\times n}$ and $B \in \mathbb{C}^{M\times N}$ is defined by

$$A \otimes B = \begin{bmatrix} a_{11}B & \dots & a_{1n}B \\ \vdots & \ddots & \vdots \\ a_{m1}B & \dots & a_{mn}B \end{bmatrix} \in \mathbb{C}^{mM\times nN}.$$

Let us mention that the Kronecker product satisfies *the mixed product property* $(A\otimes B)(C\otimes D) = (AC)\otimes(BD)$.

VIII.1 The Dirac–Pauli theorem

There can be different choices of Dirac matrices. One example is constructed in Remark VIII.3; the set of Dirac matrices could be permuted or transformed by a change of basis. Nonetheless, the analysis of Dirac-type equations is not sensitive to the choice of one set or another. Let us show that there is no dependence on which form of the Dirac matrices we choose. More generally, following **[Pau36]**, we will consider matrices $M_i \in \mathbf{GL}(N,\mathbb{C})$, $1 \le i \le \ell$, with $N \in \mathbb{N}$ and $\ell \in \mathbb{N}$, which satisfy the Clifford algebra relations

$$M_iM_j + M_jM_i = 2\delta_{ij}I_N, \qquad 1 \le i,\, j \le \ell, \qquad \ell \in \mathbb{N}, \tag{VIII.6}$$

without requiring that M_j are self-adjoint.

We will need the information about irreducible representations of the complex Clifford algebra; we closely follow **[Fed96]**.

Lemma VIII.5 (Irreducible representations of complex Clifford algebras) *Let $\ell \in \mathbb{N}$ and consider the complex Clifford algebra $\mathbf{Cl}_\ell(\mathbb{C})$, which is a unital algebra over a field of complex numbers formed by the generators e_i, $1 \le i \le \ell$, which satisfy the following relations:*

$$\{\mathrm{e}_i, \mathrm{e}_j\} = \mathrm{e}_i\mathrm{e}_j + \mathrm{e}_j\mathrm{e}_i = 2\delta_{ij}\mathbb{1}, \qquad 1 \le i,\, j \le \ell, \tag{VIII.7}$$

where $\mathbb{1}$ is the unit element in $\mathbf{Cl}_\ell(\mathbb{C})$.

(1) If $\ell = 2d$, $d \in \mathbb{N}$, then there is only one irreducible representation of $\mathbf{Cl}_\ell(\mathbb{C})$. This representation is of rank 2^d.

(2) If $\ell = 2d+1$, $d \in \mathbb{N}_0$, then there are two (non-isomorphic) irreducible representations of $\mathbf{Cl}_\ell(\mathbb{C})$. Both these representations are of rank 2^d. These two representations can be distinguished by whether the product of all Clifford algebra generators (times i),

$$\boldsymbol{\omega} = \mathrm{i}\mathrm{e}_1\mathrm{e}_2\dots\mathrm{e}_{2d+1},$$

acts by multiplication by 1 or -1 (when d is odd) or by i or $-\mathrm{i}$ (when d is even).

PROOF. Assume that

$$\ell = 2d, \qquad d \in \mathbb{N}.$$

In this case, it is well-known that there is only one irreducible representation, constructed as follows. Let e_j be the generators of the complex Clifford algebra

$\mathbf{Cl}_{2d}(\mathbb{C})$. For the convenience, we denote $\mathsf{f}_i := \mathsf{e}_{d+i}$, $1 \le i \le d$; then e_i and f_i satisfy the following relations:

$$\{\mathsf{e}_i, \mathsf{e}_j\} = \{\mathsf{f}_i, \mathsf{f}_j\} = 2\delta_{ij}\mathbb{1}, \qquad \{\mathsf{e}_i, \mathsf{f}_j\} = 0; \qquad 1 \le i, j \le d.$$

Define the elements

$$\mathsf{z}_i = \frac{1}{2}(\mathsf{e}_i + \mathrm{i}\mathsf{f}_i), \qquad \mathsf{z}_i^\star = \frac{1}{2}(\mathsf{e}_i - \mathrm{i}\mathsf{f}_i), \qquad 1 \le i \le d.$$

The elements $\mathsf{z}_i^\star$, $\mathsf{z}_i \in \mathbf{Cl}_{2d}(\mathbb{C})$ are often referred to as *the operators of creation and annihilation*; they satisfy the anticommutation relations

$$\mathsf{z}_i^2 = (\mathsf{z}_i^\star)^2 = 0, \qquad \{\mathsf{z}_i, \mathsf{z}_j^\star\} = \delta_{ij}\mathbb{1}, \qquad 1 \le i, j \le d.$$

Define the "vacuum vector" $\mathsf{p}_\emptyset = \prod_{i=1}^d (\mathsf{z}_i \mathsf{z}_i^\star) \in \mathbf{Cl}_{2d}(\mathbb{C})$, and let

$$\mathbf{S}_\emptyset = \{\mathsf{v}\mathsf{p}_\emptyset : \ \mathsf{v} \in \mathbf{Cl}_{2d}(\mathbb{C})\} \subset \mathbf{Cl}_{2d}(\mathbb{C})$$

be the left principal ideal of $\mathsf{p}_\emptyset$. Since $\mathsf{z}_i\mathsf{p}_\emptyset = 0$ for $1 \le i \le d$, the elements of the form

$$(\mathsf{z}_1^\star)^{a_1} \dots (\mathsf{z}_d^\star)^{a_d}\mathsf{p}_\emptyset, \qquad a_1, \dots, a_d \in \{0, 1\}, \tag{VIII.8}$$

form the basis of the spinor space $\mathbf{S}_\emptyset$; hence, $\dim \mathbf{S}_\emptyset = 2^d$.

As follows from the construction, the action of the algebra $\mathbf{Cl}_{2d}(\mathbb{C})$ in $\mathbf{S}_\emptyset$ is irreducible, since $\mathbf{S}_\emptyset$ is the orbit of a single element $\mathsf{p}_\emptyset \in \mathbf{S}_\emptyset$ under the action of this algebra. Moreover, the action of $\mathbf{Cl}_{2d}(\mathbb{C})$ in $\mathbf{S}_\emptyset$ is the only irreducible representation of $\mathbf{Cl}_{2d}(\mathbb{C})$. Indeed, for $\mathcal{I} \subset \{1, \dots, d\}$, we can define

$$\mathsf{p}_\mathcal{I} = \Big(\prod_{i\in\mathcal{I}} \mathsf{z}_i^\star \mathsf{z}_i\Big)\Big(\prod_{i\in\{1,\dots,d\}\setminus\mathcal{I}} \mathsf{z}_i \mathsf{z}_i^\star\Big),$$

and let $\mathbf{S}_\mathcal{I}$ be the left principal ideal of $\mathsf{p}_\mathcal{I}$. Since $\sum_{\mathcal{I}\subset\{1,\dots,d\}} \mathsf{p}_\mathcal{I} = 1$, there is a decomposition of $\mathbf{Cl}_{2d}(\mathbb{C})$ into the direct sum

$$\mathbf{Cl}_{2d}(\mathbb{C}) = \bigoplus_{\mathcal{I}\subset\{1,\dots,d\}} \mathbf{S}_\mathcal{I}.$$

One can see that the representations of $\mathbf{Cl}_{2d}(\mathbb{C})$ on each of $\mathbf{S}_\mathcal{I}$ are isomorphic. Therefore, the regular representation[1] of $\mathbf{Cl}_{2d}(\mathbb{C})$ on itself reduces to the direct sum of 2^d copies of the representation on $\mathbf{S}_\emptyset$ which we denote

$$\rho_{2d} : \mathbf{Cl}_{2d} \to \mathrm{End}(\mathbf{S}_\emptyset),$$

showing that this is the only irreducible representation.

Now we move to *Part (2)* of the lemma and consider the case

$$\ell = 2d + 1, \qquad d \in \mathbb{N}_0.$$

Let us consider the Clifford algebra $\mathbf{Cl}_{2d+1}(\mathbb{C})$. We denote its generators by e_i, $1 \le i \le d$, and f_k, $1 \le k \le d+1$:

$$\{\mathsf{e}_i, \mathsf{e}_j\} = 2\delta_{ij}\mathbb{1}, \qquad \{\mathsf{f}_k, \mathsf{f}_l\} = 2\delta_{kl}, \qquad \{\mathsf{e}_i, \mathsf{f}_k\} = 0,$$

---

[1] Recall that the regular representation contains, up to isomorphisms, all the irreducible representations, each particular representation appearing with the multiplicity equal to its dimension. See [**Ser77**].

where $1 \le i,\ j \le d$ and $1 \le k,\ l \le d+1$. Define an element of the maximal degree $2d+1$ (such elements are called *pseudoscalars*) by

$$\boldsymbol{\omega} = \mathrm{i}\left(\prod_{i=1}^{d} \mathsf{e}_i\right)\left(\prod_{k=1}^{d+1} \mathsf{f}_k\right), \qquad \boldsymbol{\omega}^2 = (-1)^{d+1}\mathsf{1}. \tag{VIII.9}$$

Let us define $\boldsymbol{\pi}^{\pm} \in \mathbf{Cl}_{2d+1}(\mathbb{C})$ by

$$\boldsymbol{\pi}^{\pm} = \begin{cases} \frac{1}{2}(\mathsf{1} \pm \boldsymbol{\omega}), & d \text{ is odd;} \\ \frac{1}{2}(\mathsf{1} \mp \mathrm{i}\boldsymbol{\omega}), & d \text{ is even.} \end{cases} \tag{VIII.10}$$

With $(\boldsymbol{\pi}^{\pm})^2 = \boldsymbol{\pi}^{\pm}$, the left multiplication by $\boldsymbol{\pi}^{\pm}$ is a projector acting on $\mathbf{Cl}_{2d+1}(\mathbb{C})$. Since all the generators of the Clifford algebra $\mathbf{Cl}_{2d+1}(\mathbb{C})$ commute with $\boldsymbol{\omega}$ (and hence with $\boldsymbol{\pi}^{\pm}$), $\mathbf{Cl}_{2d+1}(\mathbb{C})$ acts invariantly on each of $\mathbf{Cl}^{\pm}_{2d+1}(\mathbb{C}) := \boldsymbol{\pi}^{\pm}\mathbf{Cl}_{2d+1}(\mathbb{C})$, which are Clifford algebras isomorphic to the Clifford algebra $\mathbf{Cl}_{2d}(\mathbb{C})$ generated by $\{\boldsymbol{\pi}^{\pm}\mathsf{e}_i,\ \boldsymbol{\pi}^{\pm}\mathsf{f}_i\}$, $1 \le i \le d$.

Remark VIII.6 The unity elements in $\mathbf{Cl}^{\pm}_{2d+1}(\mathbb{C})$ are given by $\boldsymbol{\pi}^{\pm}$ from (VIII.10). Taking into account that $\boldsymbol{\pi}^{\pm}\boldsymbol{\omega} = \pm\boldsymbol{\pi}^{\pm}\mathsf{1}$ when d is odd and $\boldsymbol{\pi}^{\pm}\boldsymbol{\omega} = \pm\mathrm{i}\boldsymbol{\pi}^{\pm}\mathsf{1}$ when d is even, we can express $\boldsymbol{\pi}^{\pm}\mathsf{f}_{d+1}$ in terms of $\boldsymbol{\pi}^{\pm}\mathsf{e}_i$ and $\boldsymbol{\pi}^{\pm}\mathsf{f}_i$, $1 \le i \le d$:

$$\boldsymbol{\pi}^{\pm}\mathsf{f}_{d+1} = \mp\boldsymbol{\pi}^{\pm}\boldsymbol{\omega}\mathsf{f}_{d+1} = \mp\boldsymbol{\pi}^{\pm}\mathrm{i}\left(\prod_{i=1}^{d}\mathsf{e}_i\right)\left(\prod_{i=1}^{d}\mathsf{f}_i\right) = \mp\mathrm{i}\left(\prod_{i=1}^{d}\boldsymbol{\pi}^{\pm}\mathsf{e}_i\right)\left(\prod_{i=1}^{d}\boldsymbol{\pi}^{\pm}\mathsf{f}_i\right)$$

when d is odd, and

$$\boldsymbol{\pi}^{\pm}\mathsf{f}_{d+1} = \pm\mathrm{i}\boldsymbol{\pi}^{\pm}\boldsymbol{\omega}\mathsf{f}_{d+1} = \mp\boldsymbol{\pi}^{\pm}\left(\prod_{i=1}^{d}\mathsf{e}_i\right)\left(\prod_{i=1}^{d}\mathsf{f}_i\right) = \mp\left(\prod_{i=1}^{d}\boldsymbol{\pi}^{\pm}\mathsf{e}_i\right)\left(\prod_{i=1}^{d}\boldsymbol{\pi}^{\pm}\mathsf{f}_i\right)$$

when d is even.

The representation of the subalgebra with the generators e_i, f_i, $1 \le i \le d$, on each of $\mathbf{Cl}^{\pm}_{2d+1}(\mathbb{C})$ is described as above (it is 2^d copies of the irreducible representation of $\mathbf{Cl}_{2d}(\mathbb{C})$ on $\mathbf{S}_{\emptyset}$), while $\boldsymbol{\omega}$ acts on $\mathbf{Cl}^{\pm}_{2d+1}(\mathbb{C})$ by multiplication by ± 1 (when d is odd) or $\pm\mathrm{i}$ (when d is even); consequently, these two representations are not isomorphic. Therefore, the regular representation of $\mathbf{Cl}_{2d+1}(\mathbb{C})$ on itself can be decomposed into 2^d copies of two different irreducible 2^d-dimensional representations, which we denote by

$$\rho^{\pm}_{2d+1} : \ \mathbf{Cl}_{2d+1} \to \mathrm{End}(\mathbf{Cl}_{2d+1});$$

they differ by whether $\boldsymbol{\omega}$ is represented by multiplication by 1 or -1 when d is odd, or by i or $-$i when d is even. □

Theorem VIII.7 (The Dirac–Pauli theorem) *Let $\ell \in \mathbb{N}$ and let $\{M_i,\ 1 \le i \le \ell\}$ and $\{\tilde{M}_i,\ 1 \le i \le \ell\}$ be two sets of anticommuting matrices of the same size $N \in \mathbb{N}$ which satisfy the anticommutation relations*

$$M_iM_j + M_jM_i = 2\delta_{ij}I_N, \qquad \tilde{M}_i\tilde{M}_j + \tilde{M}_j\tilde{M}_i = 2\delta_{ij}I_N; \qquad 1 \le i,\ j \le \ell.$$

Assume that

$$N = 2^D n_0,$$

with n_0 an odd natural number and $D \in \mathbb{N}_0$ (note that $\ell \le 2D+1$ by Lemma VIII.5*).*

(1) If $\ell = 2d$, $d \in \mathbb{N}$, then there is $S \in \mathbf{GL}(N, \mathbb{C})$ such that

$$\tilde{M}_i = S^{-1}M_iS, \qquad 1 \le i \le \ell. \tag{VIII.11}$$

(2) *If* $\ell = 2d+1$, $d \in \mathbb{N}_0$, *then there are* S, $\Sigma \in \mathbf{GL}(N, \mathbb{C})$, *with* $\Sigma^2 = I_N$, *such that*

$$\tilde{M}_i = S^{-1} M_i S, \qquad 1 \le i \le \ell - 1; \qquad \tilde{M}_\ell = \Sigma S^{-1} M_\ell S; \tag{VIII.12}$$

$$[M_i, \Sigma] = 0, \qquad [\tilde{M}_i, \Sigma] = 0, \qquad 1 \le i \le \ell; \qquad [S, \Sigma] = 0.$$

Moreover $\sigma(\Sigma) = \{\pm 1\}$, *and the multiplicities of both eigenvalues* ± 1 *of* Σ *are multiples of* 2^d. *One can choose* $\Sigma = I_N$ *if and only if*

$$\operatorname{Tr} M_1 M_2 \dots M_\ell = \operatorname{Tr} \tilde{M}_1 \tilde{M}_2 \dots \tilde{M}_\ell.$$

(3) *If* $n_0 = 1$ *and* $D = d$, *so that* $N = 2^d$ *and* $\ell = 2d$ *or* $\ell = 2d+1$, *then the choice of* S *is unique up to a nonzero complex factor; if moreover* $\ell = 2d+1$, *then* $\Sigma = \mu I_N$ *with* $\mu \in \{\pm 1\}$.

(4) *If the matrices* M_i *and* $\tilde{M}_i$, $1 \le i \le \ell$, *are self-adjoint, then* S *in* (VIII.12) *could be chosen unitary; if, moreover,* ℓ *is odd, then* Σ *could be chosen self-adjoint.*

See [**Pau36**, **vdW74**, **Dir28**], [**Tha92**, Lemma 2.25], and also [**Kes61**, Theorem 7] for general version in odd spatial dimensions and [**Shi11**, **Shi13**] for extensions to Clifford algebras.

Remark VIII.8 Instead of (VIII.12), one could find $S, \Sigma \in \mathbf{GL}(N, \mathbb{C})$, with $\Sigma^2 = I_N$, so that

$$\tilde{M}_i = \Sigma S^{-1} M_i S; \qquad 1 \le i \le \ell; \tag{VIII.13}$$

$$[M_i, \Sigma] = 0, \qquad [\tilde{M}_i, \Sigma] = 0, \qquad 1 \le i \le \ell; \qquad [S, \Sigma] = 0.$$

Proof of Theorem VIII.7. Let $d \in \mathbb{N}_0$ be such that

$$2d \le \ell \le 2d + 1;$$

we note that our proof also applies to the case when $d = 0$, $N = 1$, when the only generator e_1 of the unital algebra $\mathbf{Cl}_1(\mathbb{C})$ is represented by $\alpha = \pm 1$.

Let us prove *Parts (1) and (2)*.

We first consider the case $D = d$, $n_0 = 1$, so that $N = 2^d$. In the case $\ell = 2d$, by Lemma VIII.5, there is only one irreducible representation of the complex Clifford algebra (VIII.7). Therefore, defining two representations ρ, $\tilde{\rho} : \mathbf{Cl}_{2d}(\mathbb{C}) \to \operatorname{End}(\mathbb{C}^N)$ by

$$\rho(\mathsf{e}_k) = M_k, \qquad \rho(\mathsf{f}_k) = M_{d+k}, \qquad 1 \le k \le d,$$

and by

$$\tilde{\rho}(\mathsf{e}_k) = \tilde{M}_k, \qquad \tilde{\rho}(\mathsf{f}_k) = \tilde{M}_{d+k}, \qquad 1 \le k \le d,$$

we conclude that there is a nondegenerate $N \times N$ matrix S such that (VIII.11) is satisfied.

If $\ell = 2d+1$, then any irreducible representation of the Clifford algebra is isomorphic to either ρ_{2d}^{+} or ρ_{2d}^{-} (see Lemma VIII.5); therefore, any two sets of the matrices $\{M_i\}$ and $\{\tilde{M}_i\}$ of size $N = 2^d$ are related by

$$\tilde{M}_i = S^{-1} M_i S, \qquad 1 \le i \le \ell \tag{VIII.14}$$

(if these sets correspond to the same representation), or

$$\tilde{M}_i = S^{-1} M_i S, \qquad 1 \le i \le \ell - 1; \qquad \tilde{M}_\ell = -S^{-1} M_\ell S \tag{VIII.15}$$

(if they correspond to different representations).

Now let us assume that the matrices M_i and $\tilde{M}_i$, $1 \le i \le \ell$, are of size $N = 2^D n_0 > 2^d$, with $D \in \mathbb{N}_0$, $D \ge d$, and $n_0 \in \mathbb{N}$ an odd number. Then $\mathbb{C}^N$ can be decomposed into the direct sum of $2^{D-d} n_0$ subspaces E_a, $1 \le a \le 2^{D-d} n_0$, each of E_a of dimension 2^d and each being an invariant subspace for all of M_i, $1 \le i \le \ell$ (for details, see [**Ser77**]). Similarly, $\mathbb{C}^N$ can be decomposed into the direct sum of $2^{D-d} n_0$ subspaces $\tilde{E}_a$, $1 \le a \le 2^{D-d} n_0$, with each of $\tilde{E}_a$ being a 2^d-dimensional invariant subspace for all of $\tilde{M}_i$, $1 \le i \le \ell$.

If $\ell = 2d$, all representations of matrices M_i in E_a and of $\tilde{M}_i$ in $\tilde{E}_a$ are isomorphic; the transformation matrix is the direct product of corresponding transformation matrices. If $\ell = 2d + 1$, then each of the representations of matrices M_i in E_a and $\tilde{M}_i$ in $\tilde{E}_a$ is either ρ^+_{2d+1} or ρ^-_{2d+1}, therefore, the representation

$$M_i\big|_{E_a}, \qquad 1 \le i \le \ell$$

is isomorphic either to

$$\tilde{M}_i\big|_{\tilde{E}_a}, \qquad 1 \le i \le 2d + 1$$

or to

$$\tilde{M}_i\big|_{\tilde{E}_a}, \quad 1 \le i \le 2d, \qquad -\tilde{M}_{2d+1}\big|_{\tilde{E}_a}.$$

While the sum of the number of copies of the representations $\rho^\pm_{2d+1}$ formed by the matrices M_i (or $\tilde{M}_i$), $1 \le i \le \ell$, equals

$$\#\rho^+_{2d+1} \; + \; \#\rho^-_{2d+1} = 2^{D-d} n_0.$$

According to Lemma VIII.5 *(2)*, the difference of the number of copies of the representations ρ^+_{2d+1} and ρ^-_{2d+1} formed by the matrices M_i, $1 \le i \le \ell$, is given by

$$\#\rho^+_{2d+1} - \#\rho^-_{2d+1} = \begin{cases} 2^{-d} \operatorname{Tr} \varOmega, & d \text{ is odd,} \\ -2^{-d} \mathrm{i} \operatorname{Tr} \varOmega, & d \text{ is even,} \end{cases}$$

where (cf. (VIII.9))

$$\varOmega = \mathrm{i} \prod_{j=1}^{2d+1} M_j;$$

there are similar formulas for the representation formed by matrices $\tilde{M}_i$, $1 \le i \le \ell$. Therefore, if $\operatorname{Tr} M_1 \dots M_\ell = \operatorname{Tr} \tilde{M}_1 \dots \tilde{M}_\ell$, then both sets of matrices have the same number of copies of irreducible representations ρ^+_{2d+1} and ρ^-_{2d+1}. Then we may take $\varSigma \in \mathbf{GL}(N, \mathbb{C})$ to be the matrix corresponding to pairwise permutations of some of the spaces E_a (thus $\varSigma^2 = I_N$) so that for each $1 \le a \le 2^{D-d} n_0$ the action of the matrices $\{M_i\}$ in E_a and $\{\tilde{M}_i\}$ in $\tilde{E}_a$ corresponds to the same representation of $\mathbf{Cl}_{2d+1}(\mathbb{C})$.

This completes the proof of *Part (1)* and *Part (2)* of Theorem VIII.7.

Part (3). Let us argue that in the case $n_0 = 1$ and $d = D$, so that $N = 2^d$ and $2d \le \ell \le 2d + 1$, the choice of S is unique up to a nonzero factor. For brevity, we only consider the case $\ell = 2d + 1$ when the relations (VIII.15) are satisfied (the relations (VIII.14) and the case $\ell = 2d$ are considered verbatim). Thus, we assume that there are S_1, $S_2 \in \mathbf{GL}(2^d, \mathbb{C})$ such that

$$\tilde{M}_i = S_1^{-1} M_i S_1, \qquad 1 \le i \le 2d; \qquad \tilde{M}_{2d+1} = -S_1^{-1} M_{2d+1} S_1;$$

$$\tilde{M}_i = S_2^{-1} M_i S_2, \qquad 1 \le i \le 2d; \qquad \tilde{M}_{2d+1} = -S_2^{-1} M_{2d+1} S_2.$$

This means that $S_1^{-1}M_iS_1 = S_2^{-1}M_iS_2$ for all $1 \le i \le 2d+1$, hence $S_2^{-1}S_1$ commutes with all of M_i, $1 \le i \le 2d+1$. Since a representation of these matrices of rank 2^d is irreducible, its center is trivial, hence $S_2^{-1}S_1 = zI_{2^d}$, with some $z \in \mathbb{C} \setminus \{0\}$.

The product $\prod_{i=1}^{2d+1} M_i$ commutes with each of M_i, $1 \le i \le 2d+1$; being in the center of the irreducible representation, it equals $z_1 I_N$, with some $z_1 \in \mathbb{C}$. Similarly, one concludes that $\prod_{i=1}^{2d+1} \tilde{M}_i = z_2 I_N$, with some $z_2 \in \mathbb{C}$. Using the relations (VIII.12), we have:

$$z_2 I_N = \prod_{i=1}^{2d+1} \tilde{M}_i = \Sigma \prod_{i=1}^{2d+1} M_i = \Sigma z_1 I_N,$$

so $\Sigma = \mu I_N$, with $\mu = z_1^{-1} z_2$. From $\Sigma^2 = I_N$ we conclude that $\mu = \pm 1$.

Part (4). Let us argue that if both M_i and $\tilde{M}_i$ are self-adjoint, then S could be chosen unitary. Let us consider the case $\ell = 2d+1$ when the relations (VIII.15) are satisfied (the case $\ell = 2d+1$ with the relations (VIII.14), as well as the case $\ell = 2d$, are considered verbatim). The hermitian conjugation of (VIII.15) yields

$$\tilde{M}_i = S^* M_i (S^{-1})^*, \qquad 1 \le i \le 2d; \qquad \tilde{M}_{2d+1} = -S^* M_{2d+1} (S^{-1})^*.$$

Due to the uniqueness of S up to a nonzero factor which we proved above, there is $z \in \mathbb{C} \setminus \{0\}$ such that

$$S^* = zS.$$

Therefore, $S_1 = \pm z^{1/2} S$ is the desired unitary matrix.

We leave the rest of the proof the reader:

Problem VIII.9 Generalize the above argument to prove that in the case $\ell = 2d+1$ the matrix Σ can be chosen self-adjoint.

This completes the proof. □

Lemma VIII.10 *Let $N = 2^D n_0$, with $n_0 \in \mathbb{N}$ an odd number and $D \in \mathbb{N}$. Let $M_i \in \mathbf{GL}(N, \mathbb{C})$, $1 \le i \le \ell$ with $\ell \le 2D$, be matrices which satisfy the anticommutation relations* (VIII.6).

(1) If $\ell = 2d$, $d \in \mathbb{N}$, $d \le D$, then the set of the matrices M_i, $1 \le i \le \ell$, could always be extended to $2D+1$ matrices.

(2) If $\ell = 2d+1$, $d \in \mathbb{N}_0$, $d \le D-1$, then the set of matrices M_i, $1 \le i \le \ell$, could be extended to a larger number of matrices if and only if

$$\operatorname{Tr} M_1 M_2 \dots M_\ell = 0. \tag{VIII.16}$$

This extension could be done up to the maximal number of matrices, $2D+1$.

Moreover, in both cases $\ell = 2d$ and $\ell = 2d+1$, this extension could be done in such a way that if M_i, $1 \le i \le \ell$, are self-adjoint, then so are M_i, $\ell+1 \le i \le 2D+1$.

Let us start with an example.

Example VIII.11 Consider the set of Dirac matrices $\left\{ \begin{bmatrix} \sigma_1 & 0 \\ 0 & \sigma_1 \end{bmatrix}, \begin{bmatrix} \sigma_2 & 0 \\ 0 & \sigma_2 \end{bmatrix} \right\}$. We can extend this set to one of the following sets of three Dirac matrices:

$$\left\{ \begin{bmatrix} \sigma_1 & 0 \\ 0 & \sigma_1 \end{bmatrix}, \quad \begin{bmatrix} \sigma_2 & 0 \\ 0 & \sigma_2 \end{bmatrix}, \quad \begin{bmatrix} \sigma_3 & 0 \\ 0 & \sigma_3 \end{bmatrix} \right\} \tag{VIII.17}$$

or

$$\left\{\begin{bmatrix}\sigma_1 & 0\\ 0 & \sigma_1\end{bmatrix},\quad \begin{bmatrix}\sigma_2 & 0\\ 0 & \sigma_2\end{bmatrix},\quad \begin{bmatrix}\sigma_3 & 0\\ 0 & -\sigma_3\end{bmatrix}\right\}. \tag{VIII.18}$$

In the first case, the product of the matrices (VIII.17) equals $\mathrm{i}I_4$; this set of the Dirac matrices can no longer be extended (one can not find a nonzero matrix anticommuting with the product of these three matrices). In the case (VIII.18), the product of the matrices equals $\mathrm{i}\begin{bmatrix} I_2 & 0\\ 0 & -I_2\end{bmatrix}$; one can further extend this set to the maximal number of Dirac matrices of size 4, e.g.

$$\left\{\begin{bmatrix}\sigma_1 & 0\\ 0 & \sigma_1\end{bmatrix},\ \begin{bmatrix}\sigma_2 & 0\\ 0 & \sigma_2\end{bmatrix},\ \begin{bmatrix}0 & \sigma_3\\ \sigma_3 & 0\end{bmatrix},\ \begin{bmatrix}0 & -\mathrm{i}\sigma_3\\ \mathrm{i}\sigma_3 & 0\end{bmatrix},\ \begin{bmatrix}\sigma_3 & 0\\ 0 & -\sigma_3\end{bmatrix}\right\},$$

that is, $\{I_2\otimes\sigma_1,\ I_2\otimes\sigma_2,\ \sigma_1\otimes\sigma_3,\ \sigma_2\otimes\sigma_3,\ \sigma_3\otimes\sigma_3\}$.

PROOF OF LEMMA VIII.10. In the case when $\ell = 2d$, $d\in\mathbb{N}$, $d\le D$ (*Part (1)* of the lemma), one can take $2d+1$ Dirac matrices α^i, $1\le i\le 2d+1$, of size 2^d, and use Theorem VIII.7 to construct $S\in\mathbf{GL}(N,\mathbb{C})$ such that

$$M_i = S^{-1}\big(\alpha^i\otimes I_{2^{D-d}}\big)S,\qquad 1\le i\le 2d.$$

Then one can define

$$M_{2d+j} = S^{-1}\big(\alpha^{2d+1}\otimes\tilde\alpha^j\big)S,\qquad 1\le j\le 2(D-d)+1, \tag{VIII.19}$$

with $\tilde\alpha^j$, $1\le j\le 2(D-d)+1$, any set of the Dirac matrices of size 2^{D-d}. If M_i, $1\le i\le\ell$, are self-adjoint, then, by Theorem VIII.7, we can choose S unitary, and then the matrices in (VIII.19) are self-adjoint.

Let us now prove *Part (2)*: the case when $\ell = 2d+1$, $d\in\mathbb{N}_0$, $d\le D-1$. Again, we take $2d+1$ Dirac matrices α^i, $1\le i\le 2d+1$, of size 2^d. By Theorem VIII.7, there are S, $\Sigma\in\mathbf{GL}(N,\mathbb{C})$, $\Sigma^2 = I_N$, such that

$$M_i = S^{-1}\big(\alpha^i\otimes I_{2^{D-d}}\big)S,\ \ 1\le i\le 2d,\quad M_\ell = \Sigma S^{-1}\big(\alpha^\ell\otimes I_{2^{D-d}}\big)S, \tag{VIII.20}$$

such that the commutation relations are satisfied:

$$[M_i,\Sigma]=0,\quad [\alpha^i\otimes I_{2^{D-d}},\Sigma]=0,\quad 1\le i\le 2d+1;\qquad [S,\Sigma]=0.$$

Note that for $1\le i\le 2d+1$, the matrix $\alpha^i\otimes I_{2^{D-d}}$ commutes with $\frac12(I_N\pm\Sigma)$, and so it is invariant on the eigenspaces of Σ corresponding to 1 and -1. Since $\{\alpha^i,\ 1\le i\le 2d+1\}$ is a set of Dirac matrices of size 2^d, the eigenspaces of Σ are direct sums of invariant subspaces of $\{\alpha^i\otimes I_{2^{D-d}},\ 1\le i\le 2d+1\}$, and thus

$$\Sigma = I_{2^d}\otimes\tilde\Sigma,\qquad M_\ell = S^{-1}\big(\alpha^\ell\otimes\tilde\Sigma\big)S \tag{VIII.21}$$

with $\tilde\Sigma\in\mathbf{GL}(2^{D-d},\mathbb{C})$, $\tilde\Sigma^2 = I_{2^{D-d}}$.

One can extend $\tilde\Sigma$ to the set of Dirac matrices $\tilde\alpha^j$, $1\le j\le 2(D-d)+1$, of size 2^{D-d}, with $\tilde\alpha^1 = \tilde\Sigma$, as long as $\operatorname{Tr}\tilde\Sigma = 0$, so that the eigenvalues $\lambda = 1$ and $\lambda = -1$ of $\tilde\Sigma$ are of the same multiplicity. It remains to use (VIII.20) and (VIII.21) to see that the trace of $\tilde\Sigma$ is proportional to (VIII.16):

$$\operatorname{Tr} M_1M_2\dots M_\ell = \operatorname{Tr}((\alpha^1\alpha^2\dots\alpha^\ell)\otimes\tilde\Sigma) = \operatorname{Tr}(cI_{2^d}\otimes\tilde\Sigma) = 2^d c\operatorname{Tr}\tilde\Sigma,$$

with $c\in\mathbb{C}$ since $\alpha^1\alpha^2\dots\alpha^\ell$ is in the center of the irreducible representation and by (VIII.9) one has $c^2 = -(\mathrm{i}\alpha^1\alpha^2\dots\alpha^\ell)^2 = (-1)^d$. Similarly to (VIII.19), one then defines matrices M_i, $2d+2\le i\le 2D+1$, by

$$M_{2d+j} = S^{-1}\big(\alpha^{2d+1}\otimes\tilde\alpha_j\big)S,\qquad 2\le j\le 2(D-d)+1. \qquad\square$$

VIII.2 Possible number of real and imaginary Dirac matrices

The construction from Remark VIII.3 shows that one can choose $d+1$ purely real and d purely imaginary Dirac matrices of size $N = 2^d$, $d \in \mathbb{N}_0$. As the matter of fact, other numbers are also possible; in this section we are going to derive a general result.

Possible numbers of purely real and purely imaginary Dirac matrices are directly related to the theory developed by Adolf Hurwitz and Johann Radon [**Hur22**, **Rad22**], who considered the following class of matrices (not necessarily with real coefficients), now called the Hurwitz–Radon matrices:

$$A_i A_j^T + A_j A_i^T = 2\delta_{ij} I_N, \qquad 1 \le i,\, j \le \nu, \tag{VIII.22}$$

where A^T denotes the *transpose* of a matrix A.

Theorem VIII.12 (Hurwitz–Radon theorem [Hur22, Rad22]) *Let* $\mathbb{F} = \mathbb{C}$ *or* $\mathbb{R}$. *There exist matrices* $A_1, \ldots, A_\nu \in \mathbf{O}(N, \mathbb{F})$ *satisfying the Hurwitz–Radon equations* (VIII.22) *if and only if* $\nu \le \rho(N)$, *where* $\rho(N)$ *are the Hurwitz–Radon numbers defined by*

$$\rho(N) = 8\alpha + 2^\beta \quad \textit{for} \quad N = 2^{4\alpha+\beta} n_0, \quad \alpha \in \mathbb{N}_0, \quad 0 \le \beta \le 3, \quad n_0 \textit{ is odd.} \tag{VIII.23}$$

Above,

$$\mathbf{O}(N, \mathbb{F}) = \{M \in \mathbf{GL}(N, \mathbb{F}) \colon \quad M^T M = M M^T = I_N\}.$$

Remark VIII.13 By [**Sha77**], the statement of Theorem VIII.12 holds for any field $\mathbb{F}$ of characteristic other than 2.

We will study the following version of the Hurwitz–Radon problem: Given $N \in \mathbb{N}$, find possible numbers, a and b, of purely real and purely imaginary matrices, respectively, which satisfy

$$M_i M_j + M_j M_i = 2\delta_{ij} I_N, \qquad 1 \le i,\, j \le \ell := a + b. \tag{VIII.24}$$

Lemma VIII.14 *The maximal number* $\nu \in \mathbb{N}$ *of real matrices of size* $N = 2^d$ *which satisfy* (VIII.22) *and the maximal number* $b \in \mathbb{N}_0$ *of self-adjoint purely imaginary matrices* M_i *of size* $N = 2^d$ *which satisfy* (VIII.24) *are related by*

$$\nu = b + 1.$$

Proof. Assume that there are self-adjoint purely imaginary matrices M_i, $1 \le i \le b$ which satisfy (VIII.24). Then

$$A_i = -\mathrm{i} M_i, \quad 1 \le i \le \nu - 1 \ \ (\text{note that } A_i^T = -A_i) \ \text{ and } \ A_\nu = I_N$$

is the set of $\nu = b+1$ real matrices which satisfy (VIII.22). Reciprocally, given real matrices A_i, $1 \le i \le \nu$, that satisfy (VIII.22), then the real matrices $B_i = A_i A_\nu^T$, $1 \le i \le \nu$, also satisfy (VIII.22). Moreover, since $B_\nu = I_N$, one can see from (VIII.22) (with $j = \nu$) that B_i, $1 \le i \le \nu - 1$ are skew-adjoint, and then $M_i = \mathrm{i} B_i$, $1 \le i \le \nu - 1$, are purely imaginary self-adjoint matrices satisfying (VIII.24). □

Definition VIII.15 (Lattice of M-matrices) For $d \in \mathbb{N}_0$, define $\Gamma_d \subset \mathbb{N}_0 \times \mathbb{N}_0$ as the set of all values $(a, b) \in \mathbb{N}_0 \times \mathbb{N}_0$ such that there are a real and b imaginary matrices M_i of size $N = 2^d$ which satisfy (VIII.24). In particular, $\Gamma_0 = \{(1, 0)\}$

(with the corresponding matrix of size $N = 1$ given by $M_1 = \pm 1$). Define the lattice of the M-matrices by

$$\Gamma = \bigcup_{d \in \mathbb{N}_0} \Gamma_d \subset \mathbb{N}_0 \times \mathbb{N}_0.$$

Theorem VIII.16 (Number of real and imaginary M-matrices) *Let $N = 2^d$, $d \in \mathbb{N}$.*

(1) There is a set of $a \in \mathbb{N}_0$ purely real and $b \in \mathbb{N}_0$ purely imaginary matrices M_i, $1 \le i \le a+b$, of size $N = 2^d$, if a, b, $d \in \mathbb{N}_0$ satisfy

$$\begin{cases} a = d+1 \mod 4, & a+b = 2d+1; \\ a = d+2 \mod 4, & a+b = 2d-1. \end{cases}$$

This set can not be enlarged by adding another matrix M_i which is purely real or purely imaginary.

(2) For $(a,b) \in \Gamma_d$, the purely real and purely imaginary matrices M_i of size 2^d, $1 \le i \le a+b$, which satisfy (VIII.24)*, could be chosen self-adjoint.*

Remark VIII.17 In particular, by Theorem VIII.16 *(1)*, the maximal number of self-adjoint purely imaginary matrices M_i of size $N = 2^d$ which satisfy (VIII.24) is given by $b = \rho(N) - 1$, with the Hurwitz–Radon numbers $\rho(N)$ from (VIII.23). By Lemma VIII.14, this means that the maximal number of real matrices of size N which satisfy (VIII.22) equals $\nu = \rho(N)$, in agreement with the Hurwitz–Radon theorem (Theorem VIII.12).

PROOF. The proof is recursive; we will generate bigger matrices from smaller ones and show that the creation process is reversible, in certain sense, allowing us to trace the maximal number of real and imaginary matrices to lower-dimensional cases. We need to start with finding the maximal number of purely imaginary matrices of size 2, 4, and 8 which satisfy (VIII.24).

Lemma VIII.18 (Maximal number of purely imaginary matrices M_i of sizes 2, 4, and 8)

(1) The number of purely imaginary matrices M_i of size N which satisfy (VIII.24) *is bounded by $N - 1$.*

(2) The maximal number of purely imaginary matrices M_i of size $N = 2$, 4, and 8 which satisfy (VIII.24) *equals* 1, 3, *and* 7, *respectively. Moreover, these matrices can be chosen self-adjoint.*

PROOF. To prove *Part (1)*, let us assume there are N purely imaginary matrices M_i, $1 \le i \le N$, of size N. Then $B_i = -\mathrm{i} M_i \in \mathbf{GL}(N, \mathbb{R})$ satisfy

$$\{B_i, B_j\} = -2\delta_{ij} I_N, \qquad 1 \le i, j \le N.$$

Then, for a fixed nonzero real vector $\xi \in \mathbb{R}^N$, the $N+1$ vectors $\xi \in \mathbb{R}^N$ and $B_i \xi \in \mathbb{R}^N$, $1 \le i \le N$, would be linearly dependent over $\mathbb{R}$: for some $c_i \in \mathbb{R}$, $0 \le i \le N$, with $\sum_{i=0}^N c_i^2 > 0$, one would have

$$0 = c_0 \xi + \sum_{i=1}^N c_i B_i \xi.$$

It follows that

$$0 = \Big(c_0 - \sum_{i=1}^{N} c_i B_i\Big)\Big(c_0\xi + \sum_{i=1}^{N} c_i B_i \xi\Big) = \Big(\sum_{i=0}^{N} c_i^2\Big)\xi,$$

in contradiction to our assumption that $\xi \in \mathbb{R}^N$ is nonzero.

To prove *Part (2)*, we will give explicit examples of imaginary matrices M_i; we will produce such matrices M_i under an additional requirement that they are self-adjoint. It will be more convenient to write an argument in terms of real matrices A_i which satisfy the Hurwitz–Radon equations (VIII.22). By Lemma VIII.14, the number ν of real matrices A_i satisfying (VIII.22) and the number b of self-adjoint purely imaginary matrices M_i satisfying the anticommutation relations (VIII.24) are related by

$$b = \nu - 1.$$

Therefore, due to the restriction proved in Lemma VIII.18 *(1)*, there is a bound $\nu \le N$, while to complete the proof of Lemma VIII.18 it remains to show that there are $\nu = N$ real matrices A_i of size N satisfying (VIII.22) when $N = 2$, 4, and 8.

Denoting the standard basis in $\mathbb{C}^N$ by $\{\boldsymbol{e}_1, \boldsymbol{e}_2, \dots, \boldsymbol{e}_N\}$, we can write A_i explicitly as follows:

• $N = 2$: $A_1 = [\boldsymbol{e}_1, \ \boldsymbol{e}_2]$, $A_2 = [\boldsymbol{e}_2, -\boldsymbol{e}_1]$, where we denote by $[\boldsymbol{v}_1, \dots, \boldsymbol{v}_N]$ the $N \times N$ matrix formed by the columns of the components of $\boldsymbol{v}_i$, $1 \le i \le N$.

• $N = 4$:

$$\text{(VIII.25)} \qquad \begin{array}{ll} A_1 = [\boldsymbol{e}_1, \ \ \boldsymbol{e}_2, \ \ \boldsymbol{e}_3, \ \ \boldsymbol{e}_4], & A_2 = [\boldsymbol{e}_2, -\boldsymbol{e}_1, \ \ \boldsymbol{e}_4, -\boldsymbol{e}_3], \\ A_3 = [\boldsymbol{e}_3, -\boldsymbol{e}_4, -\boldsymbol{e}_1, \ \ \boldsymbol{e}_2], & A_4 = [\boldsymbol{e}_4, \ \ \boldsymbol{e}_3, -\boldsymbol{e}_2, -\boldsymbol{e}_1]. \end{array}$$

• $N = 8$:

$$\text{(VIII.26)} \qquad \begin{array}{l} A_1 = [\boldsymbol{e}_1, \ \ \boldsymbol{e}_2, \ \ \boldsymbol{e}_3, \ \ \boldsymbol{e}_4, \ \ \boldsymbol{e}_5, \ \ \boldsymbol{e}_6, \ \ \boldsymbol{e}_7, \ \ \boldsymbol{e}_8], \\ A_2 = [\boldsymbol{e}_2, -\boldsymbol{e}_1, \ \ \boldsymbol{e}_4, -\boldsymbol{e}_3, \ \ \boldsymbol{e}_6, -\boldsymbol{e}_5, \ \ \boldsymbol{e}_8, -\boldsymbol{e}_7], \\ A_3 = [\boldsymbol{e}_3, -\boldsymbol{e}_4, -\boldsymbol{e}_1, \ \ \boldsymbol{e}_2, \ \ \boldsymbol{e}_7, -\boldsymbol{e}_8, -\boldsymbol{e}_5, \ \ \boldsymbol{e}_6], \\ A_4 = [\boldsymbol{e}_4, \ \ \boldsymbol{e}_3, -\boldsymbol{e}_2, -\boldsymbol{e}_1, -\boldsymbol{e}_8, -\boldsymbol{e}_7, \ \ \boldsymbol{e}_6, \ \ \boldsymbol{e}_5], \\ A_5 = [\boldsymbol{e}_5, -\boldsymbol{e}_6, -\boldsymbol{e}_7, \ \ \boldsymbol{e}_8, -\boldsymbol{e}_1, \ \ \boldsymbol{e}_2, \ \ \boldsymbol{e}_3, -\boldsymbol{e}_4], \\ A_6 = [\boldsymbol{e}_6, \ \ \boldsymbol{e}_5, \ \ \boldsymbol{e}_8, \ \ \boldsymbol{e}_7, -\boldsymbol{e}_2, -\boldsymbol{e}_1, -\boldsymbol{e}_4, -\boldsymbol{e}_3], \\ A_7 = [\boldsymbol{e}_7, -\boldsymbol{e}_8, \ \ \boldsymbol{e}_5, -\boldsymbol{e}_6, -\boldsymbol{e}_3, \ \ \boldsymbol{e}_4, -\boldsymbol{e}_1, \ \ \boldsymbol{e}_2], \\ A_8 = [\boldsymbol{e}_8, \ \ \boldsymbol{e}_7, -\boldsymbol{e}_6, -\boldsymbol{e}_5, \ \ \boldsymbol{e}_4, \ \ \boldsymbol{e}_3, -\boldsymbol{e}_2, -\boldsymbol{e}_1]. \end{array}$$

□

Now we can formulate the relations between the numbers of real and imaginary matrices M_i, which will allow us to completely describe the lattice Γ from Definition VIII.15. Essentially, the following result states that the construction in Remark VIII.4 could be reversed, given the set of the matrices $\big(\hat{M}_i\big)_{1\le i\le \hat{m}}$ of size $\hat{N} = 2^{\hat{d}}$ (satisfying (VIII.24)) and the set of matrices $\big(M_j\big)_{1\le j\le \ell}$ of smaller size $N = 2^d$, $d \le \hat{d} - 1$ (also satisfying (VIII.24)), which is maximal (so that $\ell = 2d+1$).

Proposition VIII.19 *Let d_1, $d_2 \in \mathbb{N}_0$. Assume that $(a_1, b_1) \in \Gamma_{d_1}$, with $a_1 + b_1 = 2d_1 + 1$.*

(1) Assume that $a_1 \ge 1$. Then

$$(a_2, b_2) \in \Gamma_{d_2} \quad \textit{if and only if} \quad (a_1 + a_2 - 1, b_1 + b_2) \in \Gamma_{d_1+d_2}.$$

(2) Assume that $b_1 \geq 1$. *Then*

$$(a_2, b_2) \in \Gamma_{d_2} \quad \textit{if and only if} \quad (a_1 + b_2, b_1 + a_2 - 1) \in \Gamma_{d_1+d_2}.$$

PROOF. Let there be $a_1 \in \mathbb{N}_0$ real and $b_1 \in \mathbb{N}_0$ imaginary matrices of size $N_1 = 2^{d_1}$, with $a_1 + b_1 = 2d_1 + 1$; which we denote M_j, $1 \leq j \leq a_1 + b_1$, which satisfy (VIII.24); let there be a_2 real and b_2 imaginary matrices of size $N_2 = 2^{d_2}$, which we denote $\tilde{M}^j$, $1 \leq j \leq a_2 + b_2$, which also satisfy (VIII.24).

Assume that $a_1 \geq 1$, so that at least one of the matrices M_j, $1 \leq j \leq a_1 + b_1$, has real coefficients; without loss of generality, let this matrix be $M_{a_1+b_1}$. Then the Kronecker products

$$\text{(VIII.27)} \qquad \begin{cases} \hat{M}_i = M_i \otimes I_{N_2}, & 1 \leq i \leq a_1 + b_1 - 1, \\ \hat{M}_{a_1+b_1-1+j} = M_{a_1+b_1} \otimes \tilde{M}_j, & 1 \leq j \leq a_2 + b_2 \end{cases}$$

are the matrices $\hat{M}_i$, $1 \leq i \leq a_1 + b_1 + a_2 + b_2 - 1$ of size $2^{d_1+d_2}$ which satisfy (VIII.24), such that $a_1 + a_2 - 1$ have real coefficients, while $b_1 + b_2$ have imaginary ones.

Assume that $b_1 \geq 1$. Assuming that $M_{a_1+b_1}$ has purely imaginary coefficients, the Kronecker products (VIII.27) are the matrices $\hat{M}_i$, $1 \leq i \leq a_1 + b_1 + a_2 + b_2 - 1$ of size $2^{d_1+d_2}$ which satisfy (VIII.24), such that $a_1 + b_2$ have real coefficients, while $b_1 + a_2 - 1$ have imaginary ones. This completes the proof of the easier, *only if* part. Note that the assumption $a_1 + b_1 = 2d_1 + 1$ in Proposition VIII.19 was not needed for this part of the conclusion.

To go in the other direction, we use the following two lemmata:

Lemma VIII.20 *(1) Let* M_i, $1 \leq i \leq 2d + 1$, *be the matrices of size* $N = 2^d$ *which satisfy* (VIII.24). *If there is* $A \in \operatorname{End}(\mathbb{C}^N)$ *such that* $M_i A + A M_i = 0$ *for all* $1 \leq i \leq 2d$, *then* $A = z M_{2d+1}$ *with some* $z \in \mathbb{C}$.

(2) Let M_i, $1 \leq i \leq 2d+1$, *be matrices of size* $N = 2^d$ *which satisfy* (VIII.24), *and let* $\hat{M}_i$, $1 \leq i \leq \ell$, *be matrices of size* $\hat{N} = 2^D$, *with* $D \geq d + 1$, *which also satisfy* (VIII.24). *(By Theorem* VIII.7, *one has* $\ell \leq 2D+1$.*) Assume that* $\ell \geq 2d + 1$. *If*

$$\hat{M}_i = M_i \otimes I_{2^{D-d}}, \qquad 1 \leq i \leq 2d,$$

then

$$\hat{M}_i = M_{2d+1} \otimes \tilde{M}_i, \qquad 2d + 1 \leq i \leq \ell,$$

with $\tilde{M}_i \in \operatorname{End}(\mathbb{C}^{\tilde{N}})$, $2d + 1 \leq i \leq \ell$, *the matrices of size* $\tilde{N} = 2^{D-d}$ *satisfying* (VIII.24).

PROOF. For *Part (1)*, it suffices to notice that $M_{2d+1}A$ is in the center of an irreducible representation.

To prove *Part (2)*, we first claim that each of the matrices $\hat{M}_i$, $2d + 1 \leq i \leq \ell$, anticommuting with all of $M_i \otimes I_{2^{D-d}}$, $1 \leq i \leq 2d$, has the form $M_{2d+1} \otimes B$, with some $B \in \operatorname{End}(\mathbb{C}^{\tilde{N}})$. Indeed, let us assume that $\hat{A} \in \operatorname{End}(\mathbb{C}^{\hat{N}})$ satisfies the anticommutation relations

$$\text{(VIII.28)} \qquad \hat{M}_i \hat{A} + \hat{A} \hat{M}_i = 0, \qquad 1 \leq i \leq 2d.$$

Let $\tilde{\boldsymbol{e}}_j$, $1 \leq j \leq \tilde{N} = 2^{D-d}$, be the standard basis in $\mathbb{C}^{\tilde{N}}$. Multiplying the relations (VIII.28) by $I_N \otimes \tilde{\boldsymbol{e}}_j^*$ on the left and by $I_N \otimes \tilde{\boldsymbol{e}}_k$ on the right ($1 \leq j, k \leq \tilde{N}$) and taking into account that $\hat{M}_i = M_i \otimes I_{2^{D-d}}$, we conclude that

$$M_i A_{jk} + A_{jk} M_i = 0, \qquad 1 \le i \le 2d,$$

with

$$A_{jk} = (I_N \otimes \tilde{\boldsymbol{e}}_j^*) \hat{A} (I_N \otimes \tilde{\boldsymbol{e}}_k) \in \operatorname{End}(\mathbb{C}^N).$$

By *Part (1)*, there is $z_{jk} \in \mathbb{C}$ such that $A_{jk} = z_{jk} M_{2d+1}$. It follows that

$$\hat{A} = M_{2d+1} \otimes \sum_{1 \le j,k \le \tilde{N}} (z_{jk} \tilde{\boldsymbol{e}}_j \otimes \tilde{\boldsymbol{e}}_k^*);$$

thus, $\hat{A} = M_{2d+1} \otimes B$, with some $B \in \operatorname{End}(\mathbb{C}^{\tilde{N}})$.

Applying the above consideration to each of $\hat{M}_i$, $2d+1 \le i \le \ell$, we conclude that

$$\hat{M}_i = M_{2d+1} \otimes \tilde{M}_i, \qquad 2d+1 \le i \le \ell; \qquad \tilde{M}_i \in \operatorname{End}(\mathbb{C}^{\tilde{N}}).$$

The anticommutation relations satisfied by $\hat{M}_i$ (we only consider $2d+1 \le i \le \ell$) show that $\tilde{M}_i$, $2d+1 \le i \le \ell$, satisfy the equivalent of (VIII.24):

$$\tilde{M}_i \tilde{M}_j + \tilde{M}_j \tilde{M}_i = 2\delta_{ij} I_{\tilde{N}}, \qquad 2d+1 \le i,\, j \le \ell. \qquad \square$$

Lemma VIII.21 *Let A_i, $B_i \in \operatorname{End}(\mathbb{C}^N)$, $1 \le i \le \ell$, and assume that there is $S \in \mathbf{GL}(N, \mathbb{C})$ such that $B_i = S^{-1} A_i S$, $1 \le i \le \ell$. Assume that for each $1 \le i \le \ell$, both A_i and B_i either both are purely real or purely imaginary. Then there exists $S_1 \in \mathbf{GL}(N, \mathbb{R})$ such that*

$$B_i = S_1^{-1} A_i S_1, \qquad 1 \le i \le \ell.$$

Proof. Taking a linear combination of the real and imaginary parts of the relations $SB_i = A_i S$, one has

$$(\operatorname{Re} S + z \operatorname{Im} S) B_i = A_i (\operatorname{Re} S + z \operatorname{Im} S), \qquad 1 \le i \le \ell, \quad z \in \mathbb{C}.$$

It remains to notice that there is $z_1 \in \mathbb{R}$ such that $S_1 := \operatorname{Re} S + z_1 \operatorname{Im} S$ is nonsingular (since $h(z) = \det(\operatorname{Re} S + z \operatorname{Im} S)$, $z \in \mathbb{C}$, is an entire function which is nonzero at $z = \mathrm{i}$). $\square$

Given the matrices M_i, $1 \le i \le a_1 + b_1 - 1$, of size $N_1 = 2^{d_1}$, which satisfy (VIII.24), and given the matrices $\hat{M}_i$, $1 \le i \le a_1 + b_1 + a_2 + b_2 - 1$, of size $N_1 N_2 = 2^{d_1 + d_2}$, which also satisfy (VIII.24), by Lemmata VIII.20–VIII.21, there is a nonsingular matrix $S \in \operatorname{End}(\mathbb{C}^{N_1 N_2})$ such that $S^{-1} \hat{M}_i S = M_i \otimes I_{N_2}$, $1 \le i \le a_1 + b_1 - 1$, $S^{-1} \hat{M}_i S = M_{a_1 + b_1} \otimes \tilde{M}_i$ for $a_1 + b_1 \le i \le a_1 + b_1 + a_2 + b_2 - 1$, with $\tilde{M}_i$ the matrices of size $N_2 = 2^{d_2}$ which also satisfy (VIII.24).

We now assume that for each $1 \le i \le a_1 + b_1 - 1$ the matrices $\hat{M}_i$ and M_i are either both purely real or both purely imaginary. By Theorem VIII.7, there exists a matrix $S \in \mathbf{GL}(N_1 N_2, \mathbb{C})$ such that

$$S^{-1} \hat{M}_i S = M_i \otimes I_{N_2}, \qquad 1 \le i \le a_1 + b_1 - 1 = 2d_1,$$

and by Lemma VIII.21 we can choose $S \in \mathbf{GL}(N_1 N_2, \mathbb{R})$. By Lemma VIII.20,

$$S^{-1} \hat{M}_i S = M_{a_1 + b_1} \otimes \tilde{M}_i, \qquad a_1 + b_1 \le i \le a_1 + b_1 + a_2 + b_2 - 1,$$

with $\tilde{M}_i$ the matrices of size $N_2 = 2^{d_2}$ which satisfy (VIII.24). Since $S \in \mathbf{GL}(N_1 N_2, \mathbb{R})$, the number of matrices with real and imaginary coefficients is unchanged: $\hat{M}_i$, $a_1 + b_1 \le i \le a_1 + b_1 + a_2 + b_2 - 1$, is real (imaginary, respectively) if and only if so is $M_{a_1+b_1} \otimes \tilde{M}_i$, hence if and only if so is $\tilde{M}_i$ (since $M_{a_1+b_1}$ is purely real). This completes the proof of *Part (1)* of Proposition VIII.19.

Let us prove *Part (2)*. If $b_1 \geq 1$, rearranging the matrices M_i, $1 \leq i \leq a_1 + b_1$, so that $M_{a_1+b_1}$ has purely imaginary coefficients, finding $S \in \mathbf{GL}(N_1 N_2, \mathbb{R})$ such that

$$S^{-1}\hat{M}_i S = M_i \otimes I_{N_2}$$

and defining $\tilde{M}_i$ for $i \geq a_1 + b_1$ by

$$S^{-1}\hat{M}_i S = M_{a_1+b_1} \otimes \tilde{M}_i, \qquad a_1 + b_1 \leq i \leq a_1 + b_1 + a_2 + b_2 - 1,$$

one obtains matrices $\tilde{M}_i$ of size $N_2 = 2^{d_2}$ which satisfy (VIII.24) such that a_2 of them have purely real coefficients and b_2 have purely imaginary ones.

This completes the proof of Proposition VIII.19. □

Corollary VIII.22 *Let $d \in \mathbb{N}$ and a, $b \in \mathbb{N}_0$.*

(1) $(a, b) \in \Gamma_d$ if and only if $(a + 1, b + 1) \in \Gamma_{d+1}$.
(2) Further, let $a \geq 1$. Then $(a, b) \in \Gamma_d$ if and only if $(1 + b, a - 1) \in \Gamma_d$.

PROOF. The first part follows from applying Proposition VIII.19 *(1)* to $(a, b) \in \Gamma_d$ and $(2, 1) \in \Gamma_1$. For the second part, one notices that if $(a, b) \in \Gamma_d$, then, by the first part, $(a + 1, b + 1) \in \Gamma_{d+1}$. Applying Proposition VIII.19 *(2)* to $(a + 1, b + 1) \in \Gamma_{d+1}$ and $(2, 1) \in \Gamma_1$, one concludes that $(1 + b, a - 1) \in \Gamma_d$. □

We can now complete the proof of Theorem VIII.16. Below, we will use the notation $(a, b)_\circ \in \Gamma_d$ for the sets of matrices M_i of size $N = 2^d$ that are not maximal, in the sense that they contain less than the maximal number of matrices M_i of this size which satisfy (VIII.24) (that is, one has $a + b \leq 2d$).

The sets $(2, 1) \in \Gamma_1$ (the Pauli matrices), $(0, 3)_\circ \in \Gamma_2$, and $(0, 7) \in \Gamma_3$ (the last two were constructed in Lemma VIII.18) can be used to generate sets containing all other possible numbers of real and imaginary matrices M_i which satisfy (VIII.24). These three sets are underlined on Figure VIII.1. To prove that this yields the largest possible numbers of purely real and purely imaginary matrices, we trace the construction backwards, using the reduction from Proposition VIII.19. Let us now give the details.

The lattice Γ is constructed as follows.

- $(1, 0) \in \Gamma_0$;
- $(2, 1) \in \Gamma_1$ and $(0, 3)_\circ \in \Gamma_2$ generate $(5, 0)_\circ \in \Gamma_3$ (by Proposition VIII.19 *(2)*);
- Similarly, $(2, 1) \in \Gamma_1$ and $(0, 7) \in \Gamma_3$ generate $(9, 0) \in \Gamma_4$;
- $(5, 0)_\circ$ and $(9, 0)$ generate (recursively) sets $(2d - 1, 0)_\circ \in \Gamma_d$ for $d = 3$ mod 4. We then use the reduction with the aid of $(2, 1) \in \Gamma_1$ (using Proposition VIII.19 *(2)*), obtaining $(0, 2d - 3)_\circ \in \Gamma_{d-1}$;
- $(9, 0)$ with itself generates (recursively) sets $(2d + 1, 0) \in \Gamma_d$ for $d = 0$ mod 4, and the reduction with the aid of $(2, 1) \in \Gamma_1$ (using Proposition VIII.19 *(2)*) allows us to obtain $(0, 2d - 1) \in \Gamma_{d-1}$.

$d=0$:					$(1,0)$				
$d=1$:					$\underline{(2,1)}$				
$d=2$:					$(3,2)$	$\underline{(0,3)_\circ}$			
$d=3$:				$(5,0)_\circ$	$(4,3)$	$(1,4)_\circ$	$\underline{(0,7)}$		
$d=4$:			$(9,0)$	$(6,1)_\circ$	$(5,4)$	$(2,5)_\circ$	$(1,8)$		
$d=5$:			$(10,1)$	$(7,2)_\circ$	$(6,5)$	$(3,6)_\circ$	$(2,9)$		
$d=6$:			$(11,2)$	$(8,3)_\circ$	$(7,6)$	$(4,7)_\circ$	$(3,10)$	$(0,11)_\circ$	
$d=7$:		$(13,0)_\circ$	$(12,3)$	$(9,4)_\circ$	$(8,7)$	$(5,8)_\circ$	$(4,11)$	$(1,12)_\circ$	$(0,15)$
$d=8$:	$(17,0)$	$(14,1)_\circ$	$(13,4)$	$(10,5)_\circ$	$(9,8)$	$(6,9)_\circ$	$(5,12)$	$(2,13)_\circ$	$(1,16)$
$d=9$:	$(18,1)$	$(15,2)_\circ$	$(14,5)$	$(11,6)_\circ$	$(10,9)$	$(7,10)_\circ$	$(6,13)$	$(3,14)_\circ$	$(2,17)$

FIGURE VIII.1. Possible numbers $(a,b) \in \Gamma$ of real and imaginary matrices of size 2^d satisfying (VIII.24). The notation $(a,b)_\circ$ is used when the corresponding set of numbers is "diminished", that is, $a+b$ is smaller than the maximal number $2d+1$ of Dirac matrices of size 2^d, but still can not be enlarged to the sets of either $(a+1,b)$ or $(a,b+1)$ Dirac matrices. The three underlined (a,b)-entries could be used to generate all other $d \geq 1$ entries via Proposition VIII.19.

- The above procedure gives the "boundary" of Γ: all the sets of the form $(a,0)$ and $(0,b)$ which correspond to the largest number of purely real matrices of size $N = 2^d$ and the largest number of purely imaginary matrices of this size. All other sets $(a,b) \in \Gamma$ are obtained with the aid of the Kronecker product with $(2,1)$ $(a,b) \in \Gamma_d \Rightarrow (a+1,b+1) \in \Gamma_{d+1}$ by Corollary VIII.22 *(1)*.

Possible values of $(a,b) \in \Gamma_d$ for $d \leq 9$ are given on Figure VIII.1 (we underlined the three values of (a,b) which generate the whole set Γ).

It follows that there are a purely real and b purely imaginary matrices of size $N = 2^d$ with a, b and d satisfying the following relations:

$$\begin{cases} a+b = 2d+1, & a = d+1 \mod 4; \\ a+b = 2d-1, & a = d+2 \mod 4. \end{cases}$$

Thus, we have seen that the procedure from Proposition VIII.19 allows one to construct from $(2,1)$, $(0,3)_\circ$, and $(0,7)$ all other sets $(a,b) \in \Gamma_d \setminus \{(1,0)\}$. Since on each step of this construction process at least one of the two starting sets is maximal ($2d+1$ matrices of size 2^d), the reduction from Lemma VIII.20 can be used, showing that not a single of the sets in the graph could be enlarged by adding purely real or purely imaginary matrices as long as we show that $(2,1) \in \Gamma_1$, $(0,3)_\circ \in \Gamma_2$, and $(0,7) \in \Gamma_3$ can not be enlarged by adding either purely real or purely imaginary matrices.

Let us prove that indeed there is no such enlargement. In the case $N = 2$, there is at most one purely imaginary matrix (see Lemma VIII.18). Besides, there are two purely real matrices M_i (given by the Pauli matrices); together, these three matrices correspond to the maximal number of Dirac matrices, $2d+1$, of size $N = 2^d$, with $d = 1$. Thus, $(2,1) \in \Gamma_1$ can not be enlarged. By Lemma VIII.18, there are exactly three purely imaginary matrices M_i of size 2^d with $d = 2$. There is no additional purely real matrix: otherwise, if we had $(1,3)$ in Γ_2, then the reduction from Proposition VIII.19 would allow to conclude that $(0,2)$ belonged to Γ_1, which is not the case by Lemma VIII.18. Seven purely imaginary matrices of size $N = 8$ constructed in Lemma VIII.18 correspond to the maximal number $2d+1$ of matrices M_i of size $N = 2^d$ with $d = 3$.

Finally, to prove *Part (2)* of Theorem VIII.16, it suffices to notice that if M_i and $\tilde{M}_i$ are self-adjoint, then the construction based on the Kronecker products (VIII.27) also yields self-adjoint matrices; thus, for each of $(a,b) \in \Gamma_d$, there are a purely real and b purely imaginary *self-adjoint* matrices M_i of size $N = 2^d$ satisfying (VIII.24).

This completes the proof of Theorem VIII.16. □

Remark VIII.23 The table on Figure VIII.1 is directly related to the real representations of the Clifford Algebra $\mathbf{Cl}_{p,q}(\mathbb{R})$, which is a set of generators $\mathfrak{e}_i$, $1 \le i \le p+q$, over $\mathbb{R}$, with the relations

$$\mathfrak{e}_i\mathfrak{e}_j + \mathfrak{e}_j\mathfrak{e}_i = 2\eta_{ij}\mathbb{1}, \qquad 1 \le i,\, j \le p+q,$$

where $\mathbb{1}$ is the unit element in $\mathbf{Cl}_{p,q}(\mathbb{R})$ and

$$\eta_{ij} = \begin{cases} 0, & i \ne j, \\ 1, & 1 \le i = j \le p, \\ -1, & p+1 \le i = j \le p+q. \end{cases}$$

(We note that in the real case the negative signs of η_{ij} can not be absorbed into multiplication of $\mathfrak{e}_i$, $i \ge p+1$, by an imaginary unit.) In particular, if there are p real and q imaginary Dirac matrices $M_i \in \operatorname{End}(\mathbb{R}^N)$ satisfying (VIII.24), then, denoting the real matrices by $A_i = M_i$ for $1 \le i \le p$ and defining $B_i = -\mathrm{i}M_i$ for $p+1 \le i \le p+q$, we see that

$$\mathfrak{e}_i \mapsto \begin{cases} A_i, & 1 \le i \le p, \\ B_i, & p+1 \le i \le p+q, \end{cases}$$

is the representation of $\mathbf{Cl}_{p,q}(\mathbb{R})$ in $\mathbb{R}^N$. For the complete classification of representations of Clifford algebras over $\mathbb{R}$, see e.g. [**Oku91**].

CHAPTER IX

The Soler model

The Soler model [**Iva38**, **Sol70**] describes the spinor field with the self-interaction based on the quantity $\bar{\psi}\psi = \psi^*\beta\psi$:

$$\text{(IX.1)} \qquad \mathrm{i}\partial_t\psi = D_m\psi - f(\psi^*\beta\psi)\beta\psi, \quad \psi(t,x) \in \mathbb{C}^N, \quad x \in \mathbb{R}^n, \quad n, N \in \mathbb{N},$$

where f is a continuous real-valued function with $f(0) = 0$. The quantity

$$\bar{\psi} = \psi^*\beta$$

is called the *Dirac conjugate*, with ψ^* the hermitian conjugate. This is one of the main models of the nonlinear Dirac equation, alongside with its own one-dimensional analogue, the Gross–Neveu model [**GN74**, **LG75**], and with the massive Thirring model [**Thi58**]. All these models are hamiltonian, $\mathbf{U}(1)$-invariant, and relativistically invariant.

The free Dirac operator is given by

$$\text{(IX.2)} \qquad D_m = -\mathrm{i}\boldsymbol{\alpha}\cdot\nabla + \beta m, \qquad m > 0,$$

with α^i, $1 \le i \le n$, and β mutually anticommuting self-adjoint matrices such that $D_m^2 = (-\Delta + m^2)I_N$, with I_N the $N \times N$ identity matrix. See Definition III.156. By Chapter VIII, one has $N = 2^{[(n+1)/2]}N_0$, with $N_0 \in \mathbb{N}$, and without loss of generality we assume that the matrices α^i and β have the following form:

$$\text{(IX.3)} \qquad \alpha^i = \begin{bmatrix} 0 & \sigma_i^* \\ \sigma_i & 0 \end{bmatrix}, \quad 1 \le i \le n; \qquad \beta = \begin{bmatrix} I_{N/2} & 0 \\ 0 & -I_{N/2} \end{bmatrix};$$

here σ_i, $1 \le i \le n$, are "the generalized Pauli matrices" of size $N/2$ which satisfy

$$\text{(IX.4)} \qquad \sigma_i^*\sigma_j + \sigma_j^*\sigma_i = 2I_N\delta_{ij}, \qquad \sigma_i\sigma_j^* + \sigma_j\sigma_i^* = 2I_N\delta_{ij}, \qquad 1 \le i, j \le n.$$

See Remark VIII.3 for the explicit construction. In the case $n = 3$, $N = 4$, one takes σ_i, $1 \le i \le 3$, to be the standard Pauli matrices σ_i.

Remark IX.1 The matrices σ_i in (IX.3) are not necessarily self-adjoint; for example, for $n = 4$ and $N = 4$, one can choose σ_i for $1 \le i \le 3$ to be the standard Pauli matrices, so that α^i, $1 \le i \le 3$, and β are the standard Dirac matrices, and to set $\sigma_4 = \mathrm{i}I_2$, so that

$$\text{(IX.5)} \qquad \alpha^4 = \begin{bmatrix} 0 & -\mathrm{i}I_2 \\ \mathrm{i}I_2 & 0 \end{bmatrix}.$$

The structure of the nonlinearity in (IX.1) is such that the equation is $\mathbf{U}(1)$-invariant and hamiltonian, with the Hamiltonian functional given by

$$\text{(IX.6)} \qquad E(\psi) = \int_{\mathbb{R}^n} \big(\psi^* D_m\psi - F(\psi^*\beta\psi)\big)\,dx, \quad \text{with } F(s) = \int_0^s f(t)\,dt, \quad s \in \mathbb{R}.$$

We point out that the choice of the nonlinearity based on the scalar quantity $\psi^*\beta\psi$ is motivated by Physics considerations: this quantity transforms as a scalar under Lorentz transformations (as opposed to e.g. $\psi^*\psi$ which transforms as the zero component of the Lorentz-vector corresponding to the charge-current density).

Remark IX.2 When considering in (IX.1) pure power nonlinearities, the absolute value $f(\tau) = |\tau|^\kappa$ is needed in the case when $\kappa > 0$ is not an integer. If $\kappa \in \mathbb{N}$ and is odd, then the presence or absence of the absolute value leads two different models; for instance, if $n = 3$ and $N = 4$, in the model

$$\text{(IX.7)} \qquad \mathrm{i}\partial_t\psi = D_m\psi - \psi^*\beta\psi\,\beta\psi, \qquad \psi(t,x) \in \mathbb{C}^4, \quad x \in \mathbb{R}^3,$$

the small amplitude limit (see Chapter XII) corresponding to $\omega \to -m$ is a defocusing NLS (contrary to the small amplitude limit when $\omega \to m$ which is a focusing NLS), while in the model

$$\text{(IX.8)} \qquad \mathrm{i}\partial_t\psi = D_m\psi - |\psi^*\beta\psi|\beta\psi, \qquad \psi(t,x) \in \mathbb{C}^4, \quad x \in \mathbb{R}^3,$$

such a limit is a focusing NLS (just like the small amplitude limit when $\omega \to m$). Notice that both equations (IX.8) and (IX.7) are Hamiltonian systems and both are invariant with respect to the Wigner time reversal [**BD64**, Chapter 5.4]:

$$\text{(IX.9)} \qquad \psi(t,x) \mapsto \psi_T(t,x) := \mathrm{i}\gamma^1\gamma^3\boldsymbol{K}\psi(-t,x) = -\begin{bmatrix}\sigma_2 & 0\\ 0 & \sigma_2\end{bmatrix}\boldsymbol{K}\psi(-t,x),$$

with $\boldsymbol{K} : \mathbb{C}^N \to \mathbb{C}^N$ the complex conjugation. The invariance is due to the relation

$$\psi_T(t,x)^*\beta\psi_T(t,x) = \psi(-t,x)^*\beta\psi(-t,x), \qquad \forall (t,x) \in \mathbb{R}\times\mathbb{R}^3.$$

IX.1 The well-posedness of the Cauchy problem in Sobolev spaces

Let us show that the Cauchy problem for the evolution problem (IX.1) is well-posed in $H^k(\mathbb{R}^n)$ if $k \in \mathbb{N}$ satisfies $k > n/2$. Let f be a continuous real-valued function with $f(0) = 0$ and consider the following Cauchy problem:

$$\text{(IX.10)} \qquad \begin{cases} \mathrm{i}\partial_t\psi = D_m\psi - f(\psi^*\beta\psi)\beta\psi, \\ \psi(0,x) = \psi_0(x), \end{cases}$$

with $\psi_0 \in H^k(\mathbb{R}^n)$. Below we consider mild solutions of (IX.10), that is, fixed point solutions of

$$\psi(t) = e^{-\mathrm{i}tD_m}\psi_0 + \mathrm{i}\int_0^t e^{-\mathrm{i}(t-s)D_m} f(\psi(s)^*\beta\psi(s))\beta\psi(s)\,ds,$$

with $\psi : [0,t] \mapsto H^k(\mathbb{R}^n)$. Since $k > n/2$, the integral makes sense since $\psi(t)$ is valued in $H^k(\mathbb{R}^n) \subset C(\mathbb{R}^n)$ while f is continuous.

Theorem IX.3 *Let $k \in \mathbb{N}$ satisfy $k > n/2$ and assume that f is in $C^k(\mathbb{R})$. For each $\psi_0 \in H^k(\mathbb{R}^n)$ there exist values $T^\pm_{\psi_0} \in (0,\infty]$ such that the Cauchy problem* (IX.10) *possesses a unique mild solution*

$$\psi \in C\big((-T^-_{\psi_0}, T^+_{\psi_0}), H^k(\mathbb{R}^n)\big),$$

and if $T^+_{\psi_0} < +\infty$ (respectively, $T^-_{\psi_0} < +\infty$), then

$$\lim_{t\to T^+_{\psi_0}-0} \|\psi(t)\|_{H^k} = +\infty \quad \Big(\text{ respectively, } \lim_{t\to -T^-_{\psi_0}+0} \|\psi(t)\|_{H^k} = +\infty\ \Big).$$

Moreover, ψ is in $C^1\big((-T^-_{\psi_0}, T^+_{\psi_0}), H^{k-1}(\mathbb{R}^n)\big)$ and satisfies (IX.1).

PROOF. The last assertion, as well as the blow-up alternative, are classical and their proofs are omitted. We only consider the existence and uniqueness of a local solution on a small interval of time. We emphasize that the problem is invariant with respect to time translation, hence the extension to a maximal solution is a consequence of the uniqueness.

Let us recall the Faà di Bruno formula. Let $g \in C^k(\mathbb{R}^{2N}, \mathbb{R}^{2N})$, with $N \in \mathbb{N}$. For $v \in C^k_{\mathrm{comp}}(\mathbb{R}^n)$ and $\alpha \in \mathbb{N}_0^n$, $|\alpha| \le k$,

$$\partial_x^\alpha g(v(x)) = \sum_{\substack{\ell\in\mathbb{N},\\ (\beta_1,\dots,\beta_\ell)\in(\mathbb{N}_0^n)^\ell,\\ |\beta_j|\ge 1\\ \beta_1+\dots+\beta_\ell=\alpha}} g^{(\ell)}(v(x))\Big(\otimes_{j=1}^{\ell}\partial_x^{\beta_j} v(x)\Big),$$

where

$$g^{(\ell)}(v):\ \big(\mathbb{R}^{2N}\big)^{\otimes\ell} \to \mathbb{R}^{2N}$$

is the ℓth differential of g at the point $v \in \mathbb{R}^{2N}$; the standard multi-index notations are used. Note that ℓ in this sum is necessarily smaller than or equal to $|\alpha|$. Now let $p_j = 2k/|\beta_j|$, so that $\sum 1/p_j = 1/2$; from the Hölder inequality,

$$\|\partial_x^\alpha g(v)\|_{L^2(\mathbb{R}^n)} \le \sum_{\substack{\ell\in\mathbb{N},\\ (\beta_1,\dots,\beta_\ell)\in(\mathbb{N}_0^n)^\ell,\\ |\beta_j|\ge 1\\ \beta_1+\dots+\beta_\ell=\alpha}} \|g^{(\ell)}(v)\|_{L^\infty(\mathbb{R}^n)} \prod_{j=1}^{\ell} \|\partial_x^{\beta_j} v\|_{L^{p_j}(\mathbb{R}^n)}.$$

Note that $\|g^{(\ell)}(v(\cdot))\|_{L^\infty(\mathbb{R}^n)}$ is the L^∞-norm (in $x \in \mathbb{R}^n$) of the norm of a multilinear map. Then from the Galgliardo–Nirenberg inequality (Lemma II.13 *(3)*) applied to vector-valued functions, one has:

$$\|\partial_x^{\beta_j} v\|_{L^{p_j}(\mathbb{R}^n)} \le C(n, |\beta_j|, |\alpha|)\|v\|_{H^{|\alpha|}(\mathbb{R}^n)}^{|\beta_j|/|\alpha|}\|v\|_{L^\infty(\mathbb{R}^n)}^{1-|\beta_j|/|\alpha|}.$$

As a consequence, we have the estimate

$$\|\partial_x^\alpha g(v)\|_{L^2(\mathbb{R}^n)} \le C(n,\alpha)\left(\sum_{\ell\in\mathbb{N},\,\ell\le|\alpha|} \|g^{(\ell)}(v)\|_{L^\infty(\mathbb{R}^n)}\|v\|_{L^\infty(\mathbb{R}^n)}^{\ell-1}\right)\|v\|_{H^k(\mathbb{R}^n)},$$

where

$$C(n,\alpha) := \sup_{\substack{\ell\in\mathbb{N},\\ (\beta_1,\dots,\beta_\ell)\in(\mathbb{N}_0^n)^\ell,\\ |\beta_j|\ge 1\\ \beta_1+\dots+\beta_\ell=\alpha}} \prod_{j=1}^{\ell} C(n, |\alpha|, |\beta_j|),$$

and thus

$$\|g(v)\|_{H^k(\mathbb{R}^n)} \le C(n,k)\Big(\sum_{\ell\in\mathbb{N},\,\ell\le k} \|g^{(\ell)}(v)\|_{L^\infty(\mathbb{R}^n)}\|v\|_{L^\infty(\mathbb{R}^n)}^{\ell-1}\Big)\|v\|_{H^k(\mathbb{R}^n)}.$$

Above,

$$C(n,k) = \sum_{\alpha\in\mathbb{N}_0^n,\,|\alpha|\le l} C(n,\alpha).$$

This inequality extends to $v \in H^k(\mathbb{R}^n)$, $k > n/2$, by the density argument. We will apply the above estimates to

$$g(u) = f(u^*\beta u)\beta u, \qquad u(x) \in \mathbb{C}^N \cong \mathbb{R}^{2N}, \qquad x \in \mathbb{R}^n.$$

We note that g is Lipschitz on $\mathbb{B}_R(L^\infty(\mathbb{R}^n))$, $R>0$:

$$\|g(v_2)-g(v_1)\| \le \sup_{v\in\mathbb{B}_R(L^\infty(\mathbb{R}^n))} \|g'(v)\|\|v_2-v_1\|.$$

Let $\mathcal{G}$ be the map on $L^2(\mathbb{R}^n)\cap L^\infty(\mathbb{R}^n)$ defined by

$$\mathcal{G}(\psi)(t) = e^{-\mathrm{i}tD_m}\psi_0 + \mathrm{i}\int_0^t e^{-\mathrm{i}(t-s)D_m} g(\psi(s))\,ds.$$

Let $A(s,n)$ be the best Sobolev constant in the embedding $H^k(\mathbb{R}^n)\subset L^\infty(\mathbb{R}^n)$. Pick

$$R > \|\psi_0\|_{H^k},$$

denote

$$M_1 := \sup\Big\{ \sum_{\ell\in\mathbb{N},\,\ell\le k} \|g^{(\ell)}(u)\|\|u\|^{\ell-1},\ u\in\mathbb{C}^N,\ |u|\le A(s,n)R\Big\},$$
$$M_2 := \sup\{\|g'(u)\|,\ u\in\mathbb{C}^N,\ |u|\le A(s,n)R\},$$

and let $\tau_0>0$ be such that the following two conditions are satisfied:

$$\|\psi_0\|_{H^k} + \tau_0 M_1 R < R, \tag{IX.11}$$
$$\tau_0 M_2 < 1. \tag{IX.12}$$

Lemma IX.4 *Let $s\ge 0$. The set*

$$\mathbf{K} = C\big([0,\tau_0],\overline{\mathbb{B}_R(H^s(\mathbb{R}^n))}\big), \qquad \|\cdot\|_{\mathbf{K}} = \|\cdot\|_{L^\infty([0,\tau_0],L^2(\mathbb{R}^n))}$$

is a complete metric space.

PROOF. Consider a sequence $(\psi_j)_{j\in\mathbb{N}}$ in $C\big([0,\tau_0],\overline{\mathbb{B}_R(H^s(\mathbb{R}^n))}\big)$ which is Cauchy for the norm $\|\cdot\|_{\mathbf{K}}$:

$$\forall\varepsilon>0\ \ \exists N>0,\ \forall i,\,j\in\mathbb{N},\ i,\,j\ge N \ \Rightarrow\ \sup_{t\in[0,\tau_0]}\|\psi_i(t)-\psi_j(t)\|_{H^s(\mathbb{R}^n)}\le\varepsilon.$$

For each t in $[0,\tau_0]$, the sequence $(\psi_j(t))_{j\in\mathbb{N}}$ is convergent in $H^s(\mathbb{R}^n)$ to some $\psi(t)$ in $H^s(\mathbb{R}^n)$. The latter is in $\overline{\mathbb{B}_R(H^s(\mathbb{R}^n))}$ since it is closed. Moreover,

$$\forall\varepsilon\ \ \exists N>0,\ \forall j\in\mathbb{N},\ j\ge N \ \Rightarrow\ \sup_{t\in[0,\tau_0]}\|\psi(t)-\psi_j(t)\|_{H^s(\mathbb{R}^n)}\le\varepsilon.$$

This gives the convergence to $\psi\in L^\infty\big([0,\tau_0],H^s(\mathbb{R}^n)\big)$. Since $C\big([0,\tau_0],\overline{\mathbb{B}_R(H^s(\mathbb{R}^n))}\big)$ is closed in $L^\infty\big([0,\tau_0],H^s(\mathbb{R}^n)\big)$, this concludes the proof. □

For any t in $\mathbb{R}$, $e^{-\mathrm{i}tD_m}$ is a unitary map in $H^k(\mathbb{R}^n)$. Together with (IX.11), this ensures that $\mathcal{G}$ acts invariantly in $\mathbf{K}$. Taking into account the condition (IX.12), one concludes that $\mathcal{G}$ acts as a contraction on $\mathbf{K}$; consequently, the existence and uniqueness of a fixed point of $\mathcal{G}$ in $\mathbf{K}$ follows from the Picard fixed point theorem. □

When $f(z)$ is a polynomial, the situation is simpler, allowing us to use the fractional Sobolev spaces due to the Sobolev algebra structure (see Lemma II.10).

Theorem IX.5 *Let $s \in \mathbb{R}$, $s > n/2$, and let $f \in \mathbb{R}[z]$ be a polynomial. Let $\psi_0 \in H^s(\mathbb{R}^n)$. Then there exist $T^{\pm}_{\psi_0} \in (0,\infty]$ such that the Cauchy problem* (IX.10) *possesses a unique mild solution ψ in $C\big((-T^-_{\psi_0}, T^+_{\psi_0}), H^s(\mathbb{R}^n)\big)$, and if $T^+_{\psi_0} < +\infty$ (respectively, $T^-_{\psi_0} < +\infty$), then*

$$\lim_{t\to T^+_{\psi_0}-0} \|\psi(t)\|_{H^s} = +\infty \quad \Big(\text{ respectively, } \lim_{t\to -T^-_{\psi_0}+0} \|\psi(t)\|_{H^s} = +\infty \Big).$$

Moreover, if $s \geq 1$, then ψ belongs to $C^1\big((-T^-_{\psi_0}, T^+_{\psi_0}), H^{s-1}(\mathbb{R}^n)\big)$ and satisfies (IX.1).

PROOF. The proof is similar to the one of Theorem IX.3. Instead of the Faà di Bruno formula, the Hölder inequality, and the Galgliardo–Nirenberg inequality, it is now based on Lemma II.10. □

Remark IX.6 Theorem IX.3 can be strengthened if one uses Strichartz-type estimates, see [**EV97**, **MNO03**, **MO07**], or, in the special case of Soler-type nonlinearity in lower dimensions, using the null structure. We refer to [**Bou04**, **Bou08a**, **ST10**, **Huh11**, **Can11**, **Pel11**, **Huh13a**, **Huh13b**, **Huh14**, **PS14**, **BC14**, **BH15**, **HM15**, **ZZ15**, **BH16**].

IX.2 Discrete symmetries

Let us study the discrete symmetries of the nonlinear Dirac equation (IX.1). Let $d \in \mathbb{N}$ be such that $\psi(t,x) \in \mathbb{C}^N$ with $N = 2^d$ and $x \in \mathbb{R}^n$ with $n = 2d-1$ or $n = 2d$. Assume that there are $2d+1$ Dirac matrices (α^i), $1 \leq i \leq 2d$, and β, such that α^{2i-1} with $1 \leq i \leq d$ and β are real, while α^{2i} with $1 \leq i \leq d$ are purely imaginary, so that this is in agreement with the standard case of spatial dimension $n = 3$ with $N = 4$ spinor components, and also in agreement with the construction of higher dimensional Dirac matrices from Remark VIII.3.

IX.2.1 Parity transformation. For any values of $n \in \mathbb{N}$ and $N = 2^d$, $d \in \mathbb{N}$, the linear Dirac equation

$$\mathrm{i}\partial_t\psi = D_m\psi = (-\mathrm{i}\boldsymbol{\alpha}\cdot\nabla + \beta m)\psi, \qquad \psi(t,x) \in \mathbb{C}^N, \qquad (t,x) \in \mathbb{R}\times\mathbb{R}^n,$$

is invariant with respect to the *parity transformation*:

$$\psi(t,x) \mapsto \psi_P(t,x) := \beta\psi(t,-x), \qquad (t,x) \in \mathbb{R}\times\mathbb{R}^n. \tag{IX.13}$$

IX.2.2 Time inversion. In spatial dimension $n = 2d-1$ and for spinors with $N = 2^d$ components, for any $d \in \mathbb{N}$, when an "extra" Dirac matrix α^{2d} is at our disposal (this matrix does not enter the Dirac operator D_m), the linear Dirac equation is invariant with respect to the *time-reversal transformation*,

$$\psi(t,x) \mapsto \psi_T(t,x) := \alpha^{2d}\psi(-t,x), \qquad (t,x) \in \mathbb{R}\times\mathbb{R}^n. \tag{IX.14}$$

Since

$$(\alpha^{2d}\psi(-t,x))^*\beta\alpha^{2d}\psi(-t,x) = -\psi(-t,x)^*\beta\psi(-t,x),$$

the Soler model (IX.1) is invariant with respect to this transformation if $f(\tau)$ is an even function.

IX.2.3 Charge conjugation.

- If $N = 2^d$ with d odd and either $n = 2d - 1$ or $n = 2d$, then the linear Dirac equation is invariant with respect to the transformation

$$\psi \mapsto \psi_C := -\mathrm{i}^{(d-1)d/2}\Big(\prod_{i=1}^{d}\alpha^{2i-1}\Big)\boldsymbol{K}\psi, \tag{IX.15}$$

with $\boldsymbol{K} : \mathbb{C}^N \to \mathbb{C}^N$ the complex conjugation. (A scalar factor is chosen so that the matrix $\mathrm{i}^{(d-1)d/2}\alpha^1\alpha^3\dots\alpha^{2d-1}$ is self-adjoint.) Since d is odd, one has

$$\psi_C^*\beta\psi_C = -\psi^*\beta\psi,$$

hence the Soler model (IX.1) is invariant with respect to this transformation under the additional requirement that the function $f(\tau)$ is even.

- If $N = 2^d$ with d even and $n = 2d-1$, the linear Dirac equation is invariant with respect to the transformation

$$\psi \mapsto \psi_C = -\mathrm{i}^{(d-1)d/2}\Big(\prod_{i=1}^{d-1}\alpha^{2i}\Big)\beta\boldsymbol{K}\psi. \tag{IX.16}$$

Again

$$\psi_C^*\beta\psi_C = -\psi^*\beta\psi,$$

hence (IX.1) is invariant with respect to this transformation under the additional requirement that $f(\tau)$ is even.

Remark IX.7 Note that in the standard case $n = 3$, $N = 4$, one has

$$\psi(t,x) \mapsto \psi_C(t,x) = -\mathrm{i}\alpha^2\beta\boldsymbol{K}\psi(t,x) = \mathrm{i}\gamma^2\boldsymbol{K}\psi(t,x), \tag{IX.17}$$

which is the standard charge conjugation; see [**BD64**, Section 5.2] and [**Tha92**, Section 1.4.6].

IX.3 Continuous symmetries

IX.3.1 Noether's theorem. Let us remind the basic idea of Noether's theorem that each differentiable one-parameter group of transformations of the Lagrangian results in a conservation law. For simplicity, we formulate it assuming that ψ takes values in a real Banach space E. Denote

$$\partial_\mu\psi = \frac{\partial\psi}{\partial x^\mu}, \qquad (x^\mu)_{0\le\mu\le n} = (t, x^1, \dots, x^n) \in \mathbb{R}^{n+1}.$$

Theorem IX.8 (Noether's theorem) *Let $\mathscr{L} \in C^2(E\times E^{n+1},\mathbb{R})$ be a Lagrangian density, with E a real Banach space, and let $\psi \in C^1(\mathbb{R}\times\mathbb{R}^n, E)$ be a solution to the Euler–Lagrange equation*

$$-\partial_\mu\frac{\partial\mathscr{L}(\psi,\partial_0\psi,\dots,\partial_n\psi)}{\partial(\partial_\mu\psi)} + \frac{\partial\mathscr{L}(\psi,\partial_0\psi,\dots,\partial_n\psi)}{\partial\psi} = 0, \qquad (t,x)\in\mathbb{R}\times\mathbb{R}^n.$$

Assume that $\mathscr{L}(\psi,\partial_0\psi,\dots,\partial_n\psi)$ remains invariant under the symmetry transformation

$$\psi \to \psi_s(t,x) := A(s)\psi(t,x),$$

with $A \in C^1(\mathbb{R},\mathrm{End}(E))$, $A(s)|_{s=0} = I_E$. Define the Noether current density by

$$\mathscr{J}^\mu(t,x) = \Big\langle \frac{\partial\mathscr{L}}{\partial(\partial_\mu\psi)}, \frac{\partial}{\partial s}A(s)|_{s=0}\psi\Big\rangle, \qquad (t,x)\in\mathbb{R}\times\mathbb{R}^n, \qquad 0\le\mu\le n,$$

where $\langle\cdot,\,\cdot\rangle : E^*\times E\to\mathbb{R}$*. Then there is the corresponding local conservation law*

$$\partial_\mu \mathscr{J}^\mu(t,x)=0,\qquad \forall(t,x)\in\mathbb{R}\times\mathbb{R}^n,$$

which results in the conservation law

$$\frac{d}{dt}\int_\Omega \mathscr{J}^0(t,x)\,dx=-\oint_{\partial\Omega}\mathbf{J}(t,x)\cdot\mathbf{dS},\qquad \mathbf{J}=(\mathscr{J}^1,\,\dots,\,\mathscr{J}^n),\qquad \forall t\in\mathbb{R},$$

for any bounded domain $\Omega\subset\mathbb{R}^n$ *with a sufficiently regular boundary* $\partial\Omega$*.*

Proof. We have:

$$0=\frac{d}{ds}\mathscr{L}(\psi_s,\partial_0\psi_s,\,\dots,\,\partial_n\psi_s)=\left\langle\frac{\partial\mathscr{L}}{\partial\psi},\,\frac{\partial\psi_s}{\partial s}\right\rangle+\left\langle\frac{\partial\mathscr{L}}{\partial(\partial_\mu\psi)},\,\frac{\partial(\partial_\mu\psi_s)}{\partial s}\right\rangle$$

$$=\left\langle\partial_\mu\left(\frac{\partial\mathscr{L}}{\partial(\partial_\mu\psi)}\right),\,\frac{\partial\psi_s}{\partial s}\right\rangle+\left\langle\frac{\partial\mathscr{L}}{\partial(\partial_\mu\psi)},\,\frac{\partial(\partial_\mu\psi_s)}{\partial s}\right\rangle=\partial_\mu\left\langle\frac{\partial\mathscr{L}(\psi,\partial_\mu\psi)}{\partial(\partial_\mu\psi)},\,\frac{\partial\psi_s}{\partial s}\right\rangle,$$

which at $s=0$ is equivalent to the local form of the conservation law, $\partial_\mu\mathscr{J}^\mu=0$. □

IX.3.2 Lagrangian formulation of the Soler model. We follow the convention that $0\le\mu,\nu\le n$ and $1\le i,\,j\le n$. Let γ^μ, $0\le\mu\le n$, be the $N\times N$ Dirac gamma-matrices; that is, γ^0 is self-adjoint, γ^i, $1\le i\le n$, are anti-self-adjoint, and

$$\gamma^\mu\gamma^\nu+\gamma^\nu\gamma^\mu=2g^{\mu\nu}I_N,$$

with

$$g_{\mu\nu}=g^{\mu\nu}=\mathrm{diag}[1,-1,\dots,-1],\qquad 0\le\mu,\,\nu\le n,\tag{IX.18}$$

the Minkowski metric tensor. We recall that for $\psi\in\mathbb{C}^N$, the quantity

$$\bar\psi=\psi^*\beta$$

is called the *Dirac adjoint* of ψ. Consider the Lagrangian density

$$\mathscr{L}=\bar\psi(\mathrm{i}\gamma^\mu\partial_\mu-m)\psi+F(\bar\psi\psi),\qquad \psi(t,x)\in\mathbb{C}^N,\qquad (t,x)\in\mathbb{R}\times\mathbb{R}^n,\tag{IX.19}$$

with the summation in the repeated index μ (from 0 to n), with $m>0$, $\psi\in\mathbb{C}^N$ the spinor field, and $F:\ \mathbb{R}\to\mathbb{R}$, which we assume is sufficiently smooth and satisfies $F(\tau)=o(\tau)$ as $\tau\to 0$. The Euler–Lagrange equation

$$-\partial_\mu\frac{\partial\mathscr{L}}{\partial(\partial_\mu\psi)}+\frac{\partial\mathscr{L}}{\partial\psi}=0,\qquad \psi(t,x)\in\mathbb{C}^N,\qquad (t,x)\in\mathbb{R}\times\mathbb{R}^n,\tag{IX.20}$$

obtained by taking the variation of (IX.19) with respect to $\mathrm{Re}\,\psi$ and $\mathrm{Im}\,\psi$ (or, alternatively, with respect to $\bar\psi$ considered as independent of ψ) leads to the equation

$$\mathrm{i}\partial_t\psi=D_m\psi-f(\bar\psi\psi)\beta\psi,\qquad f(\tau)=F'(\tau),\quad \tau\in\mathbb{R},\tag{IX.21}$$

where $D_m=-\mathrm{i}\alpha^i\partial_i+\beta m$ is the Dirac operator with $\alpha^i=\gamma^0\gamma^i$, $1\le i\le n$, and $\beta=\gamma^0$.

IX.3.3 Charge conservation. Let $\psi(t,x)$ be a solution to (IX.1). Due to the $\mathbf{U}(1)$-invariance of the Lagrangian, by Noether's theorem, there is a charge functional

$$Q(\psi) = \int_{\mathbb{R}^n} \psi(t,x)^* \psi(t,x)\, dx, \tag{IX.22}$$

with its value (formally) conserved in time for solutions to (IX.1) with sufficiently fast decay at infinity. This conservation takes place, for instance, if ψ_0 is in $H^s(\mathbb{R}^n)$ with $s > n/2$ when f is polynomial, or for $s = k \in \mathbb{N}$ with $k > n/2$ when f is of class C^k (considered as a map in $\mathbb{R}^2$); see Section IX.1. The local form of charge conservation is given by

$$\partial_\mu \mathscr{J}^\mu(t,x) = 0, \qquad \forall (t,x) \in \mathbb{R} \times \mathbb{R}^n, \tag{IX.23}$$

where

$$\mathscr{J} = \big(\mathscr{J}^\mu\big)_{0 \le \mu \le n}, \qquad \mathscr{J}^\mu = \bar\psi \gamma^\mu \psi = \psi^* \gamma^0 \gamma^\mu \psi, \tag{IX.24}$$

is the Minkowski vector of the charge-current density.

IX.3.4 Complex charge conservation. The Soler model is invariant under the action of a continuous symmetry group generated by the charge conjugation (see Section IX.2.3) if $n = 2d - 1$ or $n = 2d$ and $d \in \mathbb{N}$ is odd (with the charge conjugation (IX.15)) or if $n = 2d-1$ and $d \in \mathbb{N}$ is even (with the charge conjugation (IX.16)). Namely, if $B \in \mathrm{End}(\mathbb{C}^N)$ is a self-adjoint matrix such that

$$B^* B = I_N, \qquad \{B\boldsymbol{K}, D_m\} = 0,$$

where $\boldsymbol{K} : \mathbb{C}^N \to \mathbb{C}^N$ is the complex conjugation, then the Lagrangian density is invariant with respect to the symmetry transformation

$$\psi(t,x) \mapsto \psi_z(t,x) = e^{zB\boldsymbol{K}} \psi(t,x), \qquad z \in \mathbb{C}, \qquad (t,x) \in \mathbb{R} \times \mathbb{R}^n. \tag{IX.25}$$

We note that the parameter z in (IX.25) is complex-valued. One can check directly the conservation of the complex-valued *Bogoliubov charge* associated to this symmetry:

$$\Lambda = \int_{\mathbb{R}^n} \psi(t,x)^* B\boldsymbol{K} \psi(t,x)\, dx.$$

One may combine this symmetry with the unitary symmetry from Section IX.3.3 into the symmetry group $\mathbf{SU}(1,1)$:

$$\{a + b\mathbf{B}\boldsymbol{K}\colon \ a, b \in \mathbb{C},\ |a|^2 - |b|^2 = 1\} \cong \mathbf{SU}(1,1),$$

$$a + b\mathbf{B}\boldsymbol{K} \mapsto \begin{bmatrix} a & \bar b \\ b & \bar a \end{bmatrix} \in \mathbf{SU}(1,1).$$

We mention that this group appears in the Bogoliubov transformation **[Bog58]** from the Bardeen–Cooper–Schrieffer theory. More details about this symmetry and consequences for the solitary waves of the nonlinear Dirac equation are given in Chapter X.

IX.3.5 Energy-momentum conservation. Due to the invariance of the Hamiltonian with respect to spatial and temporal translations, there are associated (formally) conserved quantities called the energy and momentum. By **[BD65]**, the density of the energy-momentum tensor corresponding to the Lagrangian density $\mathscr{L}$ is given by

$$\mathscr{T}^{\mu\nu} = \frac{\partial \mathscr{L}}{\partial(\partial_\mu \psi)} g^{\nu\rho} \partial_\rho \psi - g^{\mu\nu} \mathscr{L}, \qquad 0 \le \mu, \nu \le n,$$

with $g^{\mu\nu} = \operatorname{diag}[1, -1, \dots, -1]$, $0 \le \mu, \nu \le n$ the inverse of the Minkowski tensor $g_{\mu\nu}$ from (IX.18). The components of the energy-momentum tensor

$$T^{\mu\nu} = \int_{\mathbb{R}^n} \mathscr{T}^{\mu\nu}\, dx$$

are given by

$$\text{(IX.26)} \quad T^{00} = E = \int_{\mathbb{R}^n} \mathscr{H}\, dx = \int_{\mathbb{R}^n} \Big(\bar\psi \beta(-\mathrm{i}\alpha^i \partial_{x^i} + m\beta)\psi - F(\bar\psi\psi) \Big)\, dx,$$

$$T^{0j} = T^{j0} = \mathrm{i} \int_{\mathbb{R}^n} \bar\psi \gamma^0 g^{j\rho} \partial_\rho \psi\, dx = \mathrm{i} g^{j\rho} \langle \psi, \partial_\rho \psi \rangle,$$

$$\text{(IX.27)} \qquad T^{ij} = \langle \psi, \mathrm{i}\alpha^i \partial_\rho \psi \rangle g^{\rho j} - g^{ij} L,$$

where

$$\text{(IX.28)} \qquad L = \int_{\mathbb{R}^n} \mathscr{L}\, dx.$$

Above, $\mathscr{H}$ is the density of the Hamiltonian functional corresponding to the Lagrangian density $\mathscr{L} = \bar\psi(\mathrm{i}\gamma^\mu \partial_\mu - m)\psi + F(\bar\psi\psi)$ from (IX.19):

$$\text{(IX.29)} \quad \mathscr{H} = \frac{\partial \mathscr{L}}{\partial(\partial_t \psi)} \partial_t \psi - \mathscr{L} = -\bar\psi(\mathrm{i}\gamma^i \partial_{x^i})\psi + m\bar\psi\psi - F(\bar\psi\psi),$$

with the summation in the repeated index $1 \le i \le n$.

We note that the matrix $T^{\mu\nu}$ is symmetric and real-valued. For each $0 \le \nu \le n$, there is the conserved Noether current $\big(\mathscr{T}^{\mu\nu}\big)_{0 \le \nu \le n}$. Indeed, there is the local (formal) conservation law

$$\text{(IX.30)} \qquad \partial_\mu \mathscr{T}^{\mu\nu} = 0, \qquad \forall \nu \ge 0, \quad \nu \le n,$$

which follows from the corresponding Euler–Lagrange equations (that is, (IX.1) and its hermitian adjoint) and which leads to the (formal) conservation of the Minkowski vector of the energy-momentum, $\big(T^{0\nu}\big)_{0 \le \nu \le n}$.

IX.3.6 Lorentz transformations. In this subsection, it is convenient to write

$$x = (x^0, x^1, \dots, x^n) \in \mathbb{R} \times \mathbb{R}^n, \qquad \text{with } x^0 = t.$$

The (homogeneous) group of the Lorentz transformations

$$\mathbf{O}(n, 1)$$

consists of matrices $\Lambda \in \operatorname{End}(\mathbb{R}^{n+1})$ which preserve the Lorentz metric,

$$g_{\mu\nu} = \operatorname{diag}[1, -1, \dots, -1], \qquad 0 \le \mu, \nu \le n,$$

so that $\langle \Lambda x, \Lambda y \rangle_g = \langle x, y \rangle_g$, which means that $\Lambda^* g \Lambda = g$; explicitly,

$$g_{\alpha\beta} = \Lambda^\mu_\alpha g_{\mu\nu} \Lambda^\nu_\beta, \qquad 0 \le \alpha, \beta \le n,$$

with the summation with respect to the upper and lower indices from 0 to n. Let $\Lambda = e^{sM} \in \mathbf{SO}(n,1)$, with $M = (M^{\mu}{}_{\nu})_{0\le\mu,\,\nu\le n} \in \mathbf{so}(n,1)$ and $s \in \mathbb{R}$. From $\Lambda^* g\Lambda = g$ one concludes that

$$M_{\mu\nu} = -M_{\nu\mu}, \qquad \text{where} \quad M_{\mu\nu} = g_{\mu\alpha}M^{\alpha}{}_{\nu}, \qquad 0 \le \mu,\, \nu \le n.$$

The Lagrangian of the Soler model is invariant with respect to the transformation

$$\text{(IX.31)} \quad \psi(x) \mapsto \psi_s(x) = e^{s\rho(M)}\psi(e^{sM}x), \qquad s \in \mathbb{R}, \qquad x \in \mathbb{R}\times\mathbb{R}^n,$$

where

$$\text{(IX.32)} \qquad \rho(M) = \frac{\mathrm{i}}{4}M_{\mu\nu}\sigma^{\mu\nu},$$

with $\sigma^{\mu\nu}$ given by

$$\text{(IX.33)} \qquad \sigma^{\mu\nu} = \frac{\mathrm{i}}{2}[\gamma^{\mu},\gamma^{\nu}], \qquad 0 \le \mu,\, \nu \le n.$$

We note that

$$\text{(IX.34)} \quad \gamma^0(\sigma^{\mu\nu})^*\gamma^0 = \gamma^0\frac{\mathrm{i}}{2}\big[(\gamma^{\mu})^*,(\gamma^{\nu})^*\big]\gamma^0 = \sigma^{\mu\nu}, \qquad 0 \le \mu,\, \nu \le n.$$

To verify this invariance, one needs to check that $\mathscr{L}(\psi_s)$ at the point $x \in \mathbb{R}\times\mathbb{R}^n$ equals $\mathscr{L}(\psi)$ evaluated at the point $y = e^{sM}x$. It is enough to check that

$$\big(e^{s\rho(M)}\psi(e^{sM}x)\big)^*\gamma^0(\mathrm{i}\gamma^{\alpha}\partial_{x^{\alpha}} - m)e^{s\rho(M)}\psi(e^{sM}x) = \psi(y)^*\gamma^0(\mathrm{i}\gamma^{\beta}\partial_{y^{\beta}} - m)\psi(y).$$

This would follow from the equality

$$e^{s\rho(M)^*}\gamma^0 = \gamma^0 e^{s\gamma^0\rho(M)^*\gamma^0} = \gamma^0 e^{-s\rho(M)}$$

(we used (IX.32) and (IX.34)) and the relation

$$e^{-s\rho(M)}(\mathrm{i}\gamma^{\alpha}\partial_{x^{\alpha}} - m)e^{s\rho(M)}\psi(e^{sM}x) = (\mathrm{i}\gamma^{\beta}\partial_{y^{\beta}} - m)\psi(y), \qquad y = e^{sM}x,$$

which in turn is a consequence of $(e^{-sM})^{\alpha}{}_{\beta}\partial_{x^{\alpha}} = \partial_{y^{\beta}}$ and the relation

$$\text{(IX.35)} \quad e^{-s\rho(M)}\gamma^{\alpha}e^{s\rho(M)} = (e^{-sM})^{\alpha}{}_{\beta}\gamma^{\beta}, \qquad s \in \mathbb{R}, \qquad 0 \le \alpha \le n.$$

Problem IX.9 Prove the relation (IX.35).

Hint: Show that both $Y_1(s) = e^{-sM}\begin{bmatrix}\gamma^0\\ \vdots\\ \gamma^n\end{bmatrix}$ *and* $Y_2(s) = \begin{bmatrix}e^{-s\rho(M)}\gamma^0e^{s\rho(M)}\\ \vdots\\ e^{-s\rho(M)}\gamma^n e^{s\rho(M)}\end{bmatrix}$ *satisfy the same equation* $\partial_s Y(s) = -MY(s)$, $Y \in C^1\big(\mathbb{R}, \mathbb{C}^{n+1}\otimes_{\mathbb{C}}\mathrm{End}(\mathbb{C}^N)\big)$, *with the same initial data at* $s = 0$. *Use the identity*

$$[\sigma^{\mu\nu},\gamma^{\alpha}] = 2\mathrm{i}(\gamma^{\mu}g^{\nu\alpha} - \gamma^{\nu}g^{\mu\alpha}), \qquad 0 \le \alpha,\, \mu,\, \nu \le n.$$

Remark IX.10 In particular, it follows that the quantity $\psi^*\beta\psi$ does not change under the Lorentz transformations (or, rather, behaves as a scalar-valued function of $x \in \mathbb{R}\times\mathbb{R}^n$): given ψ_s from (IX.31), one has:

$$\psi_s(x)^*\beta\psi_s(x) = \psi^*\beta\psi\big|_{y=e^{sA}x}, \qquad \forall A \in \mathbf{so}(n,1), \quad \forall s \in \mathbb{R}.$$

The quantity $\psi^*\beta\psi$ also does not change under the parity transformation (IX.13):

$$\psi_P(t,x)^*\beta\psi_P(t,x) = \psi(t,-x)^*\beta\psi(t,-x), \qquad \forall (t,x) \in \mathbb{R}\times\mathbb{R}^n.$$

In Physics terminology, such a quantity is called a *relativistic scalar*, while the Soler model (IX.1) whose Lagrangian (IX.19) is based on the quantity $\psi^*\beta\psi$ is referred to as a *model with scalar self-interaction.*

Remark IX.11 By Physics terminology, a *pseudoscalar* is a quantity that does not change under the continuous Lorentz transformation (like a relativistic scalar), but changes sign under the parity transformation $(x^1, \dots, x^n) \mapsto (-x^1, \dots, -x^n)$, while a *scalar* does not. Let $d \in \mathbb{N}$ and let α^i, $1 \le i \le 2d$, and β be the Dirac matrices of size $N = 2^d$. In odd spatial dimensions, when

$$\psi(t,x) \in \mathbb{C}^N, \qquad x \in \mathbb{R}^n, \qquad N = 2^d, \qquad n = 2d-1,$$

when the Dirac operator is given by $D_m = -\mathrm{i}\sum_{i=1}^{2d-1} \alpha^i \partial_{x^i} + \beta m$, the quantity

$$\bar\psi\Gamma\psi = \psi^*\beta\Gamma\psi, \tag{IX.36}$$

with

$$\Gamma = -\mathrm{i}^{(2d-1)(d+1)} \prod_{\mu=0}^{2d-1} \gamma^\mu, \tag{IX.37}$$

behaves like a pseudoscalar. Indeed, just like $\psi^*\beta\psi$, the quantity (IX.36) is invariant with respect to the Lorentz transformations (IX.31) (it suffices to note that Γ anticommutes with each of γ^μ, $0 \le \mu \le n$, so it commutes with $\sigma^{\mu\nu}$ and hence with $e^{s\rho(A)}$); at the same time, (IX.36) changes sign under the parity transformation (IX.13):

$$\psi_P(t,x)^*\beta\Gamma\psi_P(t,x) = -\psi(t,-x)^*\beta\Gamma\psi(t,-x), \qquad \forall (t,x) \in \mathbb{R}\times\mathbb{R}^n.$$

Let us mention that in three spatial dimensions ($d = 2$, $N = 4$, $n = 3$) one has

$$\Gamma = \mathrm{i}\gamma^0\gamma^1\gamma^2\gamma^3 = \begin{bmatrix} 0 & I_2 \\ I_2 & 0 \end{bmatrix} =: \gamma^5.$$

For more details on the Lorentz transformations in the three-dimensional case and the corresponding conserved quantities, we refer the reader to [**BD64**, **Tha92**].

IX.4 Relation to the massive Thirring model

An alternative model of the self-interacting spinor field is the massive Thirring model introduced in [**Thi58**] (known as MTM), with the vector-type self-interaction based on the scalar quantity

$$\mathscr{J}_\mu \mathscr{J}^\mu = \sum_{\mu,\,\nu=0}^{n} g_{\mu\nu}(\bar\psi\gamma^\mu\psi)(\bar\psi\gamma^\nu\psi) = (\psi^*\psi)^2 - \sum_{i=1}^{n}(\psi^*\alpha^i\psi)^2, \tag{IX.38}$$

with

$$\mathscr{J} = \big(\mathscr{J}^\mu\big)_{0\le\mu\le n}, \qquad \mathscr{J}^\mu = \bar\psi\gamma^\mu\psi = \psi^*\gamma^0\gamma^\mu\psi$$

(see (IX.24)) the Minkowski vector of the charge-current density. The Soler model and massive Thirring models coincide in two and in four spatial dimensions due to the equality of $(\bar\psi\psi)^2$ and $\mathscr{J}_\mu\mathscr{J}^\mu$ in spatial dimensions $n = 2$ and $n = 4$, as we conclude from the identities

$$(\psi^*\psi)^2 = \sum_{i=1}^{3}(\psi^*\sigma_i\psi)^2, \qquad \forall \psi \in \mathbb{C}^2, \tag{IX.39}$$

$$(\psi^*\psi)^2 = \sum_{i=1}^{4}(\psi^*\alpha^i\psi)^2 + (\psi^*\beta\psi)^2, \qquad \forall\psi \in \mathbb{C}^4, \tag{IX.40}$$

after we identify

$$\bar{\psi}\psi = \psi^*\sigma_3\psi, \qquad (\mathscr{J}^\mu)_{0\le\mu\le 2} = (\psi^*\psi, \psi^*\sigma_1\psi, \psi^*\sigma_2\psi), \qquad \psi \in \mathbb{C}^2$$

in two spatial dimensions, and

$$\bar{\psi}\psi = \psi^*\beta\psi, \quad (\mathscr{J}^\mu)_{0\le\mu\le 4} = (\psi^*\psi, \psi^*\alpha^1\psi, \psi^*\alpha^2\psi, \psi^*\alpha^3\psi, \psi^*\alpha^4\psi), \quad \psi \in \mathbb{C}^4$$

in four spatial dimensions; above, α^i, $1 \le i \le 3$, and β are the standard Dirac matrices and $\alpha^4 := \begin{bmatrix} 0 & -\mathrm{i}I_2 \\ \mathrm{i}I_2 & 0 \end{bmatrix}$ (like in the construction in Remark VIII.3).

Remark IX.12 We point out that the relations (IX.39) and (IX.40) represent a particular case ($x = y$) of the Hurwitz problem, which led to the study of the Hurwitz–Radon matrices mentioned in Section VIII.2. A sum-of-squares formula of size $[r, s, n]$ is an equation of the following type:

$$(x_1^2 + \cdots + x_r^2) \cdot (y_1^2 + \cdots + y_s^2) = z_1^2 + \cdots + z_n^2,$$

where $X = (x_1, \ldots, x_r)$ and $Y = (y_1, \ldots, y_s)$ are systems of independent variables, and each $z_k = z_k(X, Y)$ is a bilinear form in X and Y with coefficients in a given field $\mathbb{F}$. Many constructions of such formulas use coefficients only from $\{0, 1, -1\}$. For example, the $[2, 2, 2]$-formula is the identity

$$(x_1^2 + x_2^2) \cdot (y_1^2 + y_2^2) = (x_1y_1 - x_2y_2)^2 + (x_1y_2 + x_2y_1)^2, \qquad x,\, y \in \mathbb{R}^2.$$

Set $\alpha = x_1 + x_2\mathrm{i}$ and $\beta = y_1 + y_2\mathrm{i}$ and note that the norm property $|\alpha\beta| = |\alpha||\beta|$ yields the formula displayed above. Here $z_1 = x_1y_1 - x_2y_2$ and $z_2 = x_1y_2 + x_2y_1$ are bilinear forms in X and Y. The classical results by Hurwitz [**Hur98**, **Hur22**] state:

If an $[n, n, n]$-formula exists, then $n \in \{1, 2, 4, 8\}$;
An $[r, n, n]$-formula exists if and only if $r \le \rho(n)$.

Above, the Hurwitz–Radon numbers $\rho(n)$ are defined in (VIII.23). For more details, see [**Sha00**].

IX.5 Solitary waves of the nonlinear Dirac equation

Now let us consider solitary wave solutions to (IX.1).

Definition IX.13 (Solitary waves to the nonlinear Dirac equation) Solitary wave solutions (or, simply, *solitary waves*) to the nonlinear Dirac equation (IX.1) are solutions of the form

$$\psi(t, x) = \phi(x)e^{-\mathrm{i}\omega t}, \qquad \omega \in \mathbb{R}, \qquad \phi \in H^1(\mathbb{R}^n, \mathbb{C}^N). \tag{IX.41}$$

The profile $\phi \in H^1(\mathbb{R}^n, \mathbb{C}^N)$ of a solitary wave $\phi(x)e^{-\mathrm{i}\omega t}$ satisfies the stationary nonlinear Dirac equation

$$\omega\phi = D_m\phi - f(\phi^*\beta\phi)\beta\phi, \qquad \phi(x) \in \mathbb{C}^N, \qquad x \in \mathbb{R}^n. \tag{IX.42}$$

Lemma IX.14 *For any solitary wave $\phi_\omega(x)e^{-\mathrm{i}\omega t}$ with $\phi \in H^1_{1/2}(\mathbb{R}^n)$, one has*

$$\langle\phi, \alpha^i\phi\rangle = 0, \quad \textit{for all} \ \ 1 \le i \le n; \tag{IX.43}$$

$$T^{i\nu} = T^{\nu i} = \int_{\mathbb{R}^n} \mathscr{T}^{\nu i}\, dx = 0 \quad \textit{for all}\ 1 \le i \le n,\ 0 \le \nu \le n.$$

PROOF. Since the charge-current density (IX.24) of a solitary wave $\phi(x)e^{-\mathrm{i}\omega t}$ does not depend on time,

$$\mathscr{J}^\mu(t,x) = \bar\phi(x)\gamma^\mu\phi(x) = \phi(x)^*\gamma^0\gamma^\mu\phi(x), \qquad 0 \le \mu \le n,$$

we apply the local version of the charge conservation, $\partial_\mu \mathscr{J}^\mu(t,x) = 0$, to derive

$$0 = \int_{\mathbb{R}^n} x^i \partial_t \mathscr{J}^0(x)\,dx = -\sum_{j=1}^n \int_{\mathbb{R}^n} x^i \partial_j \mathscr{J}^j(x)\,dx = \int_{\mathbb{R}^n} \mathscr{J}^i(x)\,dx, \quad 1 \le i \le n.$$

Similarly, the local conservation law (IX.30) leads to vanishing of all components of the energy-momentum tensor except perhaps T^{00}. □

IX.5.1 The virial identity for the nonlinear Dirac equation. Let us derive the virial identity satisfied by the solitary wave $\psi(t,x) = \phi_\omega(x)e^{-\mathrm{i}\omega t}$, $\phi_\omega \in H^1_{1/2}(\mathbb{R}^n,\mathbb{C}^N)$. We decompose the Hamiltonian functional (IX.26) into

$$E(\psi) = K(\psi) + M(\psi) + V(\psi),$$

with

$$\begin{aligned} K(\psi) &= \int_{\mathbb{R}^n} \psi^*(-\mathrm{i}\boldsymbol{\alpha}\cdot\nabla)\psi\,dx, \\ M(\psi) &= m\int_{\mathbb{R}^n} \psi^*\beta\psi\,dx, \\ V(\psi) &= -\int_{\mathbb{R}^n} F(\bar\psi\psi)\,dx. \end{aligned} \tag{IX.44}$$

For the convenience, let us assume that $\phi_\omega \in H^1_{1/2}(\mathbb{R}^n,\mathbb{C}^N)$. Then, combining (IX.26) and (IX.43), we conclude that

$$\mathrm{i}\int_{\mathbb{R}^n} \phi_\omega(x)^*\alpha^i\partial_{x^j}\phi_\omega(x)\,dx = \delta^i_j L(\phi_\omega e^{-\mathrm{i}\omega t}), \tag{IX.45}$$

where $L(\phi_\omega e^{-\mathrm{i}\omega t})$ is defined in (IX.28). Taking the trace of (IX.45), we obtain the relation

$$K(\phi_\omega) = K(\phi_\omega e^{-\mathrm{i}\omega t}) = -nL(\phi_\omega e^{-\mathrm{i}\omega t}). \tag{IX.46}$$

We note that, given the solitary wave (IX.41), one has $E(\phi_\omega e^{-\mathrm{i}\omega t}) = E(\phi_\omega)$ since the Hamiltonian density (IX.29) does not contain the time derivatives; similarly, the values of each of the functionals Q, K, M, and V on $\phi_\omega(x)e^{-\mathrm{i}\omega t}$ and $\phi_\omega(x)$ are the same (this is not the case for L due to the time derivative in (IX.19)). Also, comparing (IX.19) and (IX.29), we obtain:

$$L(\phi_\omega e^{-\mathrm{i}\omega t}) = -E(\phi_\omega e^{-\mathrm{i}\omega t}) + \omega Q(\phi_\omega e^{-\mathrm{i}\omega t}) = -E(\phi_\omega) + \omega Q(\phi_\omega). \tag{IX.47}$$

This leads to the following *virial identity* for the nonlinear Dirac equation (cf. (V.5)):

$$\omega Q(\phi_\omega) = \frac{n-1}{n}K(\phi_\omega) + M(\phi_\omega) + V(\phi_\omega). \tag{IX.48}$$

More generally, there is the following result that mirrors Lemma V.2:

Lemma IX.15 *Let $G \in C^1(\mathbb{R},\mathbb{R})$ satisfy*

$$G(0) = 0.$$

Let $n \in \mathbb{N}$, $N \in \mathbb{N}$, $\omega \in \mathbb{R}$, *and let* $\phi_\omega \in L^\infty_{\rm loc}(\mathbb{R}^n, \mathbb{C}^N)$ *satisfy*

$$\omega\phi_\omega = D_0\phi_\omega + G'(\phi_\omega^*\beta\phi_\omega)\beta\phi_\omega \quad \text{in} \quad \mathscr{D}'(\mathbb{R}^n, \mathbb{C}^N).$$

Assume furthermore that

$$\phi_\omega \in H^1(\mathbb{R}^n, \mathbb{C}^N), \qquad G(\phi_\omega^*\beta\phi_\omega) \in L^1(\mathbb{R}^n).$$

Then ϕ_ω *satisfies the* virial identity

$$\omega \int_{\mathbb{R}^n} \phi_\omega^*\phi_\omega\, dx = \frac{n-1}{n} \int_{\mathbb{R}^n} \phi_\omega^* D_0\phi_\omega\, dx + \int_{\mathbb{R}^n} G(\phi_\omega^*\beta\phi_\omega)\, dx.$$

PROOF. The proof is similar to that of Lemma V.2. One starts with the relation

$$\begin{aligned} x\cdot\nabla\big(\omega\phi_\omega^*\phi_\omega - G(\phi_\omega^*\beta\phi_\omega)\big) &= x\cdot\big(\nabla\phi_\omega^* D_0\phi_\omega + (D_0\phi_\omega)^*\nabla\phi_\omega\big) \\ &= x^i\big(-\mathrm{i}\partial_{x^i}\phi_\omega^*\alpha^j\partial_{x^j}\phi_\omega + \mathrm{i}\partial_{x^j}\phi_\omega^*\alpha^j\partial_{x^i}\phi_\omega\big). \end{aligned}$$

Multiplying this relation by $\rho_R(x)$ with $R \geq 1$, integrating over $\mathbb{R}^n$ (as in (V.6)), and applying integration by parts in both sides, one has

$$\begin{aligned} & n\int_{\mathbb{R}^n} \rho_R(x)\big(G(\phi_\omega^*\beta\phi_\omega) - \omega\phi_\omega^*\phi_\omega\big)\, dx \\ &= \int_{\mathbb{R}^n} \rho_R(x)\big(\mathrm{i}n\phi_\omega^*\alpha^j\partial_{x^j}\phi_\omega - \mathrm{i}\delta^i_j\phi_\omega^*\alpha^j\partial_{x^i}\phi_\omega\big)\, dx + \text{ error terms}, \end{aligned}$$

with the error terms containing the gradient of ρ_R; these terms tend to zero as R tends to infinity just like in the proof of Lemma V.2 since ϕ_ω and $\nabla\phi_\omega$ are from L^2 and $G(\phi_\omega^*\beta\phi_\omega)$ is from L^1. This concludes the proof. □

IX.5.2 Existence and non-existence of solitary waves. Construction of solitary wave solutions in the Soler model has a long history. In three spatial dimensions, the solitary waves were numerically constructed by Soler [**Sol70**] and then proved to exist in [**Vaz77, CV86, Mer88, ES95**].

The construction of solitary wave solutions to the nonlinear Dirac equation in the spatial dimension $n = 1$ is explicit; see Section IX.5.3 below. The shooting method allows one to obtain solitary waves in higher dimensions.

Theorem IX.16 (Existence of solitary waves [CV86]) *Assume that* $f \in C^1(\mathbb{R}_+, \mathbb{R}_+)$ *satisfies*

$$\lim_{\tau\to 0+} f(\tau) = 0, \qquad \lim_{\tau\to+\infty} f(\tau) = +\infty.$$

Then for any $\omega \in (0, m)$ *there is a solitary wave solution* $\phi(x)e^{-\mathrm{i}\omega t}$ *with* $\phi \in C^1(\mathbb{R}^3, \mathbb{C}^4)$ *such that*

(1) $\phi(x)^*\beta\phi(x) > 0$ *for* $x \in \mathbb{R}^3$;

(2) ϕ *and* $\nabla\phi$ *have an exponential decay as* $|x| \to +\infty$;

(3) ϕ *satisfies*

$$\omega\phi = D_m\phi - f(\phi^*\beta\phi)\beta\phi, \qquad x \in \mathbb{R}^3. \tag{IX.49}$$

Above, $D_m = -\mathrm{i}\boldsymbol{\alpha}\cdot\nabla + \beta m$ with the standard choice of the Dirac matrices (VIII.2). The solution is found in the form of the Ansatz

$$\phi(x) = \begin{bmatrix} v(r)\begin{pmatrix}1\\0\end{pmatrix} \\ \mathrm{i}u(r)\begin{pmatrix}\cos\theta\\ e^{\mathrm{i}\varphi}\sin\theta\end{pmatrix}\end{bmatrix},$$

where u and v are real-valued radial functions which satisfy

$$\begin{cases}\partial_r u + \frac{n-1}{r}u = v[f(v^2-u^2)-(m-\omega)],\\ \partial_r v = u[f(v^2-u^2)-(m+\omega)],\end{cases} \tag{IX.50}$$

$n = 3$. The proof in **[CV86]** is based on the shooting method applied to the reduced system (IX.50) with $n = 3$; taking an arbitrary $n \in \mathbb{N}$, $n \geq 2$, one extends Theorem IX.16 to all dimensions.

The existence of solitary waves under slightly weaker assumptions on the nonlinearity f is proved in **[Mer88]** (also using the shooting method) and in **[ES95]** (using the variational approach). In Chapter XII we will construct solitary wave solutions with $\omega \lessapprox m$ in any spatial dimension using the bifurcation method applied to solitary wave solutions to the nonlinear Schrödinger equation; the asymptotic behavior of these solitary waves will allow us to study their spectral stability (Chapter XIII).

A consequence of Theorem VII.18 *(1)* and Lemma VII.6 is the absence of solitary waves to the nonlinear Dirac equation with $\omega \in \mathbb{R}$, $|\omega| > m$. There are different possible formulations such as the following:

Theorem IX.17 *Let $n \geq 1$. For $\omega \in \mathbb{R}\backslash[-m, m]$, there are no solutions $\phi(x)$ to* (IX.42) *such that*

$$\phi \in H^1(\mathbb{R}^n, \mathbb{C}^N)$$

and such that $F(x) := f(\phi(x)^*\beta\phi(x))$ satisfies $F \in L^n_{\mathrm{loc}}(\mathbb{R}^n)$, $F \neq 0$ almost everywhere in $x \in \mathbb{R}^n$, and

$$|F(x)| \leq \kappa\frac{\sqrt{\omega^2-m^2}}{4|\omega||x|} \qquad \textit{for } x \textit{ almost everywhere in } \Omega^n_R,$$

with some $R > 0$ and some $\kappa \in (0, 1)$.

IX.5.3 Construction of solitary waves in 1D. Let us give the explicit construction of solitary waves in the nonlinear Dirac equation with the Soler-type nonlinearity in one spatial dimension. Consider the following formulation of (IX.1) in the case $n = 1$, $N = 2$:

$$\mathrm{i}\partial_t\psi = -\mathrm{i}\alpha\partial_x\psi + g(\psi^*\beta\psi)\beta\psi, \qquad \psi(t,x) \in \mathbb{C}^2, \qquad x \in \mathbb{R}^1. \tag{IX.51}$$

As α and β, we use the Pauli matrices:

$$\alpha = -\sigma_2, \qquad \beta = \sigma_3. \tag{IX.52}$$

Noting that $\psi^*\sigma_3\psi = |\psi_1|^2 - |\psi_2|^2$, we rewrite equation (IX.51) as the following system:

$$\begin{cases}\mathrm{i}\partial_t\psi_1 = \partial_x\psi_2 + g(|\psi_1|^2-|\psi_2|^2)\psi_1,\\ \mathrm{i}\partial_t\psi_2 = -\partial_x\psi_1 - g(|\psi_1|^2-|\psi_2|^2)\psi_2.\end{cases} \tag{IX.53}$$

The next result follows from **[CV86]**.

Lemma IX.18 *Let $g \in C(\mathbb{R})$. Assume that*

$$m := g(0) > 0. \tag{IX.54}$$

Let $G(s)$ be the antiderivative of $g(s)$ such that $G(0) = 0$. Assume that for a given $\omega \in (0, m)$ there exists $\mathscr{X}_\omega > 0$ such that

$$\omega\mathscr{X}_\omega = G(\mathscr{X}_\omega), \quad \omega \neq g(\mathscr{X}_\omega), \quad \textit{and} \quad \omega s < G(s) \quad \text{for } s \in (0, \mathscr{X}_\omega). \tag{IX.55}$$

Then there is a solitary wave solution $\psi(t,x) = \phi_\omega(x)e^{-i\omega t}$, *where*

$$\phi_\omega(x) = \begin{bmatrix} v(x) \\ u(x) \end{bmatrix}, \qquad v,\ u \in H^1(\mathbb{R}), \tag{IX.56}$$

with both v *and* u *real-valued,* v *being even and* u *being odd.*

More precisely, let us define $\mathscr{X}(x)$ *and* $\mathscr{Y}(x)$ *by*

$$\mathscr{X} = v^2 - u^2, \qquad \mathscr{Y} = vu. \tag{IX.57}$$

Then $\mathscr{X}(x)$ *is the solution to*

$$\mathscr{X}'' = -\partial_{\mathscr{X}}(-2G(\mathscr{X})^2 + 2\omega^2\mathscr{X}^2), \quad \mathscr{X}(0) = \mathscr{X}_\omega, \quad \mathscr{X}'(0) = 0, \tag{IX.58}$$

and $\mathscr{Y}(x) = -\frac{1}{4\omega}\mathscr{X}'(x)$.

Remark IX.19 Condition (IX.55) imposes the restriction $\omega \le G'(0) = m$.

Proof. From (IX.53), we obtain:

$$\begin{cases} \omega v = \partial_x u + g(|v|^2 - |u|^2)v, \\ \omega u = -\partial_x v - g(|v|^2 - |u|^2)u. \end{cases} \tag{IX.59}$$

Assuming that both v and u are real-valued (this will be justified once we found real-valued v and u), we can rewrite (IX.59) as the following Hamiltonian system, with x playing the role of time:

$$\begin{cases} \partial_x u = \omega v - g(v^2 - u^2)v = \partial_v h(v,u), \\ -\partial_x v = \omega u + g(v^2 - u^2)u = \partial_u h(v,u), \end{cases} \tag{IX.60}$$

where the Hamiltonian $h(v,u)$ is given by

$$h(v,u) = \frac{\omega}{2}(v^2 + u^2) - \frac{1}{2}G(v^2 - u^2). \tag{IX.61}$$

The solitary wave corresponds to a trajectory of this Hamiltonian system such that

$$\lim_{x\to\pm\infty} v(x) = \lim_{x\to\pm\infty} u(x) = 0,$$

hence $\lim\limits_{x\to\pm\infty} \mathscr{X} = 0$. Since $G(s)$ satisfies $G(0) = 0$, we conclude that

$$h(v(x), u(x)) \equiv 0, \tag{IX.62}$$

which leads to

$$\omega(v^2 + u^2) = G(v^2 - u^2). \tag{IX.63}$$

Studying the level curves which solve this equation is most convenient in the coordinates

$$\mathscr{X} = v^2 - u^2, \qquad \mathscr{Z} = v^2 + u^2.$$

One can conclude from (IX.63) (see Figure IX.1) that solitary waves may only correspond to $|\omega| < m$, $\omega \ne 0$.

Remark IX.20 If $\omega > 0$, then there are solitary waves such that $v(x) \ne 0$ for all $x \in \mathbb{R}$ while $u(x)$ changes the sign (shifting the origin, we may assume that this happens at $x = 0$). For $\omega < 0$, the role of u and v is interchanged: for particular nonlinearities (for example, for $g(\tau)$ being even), there may be solitary waves such that $u(x) \ne 0$ for all $x \in \mathbb{R}$, while $v(x)$ changes the sign at $x = 0$. See Figure IX.1.

Remark IX.21 In the case when $G(s)$ is odd, for each solitary wave corresponding to $\omega \in \mathbb{R}$ there is a solitary wave corresponding to $-\omega$. More precisely, in this case, if $\begin{bmatrix} v(x) \\ u(x) \end{bmatrix} e^{-\mathrm{i}\omega t}$ is a solitary wave, then so is $\begin{bmatrix} u(x) \\ v(x) \end{bmatrix} e^{\mathrm{i}\omega t}$.

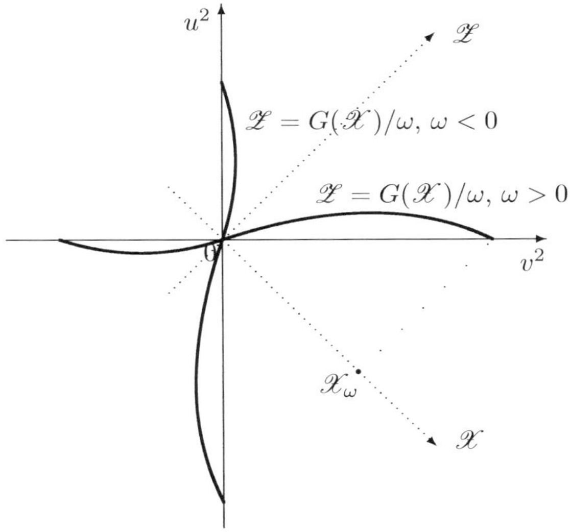

FIGURE IX.1. Solitary waves in the coordinates $\mathscr{X} = v^2 - u^2$, $\mathscr{Z} = v^2 + u^2$. Solitons with $\omega > 0$ and $\omega < 0$ correspond to the bump on the v^2 axis and to the dotted bump on the u^2 axis (respectively) in the first quadrant.

The functions $\mathscr{X}(x)$ and $\mathscr{Y}(x)$ introduced in (IX.57) are to solve

$$\text{(IX.64)} \qquad \begin{cases} \mathscr{X}' = -4\omega\mathscr{Y}, \\ \mathscr{Y}' = -(v^2 + u^2)g(\mathscr{X}) + \omega\mathscr{X} = -\frac{1}{\omega}G(\mathscr{X})g(\mathscr{X}) + \omega\mathscr{X}, \end{cases}$$

and to have the asymptotic behavior

$$\lim_{|x|\to\infty} \mathscr{X}(x) = 0, \qquad \lim_{|x|\to\infty} \mathscr{Y}(x) = 0.$$

In the second equation in (IX.64), we used the relation (IX.63). The system (IX.64) can be written as the following equation on $\mathscr{X}$:

$$\text{(IX.65)} \qquad \mathscr{X}'' = -\partial_{\mathscr{X}}(-2G(\mathscr{X})^2 + 2\omega^2\mathscr{X}^2).$$

This equation describes a particle in the potential $V_\omega(\mathscr{X}) = -2G(\mathscr{X})^2 + 2\omega^2\mathscr{X}^2$, $\mathscr{X} \in \mathbb{R}$. Due to the energy conservation (with x playing the role of time), we get:

$$\text{(IX.66)} \qquad \frac{\mathscr{X}'^2}{2} - 2G(\mathscr{X})^2 + 2\omega^2\mathscr{X}^2 = \frac{\mathscr{X}'^2}{2} + V_\omega(\mathscr{X}) = 0.$$

Using the expression for $\mathscr{X}'$ from (IX.64), the relation (IX.66) could be rewritten as

$$\begin{aligned} 0 = \frac{\mathscr{X}'^2}{2} + V_\omega(\mathscr{X}) &= 8\omega^2\mathscr{Y}^2 - 2G(\mathscr{X})^2 + 2\omega^2\mathscr{X}^2 \\ \text{(IX.67)} \qquad &= 2\omega^2\big(4v^2u^2 + (v^2 - u^2)^2\big) - 2G^2, \end{aligned}$$

which follows from (IX.63).

For a particular value of $\omega \in (0, m)$, there will be a positive solution $\mathscr{X}(x)$ such that $\lim_{x\to\pm\infty} \mathscr{X}(x) = 0$ if there exists $\mathscr{X}_\omega > 0$ so that (IX.55) is satisfied. We shift x so that $\mathscr{X}(0) = \mathscr{X}_\omega$; then $\mathscr{X}(x)$ is an even function.

Once $\mathscr{X}(x)$ is known, $\mathscr{Y}(x)$ is obtained from (IX.64), and then we can express $v(x)$, $u(x)$. □

Remark IX.22 Note that for $0 < |\omega| < 1$, the functions $v(x)$ and $u(x)$ are exponentially decaying as $|x| \to \infty$. Indeed, the exponential decay of $\mathscr{X}(x)$ could be deduced from (IX.65), which we leave as an exercise. Then the exponential decay of $\mathscr{Z}(x) = v(x)^2 + u(x)^2$ follows from the relation $\mathscr{Z} = G(\mathscr{X})/\omega$ (cf. (IX.63)).

Explicit form of the solitary waves in the cubic case. As shown in [**LG75**] for the massive Gross–Neveu model (the Soler model in 1D), in the special case of the potential

$$G(\mathscr{X}) = \mathscr{X} - \frac{\mathscr{X}^2}{2}, \qquad \mathscr{X} \in \mathbb{R}, \tag{IX.68}$$

the solitary waves can be found explicitly. Substituting $G(\mathscr{X})$ from (IX.68) into (IX.66), we get the following relation:

$$dx = -\frac{d\mathscr{X}}{2\mathscr{X}\sqrt{(1-\mathscr{X}/2)^2 - \omega^2}}. \tag{IX.69}$$

We use the substitution

$$1 - \frac{\mathscr{X}}{2} = \frac{\omega}{\cos 2\Theta}, \qquad \mathscr{X} = 2\Big(1 - \frac{\omega}{\cos 2\Theta}\Big); \tag{IX.70}$$

then

$$dx = \frac{2\frac{2\omega \sin 2\Theta}{\cos^2 2\Theta}\, d\Theta}{4(1 - \frac{\omega}{\cos 2\Theta})\sqrt{\frac{\omega^2}{\cos^2 2\Theta} - \omega^2}} = \frac{d\Theta}{\cos 2\Theta - \omega}, \tag{IX.71}$$

$$x = \frac{1}{2\sqrt{1-\omega^2}} \ln \left| \frac{\sqrt{\mu} + \tan\Theta}{\sqrt{\mu} - \tan\Theta} \right|, \tag{IX.72}$$

where

$$\mu = \frac{1-\omega}{1+\omega}. \tag{IX.73}$$

Then (IX.72) takes the form

$$(\sqrt{\mu} - \tan\Theta)e^{2x\sqrt{1-\omega^2}} = \sqrt{\mu} + \tan\Theta,$$

$$\tan\Theta(1 + e^{2x\sqrt{1-\omega^2}}) = \sqrt{\mu}(e^{2x\sqrt{1-\omega^2}} - 1),$$

which yields

$$\tan\Theta(x) = \sqrt{\mu}\tanh(x\sqrt{1-\omega^2}). \tag{IX.74}$$

Also note that

$$\mathscr{X}(x) = 2\left(1 - \frac{\omega}{\cos 2\Theta}\right) = 2\left(1 - \omega\frac{1 + \tan^2\Theta(x)}{1 - \tan^2\Theta(x)}\right), \tag{IX.75}$$

and then

$$\begin{aligned}\mathscr{Y}(x) &= -\frac{1}{4\omega}\mathscr{X}'(x) = -\frac{1}{4}\frac{2}{\cos^2 2\Theta}(-2\sin 2\Theta)\frac{d\Theta}{dx}\\ &= -\frac{1}{4}\frac{2}{\cos^2 2\Theta}(-2\sin 2\Theta)(\cos 2\Theta - \omega) = \frac{\mathscr{X}}{2}\tan 2\Theta\\ &= \frac{\mathscr{X}}{2}\frac{2\tan\Theta}{1-\tan^2\Theta} = \mathscr{X}(x)\frac{\sqrt{\mu}\tanh(x\sqrt{1-\omega^2})}{1-\mu\tanh^2(x\sqrt{1-\omega^2})}.\end{aligned}$$

Denote

$$\mathscr{Z}(x) = v^2(x) + u^2(x). \tag{IX.76}$$

Then

$$\begin{aligned}\mathscr{Z}(x) &= \frac{2}{\cos 2\Theta(x)}\left(1 - \frac{\omega}{\cos 2\Theta(x)}\right)\\ &= 2\frac{1+\tan^2\Theta(x)}{1-\tan^2\Theta(x)}\left(1 - \omega\frac{1+\tan^2\Theta(x)}{1-\tan^2\Theta(x)}\right).\end{aligned}$$

Other functions are expressed from $\mathscr{Z}$ as follows:

$$v(x) = \sqrt{\mathscr{Z}(x)}\cos\Theta(x), \qquad u(x) = -\sqrt{\mathscr{Z}(x)}\sin\Theta(x), \tag{IX.77}$$

$$\mathscr{X}(x) = \mathscr{Z}(x)\cos 2\Theta(x), \qquad \mathscr{Y}(x) = -\frac{1}{2}\mathscr{Z}(x)\sin 2\Theta(x). \tag{IX.78}$$

Remark IX.23 By (IX.74), $\tan\Theta(x)$ changes from $-\sqrt{\mu}$ to $\sqrt{\mu}$ as x changes from $-\infty$ to $+\infty$. Thus, in the limit $\omega \to 1$, when $\mu \to 0$, one has $\mathscr{X} \approx \mathscr{Z}$, while $|\mathscr{Y}| \lessapprox \mathscr{Z}\sqrt{\mu}$.

Combining equations (IX.77) with (IX.73), (IX.74) and using basic trigonometric identities, we obtain the following explicit formulae for $v(x)$ and $u(x)$:

$$v(x) = \frac{\sqrt{2(1-\omega)}}{(1-\mu\tanh^2(x\sqrt{1-\omega^2}))\cosh(x\sqrt{1-\omega^2})}, \tag{IX.79}$$

$$u(x) = \frac{\sqrt{2\mu(1-\omega)}\tanh(x\sqrt{1-\omega^2})}{(1-\mu\tanh^2(x\sqrt{1-\omega^2}))\cosh(x\sqrt{1-\omega^2})}. \tag{IX.80}$$

IX.5.4 Exponential decay of solitary waves.

Theorem IX.24 *Let $n \geq 1$ and assume that f is measurable and bounded on open sets, with $\lim_{\tau\to 0} f(\tau) = 0$. Let $\omega \in (-m, m)$ and let*

$$\phi_\omega \in L^2(\mathbb{R}^n, \mathbb{C}^N) \cap L^\infty(\mathbb{R}^n, \mathbb{C}^N)$$

be a solution to (IX.42) *which satisfies*

$$\lim_{R\to\infty} \|\phi_\omega\|_{L^\infty(\Omega^n_R, \mathbb{C}^N)} = 0, \tag{IX.81}$$

where $\Omega^n_R = \{x \in \mathbb{R}^n\colon\ |x| > R\}$. Then for any $\mu < \sqrt{m^2-\omega^2}$ one has

$$e^{\mu\langle r\rangle}\phi_\omega \in H^1(\mathbb{R}^n, \mathbb{C}^N).$$

Proof. The proof reduces to proving that $V = f(\phi_\omega^*\beta\phi_\omega)\beta$ is small at infinity compared to the Dirac operator. This, in turn, follows from the assumptions $\lim_{\tau\to 0} f(\tau) = 0$ and $\lim_{R\to\infty} \|\phi_\omega\|_{L^\infty(\Omega^n_R, \mathbb{C}^N)} = 0$. Thus, the assumption (VII.82) of Lemma VII.21 is satisfied; the application of that lemma concludes the proof. □

IX.6 Linearization at a solitary wave

To consider the linearization at the solitary wave

$$\phi_\omega(x)e^{-\mathrm{i}\omega t}, \quad \omega \in [-m,m], \quad \phi_\omega \in H^1(\mathbb{R}^n,\mathbb{C}^N)\cap L^\infty(\mathbb{R}^n,\mathbb{C}^N), \tag{IX.82}$$

we assume that in the nonlinear Dirac equation (IX.1) one has $f \in C^1(\mathbb{R})$, and write the solution in the form of the Ansatz

$$\psi(t,x) = (\phi_\omega(x) + \rho(t,x))e^{-\mathrm{i}\omega t}, \qquad \rho(t,x)\in\mathbb{C}^N, \qquad (t,x)\in\mathbb{R}\times\mathbb{R}^n.$$

The linearization at the solitary wave (IX.82) is then the linearized equation on $\rho(t,x)$, given by

$$\mathrm{i}\partial_t\rho = \mathcal{L}(\omega)\rho, \tag{IX.83}$$

where $\mathcal{L}(\omega) : L^2(\mathbb{R}^n,\mathbb{C}^N) \to L^2(\mathbb{R}^n,\mathbb{C}^N)$ is the differential operator with domain $\mathfrak{D}(\mathcal{L}(\omega)) = H^1(\mathbb{R}^n,\mathbb{C}^N)$ defined by

$$\mathcal{L}(\omega)\rho = (D_m - \omega - f\beta)\rho - 2\,\mathrm{Re}(\phi_\omega^*\beta\rho)f'\beta\phi_\omega. \tag{IX.84}$$

Above, $f = f(\tau)$ and $f' = f'(\tau)$ are evaluated at $\tau = \phi_\omega^*\beta\phi_\omega$. The operator $\mathcal{L}(\omega)$ is $\mathbb{R}$-linear but not $\mathbb{C}$-linear, because of the term with $\mathrm{Re}(\phi_\omega^*\beta\,\cdot)$. To work with $\mathbb{C}$-linear operators, we introduce the matrices representing the action of α^i with $1\le i\le n$, β, and $-\mathrm{i}$ on $\mathbb{R}^{2N}$-valued functions:

$$\begin{aligned}
\boldsymbol{\alpha}^i &= \begin{bmatrix}\mathrm{Re}\,\alpha^i & -\mathrm{Im}\,\alpha^i\\ \mathrm{Im}\,\alpha^i & \mathrm{Re}\,\alpha^i\end{bmatrix}, \quad 1\le i\le n,\\
\boldsymbol{\beta} &= \begin{bmatrix}\mathrm{Re}\,\beta & -\mathrm{Im}\,\beta\\ \mathrm{Im}\,\beta & \mathrm{Re}\,\beta\end{bmatrix},\\
\mathbf{J} &= \begin{bmatrix}0 & I_N\\ -I_N & 0\end{bmatrix},
\end{aligned} \tag{IX.85}$$

where the real part of a matrix is defined as the matrix consisting of the real part of its entries (and similarly for the imaginary part of a matrix). Since the profile of a solitary wave solution $\phi_\omega(x)e^{-\mathrm{i}\omega t}$ satisfies the stationary equation (IX.42), the profile

$$\boldsymbol{\phi}_\omega(x) = \begin{bmatrix}\mathrm{Re}\,\phi_\omega(x)\\ \mathrm{Im}\,\phi_\omega(x)\end{bmatrix} \in \mathbb{R}^{2N} \tag{IX.86}$$

satisfies the relation

$$\mathbf{L}_-(\omega)\boldsymbol{\phi}_\omega = 0, \tag{IX.87}$$

where

$$\mathbf{L}_-(\omega) = \mathbf{D}_m - \omega - f(\boldsymbol{\phi}_\omega^*\boldsymbol{\beta}\boldsymbol{\phi}_\omega)\boldsymbol{\beta}, \tag{IX.88}$$

with

$$\mathbf{D}_m = \mathbf{J}\boldsymbol{\alpha}\cdot\nabla + \boldsymbol{\beta}m \tag{IX.89}$$

representing the action of D_m on $\mathbb{C}^N$-valued functions which are considered as $\mathbb{R}^{2N}$-valued. The action of the operator (IX.84) on $\mathbb{R}^{2N}$-valued functions is represented by the operator

$$\mathbf{L}(\omega)\boldsymbol{\rho} = (\mathbf{J}\boldsymbol{\alpha}\cdot\nabla_x + \boldsymbol{\beta}m)\boldsymbol{\rho} - \omega\boldsymbol{\rho} - f\boldsymbol{\beta}\boldsymbol{\rho} - 2(\boldsymbol{\phi}_\omega^*\boldsymbol{\beta}\boldsymbol{\rho})f'\boldsymbol{\beta}\boldsymbol{\phi}_\omega, \tag{IX.90}$$

$\boldsymbol{\rho} \in L^2(\mathbb{R}^n, \mathbb{R}^{2N})$, with $f = f(\tau)$ and $f' = f'(\tau)$ evaluated at $\tau = \phi_\omega^* \beta \phi_\omega$. Recall that we assume that $f \in C^1(\mathbb{R})$ and that $\phi_\omega \in L^\infty(\mathbb{R}^n, \mathbb{C}^N)$, hence both $\mathbf{L}_-(\omega)$ and $\mathbf{L}(\omega)$ are self-adjoint operators on the domain

$$\text{(IX.91)} \qquad \mathfrak{D}(\mathbf{L}(\omega)) = H^1(\mathbb{R}^n, \mathbb{C}^{2N}) = H^1(\mathbb{R}^n, \mathbb{C} \otimes_{\mathbb{R}} \mathbb{R}^{2N}),$$

to which they are extended by $\mathbb{C}$-linearity. Thus, the linearization (IX.83) at the solitary wave takes the form

$$\text{(IX.92)} \qquad \partial_t \boldsymbol{\rho} = \mathbf{J}\mathbf{L}(\omega)\boldsymbol{\rho}, \qquad \boldsymbol{\rho}(t,x) = \begin{bmatrix} \operatorname{Re}\rho(t,x) \\ \operatorname{Im}\rho(t,x) \end{bmatrix} \in \mathbb{R}^{2N}.$$

Lemma IX.25 *The operator $\mathbf{J}\mathbf{L}(\omega)$ is closed and its spectrum $\sigma(\mathbf{J}\mathbf{L}(\omega))$ is symmetric with respect to the real and imaginary axes.*

PROOF. The closedness is immediate.

Let $\lambda \in \sigma(\mathbf{J}\mathbf{L})$. The inclusion $\bar\lambda \in \sigma(\mathbf{J}\mathbf{L})$ follows from $\mathbf{J}\mathbf{L}$ acting invariantly in the subspace $L^2(\mathbb{R}^n, \mathbb{R}^{2N}) \subset L^2(\mathbb{R}^n, \mathbb{C}^{2N})$. The inclusion $-\bar\lambda \in \sigma(\mathbf{J}\mathbf{L})$ follows from $(-\mathbf{J}\mathbf{L})^*$ being the conjugate to $\mathbf{J}\mathbf{L}$:

$$(-\mathbf{J}\mathbf{L})^* = \mathbf{L}^*(-\mathbf{J})^* = \mathbf{L}\mathbf{J} = \mathbf{J}^{-1}(\mathbf{J}\mathbf{L})\mathbf{J}. \qquad \square$$

IX.6.1 The structure of the null space. We assume that there are solitary wave solutions to (IX.21) of the form (IX.41) with $\omega \in \Omega$, where $\Omega \subset \mathbb{R}$ is an open set. Many quantities appearing below will depend on ω; we will indicate this dependence by the subscript ω. Sometimes this subscript will be omitted to shorten the notations.

Due to the $\mathbf{U}(1)$-invariance of the equation, the perturbation $\rho(t,x)$ that corresponds to the infinitesimal phase shift of the solitary wave, $\mathrm{i}\phi_\omega$, is in the kernel of the linearization $\mathbf{J}\mathbf{L}(\omega)$; as a consequence, there is a relation $\mathbf{L}(\omega)\mathbf{J}\boldsymbol{\phi}_\omega = 0$ (cf. (IX.87)), with $\boldsymbol{\phi}_\omega$ from (IX.86). Similarly, due to the translational invariance, $\partial_{x^i}\phi_\omega$ are also in the kernel of the linearized operator; as a result,

$$\text{(IX.93)} \qquad \mathbf{J}\boldsymbol{\phi}_\omega,\ \partial_{x^i}\boldsymbol{\phi}_\omega \in \mathbf{ker}(\mathbf{J}\mathbf{L}(\omega)), \qquad 1 \le i \le n.$$

One can check by direct computation that there is a Jordan block corresponding to each of these eigenvectors:

$$\text{(IX.94)} \qquad \mathbf{J}\mathbf{L}(\omega)\partial_\omega\boldsymbol{\phi}_\omega = \mathbf{J}\boldsymbol{\phi}_\omega, \qquad \mathbf{J}\mathbf{L}(\omega)\boldsymbol{\xi}^i = \partial_{x^i}\boldsymbol{\phi}_\omega, \quad 1 \le i \le n,$$

where

$$\text{(IX.95)} \qquad \boldsymbol{\xi}^i = \omega x^i \mathbf{J}\boldsymbol{\phi}_\omega - \frac{1}{2}\boldsymbol{\alpha}^i\boldsymbol{\phi}_\omega.$$

In particular, taking the derivatives in x^i, $1 \le i \le n$, and in ω of the relation $E'(\phi_\omega) = \omega Q'(\phi_\omega)$, one arrives at the relations $\mathbf{L}\partial_{x^i}\boldsymbol{\phi}_\omega = 0$ and $\mathbf{L}\partial_\omega\boldsymbol{\phi}_\omega = \boldsymbol{\phi}_\omega$.

We summarize the results on the spectral subspace of $\mathbf{J}\mathbf{L}(\omega)$ corresponding to the zero eigenvalue:

Lemma IX.26

$$\operatorname{Span}\{\mathbf{J}\boldsymbol{\phi}_\omega,\ \partial_i\boldsymbol{\phi}_\omega\colon\ 1 \le i \le n\} \subset \mathbf{ker}(\mathbf{J}\mathbf{L}(\omega)),$$

$$\operatorname{Span}\{\mathbf{J}\boldsymbol{\phi}_\omega,\ \partial_\omega\boldsymbol{\phi}_\omega,\ \partial_i\boldsymbol{\phi}_\omega,\ \boldsymbol{\alpha}^i\boldsymbol{\phi}_\omega - 2\omega x^i\mathbf{J}\boldsymbol{\phi}_\omega\colon\ 1 \le i \le n\} \subset \mathfrak{L}(\mathbf{J}\mathbf{L}(\omega)).$$

Remark IX.27 This lemma does not give the complete characterization of the kernel of $\mathbf{JL}(\omega)$; for example, there are also eigenvectors due to the rotational invariance and purely imaginary eigenvalues passing through $\lambda = 0$ at some particular values of ω [**CMKS+16**]. We also refer to the proof of Proposition XIII.26 below, which gives the dimension of the generalized null space for $\omega \lessapprox m$, showing that in this case the inclusions in Lemma IX.26 turn into equalities.

By (IX.94), there are Jordan blocks of size at least 2 corresponding to each of the vectors $\mathbf{J}\boldsymbol{\phi}_\omega$, $\partial_{x^i}\boldsymbol{\phi}_\omega$ from the null space. When two (or more) eigenvalues collide at $\lambda = 0$, at a particular value of ω, they can join one of these two types of Jordan blocks permanently residing at 0. We now consider these two events.

IX.6.2 U(1)-invariance and the Kolokolov condition. Let us revisit the Kolokolov criterion (see Section IV.2) from the point of view of the size of a particular Jordan block at $\lambda = 0$. By (IX.94), the Jordan block of $\mathbf{JL}(\omega)$ corresponding to the $\mathbf{U}(1)$-invariance is of size at least 2. The size of this Jordan block jumps up when we can solve the generalized eigenvector equation $\mathbf{JL}\boldsymbol{u} = \partial_\omega\boldsymbol{\phi}_\omega$. Since $\mathbf{L}$ is Fredholm (this follows from $0 \notin \sigma_{\mathrm{ess}}(\mathbf{L}) = \mathbb{R}\backslash(-m, m)$; see Lemma III.160), such $\boldsymbol{u}$ exists if $\partial_\omega\boldsymbol{\phi}_\omega$ is orthogonal to the null space of $(\mathbf{JL})^* = -\mathbf{LJ}$. The generalized eigenvector $\partial_\omega\boldsymbol{\phi}_\omega$ is always orthogonal to $\mathbf{J}^{-1}\partial_{x^i}\boldsymbol{\phi}_\omega \in \mathbf{ker}(\mathbf{LJ})$, $1 \le i \le n$. Indeed, we have:

$$\langle\partial_\omega\boldsymbol{\phi}_\omega, \mathbf{J}\partial_{x^i}\boldsymbol{\phi}_\omega\rangle = -\langle\boldsymbol{\phi}_\omega, \boldsymbol{\xi}^i\rangle = \frac{\langle\boldsymbol{\phi}_\omega, \boldsymbol{\alpha}^i\boldsymbol{\phi}_\omega\rangle}{2} - \omega\langle\boldsymbol{\phi}_\omega, x^i\mathbf{J}\boldsymbol{\phi}_\omega\rangle, \tag{IX.96}$$

where we used (IX.94) and self-adjointness of $\mathbf{L}$. By Lemma IX.14, the first term in the right-hand side is zero. The second term in the right-hand side is zero due to skew-symmetry of $\mathbf{J}$. Thus,

$$\langle\partial_\omega\boldsymbol{\phi}_\omega, \mathbf{J}\partial_{x^i}\boldsymbol{\phi}_\omega\rangle = 0, \qquad 1 \le i \le n. \tag{IX.97}$$

We now need to check whether $\partial_\omega\boldsymbol{\phi}_\omega$ is orthogonal to $\boldsymbol{\phi}_\omega \in \mathbf{ker}(\mathbf{LJ})$. The orthogonality condition takes the form

$$\langle\partial_\omega\boldsymbol{\phi}_\omega, \boldsymbol{\phi}_\omega\rangle = \frac{1}{2}\partial_\omega Q(\phi_\omega) = 0. \tag{IX.98}$$

This is directly related to the Kolokolov condition derived in the context of the nonlinear Schrödinger equation (see (IV.16)) and more abstract Hamiltonian systems with $\mathbf{U}(1)$-invariance [**Kol73, GSS87**].

IX.6.3 Translational invariance and the energy vanishing condition. Let us find the condition for the increase in size of the Jordan block corresponding to translational invariance [**BCS15**]. This happens if there is $\boldsymbol{\zeta}$ such that $\mathbf{JL}\boldsymbol{\zeta} = \boldsymbol{\xi}$, where $\boldsymbol{\xi} = \sum_{i=1}^n c_i\boldsymbol{\xi}^i \ne 0$ is some nontrivial linear combination of the generalized eigenvectors (IX.95). Since $\mathbf{L}$ is Fredholm, the sufficient condition is that $\boldsymbol{\xi}$ is orthogonal to vectors from $\mathbf{ker}(\mathbf{LJ})$. By (IX.96) and (IX.97), one always has

$$\Big\langle\omega x^i\mathbf{J}\boldsymbol{\phi}_\omega - \frac{1}{2}\boldsymbol{\alpha}^i\boldsymbol{\phi}_\omega, \boldsymbol{\phi}_\omega\Big\rangle = 0, \qquad 1 \le i \le n, \tag{IX.99}$$

ensuring orthogonality of $\boldsymbol{\xi}$ to $\boldsymbol{\phi}_\omega \in \mathbf{ker}(\mathbf{LJ})$.

Now we need to ensure orthogonality of $\boldsymbol{\xi}$ to all of $\mathbf{J}^{-1}\partial_{x^i}\boldsymbol{\phi}_\omega \in \mathbf{ker}(\mathbf{LJ})$, $1 \le i \le n$. We may write this condition in the form

$$\det C^i_j(\omega) = 0, \qquad C^i_j(\omega) := -2\langle\boldsymbol{\xi}^i, \mathbf{J}\partial_{x^j}\boldsymbol{\phi}_\omega\rangle, \qquad 1 \le i, j \le n. \tag{IX.100}$$

Substituting $\boldsymbol{\xi}^i$ from (IX.95), we have:

$$C^i_j(\omega) = \langle \boldsymbol{\alpha}^i \boldsymbol{\phi}_\omega - 2\omega x^i \mathbf{J}\boldsymbol{\phi}_\omega, \mathbf{J}\partial_{x^j}\boldsymbol{\phi}_\omega \rangle, \qquad 1 \le i, j \le n. \tag{IX.101}$$

Since

$$\langle 2x^i \mathbf{J}\phi_\omega, \mathbf{J}\partial_{x^j}\boldsymbol{\phi}_\omega \rangle = \int_{\mathbb{R}^n} x^i \partial_{x^j} (\boldsymbol{\phi}_\omega^* \boldsymbol{\phi}_\omega)\, dx = -Q(\phi_\omega)\delta^i_j, \qquad 1 \le i, j \le n,$$

we rewrite (IX.101) as

$$C^i_j(\omega) = \langle \boldsymbol{\alpha}^i \boldsymbol{\phi}_\omega, \mathbf{J}\partial_{x^j}\boldsymbol{\phi}_\omega \rangle + \omega Q(\phi_\omega)\delta^i_j, \qquad 1 \le i, j \le n. \tag{IX.102}$$

Using (IX.45), (IX.47), and (IX.102), we get $C^i_j(\omega) = E(\phi_\omega)\delta^i_j$, with $E(\phi_\omega)$ and $Q(\phi_\omega)$ the values of the energy functionals (IX.26) and the charge functional (IX.22) at the solitary wave $\phi_\omega(x)e^{-\mathrm{i}\omega t}$ (we notice that these functionals do not contain the time derivatives, hence $E(\phi_\omega) = E(\phi_\omega e^{-\mathrm{i}\omega t})$, $Q(\phi_\omega) = Q(\phi_\omega e^{-\mathrm{i}\omega t})$). Thus, the condition (IX.100) for the increase of the size of the Jordan block in the nonlinear Dirac equation is equivalent to

$$E(\phi_\omega) = 0. \tag{IX.103}$$

We mention that, by **[BCS15]**, this condition is satisfied for the massive Thirring model in one dimension, for any pure power nonlinearity other than cubic, at some $\omega \in (-m, 0)$.

Remark IX.28 In the case of the nonlinear Schrödinger equation (V.1) with any nonlinearity $f \in C(\mathbb{R})$, the corresponding Jordan block is always of constant size; this follows from equations (V.86), (V.87), and (V.88) in the proof of Lemma V.27.

The Dirac equation with pure power nonlinearity does not have such a degeneracy for $\omega > 0$:

Lemma IX.29 *Let $\kappa > 0$ and let $F(\tau) = c|\tau|^{\kappa+1}$, with $c > 0$, and assume that $\kappa \le 2/(n-2)$ if $n \ge 3$. If $\phi_\omega(x)e^{-\mathrm{i}\omega t}$, $\phi_\omega \in H^1(\mathbb{R}^n, \mathbb{C}^N) \cap L^\infty_{\mathrm{loc}}(\mathbb{R}^n, \mathbb{C}^N)$, $\omega > 0$, is a solitary wave solution to* (IX.1), *then one has*

$$E(\phi_\omega) > 0. \tag{IX.104}$$

PROOF. Multiplying (IX.42) on the left by ϕ_ω^* and integrating, we arrive at the relation

$$\omega Q = K + M + (\kappa + 1)V,$$

with Q, K, M, and V from (IX.44). This relation and the virial identity from Lemma IX.15 yield the equality $\frac{1}{n}K(\phi_\omega) = -\kappa V(\phi_\omega)$. With $V = -\int_{\mathbb{R}^n} F(\bar{\phi}_\omega \phi_\omega)\, dx \le 0$ and $\omega > 0$, one arrives at $E = \omega Q - \kappa V > 0$. We note that $F(\bar{\phi}_\omega \phi_\omega) \in L^1(\mathbb{R}^n)$ due to the assumption that $\kappa \le 2/(n-2)$ when $n \ge 3$. □

IX.6.4 Contribution into the kernel due to the rotational invariance. Using the expression for the Lorentz transformation of $\psi(t,x)$ from (IX.31), one can compute the vectors from the null space of the linearization operator $\mathcal{L}(\omega)$ from (IX.84) which are contributed by the rotational symmetry:

$$\begin{aligned} &\frac{\partial}{\partial s}\Big|_{s=0} e^{s\rho(R)}\phi_\omega(e^{sR}x) \\ &= \rho(R)\phi_\omega(x) + \nabla\phi_\omega(x)Rx = \frac{\mathrm{i}}{4}R_{ij}\sigma^{ij}\phi_\omega(x) + R^i{}_j x^j \partial_{x^i}\phi_\omega(x) \end{aligned}$$

with $R = (R^i{}_j)_{1 \le i,j \le n} \in \mathbf{so}(n)$, $R_{ij} = g_{ik}R^k{}_j = -R^i{}_j$, $1 \le i, j \le n$ and with $\rho(R) = \frac{\mathrm{i}}{4}R_{ij}\sigma^{ij}$ (see (IX.32)).

Rotational symmetry in three spatial dimensions. Let us study these vectors in more detail in the case of three spatial dimensions, with the standard choice of the Dirac matrices. Denote

$$\varSigma_i = \mathrm{diag}[\sigma_i, \sigma_i], \quad \mathbf{\Sigma}_i = \begin{bmatrix} \mathrm{Re}\,\varSigma_i & -\mathrm{Im}\,\varSigma_i \\ \mathrm{Im}\,\varSigma_i & \mathrm{Re}\,\varSigma_i \end{bmatrix}, \qquad 1 \le i \le 3.$$

Then the rotational symmetry leads to the following vectors in the null space of the linearization $\mathbf{L}(\omega)$ from (IX.90) ($\mathbb{C}$-linear version of $\mathcal{L}(\omega)$) at the solitary wave:

(IX.105) $$\mathbf{\Theta}_i = -\mathbf{J}\mathbf{\Sigma}_i \boldsymbol{\phi}_\omega + 2\epsilon_{ijk} x^j \partial_{x^k} \boldsymbol{\phi}_\omega \in \mathbf{ker}(\mathbf{J}\mathbf{L}(\omega)), \qquad 1 \le i \le 3.$$

Above, ϵ_{ijk} are the Levi-Civita symbols, skew-symmetric in each pair of the indices and with $\epsilon_{1\,2\,3} = 1$.

One can check that the generalized eigenvector $\partial_\omega \boldsymbol{\phi}_\omega$ is orthogonal to $\mathbf{J}\mathbf{\Theta}_i$, $1 \le i \le 2$, while the condition for its orthogonality to $\mathbf{J}\mathbf{\Theta}_i$ with $i = 3$, is given by the Kolokolov condition

$$\langle \partial_\omega \boldsymbol{\phi}_\omega, \mathbf{J}\mathbf{\Theta}_3 \rangle = \langle \partial_\omega \boldsymbol{\phi}_\omega, \boldsymbol{\phi}_\omega \rangle.$$

One can also check that the generalized eigenvectors $\boldsymbol{\xi}^i$, $1 \le i \le 3$ (see (IX.95)), are always orthogonal to $\mathbf{J}\mathbf{\Theta}_j$, $1 \le j \le 3$:

$$\Big\langle \omega x^i \mathbf{J}\boldsymbol{\phi}_\omega - \frac{1}{2}\alpha^i \boldsymbol{\phi}_\omega, \mathbf{J}\mathbf{\Theta}_j \Big\rangle = 0, \qquad 1 \le i,\, j \le 3.$$

Therefore, the presence of these eigenvectors in the kernel of $\mathbf{J}\mathbf{L}$ in the (3+1)D case does not affect the size of the Jordan blocks associated with the unitary invariance and the translational invariance; these sizes are completely characterized by the conditions (IX.98) and (IX.103).

There are no new Jordan blocks associated to $\mathbf{\Theta}_i$. For example, for the standard Ansatz

(IX.106) $$\phi_\omega(x) = \begin{bmatrix} v(r) \begin{pmatrix} 1 \\ 0 \end{pmatrix} \\ \mathrm{i}u(r) \begin{pmatrix} \cos\theta \\ e^{\mathrm{i}\varphi}\sin\theta \end{pmatrix} \end{bmatrix},$$

one has $\mathbf{\Theta}_3 = -\mathbf{J}\boldsymbol{\phi}_\omega$, $\mathbf{\Theta}_1 = \mathbf{J}\mathbf{\Theta}_2$, so that

(IX.107) $$\langle \mathbf{\Theta}_1, \mathbf{J}\mathbf{\Theta}_2 \rangle = \langle \mathbf{\Theta}_1, \mathbf{\Theta}_1 \rangle > 0.$$

As a consequence, the Jordan block corresponding to $\mathbf{\Theta}_3$ is the same as the one corresponding to the unitary invariance (whose size is controlled by the Kolokolov condition (IX.98)), and there is no Jordan block corresponding to $\mathbf{\Theta}_1$ since by (IX.107) it is not orthogonal to $\mathbf{ker}((\mathbf{J}\mathbf{L})^*)$ which contains, in particular, $\mathbf{J}^{-1}\mathbf{\Theta}_2$. Similarly, there is no Jordan block corresponding to $\mathbf{\Theta}_2$ since it is not orthogonal to $\mathbf{J}^{-1}\mathbf{\Theta}_1 \in \mathbf{ker}((\mathbf{J}\mathbf{L})^*)$.

Remark IX.30 In the $(2+1)$D case, the story is similar: the eigenvector from the null space which corresponds to the infinitesimal rotation coincides with $\mathbf{J}\boldsymbol{\phi}_\omega$, the eigenvector which corresponds to the unitary symmetry. The size of the corresponding Jordan block jumps up (indicating collision of eigenvalues at the origin) if and only if the Kolokolov condition $\partial_\omega Q(\phi_\omega) = 0$ is satisfied.

CHAPTER X

Bi-frequency solitary waves

The Soler model shares the symmetry features with its more physically relevant counterpart, Dirac–Klein–Gordon system (the Dirac equation with the Yukawa self-interaction, which is also based on the scalar quantity $\psi^*\beta\psi$):

$$\text{(X.1)} \qquad \begin{cases} \mathrm{i}\partial_t\psi = D_m\psi - \Phi\beta\psi, \\ (\partial_t^2 - \Delta + M^2)\Phi = \psi^*\beta\psi, \end{cases} \qquad \psi(t,x) \in \mathbb{C}^N, \quad \Phi(t,x) \in \mathbb{R}, \quad x \in \mathbb{R}^n,$$

where $M > 0$ is the mass of the particles described by the scalar field Φ. It was noticed by Alberto Galindo in **[Gal77]** that both the Soler model (IX.1) and the Dirac–Klein–Gordon model (X.1) have not only the usual $\mathbf{U}(1)$ symmetry, but also the $\mathbf{SU}(1,1)$ symmetry, which is essentially equivalent to the Bogoliubov transformation **[Bog58]** from the Bardeen–Cooper–Schrieffer theory, and in particular also contains the complex conjugation. In the three-dimensional case ($n = 3$, $N = 4$) with the standard choice of the Dirac matrices,

$$\text{(X.2)} \qquad \alpha^i = \begin{bmatrix} 0 & \sigma_i \\ \sigma_i & 0 \end{bmatrix}, \quad 1 \le i \le 3, \qquad \beta = \begin{bmatrix} I_2 & 0 \\ 0 & -I_2 \end{bmatrix},$$

with σ_i the Pauli matrices, this symmetry group takes the form

$$\text{(X.3)} \qquad G_{\mathrm{Bogoliubov}} = \left\{ a + b\mathrm{i}\gamma^2\boldsymbol{K}\colon \quad a,\, b \in \mathbb{C}, \ |a|^2 - |b|^2 = 1 \right\} \cong \mathbf{SU}(1,1),$$

where $\boldsymbol{K} : \mathbb{C}^N \to \mathbb{C}^N$ is the antilinear operator of complex conjugation; the group isomorphism is given by

$$a + b\mathrm{i}\gamma^2\boldsymbol{K} \ \mapsto \ \begin{bmatrix} a & b \\ \bar{b} & \bar{a} \end{bmatrix}.$$

By Noether's theorem (Theorem IX.8), each continuous symmetry group leads to a conservation law. Moreover, the Bogoliubov group, when applied to standard solitary waves $\phi(x)e^{-\mathrm{i}\omega t}$, produces bi-frequency solitary waves of the form

$$\text{(X.4)} \qquad a\phi(x)e^{-\mathrm{i}\omega t} + b\phi_C(x)e^{\mathrm{i}\omega t}, \qquad a,\, b \in \mathbb{C}, \qquad |a|^2 - |b|^2 = 1,$$

with $\phi_C = \mathrm{i}\gamma^2\boldsymbol{K}\phi$ the *charge conjugate* of ϕ from Quantum Electrodynamics; here $\mathrm{i}\gamma^2\boldsymbol{K}$ is one of the infinitesimal generators of $\mathbf{SU}(1,1)$.

Chadam and Glassey **[CG74]** noted an interesting feature of the Dirac–Klein–Gordon model: under the standard choice of 4×4 Dirac matrices (X.2), as long as the solution $\psi(t,x)$ is sufficiently regular, there is the conservation of the quantity

$$\text{(X.5)} \qquad \int_{\mathbb{R}^3} \left(|\psi_1 - \bar{\psi}_4|^2 + |\psi_2 + \bar{\psi}_3|^2 \right) dx.$$

As a consequence, if (X.5) is zero at some and hence at all moments of time, then $|\psi_1| = |\psi_4|$ and $|\psi_2| = |\psi_3|$ for almost all x and t, hence $\psi^*\beta\psi = 0$ (almost everywhere), meaning that the self-interaction plays no role in the evolution and that as the matter of fact $\psi(t,x)$ satisfies the linear equation (without self-interaction). A similar analysis in the context of the Soler model (IX.1) was performed in [**OY04**]. The relation of the conservation of the quantity (X.5) to the $\mathbf{SU}(1,1)$ symmetry of the corresponding Lagrangian based on combinations of $\psi^* D_m\psi$ and $\psi^*\beta\psi$ was noticed by Galindo [**Gal77**]. We will clarify all these issues in Section X.1.

In Section X.2, we will show that, as the matter of fact, the manifold of bi-frequency solitary waves in dimensions $n \geq 3$ is larger than the one obtained by applying the Bogoliubov transformation to one-frequency solitary waves. These bi-frequency waves are of a particular importance to us. We will show in Chapter XIII that they play an important role in the proof of spectral stability of weakly relativistic one-frequency solitary waves, allowing us to compute the multiplicity of eigenvalues $\pm 2\omega\mathrm{i}$ of the linearization operator.

Moreover, when proving the asymptotic stability of one-frequency solitary waves in the future, one will have to do this with respect to the manifold of bi-frequency solitary waves: a bi-frequency solitary wave, being itself an exact solution, if considered as a small perturbation of a one-frequency solitary wave, can not converge to a nearby one-frequency solitary wave.

X.1 The Bogoliubov symmetry and associated charges

Let D_m be the Dirac operator from Definition III.156 and assume that $B \in \mathrm{End}(\mathbb{C}^N)$ is a matrix which satisfies

$$\{B\boldsymbol{K}, D_m\} = 0, \qquad B\boldsymbol{K} = \boldsymbol{K}B^*, \qquad B^*B = I_N. \tag{X.6}$$

The relations (X.6) imply that

$$(B\boldsymbol{K})^2 = \boldsymbol{K}B^*B\boldsymbol{K} = I_N,$$

$$(\beta B)^T = \boldsymbol{K}(\beta B)^*\boldsymbol{K} = \boldsymbol{K}B^*\beta\boldsymbol{K} = B\boldsymbol{K}\beta\boldsymbol{K} = -\beta B\boldsymbol{K}^2 = -\beta B. \tag{X.7}$$

Remark X.1 One can think of the complex conjugation $\boldsymbol{K}$ as self-adjoint in the sense that

$$\mathrm{Re}\left((\boldsymbol{K}\psi)^*\theta\right) = \mathrm{Re}\,\overline{(\psi^*\boldsymbol{K}\theta)} = \mathrm{Re}\left(\psi^*\boldsymbol{K}\theta\right).$$

We can summarize [**CG74, OY04**] as follows:

Lemma X.2 *If the solution to* (IX.1) *satisfies* $B\boldsymbol{K}\psi|_{t=0} = z\psi|_{t=0}$, *for some* $z \in \mathbb{C}$, $|z| = 1$, *then* $B\boldsymbol{K}\psi = z\psi$ *for all* $t \in \mathbb{R}$, *and moreover* $\psi^*\beta\psi = 0$.

PROOF. The first statement is an immediate consequence of the fact that $B\boldsymbol{K}$ commutes with the flow of the equation. If $\psi(t,x)$ satisfies $\mathrm{i}\partial_t\psi = D_m\psi - f(\psi^*\beta\psi)\beta\psi$, then

$$B\boldsymbol{K}\partial_t\psi = B\boldsymbol{K}(-\mathrm{i}(D_m-\beta f))\psi = (-\mathrm{i}(D_m-\beta f))B\boldsymbol{K}\psi = z(-\mathrm{i}D_m+\mathrm{i}\beta f)\psi = z\partial_t\psi.$$

Finally, since $B\boldsymbol{K}\psi = z\psi$, one has:

$$z\psi^*\beta\psi = \psi^*\beta B\boldsymbol{K}\psi = (\boldsymbol{K}\psi)^T\beta B\boldsymbol{K}\psi = -(\boldsymbol{K}\psi)^T\beta B\boldsymbol{K}\psi = 0.$$

We took into account that $(\beta B)^T = -\beta B$ by (X.7). □

As in [**Gal77**], the Lagrangian of the Soler model (IX.1) with the density

$$\mathscr{L} = \psi^* D_m\psi + F(\psi^*\beta\psi),$$

where $\psi(t,x) \in \mathbb{C}^N$, is invariant under the action of the continuous symmetry group

$$g \in G_{\text{Bogoliubov}}, \qquad g: \ \psi \mapsto (a + bB\boldsymbol{K})\psi, \qquad |a|^2 - |b|^2 = 1$$

(cf. (X.3)). Noether's theorem (Theorem IX.8) leads to the conservation of the standard charge $Q = \int_{\mathbb{R}^3} \psi^*\psi\,dx$ corresponding to the standard charge-current density $\psi\gamma^0\gamma^\mu\psi$ (note that the unitary group is a subgroup of $\mathbf{SU}(1,1)$), and the complex-valued Bogoliubov charge $\Lambda = \int_{\mathbb{R}^n} \psi^* B\boldsymbol{K}\psi\,dx$ corresponding to the complex-valued current density $\psi^*\gamma^0\gamma^\mu B\boldsymbol{K}\psi$. Now Galindo's observation **[Gal77]** could be stated as follows.

Lemma X.3 (The Bogoliubov $\mathbf{SU}(1,1)$ symmetry and the charge conservation) *Let $B \in \operatorname{End}(\mathbb{C}^N)$ satisfy* (X.6). *Then the Soler model* (IX.1) *has a continuous symmetry group, $G_{\text{Bogoliubov}} \cong \mathbf{SU}(1,1)$, in the following sense:*

(1) The Hamiltonian density

$$\mathscr{H} = \psi^* D_m \psi - F(\psi^*\beta\psi), \tag{X.8}$$

where $F(\tau) = \int_0^\tau f(t)\,dt$, $\tau \in \mathbb{R}$, satisfies

$$\mathscr{H}(g\psi) = \mathscr{H}(\psi), \qquad \forall g \in G_{\text{Bogoliubov}}.$$

(2) Moreover, let $k \in \mathbb{N}$ satisfy $k > n/2$ and assume that $f \in C^k(\mathbb{R},\mathbb{R})$. Let $\psi(t,x) \in \mathbb{C}^N$ be a solution to (IX.1) *with $\psi|_{t=0} \in H^k(\mathbb{R},\mathbb{C}^N)$. Then, for any $g \in G_{\text{Bogoliubov}}$, $g\psi(t,x)$ is also a solution to* (IX.1).

(3) Moreover, for this solution $\psi(t)$, the following quantities are conserved:

$$Q(\psi) = \langle \psi, \psi \rangle = \int_{\mathbb{R}^n} \psi(t,x)^*\psi(t,x)\,dx,$$

$$\Lambda(\psi) = \langle \psi, B\boldsymbol{K}\psi \rangle = \int_{\mathbb{R}^n} B_{jk}\overline{\psi_j(t,x)}\;\overline{\psi_k(t,x)}\,dx.$$

PROOF. We notice that $\overline{\varphi^*\boldsymbol{K}\rho} = (\boldsymbol{K}\varphi)^*\rho$, for all φ and ρ in $\mathbb{C}^N$, hence

$$\operatorname{Re}\{\varphi^*\boldsymbol{K}\rho\} = \operatorname{Re}\{(\boldsymbol{K}\varphi)^*\rho\},$$

resulting in

$$\begin{aligned}(g\psi)^*\beta g\psi &= \operatorname{Re}\{\psi^*(\bar{a} + \boldsymbol{K}B^*\bar{b})\beta(a + bB\boldsymbol{K})\psi\} \\ &= \operatorname{Re}\{\psi^*(\bar{a} + bB\boldsymbol{K})(a - bB\boldsymbol{K})\beta\psi\} = \psi^*\beta\psi.\end{aligned} \tag{X.9}$$

Now the invariance of the Hamiltonian density follows from

$$(g\psi)^* D_m g\psi = \operatorname{Re}\{\psi^*(\bar{a} + bB\boldsymbol{K})(a - bB\boldsymbol{K})D_m\psi\} = \psi^* D_m\psi$$

and from $F((g\psi)^*\beta g\psi) = F(\psi^*\beta\psi)$.

Under the assumptions of Theorem IX.3, for any ψ_0 in $H^k(\mathbb{R}^n,\mathbb{C}^N)$, there exists a maximal solution ψ to (IX.1) with the initial data $\psi|_{t=0} = \psi_0$:

$$\psi \in C^0\big((-T^-_{\psi_0}, T^+_{\psi_0}), H^k(\mathbb{R}^n,\mathbb{C}^N)\big) \cap C^1\big((-T^-_{\psi_0}, T^+_{\psi_0}), H^{k-1}(\mathbb{R}^n,\mathbb{C}^N)\big).$$

Let $g = a + bB\boldsymbol{K}$, $|a|^2 - |b|^2 = 1$. Since $B\boldsymbol{K}$ anti-commutes with both i and D_m, one has

$$\mathrm{i}\partial_t(B\boldsymbol{K}\psi) = (D_m - f(\psi^*\beta\psi)\beta)(B\boldsymbol{K}\psi).$$

Taking the linear combination with (IX.1), we arrive at

$$\mathrm{i}\partial_t(a + bB\boldsymbol{K})\psi = (D_m - f(\psi^*\beta\psi)\beta)((a + bB\boldsymbol{K})\psi).$$

It remains to use (X.9).

By Noether's theorem (Theorem IX.8), the invariance under the action of a continuous group results in the conservation laws. Let us check the conservation of the complex charge Λ. Writing $f = f(\psi^*\beta\psi)$, we have:

$$\partial_t\Lambda(\psi) = \langle -\mathrm{i}(D_m - f\beta)\psi, B\boldsymbol{K}\psi\rangle + \langle\psi, B\boldsymbol{K}(-\mathrm{i}(D_m - f\beta)\psi)\rangle$$
$$= \langle\psi, \mathrm{i}(D_m - f\beta)B\boldsymbol{K}\psi\rangle + \langle\psi, \mathrm{i}B\boldsymbol{K}(D_m - f\beta)\psi\rangle = 0.$$

In the last relation, we took into account the anticommutation relations from (X.6). We also note that for the densities, we have

$$\partial_t(\psi^* B\boldsymbol{K}\psi) = -(\alpha^j\partial_j\psi)^* B\boldsymbol{K}\psi - \psi^* B\boldsymbol{K}\alpha^j\partial_j\psi = -\partial_{x^j}(\psi^*\alpha^j\psi B\boldsymbol{K}\psi),$$

showing that the Minkowski vector of the Bogoliubov charge-current density is given by

$$\psi(t,x)^*\gamma^0\gamma^\mu B\boldsymbol{K}\psi(t,x), \qquad 0 \le \mu \le n. \qquad \square$$

Remark X.4 Three conserved quantities, the standard charge and the real and imaginary parts of the complex charge, correspond to $\dim_{\mathbb{R}} \mathbf{SU}(1,1) = 3$.

Example X.5 For $N = 2$ and $n \le 2$, with $D_m = -\mathrm{i}\sum_{i=1}^n \sigma_i\partial_{x^i} + \sigma_3 m$, one takes $B = \sigma_1$ (σ_j being the Pauli matrices);

$$B\boldsymbol{K}\psi = \sigma_1\boldsymbol{K}\psi =: \psi_C, \qquad \psi \in \mathbb{C}^2.$$

The conserved quantity is

$$\Lambda = \int_{\mathbb{R}} \psi^*\sigma_1\boldsymbol{K}\psi\,dx = 2\int_{\mathbb{R}} \bar\psi_1\bar\psi_2\,dx.$$

It follows that the charge can be decomposed into

$$Q = Q_- + Q_+,$$

with both

$$Q_\pm := \frac{1}{2}(Q \pm \operatorname{Re}\Lambda) = \frac{1}{2}\int_{\mathbb{R}} (|\psi_1|^2 + |\psi_2|^2 \pm 2\operatorname{Re}\psi_1\psi_2)\,dx = \frac{1}{2}\int_{\mathbb{R}} |\psi_1 \pm \bar\psi_2|^2\,dx$$

conserved in time; so, if at $t = 0$ one has $\bar\psi_2 = \psi_1$ (or, similarly, if $\bar\psi_2 = -\psi_1$), then these relations persist for all times (hence $\psi^*\sigma_3\psi = |\psi_1|^2 - |\psi_2|^2 = 0$ for all times) due to the conservation of $Q_\pm$.

Example X.6 In the case $n = 3$, $N = 4$, using the standard choice of the Dirac matrices (X.2), one can take $B = \mathrm{i}\gamma^2$, so that

$$B\boldsymbol{K}\psi = \mathrm{i}\gamma^2\boldsymbol{K}\psi =: \psi_C, \qquad \psi \in \mathbb{C}^4.$$

Then the quantity

$$\Lambda(\psi) = \int_{\mathbb{R}^3} \psi^*\mathrm{i}\gamma^2\boldsymbol{K}\psi\,dx = 2\int_{\mathbb{R}^3} (\bar\psi_1\bar\psi_4 - \bar\psi_2\bar\psi_3)\,dx$$

is conserved, hence so are

$$Q_\pm := \frac{1}{2}(Q \pm \operatorname{Re}\Lambda) = \frac{1}{2}\int_{\mathbb{R}^3} \left(|\psi_1 \pm \bar\psi_4|^2 + |\psi_2 \mp \bar\psi_3|^2\right)dx.$$

Thus, if at some moment of time one has $\bar\psi_4 = \psi_1$ and $\bar\psi_3 = -\psi_2$ (or, similarly, if $\bar\psi_4 = -\psi_1$ and $\bar\psi_3 = \psi_2$), then this relation persists for all times (hence $\psi^*\beta\psi = 0$) due to the conservation of $Q_\pm$. Note that $2Q_-$ coincides with (X.5), so our conclusions are in agreement with **[CG74, OY04]**.

Remark X.7 For $n = 4$ and $N = 4$ (cf. Remark IX.1), there is no B satisfying (X.6) and thus no $\mathbf{SU}(1,1)$ symmetry.

Lemma X.8 (Charge transformation under the action of $\mathbf{SU(1,1)}$) *Let $a, b \in \mathbb{C}$, $|a|^2 - |b|^2 = 1$, so that $g = a + bB\boldsymbol{K} \in G_{\mathrm{Bogoliubov}}$. Then $g\psi = (a + bB\boldsymbol{K})\psi$ satisfies*

$$Q((a + bB\boldsymbol{K})\psi) = (|a|^2 + |b|^2)Q(\psi) + 2\,\mathrm{Re}\{\bar{a}b\Lambda(\psi)\}, \tag{X.10}$$

$$\Lambda(a + bB\boldsymbol{K}\psi) = \bar{a}^2\Lambda(\psi) + 2\bar{a}\bar{b}Q(\psi) + \bar{b}^2\overline{\Lambda(\psi)}. \tag{X.11}$$

The quantity $Q^2 - |\Lambda|^2$ is invariant under the action of $\mathbf{SU}(1,1)$:

$$Q(g\psi)^2 - |\Lambda(g\psi)|^2 = Q(\psi)^2 - |\Lambda(\psi)|^2. \tag{X.12}$$

PROOF. For the charge density of $g\psi = (a + bB\boldsymbol{K})\psi$, one has

$$(g\psi)^*g\psi = \mathrm{Re}\{\psi^*(\bar{a} + bB\boldsymbol{K})(a + bB\boldsymbol{K})\psi\} = \mathrm{Re}\{\psi^*(|a|^2 + |b|^2 + 2\bar{a}bB\boldsymbol{K})\psi\}$$

$$= (|a|^2 + |b|^2)\psi^*\psi + 2\,\mathrm{Re}\{\bar{a}b\psi^*B\boldsymbol{K}\psi\}.$$

For the complex charge density of $g\psi$, using the identity $(B\boldsymbol{K}u)^*B\boldsymbol{K}v = \overline{u^*v}$, $\forall u, v \in \mathbb{C}^N$ which follows from (X.6), one has

$$(g\psi)^*B\boldsymbol{K}g\psi = ((a + bB\boldsymbol{K})\psi)^*B\boldsymbol{K}(a + bB\boldsymbol{K})\psi$$

$$= \bar{a}\psi^*(\bar{a}B\boldsymbol{K} + \bar{b})\psi + \bar{b}\,\overline{\psi^*(a + bB\boldsymbol{K})\psi} = \bar{a}^2\psi^*B\boldsymbol{K}\psi + 2\bar{a}\bar{b}\psi^*\psi + \bar{b}^2\overline{\psi^*B\boldsymbol{K}\psi}.$$

The integration of the above charge densities leads to (X.11). The relation (X.12) is verified by the explicit computation. □

X.2 Bi-frequency solitary waves

Let $n \in \mathbb{N}$, $N \in \mathbb{N}$. We assume that the Dirac matrices are given by (IX.3) in terms of the generalized Pauli matrices σ_i from (IX.4), and consider the solitary wave solution $\phi_\xi(x)e^{-\mathrm{i}\omega t}$, $\omega \in [-m, m]$, with

$$\phi_\xi(x) = \begin{bmatrix} v(r)\xi \\ \mathrm{i}u(r)\sigma_r\xi \end{bmatrix}, \qquad \xi \in \mathbb{C}^{N/2}, \quad |\xi| = 1, \qquad r = |x|, \qquad x \in \mathbb{R}^n$$

(above, we do not indicate the dependence of ϕ, v, and u of ω), and with $v(r)$ and $u(r)$ real-valued functions which satisfy

$$\begin{cases} \omega v = \partial_r u + \frac{n-1}{r}u + (m - f(v^2 - u^2))v, \\ \omega u = -\partial_r v - (m - f(v^2 - u^2))u, \end{cases} \qquad r > 0; \qquad \lim_{r\to 0} u(r,\omega) = 0. \tag{X.13}$$

We assume that the functions $v(r)$ and $u(r)$ satisfy

$$\sup_{r\ge 0} |u(r)/v(r)| < 1, \tag{X.14}$$

which is true in particular for small amplitude solitary waves with $\omega \lessapprox m$ which we will construct in Chapter XII.

We denote

$$\sigma_r = \frac{x\cdot\boldsymbol{\sigma}}{r}, \qquad x \in \mathbb{R}^n \setminus \{0\}, \qquad r = |x|. \tag{X.15}$$

The relations (IX.4) imply that $\sigma_r\sigma_r^* = I_{N/2}$.

Lemma X.9 *Let $n \in \mathbb{N}$, $\omega \in [-m, m]$. If $v(r)$, $u(r)$ are real-valued functions which solve* (X.13)*, so that for any $\xi \in \mathbb{C}^{N/2}$, $|\xi| = 1$, the function*

$$\psi(t, x) = \phi_\xi(x) e^{-\mathrm{i}\omega t}, \tag{X.16}$$

with

$$\phi_\xi(x) = \begin{bmatrix} v(r)\xi \\ \mathrm{i}u(r)\sigma_r\xi \end{bmatrix}, \qquad r = |x|, \tag{X.17}$$

is a solitary wave solution to the nonlinear Dirac equation (IX.1)*, then for any Ξ, $\mathrm{H} \in \mathbb{C}^{N/2} \setminus \{0\}$, $|\Xi|^2 - |\mathrm{H}|^2 = 1$, the function*

$$\begin{aligned} \theta_{\Xi,\mathrm{H}}(t, x) &= \phi_\Xi(x) e^{-\mathrm{i}\omega t} + \chi_\mathrm{H}(x) e^{\mathrm{i}\omega t} \\ &= |\Xi| \phi_\xi(x) e^{-\mathrm{i}\omega t} + |\mathrm{H}| \chi_\eta(x) e^{\mathrm{i}\omega t}, \qquad \xi = \frac{\Xi}{|\Xi|}, \quad \eta = \frac{\mathrm{H}}{|\mathrm{H}|}, \end{aligned} \tag{X.18}$$

with

$$\chi_\eta(x) = \begin{bmatrix} -\mathrm{i}u(r)\sigma_r^*\eta \\ v(r)\eta \end{bmatrix}, \qquad r = |x|, \tag{X.19}$$

is a solution to (IX.1).

PROOF. The statement of the lemma is verified by the direct substitution. First one checks that

$$\theta_{\Xi,\mathrm{H}}^* \beta \theta_{\Xi,\mathrm{H}} = |\Xi|^2(v^2 - u^2) + |\mathrm{H}|^2(u^2 - v^2) = v^2 - u^2 = \phi_\xi^* \beta \phi_\xi.$$

It remains to prove that the relation $\omega\phi_\xi = D_m\phi_\xi - \beta f \phi_\xi$ implies the relation

$$-\omega\chi_\eta = D_m\chi_\eta - \beta f \chi_\eta,$$

with $\chi_\eta(x)$ defined in (X.19), for any $\eta \in \mathbb{C}^{N/2}$, $|\eta| = 1$. With D_m built with the Dirac matrices from (IX.3), the above relations on ϕ_ξ and χ_η are written explicitly as follows:

$$\begin{aligned} \omega \begin{bmatrix} v\xi \\ \mathrm{i}\sigma_r u\xi \end{bmatrix} &= \begin{bmatrix} 0 & -\mathrm{i}\boldsymbol{\sigma}^* \cdot \nabla \\ -\mathrm{i}\boldsymbol{\sigma} \cdot \nabla & 0 \end{bmatrix} \begin{bmatrix} v\xi \\ \mathrm{i}\sigma_r u\xi \end{bmatrix} + (m - f) \begin{bmatrix} v\xi \\ -\mathrm{i}\sigma_r u\xi \end{bmatrix}, \\ -\omega \begin{bmatrix} -\mathrm{i}\sigma_r^* u\eta \\ v\eta \end{bmatrix} &= \begin{bmatrix} 0 & -\mathrm{i}\boldsymbol{\sigma}^* \cdot \nabla \\ -\mathrm{i}\boldsymbol{\sigma} \cdot \nabla & 0 \end{bmatrix} \begin{bmatrix} -\mathrm{i}\sigma_r^* u\eta \\ v\eta \end{bmatrix} + (m - f) \begin{bmatrix} -\mathrm{i}\sigma_r^* u\eta \\ -v\eta \end{bmatrix}; \end{aligned}$$

each of these relations is equivalent to the system (X.13) once we take into account that $\boldsymbol{\sigma} \cdot \nabla v = \sigma_r \partial_r v$, $\boldsymbol{\sigma}^* \cdot \nabla v = \sigma_r^* \partial_r v$, and

$$(\boldsymbol{\sigma}^* \cdot \nabla)\sigma_r u = (\boldsymbol{\sigma}^* \cdot \nabla)\sigma_r \partial_r U = (\boldsymbol{\sigma}^* \cdot \nabla)(\boldsymbol{\sigma} \cdot \nabla)U = \Delta U = \partial_r u + \frac{n-1}{r} u,$$

where we introduced $U(r) = \int_0^r u(s)\,ds$ and used the identity $(\boldsymbol{\sigma}^* \cdot \nabla)(\boldsymbol{\sigma} \cdot \nabla) = I_{N/2}\Delta$ which follows from (IX.4); similarly, one has $(\boldsymbol{\sigma} \cdot \nabla)\sigma_r^* u = \partial_r u + \frac{n-1}{r} u$. □

Remark X.10 By Lemma X.9, for $N \geq 4$, there is a larger symmetry group,

$$\mathbf{SU}(N/2, N/2),$$

with a faithful action on the set of bi-frequency solitary waves, which is present at the level of bi-frequency solitary wave solutions while being absent at the level of the Lagrangian.

Remark X.11 We note that if $f(\tau)$ in (IX.1) is even, then $\theta_{\Xi,\mathrm{H}}(t, x)$ given by (X.18) with Ξ, $\mathrm{H} \in \mathbb{C}^{N/2}$ such that $|\Xi|^2 - |\mathrm{H}|^2 = -1$ is also a solitary wave solution.

Remark X.12 We point out that the asymptotic stability of standard, *one-frequency* solitary waves can only hold with respect to the whole manifold of solitary wave solutions (X.18), which includes both one-frequency and bi-frequency solitary waves: if a small perturbation of a one-frequency solitary wave is a bi-frequency solitary wave, which is an exact solution, then convergence to the set of one-frequency solitary waves is out of question. In this regard, we recall that the asymptotic stability results [**BC12b**, **PS12**, **CPS17**] were obtained under certain restrictions on the class of perturbations. It turns out that these restrictions were sufficient to remove not only translations, but also the perturbations in the directions of bi-frequency solitary waves; this is exactly why the proof of asymptotic stability of the set of *one-frequency* solitary waves with respect to such class of perturbations was possible at all.

We define the solitary manifold of one- and bi-frequency solutions of the form (X.18) corresponding to some particular value of ω by

$$\text{(X.20)} \qquad \mathscr{M} = \Big\{\theta_{\Xi,\mathrm{H}}(t,x): \qquad \Xi,\ \mathrm{H} \in \mathbb{C}^{N/2}, \quad |\Xi|^2 - |\mathrm{H}|^2 = 1\Big\}.$$

Remark X.13 We do not consider translations and Lorentz boosts, thus preserving the spatial location of the solitary wave.

Now let $\mathcal{O} = \mathcal{O}(\phi e^{-\mathrm{i}\omega t})$ be the orbit of $\phi e^{-\mathrm{i}\omega t}$ under the action of the available symmetry groups, $G_{\mathrm{Bogoliubov}}$ defined in (X.3) and $\mathbf{SO}(n)$:

$$\mathcal{O} = \left\{rg\left(\phi e^{-\mathrm{i}\omega t}\right): \quad r \in \mathbf{SO}(n), \quad g \in G_{\mathrm{Bogoliubov}}\right\} \subset \mathscr{M}.$$

In lower spatial dimensions $n \le 2$, when $N = 2$, with

$$D_m = \begin{cases} \mathrm{i}\sigma_2\partial_x + \sigma_3 m, & n = 1, \\ -\mathrm{i}\sigma_1\partial_{x^1} - \mathrm{i}\sigma_2\partial_{x^2} + \sigma_3 m, & n = 2, \end{cases}$$

the orbit $\mathcal{O}$ is given by

$$\mathcal{O} = \{\ g(\phi(x)e^{-\mathrm{i}\omega t}): \quad g \in G_{\mathrm{Bogoliubov}}\ \},$$

with $G_{\mathrm{Bogoliubov}} = \{a + b\sigma_1\boldsymbol{K},\ a,\ b \in \mathbb{C},\ |a|^2 - |b|^2 = 1\}$; thus, in spatial dimension $n = 1$, $\mathcal{O}$ coincides with the solitary manifold

$$\mathscr{M} = \left\{a \begin{bmatrix} v(x) \\ u(x) \end{bmatrix} e^{-\mathrm{i}\omega t} + b \begin{bmatrix} u(x) \\ v(x) \end{bmatrix} e^{\mathrm{i}\omega t}: \quad a,\ b \in \mathbb{C},\ |a|^2 - |b|^2 = 1\right\},$$

where $(v(x), u(x))$ is a solution to the system (IX.59). There is a similar situation in spatial dimension $n = 2$: again, one has $\mathscr{M} = \mathcal{O}$. In general, though, the solitary manifold $\mathscr{M}$ can be larger than $\mathcal{O}$. For example, in three spatial dimensions, $n = 3$ and $N = 4$, the solitary manifold $\mathscr{M}$ is larger than the orbit $\mathcal{O}$ of $\phi(x)e^{-\mathrm{i}\omega t}$ under the action of the available symmetry groups: spatial rotations $\mathbf{SO}(n)$ and the Bogoliubov group $G_{\mathrm{Bogoliubov}}$ given by elements of the form $a + b\mathrm{i}\gamma^2\boldsymbol{K}$, where $a,\ b \in \mathbb{C}$, $|a|^2 - |b|^2 = 1$. Indeed, one has $\mathcal{O} \subsetneq \mathscr{M}$, since

$$\dim_{\mathbb{R}} \mathcal{O} = 5 < \dim_{\mathbb{R}} \mathbf{SO}(3) + \dim_{\mathbb{R}} G_{\mathrm{Bogoliubov}} = 3 + 3,$$

$$\dim_{\mathbb{R}} \mathscr{M} = \dim_{\mathbb{R}}\{(\Xi, \mathrm{H}) \in \mathbb{C}^{N/2} \times \mathbb{C}^{N/2}: \quad |\Xi|^2 - |\mathrm{H}|^2 = 1\} = 2N - 1 = 7.$$

Note that in the above inequality for $\dim_{\mathbb{R}} \mathcal{O}$ one has "strictly smaller", since the generator corresponding to the standard $\mathbf{U}(1)$-invariance which enters the Lie algebra of $G_{\mathrm{Bogoliubov}}$ also coincides with the generator of rotation around z-axis if we choose $\phi = \phi_\xi$ (see (X.17)) with $\xi = \boldsymbol{e}_1$ being the first vector of the standard basis in $\mathbb{C}^2$.

In the case of four spatial dimensions, with $n = 4$ and $N = 4$, the symmetry group simplifies to $\mathbf{U}(1)$ since the generator $\mathrm{i}\gamma^2\boldsymbol{K}$ is no longer available: the Dirac operator $D_m = -\mathrm{i}\boldsymbol{\alpha}\cdot\nabla + \beta m$ now contains the summand $-\mathrm{i}\alpha^4\partial_{x^4} = \beta\gamma^5\partial_{x^4}$ (with $\alpha^4 := \begin{bmatrix} 0 & -\mathrm{i}I_2 \\ \mathrm{i}I_2 & 0 \end{bmatrix} = -\mathrm{i}\beta\gamma^5$), which breaks the anticommutation of D_m with $\mathrm{i}\gamma^2\boldsymbol{K}$:

$$\{-\mathrm{i}\alpha^4, \mathrm{i}\alpha^2\beta\boldsymbol{K}\} = \{\beta\gamma^5, \mathrm{i}\alpha^2\beta\boldsymbol{K}\} = \beta\gamma^5\mathrm{i}\alpha^2\beta\boldsymbol{K} + \mathrm{i}\alpha^2\beta\boldsymbol{K}\beta\gamma^5$$
$$= \mathrm{i}\gamma^5\alpha^2\boldsymbol{K} + \mathrm{i}\alpha^2\gamma^5\boldsymbol{K} = 2\mathrm{i}\alpha^2\gamma^5\boldsymbol{K} \ne 0.$$

Again, $\mathscr{O} \subsetneq \mathscr{M}$ since

$$\dim_{\mathbb{R}}\mathscr{O} \le \dim_{\mathbb{R}}\mathbf{SO}(4) + \dim_{\mathbb{R}} G_{\mathrm{Bogoliubov}} \le 6 + 1, \qquad \dim_{\mathbb{R}}\mathscr{M} = 2N - 1 = 7,$$

with the Lie algebras of $\mathbf{SO}(4)$ and $G_{\mathrm{Bogoliubov}}$ sharing one element (a generator of the standard $\mathbf{U}(1)$-symmetry). Moreover, the action of $\mathbf{SO}(4)$ ($\dim_{\mathbb{R}} = 6$) on $\mathbb{C}^2$ ($\dim_{\mathbb{R}} = 4$) is not *faithful*; the orbit of an element $\xi \in \mathbb{C}^2$ under the action of $\mathbf{SO}(4)$ is only three-dimensional. As a result, in the case $n = 4$, $N = 4$, one has

$$\dim_{\mathbb{R}}\mathscr{O} = 3, \qquad \dim_{\mathbb{R}}\mathscr{M} = 2N - 1 = 7.$$

Remark X.14 We can rephrase the above situation in the following way. When moving from $n = 3$ to $n = 4$, additional rotations in $\mathbb{R}^4$ do not add to the orbit of $\xi \in \mathbb{C}^2$ which has already been of maximal dimension when $n = 3$ (which equals three: it is the real dimension of the unit sphere in $\mathbb{C}^2$), while the loss of the generator $B\boldsymbol{K}$ from the Bogoliubov group led to the loss of two real dimensions of the orbit $\mathscr{O}$.

Remark X.15 Let us briefly discuss the pseudo-scalar theories in spatial dimension $n = 3$. Instead of the (scalar) Yukawa interaction, given by the term $\phi\,\psi^*\beta\psi$ in the Lagrangian, one can consider pseudoscalar interaction, introducing the term $\phi\,\psi^*\mathrm{i}\beta\gamma^5\psi$, which we write as $-\phi\psi^*\alpha^4\psi$ with

$$\alpha^4 = -\mathrm{i}\beta\gamma^5 = \begin{bmatrix} 0 & -\mathrm{i}I_2 \\ \mathrm{i}I_2 & 0 \end{bmatrix}.$$

The Bogoliubov symmetry $\mathbf{SU}(1,1)$ is no longer present in a model with such an interaction. For $g = a + b\mathrm{i}\gamma^2\boldsymbol{K}$, $|a|^2 - |b|^2 = 1$, one has:

$$\begin{aligned}
(g\psi)^*\alpha^4(g\psi) &= \mathrm{Re}\left[(g\psi)^*\alpha^4(g\psi)\right] = \mathrm{Re}\left[((a + b\mathrm{i}\gamma^2\boldsymbol{K})\psi)^*\alpha^4(a + b\mathrm{i}\gamma^2\boldsymbol{K})\psi\right] \\
&= \mathrm{Re}\left[\psi^*(\bar a + \boldsymbol{K}\mathrm{i}\bar b\gamma^2)\alpha^4(a + b\mathrm{i}\gamma^2\boldsymbol{K})\psi\right] = \mathrm{Re}\left[\psi^*(\bar a + b\mathrm{i}\gamma^2\boldsymbol{K})\alpha^4(a + b\mathrm{i}\gamma^2\boldsymbol{K})\psi\right] \\
&= \mathrm{Re}\left[\psi^*\alpha^4(\bar a + b\mathrm{i}\gamma^2\boldsymbol{K})(a + b\mathrm{i}\gamma^2\boldsymbol{K})\psi\right] \\
&= \mathrm{Re}\left[\psi^*\alpha^4(|a|^2 + |b|^2 + 2\mathrm{i}\bar a b\gamma^2\boldsymbol{K})\psi\right] = \psi^*\alpha^4(|a|^2 + |b|^2)\psi,
\end{aligned}$$

which in general is different from $\psi^*\alpha^4\psi$. Above, in the last equality, we took into account that the matrix $\alpha^4\gamma^2 = \begin{bmatrix} \mathrm{i}\sigma_2 & 0 \\ 0 & \mathrm{i}\sigma_2 \end{bmatrix}$ is antisymmetric, hence

$$\psi^*\alpha^4\gamma^2\boldsymbol{K}\psi = (\boldsymbol{K}\psi)^T\alpha^4\gamma^2\boldsymbol{K}\psi = 0.$$

CHAPTER XI

Bifurcations of eigenvalues from the essential spectrum

The linearization of (IX.1) at a solitary wave solution $\psi(t,x) = \phi_\omega(x)e^{-i\omega t}$ is represented by a nonselfadjoint operator of the form

$$(\text{XI.1}) \qquad J(D_m - \omega + V(x,\omega)), \qquad \text{with } J \text{ skew-adjoint, } J^2 = -1,$$

where the matrix J commutes with D_m but not necessarily with the potential $V(x,\omega)$. We say that the solitary wave is spectrally stable if the spectrum of this linearization operator has no points with positive real part. The spectral stability is the weakest type of stability; it does not necessarily lead to actual, dynamical one. The essential spectrum is easy to analyze: the application of Weyl's theorem (see e.g. Theorem III.135) shows that the essential spectrum of the operator corresponding to the linearization at a solitary wave starts at $\pm(m-|\omega|)i$ and extends to $\pm\infty i$. Thus, the spectral stability of the corresponding solitary wave would be a corollary of the absence of eigenvalues with positive real part in the spectrum of $J(D_m - \omega + V(\omega))$ in (XI.1). The major difficulties in identifying the point spectrum $\sigma_p\big(J(D_m - \omega + V(\omega))\big)$ are due to the spectrum of D_m extending to both $\pm\infty$; this prevents us from using standard tools developed in the NLS context.

In the absence of linear stability (that is when the linearized system is not dynamically stable), one expects to be able to prove *orbital instability*, in the sense of [**GSS87**, **DBRN19**, **DBGRN15**]. In [**GO12**], this instability is proved in the context of the nonlinear Schrödinger equation; such results are still absent for the nonlinear Dirac equation.

Since the isolated eigenvalues depend continuously on the perturbation, it is convenient to trace the location of "unstable" eigenvalues (eigenvalues with positive real part) considering ω as a parameter. One wants to know how and when the "unstable" eigenvalues may emerge from the imaginary axis, particularly from the essential spectrum; that is, at which critical values of ω the solitary waves start developing an instability. Below, we describe the possible scenarios.

Instability scenario 1: collision of eigenvalues. For the nonlinear Schrödinger equation (Section IV.2), the Kolokolov stability criterion ([**Kol73**]) keeps track of the collision of purely imaginary eigenvalues at the origin and a subsequent birth of a positive and a negative eigenvalue. When $\partial_\omega Q(\omega) < 0$, with $Q(\omega) = \|\phi_\omega\|^2_{L^2}$ being the charge of the solitary wave (V.69) (charge-subcritical $\kappa < \frac{2}{n}$), then the linearization at a solitary wave has purely imaginary spectrum; when $\partial_\omega Q(\omega) > 0$ (charge-supercritical $\kappa < \frac{2}{n}$), there are two real (one positive, one negative) eigenvalues of the linearization operator. The vanishing of the quantity $\partial_\omega Q(\omega)$ at some value of ω indicates the moment of the collision of eigenvalues, when the Jordan block corresponding to the zero eigenvalue has a jump of two in

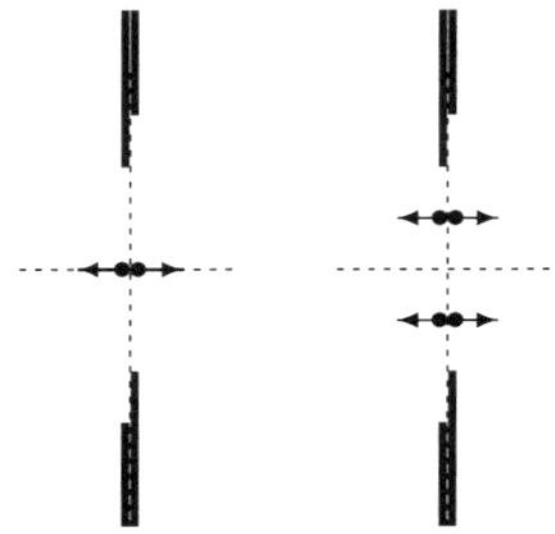

FIGURE XI.1. When the frequency ω of the solitary wave $\phi_\omega e^{-i\omega t}$ changes, the "unstable", positive-real-part eigenvalues in the linearized equation could be born from the collisions of discrete imaginary eigenvalues

its size. A nice feature of the linearization at a ground state solitary wave in the nonlinear Schrödinger equation is that its spectrum belongs to the imaginary axis, with some eigenvalues possibly located on the real axis; thus, the collision of eigenvalues at $\lambda = 0$ is the only way how the spectral instability could develop. In the NLD context, such a collision does not necessarily occur at $\lambda = 0$; both situations as in Fig. XI.1 are possible.

In NLD the collision of eigenvalues at the origin and a subsequent transition to instability is characterized not only by the Vakhitov–Kolokolov condition $dQ/d\omega = 0$ (see Section IX.6.2), but also by the condition $E(\omega) = 0$ (see Section IX.6.3), where E is the value of the energy functional on the corresponding solitary wave.

The eigenvalues with positive real part could also be born from the collision of purely imaginary eigenvalues at some point in the spectral gap but away from the origin. It was recently observed this scenario in the cubic Soler model in two spatial dimensions [**CMKS$^+$16**] but there is no criterion for such a collision of eigenvalues.

Instability scenario 2: bifurcations from the essential spectrum. The most peculiar feature of the linearization at a solitary wave in the NLD context is the possibility of bifurcations of eigenvalues with nonzero real part off the imaginary axis, out of the bulk of the essential spectrum.

In the present chapter, we give a thorough analytical study of eigenvalues of the Dirac operators, focusing on whether and how such eigenvalues can bifurcate from the essential spectrum. Generalizing the Jensen–Kato approach [**JK79**] to the context of the Dirac operators, we show in Theorem XI.1, that for $|\omega| < m$ the bifurcations from the essential spectrum are only possible from embedded eigenvalues (Fig. XI.2, center), with the following exceptions: the bifurcation could start at the embedded thresholds located at $\pm i(m + |\omega|)$ (Fig. XI.2, left), or they could start at $\lambda = \pm im$ when $\omega = 0$ (Fig. XI.2, right; this situation correspond to the collision of thresholds). Indeed, bifurcations from the embedded thresholds have been observed in a one-dimensional NLD-type model of coupled-mode equations [**BPZ98, CP06**]. The bifurcations from the collision of thresholds at $\pm im$ (when $\omega = 0$) were demonstrated in [**KS02**] in the context of the perturbed massive Thirring model.

As to the bifurcations from the embedded eigenvalues before the embedded thresholds, as in Fig. XI.2 (center), we do not have any such examples in the NLD

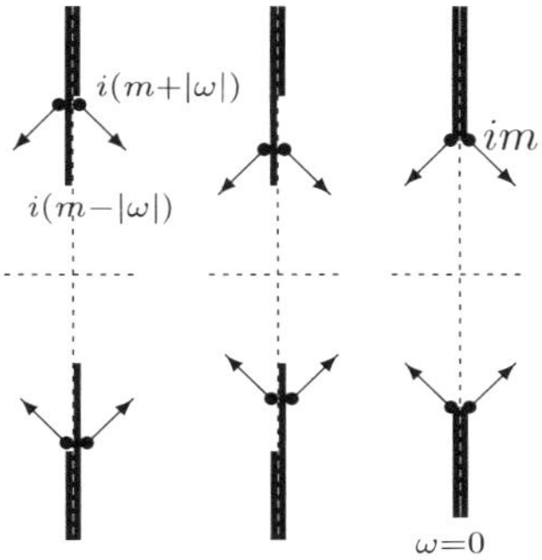

FIGURE XI.2. Theoretically, when $|\omega| < m$, the nonzero-real-part eigenvalues could be born from the embedded thresholds at $\pm i(m+|\omega|)$, from the embedded eigenvalue in the bulk of the essential spectrum between the threshold and the embedded threshold, and from the collision of the thresholds at $\pm im$ when $\omega = 0$.

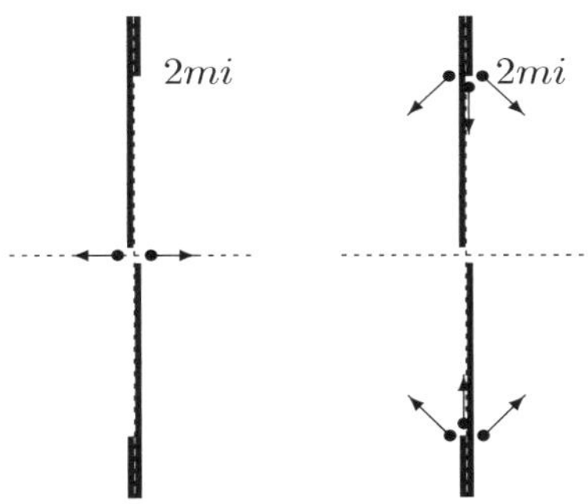

FIGURE XI.3. Bifurcations from $\lambda = 0$ and hypothetical bifurcations from $\lambda = \pm 2mi$ in the nonrelativistic limit, $\omega \lesssim m$. The nonzero-real-part eigenvalues could be present in the spectrum of the linearization at a solitary wave $\phi_\omega e^{-i\omega t}$ for ω arbitrarily close to m; these eigenvalues would have to be located near $\lambda = 0$ or near the embedded threshold at $\lambda = \pm 2mi$.

context, although such examples could be produced for Dirac operators of the form (XI.1) (with V kept self-adjoint).

Instability scenario 3: bifurcations from the nonrelativistic limit. The nonzero-real-part eigenvalues could be present in the spectrum of the linearization operators at small amplitude solitary waves for all $\omega \lesssim m$, being born "from the nonrelativistic limit".

This situation is analyzed in Chapter XIII. It is shown that under very mild assumptions, the bifurcations of eigenvalues for ω departing from $\pm m$ are only possible from the thresholds $\lambda = 0$ and $\lambda = \pm 2mi$; see Fig. XI.3.

For this scenario to be considered (Chapter XIII), we have to analyze first the nonrelativistic limit at the level of the solitary waves in Chapter XII.

XI.1 Bifurcations of eigenvalues from the essential spectrum

Theorem XI.1 (Bifurcations from the essential spectrum) *Let $n \geq 1$, $m > 0$, and let*

$$D_m = -\mathrm{i}\boldsymbol{\alpha} \cdot \nabla + \beta m : \ L^2(\mathbb{R}^n, \mathbb{C}^N) \to L^2(\mathbb{R}^n, \mathbb{C}^N), \qquad \mathfrak{D}(D_m) = H^1(\mathbb{R}^n, \mathbb{C}^N),$$

be the Dirac operator. Let $J \in \mathrm{End}(\mathbb{C}^N)$ be skew-adjoint and invertible, such that

$$J^2 = -I_N, \qquad [J, D_m] = 0.$$

Let $(\omega_j)_{j\in\mathbb{N}}$, $\omega_j \in [-m, m]$, be a sequence with $\lim_{j\to\infty} \omega_j = \omega_0 \in [-m, m]$, and assume that V is hermitian and that there is $\varepsilon > 0$ such that

$$\text{(XI.2)} \qquad \begin{cases} \|\langle r\rangle^{1+\varepsilon} V(\omega_0)\|_{L^\infty(\mathbb{R}^n,\mathrm{End}(\mathbb{C}^N))} < \infty, \\ \lim_{j\to\infty} \|\langle r\rangle^{1+\varepsilon} \left(V(\omega_j) - V(\omega_0)\right)\|_{L^\infty(\mathbb{R}^n,\mathrm{End}(\mathbb{C}^N))} = 0, \end{cases}$$

where $\|\langle r\rangle^{1+\varepsilon} V(\omega)\| = \|\langle\cdot\rangle^{1+\varepsilon} V(\cdot, \omega)\|$. Let

$$L(\omega) = D_m - \omega + V(x, \omega)$$

and let $\lambda_j \in \sigma_{\mathrm{p}}(JL(\omega_j))$, $j \in \mathbb{N}$ be a sequence such that

$$\mathrm{Re}\,\lambda_j \neq 0 \quad \forall j \in \mathbb{N}, \qquad \lambda_j \underset{j\to\infty}{\longrightarrow} \lambda_0 \in \mathrm{i}\mathbb{R}, \qquad \lambda_0 \neq \pm\mathrm{i}(m + |\omega_0|).$$

If $\omega_0 = \pm m$, additionally assume that

$$\text{(XI.3)} \qquad \lambda_0 \neq 0.$$

Then

$$\lambda_0 \in \sigma_{\mathrm{p}}(JL(\omega_0)).$$

Combining Theorem XI.1 with the absence of embedded eigenvalues stated in Theorem VII.18, we arrive at the following result:

Corollary XI.2 *For the linearizations at solitary waves, there are no bifurcations of eigenvalues from the essential spectrum beyond the embedded thresholds $\pm\mathrm{i}(m + |\omega|)$.*

Remark XI.3 The conclusion of Theorem XI.1 is trivial when V depends continuously on ω and if $\lambda_0 \in \mathrm{i}\mathbb{R}$ with $|\lambda_0| < m - |\omega_0|$, so that λ_0 is not in the essential spectrum, and the inclusion $\lambda_0 \in \sigma_{\mathrm{p}}(JL(\omega_0))$ follows from the continuous dependence of isolated eigenvalues on a parameter ω.

Remark XI.4 If JL corresponds to the linearization at solitary waves, then, due to the exponential decay of ϕ_ω (cf. Theorem IX.24), the condition (XI.2) is trivially satisfied for any $\omega_0 \in (-m, m)$ (and with any $\varepsilon > 0$).

XI.2 Bifurcation of eigenvalues before the embedded threshold

Let us consider bifurcations from the interval of the imaginary axis between the embedded thresholds, proving Theorem XI.1 for the case $|\lambda_0| < m + |\omega_0|$.

We start with the following elementary result.

Lemma XI.5 *Let $J \in \mathrm{End}(\mathbb{C}^N)$ be skew-adjoint and invertible and let L be self-adjoint on $L^2(\mathbb{R}^n, \mathbb{C}^N)$. If $\lambda \in \sigma_{\mathrm{p}}(JL) \setminus \mathrm{i}\mathbb{R}$ and ζ is a corresponding eigenvector, then*

$$\langle \zeta, L\zeta\rangle = 0, \qquad \langle \zeta, J^{-1}\zeta\rangle = 0.$$

PROOF. One has $JL\zeta = \lambda\zeta$, $L\zeta = \lambda J^{-1}\zeta$, hence

$$\langle\zeta, L\zeta\rangle = \lambda\langle\zeta, J^{-1}\zeta\rangle. \tag{XI.4}$$

Since $\langle\zeta, L\zeta\rangle \in \mathbb{R}$ and $\langle\zeta, J^{-1}\zeta\rangle \in \mathrm{i}\mathbb{R}$, the condition $\operatorname{Re}\lambda \neq 0$ implies that both sides in (XI.4) are equal to zero. □

Lemma XI.6 *Let $n \geq 1$ and $m > 0$. Let $J \in \operatorname{End}(\mathbb{C}^N)$ be skew-adjoint and invertible, such that $J^2 = -I_N$, $[J, D_m] = 0$. Let $\omega_j \in (-m, m)$, $\omega_j \underset{j\to\infty}{\longrightarrow} \omega_0 \in [-m, m]$. Let*

$$L(\omega) = D_m - \omega + V(\omega), \qquad \omega \in [-m, m],$$

where $V(\omega) \in L^\infty(\mathbb{R}^n, \operatorname{End}(\mathbb{C}^N))$ is hermitian, with $\varepsilon > 0$ such that the assumption (XI.2) is satisfied. Let $\lambda_j \in \sigma_{\mathrm{d}}(JL(\omega_j))$ be a sequence such that

$$\lambda_j \underset{j\to\infty}{\longrightarrow} \lambda_0 \in \mathrm{i}\mathbb{R}, \qquad |\lambda_0| < m + |\omega_0|, \tag{XI.5}$$

with

$$\operatorname{Re}\lambda_j \neq 0, \quad \forall j \in \mathbb{N}. \tag{XI.6}$$

If $\omega_0 = \pm m$, additionally assume that

$$\lambda_0 \neq 0. \tag{XI.7}$$

Then

$$\lambda_0 \in \sigma_{\mathrm{p}}(JL(\omega_0)). \tag{XI.8}$$

PROOF. Let $(\zeta_j)_{j\in\mathbb{N}}$ be a sequence of unit eigenvectors associated with eigenvalues λ_j, so that $JL(\omega_j)\zeta_j = \lambda_j\zeta_j$. It follows that

$$(D_m - \omega_j + \lambda_j J)\zeta_j = -V(\omega_j)\zeta_j. \tag{XI.9}$$

Let $\Pi^\pm$ (cf. (VII.92)) be the projectors onto eigenspaces of J corresponding to $\pm\mathrm{i} \in \sigma(J)$, respectively. We denote $\zeta_j^\pm = \Pi^\pm\zeta_j$. By (XI.6), applying Lemma XI.5, we conclude that $0 = \langle\zeta_j, J\zeta_j\rangle = \mathrm{i}\|\zeta_j^+\|^2 - \mathrm{i}\|\zeta_j^-\|^2$, $j \in \mathbb{N}$, while $1 = \|\zeta_j\|^2 = \|\zeta_j^+\|^2 + \|\zeta_j^-\|^2$, $j \in \mathbb{N}$; we conclude that $\|\zeta_j^\pm\| = 1/\sqrt{2}$. Applying $\Pi^\pm$ to (XI.9), we have:

$$\big(D_m - \omega_j + \mathrm{i}\lambda_j\big)\zeta_j^+ = -\Pi^+ V(\omega_j)\zeta_j, \tag{XI.10}$$

$$\big(D_m - \omega_j - \mathrm{i}\lambda_j\big)\zeta_j^- = -\Pi^- V(\omega_j)\zeta_j. \tag{XI.11}$$

Above, we took into account that $[J, D_m] = 0$, hence the projections $\Pi^\pm$ also commute with D_m.

If the condition (XI.5) (as well as (XI.7) when $\omega_0 = \pm m$) is satisfied, then either $\omega_0 + \mathrm{i}\lambda_0 \in (-m, m)$ or $\omega_0 - \mathrm{i}\lambda_0 \in (-m, m)$. Both cases are considered similarly; for definiteness, we will assume that

$$\omega_0 - \mathrm{i}\lambda_0 \in (-m, m).$$

In this case, without loss of generality, we may also assume that

$$\omega_j - \mathrm{i}\lambda_j \in (-m, m), \qquad \forall j \in \mathbb{N}. \tag{XI.12}$$

By (XI.2), the right-hand side of (XI.10) belongs to L^2_s for any $s \leq 1 + \varepsilon$. Due to (XI.12), we may apply Lemma VI.27 to (XI.10), concluding that there is $s \in (0, 1)$ such that $\|\zeta_j^+\|_{H^1_s}$ are uniformly bounded when j is large enough (so that $\mathrm{i}\lambda_j + \omega_j$ are sufficiently close to $\mathrm{i}\lambda_0 + \omega_0$). Thus the sequence $(\zeta_j^+)_{j\in\mathbb{N}}$ is precompact in

$L^2(\mathbb{R}^n, \mathbb{C}^N)$, and we can choose a subsequence which converges to some vector $\zeta_0^+ \in L^2(\mathbb{R}^n, \mathbb{C}^N)$ of norm $\|\zeta_0^+\| = \lim_{j\to\infty} \|\zeta_j^+\| = 1/\sqrt{2}$. At the same time, any subsequence of the bounded sequence $(\zeta_j^-)_{j\in\mathbb{N}}$ also contains a weakly convergent subsequence. We conclude that there is a subsequence $(\zeta_j)_{j\in\mathbb{N}}$ which has a nonzero weak limit; this limit is necessarily an eigenvector of $JL(\omega_0)$ corresponding to λ_0. □

XI.3 Bifurcation of eigenvalues beyond the embedded thresholds

We now turn to the proof of the limiting absorption principle for the linearized operator in a neighborhood of any purely imaginary point beyond the embedded thresholds $\pm\mathrm{i}(m + |\omega|)$, proving Theorem XI.1 for the case $|\lambda_0| > m + |\omega_0|$. In that respect we closely follow the strategy initiated in a work by Jensen and Kato **[JK79]** which is related to the approach by **[Agm75]**. We start with the following identity:

$$J(D_m - \omega + V(\omega)) - \lambda = (J(D_m - \omega) - \lambda)\left(I + \left(J(D_m - \omega) - \lambda\right)^{-1} JV(\omega)\right),$$

$\lambda \in \mathbb{C} \setminus \sigma(J(D_m - \omega))$. After diagonalizing J (which commutes with D_m), Lemma VI.24 provides the limiting absorption principle for $J(D_m - \omega) - \lambda$; hence, our task reduces to proving that the operator $A(\lambda, \omega) = I + \left(J(D_m - \omega) - \lambda\right)^{-1} JV(\omega)$,

$$A(\lambda, \omega) : L^2_{-s}(\mathbb{R}^n, \mathbb{C}^N) \to L^2_{-s}(\mathbb{R}^n, \mathbb{C}^N), \qquad s > 1/2,$$

has an inverse which is bounded uniformly in λ, with $\operatorname{Re}\lambda > 0$, in the vicinity of any particular point

$$(\omega_0, \lambda_0), \qquad \omega_0 \in [-m, m], \qquad \lambda_0 \in \mathrm{i}\mathbb{R}, \qquad |\lambda_0| > \mathrm{i}(m + |\omega_0|)$$

where we know that A is invertible, and also proving that $A(\omega_0, \lambda_0)$ is not invertible if and only if λ_0 is an eigenvalue of $J(D_m - \omega_0 + V(\omega_0))$.

Proposition XI.7 *Let $V : \mathbb{R}^n \to \operatorname{End}(\mathbb{C}^N)$ be measurable, hermitian-valued, and assume that there are $\varepsilon > 0$ and $C > 0$ such that*

$$\text{(XI.13)} \qquad \|\langle r\rangle^{1+\varepsilon} V\|_{L^\infty(\mathbb{R}^n, \operatorname{End}(\mathbb{C}^N))} < C.$$

Let $\omega \in [-m, m]$, $\lambda \in \mathrm{i}\mathbb{R}$, $|\lambda| > m + |\omega|$, $s \in (1/2, (1+\varepsilon)/2)$. Then either the operator

$$A = I + \left(J(D_m - \omega) - \lambda\right)^{-1} JV, \qquad A : H^{1/2}_{-s}(\mathbb{R}^n, \mathbb{C}^N) \to H^{1/2}_{-s}(\mathbb{R}^n, \mathbb{C}^N)$$

is invertible, or there is a nonzero function $F \in \mathbf{ker}(A)$, $F \in L^2(\mathbb{R}^n, \mathbb{C}^N)$.

Proof. By (XI.13), for any $s < (1+\varepsilon)/2$, the map $V : u \mapsto Vu$ is bounded from $H^{1/2}_{-s}$ to L^2_s. Due to the limiting absorption principle for the Dirac operator (cf. Lemma VI.24), the resolvent

$$R_0(\lambda) := (J(D_m - \omega) - \lambda)^{-1}, \qquad \lambda \in \mathbb{C}, \quad \operatorname{Re}\lambda > 0,$$

can be extended onto the closure of the right half-plane, excluding arbitrarily small open neighborhoods of $\pm\mathrm{i}(m\pm\omega)$ (we keep the same notation R_0 for this extension), so that for any $s > 1/2$ one has

$$R_0(\lambda) \in \mathscr{B}(L^2_s(\mathbb{R}^n, \mathbb{C}^N), H^1_{-s}(\mathbb{R}^n, \mathbb{C}^N)), \qquad \lambda \in \mathbb{C}, \ \operatorname{Re}\lambda \geq 0, \ \lambda \neq \pm\mathrm{i}(m \pm \omega).$$

Due to the decay of V, the operator $A-I$ is compact in L^2_{-s} if $s\in\big(1/2,(1+\varepsilon)/2\big)$. Hence, by the Fredholm alternative, A is invertible in L^2_{-s} if and only if its null space

$$\mathfrak{M}_s := \mathbf{ker}\left(A|_{L^2_{-s}}\right)$$

is trivial.

Let us consider the Fredholm operator

$$B = 1 + VJ\big(J(D_m-\omega)+\lambda\big)^{-1}$$

on $H_s^{-1/2}(\mathbb{R}^n,\mathbb{C}^N)$, with $s\in\big(1/2,(1+\varepsilon)/2\big)$. We denote its null space by

$$\mathfrak{N}_s := \mathbf{ker}\left(B|_{L^2_s}\right).$$

Being compact perturbations of the identity, both A and B are Fredholm operators of index zero. Proceeding as in [**JK79**, Section 3], we notice that the finite-dimensional spaces $\mathfrak{M}_s$ and $\mathfrak{N}_s$ are respectively non-decreasing and non-increasing as s grows. Since $A|_{H^{1/2}_{-s}}$ and $B|_{H_s^{-1/2}}$ are mutually adjoint,

$$\begin{aligned} 0=\operatorname{ind} A|_{H^{1/2}_{-s}} &= \dim \mathbf{ker}(A|_{L^2_{-s}}) - \dim \mathbf{coker}(A|_{L^2_{-s}}) \\ &= \dim \mathbf{ker}(A|_{L^2_{-s}}) - \dim \mathbf{ker}(B|_{L^2_s}) \end{aligned}$$

is a non-decreasing function of $s\in\big(1/2,(1+\varepsilon)/2\big)$, hence $\dim\mathfrak{M}_s=\dim\mathfrak{N}_s$ does not depend on $s\in\big(1/2,(1+\varepsilon)/2\big)$. We conclude that the spaces $\mathfrak{M}_s$, $\mathfrak{N}_s$ do not depend on s for $s\in\big(1/2,(1+\varepsilon)/2\big)$; we will denote these spaces by $\mathfrak{M}$ and $\mathfrak{N}$, respectfully.

One key fact is the following lemma.

Lemma XI.8 *Let $s>1/2$, $\omega\in[-m,m]$, $\lambda\in\mathrm{i}\mathbb{R}$, $|\lambda|>m+|\omega|$. Then*

$$(J(D_m-\omega)-\lambda)\,R_0(\lambda)v=v,\qquad \forall v\in H_s^{-1/2}(\mathbb{R}^n,\mathbb{C}^N).$$

Proof. This is an adaptation of [**JK79**, Lemma 2.4]. Fix $v\in H_s^{-1/2}(\mathbb{R}^n,\mathbb{C}^N)$. We note that for $\lambda\in\mathrm{i}\mathbb{R}$ one has

$$(J(D_m-\omega)-\lambda)^* = -(D_m-\omega)J-\bar\lambda = -(J(D_m-\omega)-\lambda),\qquad \forall\lambda\in\mathrm{i}\mathbb{R},$$

where we took into account that $[J,D_m]=0$. Therefore, for any $\varphi\in C^\infty_{\mathrm{comp}}(\mathbb{R}^n,\mathbb{C}^N)$, since $R_0(\lambda)v\in H^{1/2}_{-s}$, we have:

$$\langle (J(D_m-\omega)-\lambda)R_0(\lambda)v,\varphi\rangle = -\langle R_0(\lambda)v,(J(D_m-\omega)-\lambda)\varphi\rangle.$$

Using $R_0(\lambda)^*=-R_0(\lambda)$ for $\lambda\in\mathrm{i}\mathbb{R}$, we then write

$$\big\langle (J(D_m-\omega)-\lambda)R_0(\lambda)v,\varphi\big\rangle = \big\langle v,R_0(\lambda)(J(D_m-\omega)-\lambda)\varphi\big\rangle = \langle v,\varphi\rangle,$$

where the identity

$$R_0(\lambda)(J(D_m-\omega)-\lambda)\varphi=\varphi,\qquad \forall\varphi\in C^\infty_{\mathrm{comp}}(\mathbb{R}^n,\mathbb{C}^N)$$

follows from the same identity on the Fourier transform side. This completes the proof. □

We deduce from Lemma XI.8 that any $u \in \mathfrak{M}$ satisfies

$$\big(J(D_m - \omega + V) - \lambda\big)u = 0.$$

In the following, we argue that there is the inclusion

$$\mathfrak{M} \subset L^2(\mathbb{R}^n, \mathbb{C}^N), \tag{XI.14}$$

which would conclude the proof of Proposition XI.7.

The inclusion (XI.14) is proved using the following three complementary results.

Lemma XI.9 *Let $s > 1/2$, $\epsilon > 0$. Let $\Lambda \in \mathbb{R}$, $|\Lambda| > m$. If $f \in H_s^{-1/2}(\mathbb{R}^n, \mathbb{C}^N)$, then*

$$\lim_{\epsilon \to 0+} \operatorname{Im} \big\langle f, (D_m - \Lambda - \mathrm{i}\epsilon)^{-1} f \big\rangle = \frac{\pi |\Lambda|}{\sqrt{\Lambda^2 - m^2}} \int_{\xi^2 + m^2 = \Lambda^2} |\tau P^+ \hat{f}(\xi)|^2 \, d\sigma(\xi),$$

where τ denotes the trace operator on Sobolev space $H^\mu(\mathbb{R}^n, \mathbb{C}^N)$ of order $\mu > 1/2$, $\hat{\ }$ is the unitary Fourier transform on tempered distributions, $d\sigma$ is the induced measure on the surface $\xi^2 + m^2 = \Lambda^2$, and $P^\pm = \frac{1}{2}\Big(1 \pm \frac{d_m(\xi)}{\sqrt{\xi^2+m^2}}\Big)$ are the projectors onto positive and negative eigenvalues, $\pm\sqrt{\xi^2 + m^2}$, of the symbol $d_m(\xi) = \boldsymbol{\alpha} \cdot \xi + \beta m$.

PROOF. First we notice that for $f \in H_s^{-1/2}(\mathbb{R}^n, \mathbb{C}^N)$, one has:

$$\begin{aligned} \big\langle f, (D_m - \Lambda - \mathrm{i}\epsilon)^{-1} f \big\rangle &= \int_{\mathbb{R}^n} (\boldsymbol{\alpha} \cdot \xi + \beta m - \Lambda - \mathrm{i}\epsilon)^{-1} \, |\hat{f}(\xi)|^2 \, d\xi \\ &= \int_{\mathbb{R}^n} \frac{|P^+(\xi)\hat{f}(\xi)|^2 \, d\xi}{\sqrt{\xi^2 + m^2} - \Lambda - \mathrm{i}\epsilon} + \int_{\mathbb{R}^n} \frac{|P^-(\xi)\hat{f}(\xi)|^2 \, d\xi}{-\sqrt{\xi^2 + m^2} - \Lambda - \mathrm{i}\epsilon}, \end{aligned}$$

and hence

$$\begin{aligned} &\operatorname{Im} \big\langle f, (D_m - \Lambda - \mathrm{i}\epsilon)^{-1} f \big\rangle \\ &= \int_{\mathbb{R}^n} \frac{\epsilon |P^+(\xi)\hat{f}(\xi)|^2 \, d\xi}{(\sqrt{\xi^2 + m^2} - \Lambda)^2 + \epsilon^2} + \int_{\mathbb{R}^n} \frac{\epsilon |P^-(\xi)\hat{f}(\xi)|^2 \, d\xi}{(\sqrt{\xi^2 + m^2} + \Lambda)^2 + \epsilon^2}. \end{aligned} \tag{XI.15}$$

Let us assume that $\Lambda > m$. In the limit $\epsilon \to 0+$, the second integral in the right-hand side of (XI.15) tends to 0. The first integral can be written as

$$\begin{aligned} &\int_{\mathbb{R}^n} \frac{2\epsilon}{(\sqrt{\xi^2 + m^2} - \Lambda)^2 + \epsilon^2} |P^+(\xi)\hat{f}(\xi)|^2 \, d\xi \\ &= \int_{\mathbb{S}^{n-1}} \int_{\mathbb{R}^+} \frac{2\epsilon}{(\sqrt{k^2 + m^2} - \Lambda)^2 + \epsilon^2} |P^+(k\omega)\hat{f}(k\omega)|^2 \, k^{n-1} \, dk \, d\omega \\ &= \int_{\mathbb{S}^{n-1}} \int_m^\infty \Big| P^+(\omega\sqrt{\lambda^2 - m^2})\hat{f}(\omega\sqrt{\lambda^2 - m^2}) \Big|^2 \frac{2\epsilon \, (\lambda^2 - m^2)^{(n-2)/2} \lambda \, d\lambda \, d\omega}{(\lambda - \Lambda)^2 + \epsilon^2}, \end{aligned}$$

which converges, in the limit $\epsilon \to 0+$, to

$$\begin{aligned} &2\pi(\Lambda^2 - m^2)^{\frac{n-2}{2}} \Lambda \int\limits_{\mathbb{S}^{n-1}} \Big| P^+(\omega\sqrt{\Lambda^2 - m^2})\hat{f}(\omega\sqrt{\Lambda^2 - m^2}) \Big|^2 \, d\omega \\ &= \frac{2\pi\Lambda}{\sqrt{\Lambda^2 - m^2}} \int\limits_{\xi^2 + m^2 = \Lambda^2} |\tau(P^+(\xi)\hat{f}(\xi)|^2 \, d\sigma. \end{aligned}$$

This proves the required identity in the case $\Lambda > m$. The case $\Lambda < -m$ is considered similarly. □

The second result we need is directly inspired by [**CPV05**, Proof of Lemma 2.4].

Lemma XI.10 *Assume that V is hermitian and there are $\varepsilon > 0$ and $C > 0$ such that*

$$\|\langle r\rangle^{1+\varepsilon} V\|_{L^\infty(\mathbb{R}^n,\mathrm{End}(\mathbb{C}^N))} < C.$$

Let $s \in \big(1/2, (1+\varepsilon)/2\big)$, $\omega \in [-m, m]$. Let $\lambda \in \mathrm{i}\mathbb{R}$, $|\lambda| > m + |\omega|$.

Then for any $F \in H^{1/2}_{-s}(\mathbb{R}^n, \mathbb{C}^N)$ such that

$$J(D_m - \omega + V)F = \lambda F,$$

the function $G := (D_m - \omega + J\lambda)F = -VF$ satisfies

$$\int_{\sqrt{\xi^2+m^2}=|\omega\mp\mathrm{i}\lambda|} \Big|\Pi^\pm\tau\big(P^+(\xi)\hat{G}(\xi)\big)\Big|^2\,d\sigma = 0,$$

where $\Pi^\pm = \frac{1}{2}(1 \mp \mathrm{i}J)$ is the projector onto eigenspaces of J corresponding to $\pm\mathrm{i} \in \sigma(J)$ (cf. (VII.92)).

Proof. We assume that $\lambda = \mathrm{i}\Lambda$, with $\Lambda > m + |\omega|$. (The case $\lambda = -\mathrm{i}\Lambda$ is considered verbatim.)

Applying the spectral projectors $\Pi^\pm$ to the relation

$$G = (D_m - \omega + J\lambda)F$$

and denoting $F^\pm = \Pi^\pm F$, $G^\pm = \Pi^\pm G$, we have:

$$G^\pm = (D_m - \omega \pm \mathrm{i}\lambda)F^\pm = (D_m - \omega \mp \Lambda)F^\pm = -\Pi^\pm VF.$$

One has:

$$\begin{aligned}
\lim_{\epsilon\to 0+}\langle G^\pm, (D_m - \omega \mp \Lambda - \mathrm{i}\epsilon)^{-1}G^\pm\rangle &= -\lim_{\epsilon\to 0+}\langle G^\pm, (D_m - \omega \mp \Lambda - \mathrm{i}\epsilon)^{-1}\Pi^\pm VF\rangle \\
&= -\big\langle F^\pm, \Pi^\pm VF\big\rangle = -\langle F^\pm, VF\rangle.
\end{aligned}$$

Summing up the expressions corresponding to $\pm$ signs, taking the imaginary part, and applying Lemma XI.9 leads to

$$\begin{aligned}
0 = -\operatorname{Im}\langle F, VF\rangle &= -\operatorname{Im}\big(\langle F^+, VF\rangle + \langle F^-, VF\rangle\big) \\
&= \sum_\pm \lim_{\epsilon\to 0+}\operatorname{Im}\langle G^\pm, (D_m - \omega \mp \Lambda - \mathrm{i}\epsilon)^{-1}G^\pm\rangle \\
&= \frac{\pi|\omega+\Lambda|}{\sqrt{(\omega+\Lambda)^2 - m^2}} \int_{\sqrt{\xi^2+m^2}=|\omega+\Lambda|} \Big|\Pi^+\tau\big(P^+(\xi)\hat{G}(\xi)\big)\Big|^2\,d\sigma(\xi) \\
&\quad + \frac{\pi|\omega-\Lambda|}{\sqrt{(\omega-\Lambda)^2 - m^2}} \int_{\sqrt{\xi^2+m^2}=|\omega-\Lambda|} \Big|\Pi^-\tau\big(P^+(\xi)\hat{G}(\xi)\big)\Big|^2\,d\sigma(\xi).
\end{aligned}$$

The assumption that V is hermitian was used in the very first equality.

Since both the coefficients and the integrals in the right-hand side are positive, the conclusion follows. □

The last step is needed to exclude non-square-integrable resonances. It is directly inspired by [**BG87**, Theorem 2].

Lemma XI.11 *Let $s > 1/2$. Let $\omega \in [-m, m]$. Let $\lambda = \mathrm{i}\Lambda \in \mathrm{i}\mathbb{R}$ with $\Lambda > m+|\omega|$. If $F \in \mathscr{S}'(\mathbb{R}^n, \mathbb{C}^N)$ is such that*

$$[J(D_m - \omega) - \lambda]F \in L^2_s(\mathbb{R}^n, \mathbb{C}^N)$$

and if $(1 \pm \mathrm{i}J)P^+(\xi)\big(J(d_m(\xi) - \omega) - \lambda\big)\hat{F}$ (with $d_m(\xi) = \boldsymbol{\alpha} \cdot \xi + \beta m$) vanish on the spheres $\sqrt{\xi^2 + m^2} = \Lambda \pm \omega$, respectively, then

$$\|F\|_{s-1} \le C\|[J(D_m - \omega) - \lambda]F\|_s, \tag{XI.16}$$

for some constant $C > 0$ depending on s, λ and ω only.

Proof. The proof of [**BG87**, Theorem 2] works with a straightforward adaptation of the key [**BG87**, Lemma 5], which is a consequence of [**Agm75**, Appendix B] which in turn is valid in any dimension; the assumptions needed to apply it are the ones stated in the condition of the Lemma. □

Remark XI.12 In [**BG87**], Berthier and Georgescu proved a result similar to (XI.16) under the L^1_{loc} assumption on the Fourier transform of F. Such an assumption provides that $(1 \pm \mathrm{i}J)P^+(\xi)\big(J(d_m(\xi) - \omega) - \lambda\big)\hat{F}$ vanish on the spheres $\sqrt{\xi^2 + m^2} = \Lambda \pm \omega$, respectively.

Lemmata XI.9, XI.10 and XI.11 complete the proof of the inclusion (XI.14). Proposition XI.7 follows from (XI.14) and Lemma XI.8. □

Proposition XI.13 *Let $V : \mathbb{R}^n \times [-m, m] \to \mathrm{End}(\mathbb{C}^N)$ be measurable, and assume that there are $C > 0$ and $\varepsilon > 0$ such that (XI.2) is satisfied. Let $\omega_0 \in [-m, m]$ and let*

$$\lambda_0 \in \mathrm{i}\mathbb{R}, \qquad |\lambda_0| > m + |\omega_0|, \qquad \lambda_0 \notin \sigma_{\mathrm{p}}\big(J(D_m - \omega + V(\omega))\big).$$

Then for any $s \in \big(1/2, (1+\varepsilon)/2\big)$ there exist an open neighborhood $I \subset [-m, m]$ of ω_0 (I is a one-sided neighborhood of ω_0 if $\omega_0 = \pm m$) and an open neighborhood $U \subset \mathbb{C}$ of λ_0 such that for $\omega \in I$ the resolvent of $J(D_m - \omega + V(\omega))$ at $\lambda \in \overline{U} \setminus \mathrm{i}\mathbb{R}$ extends to a continuous mapping

$$\big(J(D_m - \omega + V(\omega)) - \lambda\big)^{-1} : H^{-1/2}_s(\mathbb{R}^n, \mathbb{C}^N) \to H^{1/2}_{-s}(\mathbb{R}^n, \mathbb{C}^N),$$

which is bounded uniformly in $\lambda \in \overline{U} \setminus \mathrm{i}\mathbb{R}$.

Proof. If $\lambda_0 \notin \sigma_{\mathrm{p}}\big(J(D_m - \omega_0 + V(\omega_0))\big)$, then, by Proposition XI.7, the operator

$$A(\omega, \lambda) = 1 + \big(J(D_m - \omega) - \lambda\big)^{-1} JV(\omega),$$
$$A(\omega, \lambda) : H^{1/2}_{-s}(\mathbb{R}^n, \mathbb{C}^N) \to H^{1/2}_{-s}(\mathbb{R}^n, \mathbb{C}^N)$$

is invertible at (ω_0, λ_0). By the limiting absorption principle (Lemma VI.24), for any $s > 1/2$, there is an open neighborhood $I \subset [-m, m]$ of ω_0 (I is a one-sided neighborhood if $\omega_0 = \pm m$) and an open neighborhood $U \subset \mathbb{C}$ of λ_0 such that $R_0(\lambda) = (J(D_m - \omega) - \lambda)^{-1}$ remains continuous in the $H^{-1/2}_s \to H^{1/2}_{-s}$ operator topology for $\omega \in I$, $\lambda \in U$, $\mathrm{Re}\,\lambda \ge 0$. Similarly, $V(\omega) : H^{1/2}_{-s} \to L^2_s$ remains continuous in the corresponding operator topology for $\omega \in I \cap [-m, m]$ (cf. (XI.2)). By continuity in ω and λ, the operator $A(\omega, \lambda)$ is continuous in the $H^{1/2}_{-s} \to H^{1/2}_{-s}$ operator topology, remaining invertible for (ω, λ) in an open neighborhood of (ω_0, λ_0), with $\mathrm{Re}\,\lambda \ge 0$. □

Now we can finish the proof of Theorem XI.1.

Lemma XI.14 *Let $(\omega_j)_{j\in\mathbb{N}}$, $\omega_j \in (-m,m)$, $\omega_j \to \omega_0 \in [-m,m]$. Assume that*

$$\lambda_0 \in \mathrm{i}\mathbb{R}, \qquad |\lambda_0| > m + |\omega_0|, \qquad \lambda_0 \notin \sigma_{\mathrm{p}}\big(J(D_m - \omega_0 + V(\omega_0))\big).$$

Then there is no sequence $\lambda_j \in \sigma_{\mathrm{p}}\big(J(D_m - \omega_j + V(\omega_j))\big)$ such that $\lambda_j \to \lambda_0 \in \mathrm{i}\mathbb{R}$.

Proof. We use the last part of the proof of Proposition XI.13 to argue by contradiction. Indeed, let $(\omega_j)_{j\in\mathbb{N}}$, $\omega_j \in (-m,m)$, $\omega_j \to \omega_0 \in [-m,m]$, and assume that

$$\lambda_j \in \sigma_{\mathrm{p}}\big(J(D_m - \omega_j + V(\omega_j))\big), \qquad \lambda_j \to \lambda_0 \in \mathrm{i}\mathbb{R}, \qquad |\lambda_0| > m + |\omega_0|.$$

Fix s, s' such that

$$1/2 < s < s' < (1+\varepsilon)/2,$$

with $\varepsilon > 0$ from (XI.2). Since the operator

$$\big(J(D_m - \omega_j) - \lambda_j\big)^{-1} JV(\omega_j)$$

is bounded from $L^2_{-s'}$ to $H^{1/2}_{-s}$ uniformly in $j \in \mathbb{N}$, while the latter embeds compactly into $L^2_{-s'}$, we conclude that any sequence of eigenvectors (normalized in $L^2_{-s'}$) associated to $\lambda_j \in \sigma_{\mathrm{p}}\big(J(D_m - \omega_j + V(\omega_j))\big)$ is compact in $L^2_{-s'}$, converging to a nonzero vector from $H^{1/2}_{-s}$, leading to a contradiction with the operator

$$A(\lambda_0) = 1 + \big(J(D_m - \omega_0) - \lambda_0\big)^{-1} JV(\omega_0) : \ H^{1/2}_{-s}(\mathbb{R}^n, \mathbb{C}^N) \to H^{1/2}_{-s}(\mathbb{R}^n, \mathbb{C}^N)$$

being invertible by Proposition XI.13. □

This finishes the proof of Theorem XI.1.

CHAPTER XII

Nonrelativistic asymptotics of solitary waves

In this chapter, we use the bifurcation approach to construct solitary wave solutions in the Soler model

$$\mathrm{i}\partial_t\psi = D_m\psi - f(\psi^*\beta\psi)\beta\psi, \qquad \psi(t,x)\in\mathbb{C}^N, \quad x\in\mathbb{R}^n, \quad n\in\mathbb{N},$$

with $m>0$; we closely follow **[BC17]**. We will assume that the nonlinearity is represented by $f(\tau)\sim|\tau|^\kappa$, with $\kappa>0$ small. In this case, we will obtain solitary wave solutions in the nonrelativistic limit,

$$\phi_\omega(x)e^{-\mathrm{i}\omega t}, \qquad \phi_\omega\in H^1(\mathbb{R}^n,\mathbb{C}^N), \qquad \omega\lessapprox m$$

($\omega\in(m-\varepsilon,m)$ for some $\varepsilon>0$ small enough), building them as bifurcations from solitary waves of nonlinear Schrödinger equation; the construction provides description of solitary waves which we will need for the analysis of their spectral stability (presence or absence of eigenvalues with positive real part in the spectrum of the linearization at a solitary wave).

A perturbation method for the construction of solitary waves in the nonlinear Dirac equation was used in **[Oun00]**. This work was later followed in **[Gua08, CGG14]** and also generalized to the Einstein–Dirac and Einstein–Dirac–Maxwell systems **[RN10b, Stu10, RN10a]** and to the Dirac–Maxwell system **[CS18]**. Our aim here is to make the perturbative approach of **[Oun00]** rigorous for the important case of lower order nonlinearities. The usefulness of such an approach is that it gives the asymptotics of solitary waves needed for the study of their stability properties.

Most common models considered by physicists and chemists (e.g. **[Rañ83a]**) are based on the pure power nonlinearity $f(\tau)=|\tau|^\kappa$, $\kappa>0$, often cubic ($\kappa=1$) and quintic ($\kappa=2$). We will need the pure power case for the application to stability of weakly relativistic solitary waves in the nonlinear Dirac equation. As we already mentioned, there have been several implementations of constructing solitary waves via the bifurcation method for such models, but these approaches did not allow one to handle the low regularity case, such as $f(\tau)=|\tau|^\kappa$, with $\kappa\in(0,1)$, when $f(\tau)$ is no longer differentiable at $\tau=0$, hence the derivative of f could contribute a singularity if the Lorentz scalar $\phi_\omega^*\beta\phi_\omega$ vanished. On the other hand, this low regularity case corresponds to the interesting "Schrödinger charge-subcritical" case, when $\kappa\in(0,2/n)$ (with $n\geq 2$), so that the "groundstate" solitary waves for NLS are stable (groundstate is understood in the sense of **[BL83a]**: it is a strictly positive, spherically symmetric, decaying solution to the stationary NLS). With these values of κ, one can compare stability properties in both models, pushing further the discussion from **[CMKS$^+$16]**. We overcome the difficulties resulting from the low regularity of f in the nonrelativistic limit $\omega\lessapprox m$, constructing solitary waves for arbitrary f with $f(\tau)\sim|\tau|^\kappa$ with $\kappa>0$ arbitrarily small. The main points are to base the construction on the Schauder fixed point theorem (instead of the contraction mapping principle which is not available to us when $f(\tau)$ is not Lipschitz) and to prove that $\phi_\omega(x)^*\beta\phi_\omega(x)$ is bounded from below by $c\phi_\omega(x)^*\phi_\omega(x)$ with some $c \in (0,1)$, for ω sufficiently close to m. In the case when f is differentiable away from the origin, we will additionally prove uniqueness of ϕ_ω (up to the symmetry transformations) and also its differentiability with respect to ω.

In the present chapter, we prove the existence of solitary waves for the case of a continuous nonlinearity (Theorem XII.1) and then derive the improvement for the case of nonlinearity differentiable everywhere except perhaps at zero (Theorem XII.3).

XII.1 Main results

We consider the nonlinear Dirac equation

$$\text{(XII.1)} \qquad \mathrm{i}\partial_t\psi = D_m\psi - f(\psi^*\beta\psi)\beta\psi, \qquad \psi(t,x)\in\mathbb{C}^N, \qquad x\in\mathbb{R}^n,$$

where D_m is the Dirac operator (see (IX.2)). We assume that the nonlinearity is represented by the function $f\in C(\mathbb{R})$, $f(0)=0$, satisfying

$$\text{(XII.2)} \qquad f\in C(\mathbb{R}), \qquad f(\tau)=|\tau|^\kappa + o(|\tau|^\kappa), \qquad \tau\in\mathbb{R},$$

with some $\kappa>0$. In the nonrelativistic limit $\omega \lessapprox m$, the solitary waves to nonlinear Dirac equation could be obtained as bifurcations of the solitary wave solutions $\phi_\omega(x)e^{-\mathrm{i}\omega t}$ to the nonlinear Schrödinger equation with pure power nonlinearity,

$$\text{(XII.3)} \qquad \mathrm{i}\partial_t\psi = -\frac{1}{2m}\Delta\psi - |\psi|^{2\kappa}\psi, \qquad \psi(t,x)\in\mathbb{C}, \qquad x\in\mathbb{R}^n,$$

with the same value of $\kappa>0$ as in (XII.2). For the applications to the nonlinear Dirac equation, it is very convenient to keep the factor $1/(2m)$ at the Laplace operator and to consider the solitary wave with $\omega=-1/(2m)$. We assume that

$$n\in\mathbb{N}, \qquad m>0, \qquad \kappa>0; \qquad \kappa<2/(n-2) \quad \text{if} \quad n\ge 3.$$

By the results of Chapter V, the stationary nonlinear Schrödinger equation

$$\text{(XII.4)} \qquad -\frac{1}{2m}u = -\frac{1}{2m}\Delta u - |u|^{2\kappa}u, \qquad u(x)\in\mathbb{R}, \qquad x\in\mathbb{R}^n,$$

has a strictly positive spherically symmetric exponentially decaying solution u_κ, called the groundstate. By Theorem V.7, $u_\kappa\in H^1_r(\mathbb{R}^n)\cap C^2(\mathbb{R}^n)$; as the matter of fact, by Lemma V.24,

$$\text{(XII.5)} \qquad u_\kappa\in H^s_r(\mathbb{R}^n)\cap C^2(\mathbb{R}^n), \qquad \forall s<n/2+2.$$

By Lemma V.19, there are $0<c_{n,\kappa}<C_{n,\kappa}<\infty$ such that

$$\text{(XII.6)} \qquad c_{n,\kappa}\langle x\rangle^{-(n-1)/2}e^{-|x|} \le u_\kappa(x) \le C_{n,\kappa}\langle x\rangle^{-(n-1)/2}e^{-|x|}, \qquad x\in\mathbb{R}^n.$$

The linearization at the NLS solitary wave $u_\kappa(x)e^{-\mathrm{i}\omega t}$, $\omega=-\frac{1}{2m}$, is given by

$$\text{(XII.7)} \qquad \partial_t\boldsymbol{\rho} = \begin{bmatrix} 0 & \mathfrak{l}_- \\ -\mathfrak{l}_+ & 0\end{bmatrix}\boldsymbol{\rho}, \qquad \boldsymbol{\rho}(t,x)\in\mathbb{C}^2, \qquad x\in\mathbb{R}^n,$$

where $\mathfrak{l}_\pm$ are defined by

$$\text{(XII.8)} \qquad \mathfrak{l}_- = \frac{1}{2m} - \frac{\Delta}{2m} - u_\kappa^{2\kappa}, \qquad \mathfrak{l}_+ = \frac{1}{2m} - \frac{\Delta}{2m} - (1+2\kappa)u_\kappa^{2\kappa},$$

with the domain $\mathfrak{D}(\mathfrak{l}_\pm) = H^2(\mathbb{R}^n)$. (We recall that the operator of multiplication by $u_\kappa^{2\kappa}$ is a continuous linear operator in $L^2(\mathbb{R}^n)$ by Lemma V.24 *(3)*.)

Later we will need the following relations:

$$\text{(XII.9)} \qquad \mathfrak{l}_+\theta(x) = u_\kappa(x), \qquad \theta(x) := -\frac{m}{\kappa}u_\kappa - mx\cdot\nabla_x u_\kappa, \qquad x\in\mathbb{R}^n$$

(cf. (V.78), (V.80)). By Lemma V.27, if $\kappa = 2/n$, then there are α, $\beta \in H^2_r(\mathbb{R}^n)$ such that

$$\text{(XII.10)} \qquad \mathfrak{l}_-\alpha(x) = \theta(x), \qquad \mathfrak{l}_+\beta(x) = \alpha(x), \qquad x\in\mathbb{R}^n.$$

It will often be convenient to extend functions of $r = |x| \ge 0$ by symmetry to functions of $t\in\mathbb{R}$. For the future use, we set

$$\text{(XII.11)} \qquad \hat{V}(t) := u_\kappa(|t|), \qquad \hat{U}(t) := -\frac{1}{2m}\hat{V}'(t), \qquad t\in\mathbb{R},$$

where u_κ is considered as a function of $r = |x|$, $x\in\mathbb{R}^n$. Note that the inclusion $u_\kappa\in C^2(\mathbb{R}^n)$ implies that $\hat{V}\in C^2(\mathbb{R})$ and $\hat{U}\in C^1(\mathbb{R})$. By (XII.4), the functions $\hat{V}$ and $\hat{U}$ (which are even and odd, respectively) satisfy

$$\text{(XII.12)} \qquad \frac{1}{2m}\hat{V} + \partial_t\hat{U} + \frac{n-1}{t}\hat{U} = |\hat{V}|^{2\kappa}\hat{V}, \qquad \partial_t\hat{V} + 2m\hat{U} = 0, \qquad t\in\mathbb{R},$$

where $\hat{U}(t)/t$ at $t=0$ is understood in the limit sense (see (XII.11)):

$$\lim_{t\to 0}\hat{U}(t)/t = \hat{U}'(0) = -\frac{1}{2m}\hat{V}''(0).$$

We will obtain the solitary wave solutions to (XII.1) as bifurcations from $(\hat{V}, \hat{U})$.

When constructing solitary waves, we will need the following Banach spaces:

$$\text{(XII.13)} \qquad \begin{aligned} &\mathbf{X} = L^2(\mathbb{R}, |t|^{n-1}\,dt;\,\mathbb{C})\cap L^\infty(\mathbb{R};\,\mathbb{C}),\\ &\text{with} \quad \|\cdot\|_{\mathbf{X}} = c\left(\|\cdot\|_{L^2(\mathbb{R},|t|^{n-1}\,dt;\,\mathbb{C})} + \|\cdot\|_{L^\infty(\mathbb{R};\,\mathbb{C})}\right);\end{aligned}$$

$$\text{(XII.14)} \qquad \mathbf{Y} = H^1(\mathbb{R}, \langle t\rangle^{n-1}\,dt;\,\mathbb{C}) = H^1_{(n-1)/2}(\mathbb{R};\,\mathbb{C})\subset\mathbf{X}.$$

The space $\mathbf{Y}$ is equipped with the standard norm of $H^1_s(\mathbb{R})$, $s = (n-1)/2$, while the constant $c>0$ in (XII.13) is chosen so that

$$\text{(XII.15)} \qquad \|\xi\|_{\mathbf{X}} \le \|\xi\|_{\mathbf{Y}}, \qquad \forall\xi\in\mathbf{Y}.$$

We note that both $\mathbf{X}$ and $\mathbf{Y}$ are algebras: there is $C>0$ such that

$$\text{(XII.16)} \qquad \|\xi\eta\|_{\mathbf{X}} \le C\|\xi\|_{\mathbf{X}}\|\eta\|_{\mathbf{X}}, \qquad \forall\xi,\,\eta\in\mathbf{X};$$

$$\text{(XII.17)} \qquad \|\xi\eta\|_{\mathbf{Y}} \le C\|\xi\|_{\mathbf{Y}}\|\eta\|_{\mathbf{Y}}, \qquad \forall\xi,\,\eta\in\mathbf{Y}.$$

Abusing notations, for $\psi = \begin{bmatrix}\psi_1\\ \psi_2\end{bmatrix}$ with ψ_1, $\psi_2\in\mathbf{X}$, we also denote

$$\|\psi\|_{\mathbf{X}} = \sqrt{\|\psi_1\|^2_{\mathbf{X}} + \|\psi_2\|^2_{\mathbf{X}}},$$

and similarly in the case of $\mathbf{Y}$ instead of $\mathbf{X}$.

The space

$$\text{(XII.18)} \qquad H^1_{e,o}(\mathbb{R}, |t|^{n-1}\,dt;\,\mathbb{C}^2) := H^1_{\mathrm{even}}(\mathbb{R}, |t|^{n-1}\,dt;\mathbb{C})\times H^1_{\mathrm{odd}}(\mathbb{R}, |t|^{n-1}\,dt;\,\mathbb{C})$$

denotes the subspace of $\mathbb{C}^2$-valued functions on $\mathbb{R}$ such that the first component is even and the second is odd. Similarly, we denote

$$\begin{aligned} \mathbf{X}_{e,o} &:= L^2_{e,o}(\mathbb{R}, |t|^{n-1}\,dt;\, \mathbb{C}^2) \cap L^\infty(\mathbb{R};\, \mathbb{C}^2), \\ \mathbf{Y}_{e,o} &:= H^1_{e,o}(\mathbb{R}, \langle t\rangle^{n-1}\,dt;\, \mathbb{C}^2). \end{aligned} \tag{XII.19}$$

Theorem XII.1 *Let $n \in \mathbb{N}$, $N = 2^{[(n+1)/2]}$, and assume that $f \in C(\mathbb{R})$ and that there is $\kappa > 0$ such that*

$$|f(\tau) - |\tau|^\kappa| \le o(|\tau|^\kappa), \qquad |\tau| \le 1. \tag{XII.20}$$

If $n \ge 3$, we additionally assume that $\kappa < 2/(n-2)$.

(1) *There is*

$$\omega_0 \in \left(\frac{m}{2}, m\right) \tag{XII.21}$$

such that for all $\omega \in (\omega_0, m)$ there are solitary wave solutions $\phi_\omega(x)e^{-\mathrm{i}\omega t}$ to (XII.1) *with $\phi_\omega \in H^1(\mathbb{R}^n, \mathbb{C}^N)$, with*

$$\phi_\omega(x) = \begin{bmatrix} v(r,\omega)\xi \\ \mathrm{i}u(r,\omega)\frac{x}{r}\cdot\boldsymbol{\sigma}\,\xi \end{bmatrix}, \qquad r = |x|, \qquad \xi \in \mathbb{C}^{N/2}, \quad |\xi| = 1, \tag{XII.22}$$

$$\lim_{r\to 0} u(r,\omega) = 0. \tag{XII.23}$$

Moreover, if we express

$$\begin{aligned} v(r,\omega) = \epsilon^{\frac{1}{\kappa}} V(\epsilon r, \epsilon), \qquad & u(r,\omega) = \epsilon^{1+\frac{1}{\kappa}} U(\epsilon r, \epsilon), \\ \epsilon = \sqrt{m^2 - \omega^2} > 0, \qquad & r \ge 0, \end{aligned} \tag{XII.24}$$

decomposing

$$\begin{aligned} V(t,\epsilon) = \hat{V}(t) + \tilde{V}(t,\epsilon), \qquad & U(t,\epsilon) = \hat{U}(t) + \tilde{U}(t,\epsilon), \\ & t \in \mathbb{R}, \quad \epsilon > 0, \end{aligned} \tag{XII.25}$$

with $\hat{V}(t)$, $\hat{U}(t)$ defined in (XII.12)*, then there is $\gamma > 0$ such that $\tilde{V}(t,\epsilon)$, $\tilde{U}(t,\epsilon)$ satisfy*

$$\lim_{\epsilon\to 0+} \left\| e^{\gamma\langle t\rangle} \begin{bmatrix} \tilde{V}(\cdot,\epsilon) \\ \tilde{U}(\cdot,\epsilon) \end{bmatrix} \right\|_{H^1(\mathbb{R},\mathbb{C}^2)} = 0, \tag{XII.26}$$

with $\langle t\rangle$ the operator of multiplication by $(1+t^2)^{1/2}$. Moreover, $V(t,\epsilon)$ and $U(t,\epsilon)$ are differentiable in $t \in \mathbb{R}$, and there is $C > 0$ such that

$$|\partial_t V(t,\epsilon)| + |\partial_t U(t,\epsilon)| \le C, \qquad \forall t \in \mathbb{R}, \qquad \forall \epsilon \in (0, \epsilon_0),$$

where $\epsilon_0 := \sqrt{m^2 - \omega_0^2} > 0$.

(2) *There is $\epsilon_1 \in (0, \epsilon_0)$ such that*

$$\epsilon_1 |U(t,\epsilon)| \le \frac{1}{2}|V(t,\epsilon)|, \qquad \forall t \in \mathbb{R}, \quad \forall \epsilon \in (0, \epsilon_1), \tag{XII.27}$$

$$\begin{aligned} \phi_\omega(x)^*\beta\phi_\omega(x) \ge |\phi_\omega(x)|^2/2, \qquad & \omega = \sqrt{m^2 - \epsilon^2}, \\ \forall x \in \mathbb{R}^n, \qquad & \forall \epsilon \in (0, \epsilon_1). \end{aligned} \tag{XII.28}$$

(3) One has

$$|\tilde{V}(t,\epsilon)| + |\tilde{U}(t,\epsilon)| \le o(1)\hat{V}(t), \qquad \forall t\in\mathbb{R}, \quad \forall \epsilon\in(0,\epsilon_1), \tag{XII.29}$$

where $o(1)$ *is with respect to* ϵ *uniformly in* t *(so that* $\sup_{t\in\mathbb{R}}|o(1)|\to 0$ *as* $\epsilon\to 0$*), and there is* $b_0>0$ *such that*

$$|V(t,\epsilon)| + |U(t,\epsilon)| \le b_0\langle t\rangle^{-(n-1)/2}e^{-|t|}, \quad \forall t\in\mathbb{R}, \quad \forall\epsilon\in(0,\epsilon_1). \tag{XII.30}$$

(4) The solitary waves satisfy

$$\|\phi_\omega\|_{L^\infty(\mathbb{R}^n,\mathbb{C}^N)} = O\big(\epsilon^{\frac{1}{\kappa}}\big), \qquad \|\phi_\omega\|_{L^2(\mathbb{R}^n,\mathbb{C}^N)} = O\big(\epsilon^{\frac{1}{\kappa}-\frac{2}{n}}\big); \quad \omega \lessapprox m. \tag{XII.31}$$

(5) Assume, moreover, that there is $K>\kappa$ *such that*

$$|f(\tau)-|\tau|^\kappa| = O(|\tau|^K), \qquad |\tau|\le 1. \tag{XII.32}$$

Then there are b_1, $b_2>0$ *such that* $\tilde{V}(t,\epsilon)$, $\tilde{U}(t,\epsilon)$ *satisfy*

$$\left\| e^{\gamma\langle t\rangle}\begin{bmatrix}\tilde{V}(\cdot,\epsilon)\\ \tilde{U}(\cdot,\epsilon)\end{bmatrix}\right\|_{H^1(\mathbb{R},\mathbb{C}^2)} \le b_1\epsilon^{2\varkappa}, \qquad \epsilon\in(0,\epsilon_1) \tag{XII.33}$$

and

$$|\tilde{V}(t,\epsilon)| + |\tilde{U}(t,\epsilon)| \le b_2\epsilon^{2\varkappa}\hat{V}(t), \qquad \forall t\in\mathbb{R}, \quad \forall\epsilon\in(0,\epsilon_1), \tag{XII.34}$$

with

$$\varkappa = \min\Big(1, \frac{K}{\kappa}-1\Big). \tag{XII.35}$$

Remark XII.2 We expect that, for solitary wave solutions $\phi_\omega(x)e^{-\mathrm{i}\omega t}$ to (XII.1), the profiles $\phi_\omega\in H^1(\mathbb{R}^n,\mathbb{C}^N)$ are continuous and thus all solitary waves of the form (XII.22) satisfy the condition (XII.23) Below, we only prove that (XII.23) is satisfied by the family constructed in Theorem XII.1.

Theorem XII.1 *(1)* is proved in Section XII.2. The positivity of $\phi_\omega^*\beta\phi_\omega$ (stated in Theorem XII.1 *(2)*) and the asymptotics of solitary waves (Theorem XII.1 *(3)*) are in Section XII.3. The asymptotics stated in Theorem XII.1 *(4)* follow from the estimates in Theorem XII.1 *(1)* and *(2)*. The error estimates from Theorem XII.1 *(5)* are proved in Section XII.4.

Theorem XII.3 *Let* $n\in\mathbb{N}$, $N=2^{[(n+1)/2]}$, *and assume that* $f\in C^1(\mathbb{R}\setminus\{0\})\cap C(\mathbb{R})$ *and that there are* $\kappa>0$ *and* $K>\kappa$ *such that*

$$|f(\tau)-|\tau|^\kappa| = O(|\tau|^K), \qquad |\tau|\le 1; \tag{XII.36}$$

$$|\tau f'(\tau)-\kappa|\tau|^\kappa| = O(|\tau|^K), \qquad |\tau|\le 1. \tag{XII.37}$$

If $n\ge 3$, *we additionally assume that* $\kappa<2/(n-2)$. *We can take* $\epsilon_1>0$ *in Theorem* XII.1 *smaller if necessary so that for* $\omega=\sqrt{m^2-\epsilon^2}$, $\epsilon\in(0,\epsilon_1)$, *the functions* $\phi_\omega(x)$, $\tilde{V}(t,\epsilon)$, *and* $\tilde{U}(t,\epsilon)$ *from Theorem* XII.1 *(1) (cf.* (XII.22)–(XII.26)*) are unique and satisfy the following additional properties.*

(1) One has $\phi_\omega\in H^2(\mathbb{R}^n,\mathbb{C}^N)$. *The map*

$$\omega\mapsto\phi_\omega\in H^1(\mathbb{R}^n,\mathbb{C}^N)$$

is C^1, *with* $\partial_\omega\phi_\omega\in H^1(\mathbb{R}^n,\mathbb{C}^N)$. *Moreover,* $\partial_\epsilon\tilde{W}(\cdot,\epsilon)\in H^1(\mathbb{R},\mathbb{C}^2)$, *with*

$$\|e^{\gamma\langle t\rangle}\partial_\epsilon\tilde{W}(\cdot,\epsilon)\|_{H^1(\mathbb{R},\mathbb{C}^2)} = O(\epsilon^{2\varkappa-1}), \qquad \epsilon\in(0,\epsilon_1), \tag{XII.38}$$

where $\tilde{W}(t,\epsilon) = \begin{bmatrix}\tilde{V}(t,\epsilon)\\ \tilde{U}(t,\epsilon)\end{bmatrix}$, *and there is* $c > 0$ *such that*

(XII.39) $$\|\partial_\omega \phi_\omega\|_{L^2}^2 = c\epsilon^{-n+\frac{2}{\kappa}}(1 + O(\epsilon^{2\varkappa})), \quad \omega = \sqrt{m^2 - \epsilon^2}, \quad \epsilon \in (0, \epsilon_1).$$

(2) Additionally, assume that κ, K *from* (XII.36) *and* (XII.37) *satisfy either*

(XII.40) $$0 < \kappa < 2/n, \qquad K > \kappa,$$

or

(XII.41) $$\kappa = 2/n, \qquad K > 4/n.$$

Then there is $\omega_* \in (0, m)$ *such that* $\partial_\omega Q(\omega) < 0$ *for all* $\omega \in (\omega_*, m)$.

If

(XII.42) $$\kappa > 2/n,$$

then there is $\omega_* \in (0, m)$ *such that* $\partial_\omega Q(\omega) > 0$ *for all* $\omega \in (\omega_*, m)$.

Remark XII.4 By **[BL83b]**, the pure power stationary Schrödinger equation (XII.4) with $n \geq 3$ has infinitely many distinct radial solutions, and one expects that there is a family of solitary waves of (XII.1) bifurcating from any of these radial solutions (similarly to what we state in Theorem XII.1).

Remark XII.5 Let us also consider the following question: *Given a sequence of solitary wave solutions corresponding to* $\omega_j \to m$, *does this sequence (up to symmetries and extraction of a subsequence) always converge to a solution of a nonlinear Schrödinger equation, in the sense of Theorem* XII.1*?* The answer to this question is negative in general. One obstacle can be illustrated as follows. In particular, in dimension $n = 3$, according to **[ES95**, Theorem 1], there are solitary wave solutions to (XII.1) for the pure power nonlinearity with $\kappa = 1$ or any real $\kappa \geq 2$ (so that $|\tau|^{\kappa+1}$ remains C^2 at $\tau = 0$, meeting the assumptions of **[ES95]**); see also earlier works **[CV86, Mer88]**. On the other hand, if $\kappa \geq \frac{2}{n-2} = 2$, the nonrelativistic limit can not converge to a stationary solution of the nonlinear Schrödinger equation, which does not exist for such values of κ.

We do not know whether in the case $0 < \kappa < \frac{2}{n-2}$ any sequence of solitary waves ϕ_ω, $\omega \lessapprox m$, could be obtained as a bifurcation from an NLS solitary wave.

Theorem XII.3 *(1)* is proved in Section XII.5. The Kolokolov condition in the critical case (Theorem XII.3 *(2)*) is analyzed in Section XII.6.

XII.2 Solitary waves in the nonrelativistic limit: the case $f \in C$

In this section, we prove Theorem XII.1, constructing a particular family of solitary waves bifurcating from solitary waves of the nonlinear Schrödinger equation.

First of all, we need to rewrite the assumption $f(\tau) = |\tau|^\kappa + o(|\tau|^\kappa)$ in a more convenient form. For $\hat{V}$, $\hat{U}$ from (XII.11), let us denote

(XII.43) $$\Lambda_\kappa := \sup_{x\in\mathbb{R}^n} |\hat{V}(x)| + m \sup_{x\in\mathbb{R}^n} |\hat{U}(x)| < \infty.$$

We focus on solitary waves with $\tilde{V}(t,\epsilon)$, $\tilde{U}(t,\epsilon)$ from (XII.25), with $\epsilon = \sqrt{m^2 - \omega^2}$, which satisfy

(XII.44) $$|\tilde{V}(t,\epsilon)| + m|\tilde{U}(t,\epsilon)| < \Lambda_\kappa, \qquad \forall t \in \mathbb{R};$$

we will see below that this imposes certain smallness assumptions onto $\epsilon > 0$. It follows from (XII.43) and (XII.44) that for these values of $\epsilon > 0$ one has

$$\text{(XII.45)} \qquad \begin{aligned} &|V(t,\epsilon)| \le 2\Lambda_\kappa, \qquad m|U(t,\epsilon)| \le 2\Lambda_\kappa, \qquad \forall t \in \mathbb{R}; \\ &|V(t,\epsilon)^2 - \epsilon^2 U(t,\epsilon)^2| < 4\Lambda_\kappa^2, \qquad \forall t \in \mathbb{R}. \end{aligned}$$

We are going to build small amplitude solitary waves, hence the proof below would not be affected by a change of the nonlinearity $f(\tau)$ outside of an open neighborhood of $\tau = 0$; thus, by (XII.20), we could assume that

$$\text{(XII.46)} \qquad |f(\tau)| \le 2|\tau|^\kappa, \qquad \tau \in \mathbb{R},$$

and that

$$\text{(XII.47)} \qquad |f(\tau) - |\tau|^\kappa| \le |\tau|^\kappa H(\tau), \qquad \tau \in \mathbb{R},$$

where $H \in C(\mathbb{R})$ is monotonically increasing for $\tau \ge 0$, with $H(0) = 0$. It is convenient to define

$$\text{(XII.48)} \qquad h(\epsilon) := \max\Big(H(\epsilon^{2/\kappa} 4\Lambda_\kappa^2),\, \epsilon^{2\kappa},\, \epsilon^2\Big).$$

Note that, by (XII.45),

$$\text{(XII.49)} \qquad H(v^2 - u^2) = H(\epsilon^{2/\kappa}(V^2 - \epsilon^2 U^2)) \le H\big(\epsilon^{2/\kappa} 4\Lambda_\kappa^2\big) \le h(\epsilon);$$

from (XII.47) and (XII.49) we obtain the following estimate for later use:

$$\text{(XII.50)} \qquad \big|f\big(\epsilon^{2/\kappa}(V^2 - \epsilon^2 U^2)\big) - \epsilon^2 |V^2 - \epsilon^2 U^2|^\kappa\big| \le C\epsilon^2 |V^2 - \epsilon^2 U^2|^\kappa h(\epsilon).$$

Now we are ready to start the proof of Theorem XII.1. We extend the argument of [**CGG14**, Section 4.2]. As in Section IX.5.2, substituting the Ansatz (XII.22) into the nonlinear Dirac equation (XII.1) gives the system

$$\text{(XII.51)} \qquad \begin{cases} \partial_r u + \frac{n-1}{r} u + (m - \omega) v = f(v^2 - u^2) v, \\ \partial_r v + (m + \omega) u = f(v^2 - u^2) u, \end{cases} \qquad r > 0,$$

for the pair of real-valued functions $v = v(r,\omega)$, $u = u(r,\omega)$. We will always impose the condition

$$\text{(XII.52)} \qquad \lim_{r \to 0} u(r,\omega) = 0$$

(cf. (XII.23)); this allows us to extend $v(r,\omega)$ and $u(r,\omega)$ continuously onto $\mathbb{R}$ so that v is even and u is odd:

$$\text{(XII.53)} \qquad v(x,\omega) = v(-x,\omega), \qquad u(x,\omega) = -u(-x,\omega), \qquad x \le 0.$$

Then (XII.51) extends onto the whole real axis:

$$\text{(XII.54)} \qquad \begin{cases} \partial_x u + \frac{n-1}{x} u + (m - \omega) v = f(v^2 - u^2) v, \\ \partial_x v + (m + \omega) u = f(v^2 - u^2) u, \\ u\big|_{x=0} = 0, \end{cases} \qquad x \in \mathbb{R}.$$

In (XII.54), the term $\frac{u(x,\omega)}{x}$ at $x = 0$ is understood as the limit $\lim_{x \to 0} \frac{u(x,\omega)}{x} = \partial_x u(0,\omega)$.

By (XII.54), V and U from (XII.24) are to satisfy

$$\text{(XII.55)} \qquad \begin{cases} \epsilon^2\Big(\partial_t U + \dfrac{n-1}{t}U\Big) + (m-\omega)V = fV, \\ \partial_t V + (\omega+m)U = fU, \\ U\big|_{t=0} = 0, \end{cases} \qquad t \in \mathbb{R},$$

with $t = \epsilon x$ and with

$$\text{(XII.56)} \qquad f = f\big(\epsilon^{2/\kappa}\big(V(t,\epsilon)^2 - \epsilon^2 U(t,\epsilon)^2\big)\big).$$

According to (XII.53), $V(t,\epsilon)$ is even in $t \in \mathbb{R}$ and $U(t,\epsilon)$ is odd. We recall that the term $U(t,\epsilon)/t$ at $t=0$ is understood as the limit $\lim_{t\to 0} U(t,\epsilon)/t = \partial_t U(0,\epsilon)$. We rewrite the system (XII.55) as

$$\text{(XII.57)} \qquad \begin{cases} \partial_t U + \dfrac{n-1}{t}U + \dfrac{1}{m+\omega}V = \dfrac{f}{\epsilon^2}V, \\ \partial_t V + (m+\omega)U = fU, \\ U\big|_{t=0} = 0, \end{cases} \qquad t \in \mathbb{R}.$$

We note that the system (XII.12) corresponds to the limit of (XII.57) as $\epsilon \to 0$ (that is, $\omega \to m$) after the substitution (XII.25). For sufficiently small $\epsilon > 0$, we will construct the solution (V, U) as a bifurcation from $(\hat{V}, \hat{U})$.

Substituting $V(t,\epsilon) = \hat{V}(t) + \tilde{V}(t,\epsilon)$ and $U(t,\epsilon) = \hat{U}(t) + \tilde{U}(t,\epsilon)$ into (XII.57) and then subtracting equations (XII.12), we arrive at

$$\text{(XII.58)} \qquad \begin{cases} (\partial_t + \frac{n-1}{t})\tilde{U} + \frac{1}{m+\omega}\tilde{V} = (1+2\kappa)|\hat{V}|^{2\kappa}\tilde{V} - G_1(\epsilon,\tilde{V},\tilde{U}), \\ \partial_t \tilde{V} + (m+\omega)\tilde{U} = G_2(\epsilon,\tilde{V},\tilde{U}), \end{cases}$$

$t \in \mathbb{R}$, $\epsilon > 0$, where

$$\text{(XII.59)} \qquad G_1(\epsilon,\tilde{V},\tilde{U}) = -\epsilon^{-2} f\big(\epsilon^{2/\kappa}(V^2 - \epsilon^2 U^2)\big)V + \hat{V}^{2\kappa}\hat{V} + (1+2\kappa)\hat{V}^{2\kappa}\tilde{V} + \left(\frac{1}{m+\omega} - \frac{1}{2m}\right)\hat{V},$$

$$\text{(XII.60)} \qquad G_2(\epsilon,\tilde{V},\tilde{U}) = f\big(\epsilon^{2/\kappa}(V^2 - \epsilon^2 U^2)\big)U + (m-\omega)\hat{U}, \qquad \omega = \sqrt{m^2 - \epsilon^2},$$

with (XII.25) giving the relations between V, U and $\tilde{V}$, $\tilde{U}$. Let us denote

$$\text{(XII.61)} \qquad \tilde{W} = \begin{bmatrix} \tilde{V} \\ \tilde{U} \end{bmatrix}, \qquad G(\epsilon,\tilde{W}) = \begin{bmatrix} G_1(\epsilon,\tilde{V},\tilde{U}) \\ G_2(\epsilon,\tilde{V},\tilde{U}) \end{bmatrix},$$

and introduce the operator

$$\text{(XII.62)} \qquad A(\epsilon) = \begin{bmatrix} -\frac{1}{m+\omega} + (1+2\kappa)|\hat{V}|^{2\kappa} & -\partial_t - \frac{n-1}{t} \\ \partial_t & m+\omega \end{bmatrix}, \qquad \omega = \sqrt{m^2 - \epsilon^2},$$

defined for $\epsilon \in [0, m]$, with the domain

$$\mathfrak{D}(A(\epsilon)) = H^1_{e,o}(\mathbb{R}, |t|^{n-1}\,dt; \mathbb{C}^2),$$

where $H^1_{e,o}(\mathbb{R}, |t|^{n-1}\,dt; \mathbb{C}^2)$ is defined in (XII.18). Note that

$$A(\epsilon): \; H^1_{e,o}(\mathbb{R}, |t|^{n-1}\,dt; \mathbb{C}^2) \to L^2_{e,o}(\mathbb{R}, |t|^{n-1}\,dt; \mathbb{C}^2).$$

Now the system (XII.58) takes the form

$$\text{(XII.63)} \qquad A(\epsilon)\tilde{W}(t,\epsilon) = G(\epsilon, \tilde{W}(t,\epsilon)), \qquad t \in \mathbb{R}, \qquad \epsilon > 0.$$

Lemma XII.6 *For any $\epsilon \in [0, m]$ the differential operator $A(\epsilon)$ defined on the domain $\mathfrak{D}(A(\epsilon)) = H^1_{e,o}(\mathbb{R}, |t|^{n-1}\, dt; \mathbb{C}^2)$ is self-adjoint.*

PROOF. The case $n = 1$ is immediate.

Let now $n \geq 2$. For each $\xi \in \mathbb{C}^{N/2}$ with $|\xi| = 1$, the space

$$L^2_\xi(\mathbb{R}^n, \mathbb{C}^N) := \left\{ \phi(x) = \begin{bmatrix} v(r)\xi \\ \mathrm{i}u(r)\frac{x}{r} \cdot \boldsymbol{\sigma}\, \xi \end{bmatrix},\ (v, u) \in L^2(\mathbb{R}_+, r^{n-1}\, dr; \mathbb{C}^2) \right\}$$

is a closed invariant subspace for D_m with domain

$$H^1_\xi(\mathbb{R}^n, \mathbb{C}^N) := \Big\{ \phi(x) = \begin{bmatrix} v(r)\xi \\ \mathrm{i}u(r)\frac{x}{r} \cdot \boldsymbol{\sigma}\, \xi \end{bmatrix},\ (v, u) \in L^2(\mathbb{R}_+, r^{n-1}\, dr; \mathbb{C}^2),$$
$$\Big(\partial_r u + \frac{n-1}{r} u,\ \partial_r v\Big) \in L^2(\mathbb{R}_+, r^{n-1}\, dr; \mathbb{C}^2)\Big\}.$$

Due to this invariance, the restriction of D_m to $L^2_\xi(\mathbb{R}^n, \mathbb{C}^N)$ is self-adjoint. Indeed, D_m acting on $L^2_\xi(\mathbb{R}^n, \mathbb{C}^N)$ is symmetric and the domain of its adjoint is the orthogonal projection of $H^1(\mathbb{R}^n, \mathbb{C}^N)$ onto $L^2_\xi(\mathbb{R}^n, \mathbb{C}^N)$; that is, $H^1_\xi(\mathbb{R}^n, \mathbb{C}^N)$.

The map $E :\ H^1_\xi(\mathbb{R}^n, \mathbb{C}^N) \to H^1_{e,o}(\mathbb{R}, |t|^{n-1} dt;\, \mathbb{C}^2)$,

$$E :\ (v(t), u(t)) \mapsto \begin{cases} \frac{1}{\sqrt{2}}(v(t), u(t)) & \text{if } t > 0, \\ \frac{1}{\sqrt{2}}(v(-t), -u(-t)) & \text{if } t < 0 \end{cases}$$

is well-defined and isometric. It ensures that

$$\begin{bmatrix} -m & -\partial_t - \frac{n-1}{t} \\ \partial_t & m \end{bmatrix},$$

with domain $H^1_{e,o}(\mathbb{R}, |t|^{n-1}\, dt; \mathbb{C}^2)$, is self-adjoint. Since $A(\epsilon)$ is a symmetric bounded perturbation of the above operator, the Kato–Rellich theorem (Theorem III.111) concludes the proof. □

We notice that the essential spectrum of $A(\epsilon)$, $\epsilon \in [0, m]$ with $\hat{V}$ substituted by zero is $(-\infty, -\frac{1}{m+\omega}] \cup [m+\omega, +\infty)$ (see e.g. [**Wei82**, Satz 2.1] or [**Tha92**, Theorem 4.18]). Applying Lemma III.160 and Weyl's theorem (Theorem III.135), we deduce that the essential spectrum of $A(\epsilon)$ is also given by

$$\sigma_{\mathrm{ess}}\big(A(\epsilon)\big) = \Big(-\infty, -\frac{1}{m+\omega}\Big] \cup \Big[m + \omega, +\infty\Big), \qquad \epsilon \in [0, m].$$

Since the inclusion $\begin{bmatrix} \xi \\ \eta \end{bmatrix} \in \mathbf{ker}(A(0))$ would lead to $\eta(t) = -\frac{1}{2m}\xi'(t)$ for $t \in \mathbb{R}$ and then to $\xi(|x|) \in \mathbf{ker}(\mathfrak{l}_+)$, with $\mathfrak{l}_+$ defined in (XII.8) and $x \in \mathbb{R}^n$, while the restriction of $\mathfrak{l}_+$ to spherically symmetric functions has zero kernel (see Lemma V.26 *(2)*), we see that

$$\mathbf{ker}\,\big(A(0)|_{H^1_{e,o}(\mathbb{R}, |t|^{n-1}\, dt;\, \mathbb{C}^2)}\big) = \{0\}.$$

Thus, $\lambda = 0$ does not belong to the spectrum of $A(0)|_{L^2_{e,o}}$, hence $A(0)^{-1}$ is bounded from $L^2_{e,o}(\mathbb{R}, |t|^{n-1}\, dt;\, \mathbb{C}^2)$ to $H^1_{e,o}(\mathbb{R}, |t|^{n-1}\, dt;\, \mathbb{C}^2)$. By continuity in ϵ in the norm resolvent sense (see Definition III.143), there is $\epsilon_0 > 0$ such that the mapping

$$A(\epsilon)^{-1} :\ L^2_{e,o}(\mathbb{R}, |t|^{n-1}\, dt;\, \mathbb{C}^2) \to H^1_{e,o}(\mathbb{R}, |t|^{n-1}\, dt;\, \mathbb{C}^2), \qquad \epsilon \in [0, \epsilon_0] \tag{XII.64}$$

is continuous, with the norm bounded uniformly in $\epsilon \in [0, \epsilon_0]$.

We actually need a stronger statement on continuity of $A(\epsilon)^{-1}$ in the spaces

$$\mathbf{X}_{e,o} = L^2_{e,o}(\mathbb{R}, |t|^{n-1}\,dt;\,\mathbb{C}^2) \cap L^\infty(\mathbb{R};\,\mathbb{C}^2), \qquad \mathbf{Y}_{e,o} = H^1_{e,o}(\mathbb{R}, \langle t\rangle^{n-1}\,dt;\,\mathbb{C}^2),$$

from (XII.13) and (XII.14), with the norms

$$\|(\xi_1,\xi_2)\|^2_{\mathbf{X}_{e,o}} = \|\xi_1\|^2_{\mathbf{X}} + \|\xi_2\|^2_{\mathbf{X}} \quad \text{for} \quad (\xi_1,\xi_2) \in \mathbf{X}_{e,o}$$

and

$$\|(\xi_1,\xi_2)\|^2_{\mathbf{Y}_{e,o}} = \|\xi_1\|^2_{\mathbf{Y}} + \|\xi_2\|^2_{\mathbf{Y}} \quad \text{for} \quad (\xi_1,\xi_2) \in \mathbf{Y}_{e,o},$$

where

$$\|\cdot\|_{\mathbf{X}} = c\left(\|\cdot\|_{L^2(\mathbb{R},|t|^{n-1}\,dt;\,\mathbb{C})} + \|\cdot\|_{L^\infty(\mathbb{R};\,\mathbb{C})}\right), \quad c>0; \quad \|\cdot\|_{\mathbf{Y}} = \|\cdot\|_{H^1_{(n-1)/2}(\mathbb{R})}.$$

Abusing the notations, we will denote these norms by $\|\cdot\|_{\mathbf{X}}$ and $\|\cdot\|_{\mathbf{Y}}$, respectively.

Lemma XII.7 *The restriction of the mapping* (XII.64) *onto*

$$\mathbf{X}_{e,o} \subset L^2_{e,o}(\mathbb{R}, |t|^{n-1}\,dt;\mathbb{C}^2)$$

defines a continuous map

$$A(\epsilon)^{-1}:\ \mathbf{X}_{e,o} \to \mathbf{Y}_{e,o}, \qquad \epsilon \in [0,\epsilon_0], \tag{XII.65}$$

with the norm bounded uniformly in $\epsilon \in [0,\epsilon_0]$.

Proof. The uniform boundedness in ϵ will follow from the resolvent identity. Indeed, assume that the map

$$A(\epsilon)^{-1}:\ \mathbf{X}_{e,o} \to \mathbf{Y}_{e,o},$$

is well-defined and bounded for each ϵ in $[0,\epsilon_0]$. Then, from the resolvent identity, since the coefficients are continuous in ϵ, we deduce that this map is continuous in ϵ.

Due to the boundedness of the mapping (XII.64), we already know that the solution of

$$A(\epsilon)\begin{bmatrix} v \\ u \end{bmatrix} = \begin{bmatrix} b \\ a \end{bmatrix} \in L^2_{e,o}(\mathbb{R}, |t|^{n-1}\,dt;\,\mathbb{C}^2) \cap L^\infty(\mathbb{R};\,\mathbb{C}^2) \tag{XII.66}$$

from $L^2_{e,o}(\mathbb{R}, |t|^{n-1}\,dt;\,\mathbb{C}^2)$ satisfies $\begin{bmatrix} v \\ u \end{bmatrix} \in H^1_{e,o}(\mathbb{R}, |t|^{n-1}\,dt;\,\mathbb{C}^2)$.

In the case $n=1$, we are done.

In the case $n \ge 2$, we proceed as follows. We already know that

$$\begin{bmatrix} v \\ u \end{bmatrix} \in H^1_{e,o}(\mathbb{R}, |t|^{n-1}\,dt;\,\mathbb{C}^2) \subset H^1_{e,o}(\mathbb{R}\setminus[-1,1], \langle t\rangle^{n-1}\,dt;\mathbb{C}^2).$$

It suffices to prove that $\begin{bmatrix} v \\ u \end{bmatrix}$ also satisfies

$$\begin{bmatrix} v \\ u \end{bmatrix} \in L^\infty([-1,1];\,\mathbb{C}^2), \qquad \partial_t \begin{bmatrix} v \\ u \end{bmatrix} \in L^\infty([-1,1];\,\mathbb{C}^2),$$

with the norms bounded by $\left\|\begin{bmatrix} b \\ a \end{bmatrix}\right\|_{L^2(\mathbb{R},|t|^{n-1}\,dt;\,\mathbb{C}^2)} + \left\|\begin{bmatrix} b \\ a \end{bmatrix}\right\|_{L^\infty(\mathbb{R};\,\mathbb{C}^2)}$ (times a constant factor). Equation (XII.66) can be written out as the following system:

$$\begin{cases} \left((1+2\kappa)\hat{V}(t)^{2\kappa} - \frac{1}{m+\omega}\right)v(t) - \partial_t u - \frac{n-1}{t}u(t) = b(t), \\ \partial_t v + (m+\omega)u(t) = a(t). \end{cases} \tag{XII.67}$$

From $\begin{bmatrix} v \\ u \end{bmatrix} \in H^1_{e,o}(\mathbb{R}, |t|^{n-1}\,dt; \mathbb{C}^2)$ we deduce that

$$v,\ u \in C(\mathbb{R} \setminus \{0\}),$$

and that $|t|^{\frac{n-1}{2}} v \in L^\infty([-1,1])$ and $|t|^{\frac{n-1}{2}} u \in L^\infty([-1,1])$, as a consequence of the Sobolev inequality and the Hardy inequality (see e.g. [**Ste70**, Appendix A.4] for the latter), and moreover

$$\begin{aligned} \| |t|^{\frac{n-1}{2}} v\|_{L^\infty([-1,1])} + \| |t|^{\frac{n-1}{2}} u\|_{L^\infty([-1,1])} &\le C \left\| \begin{bmatrix} v \\ u \end{bmatrix} \right\|_{H^1(\mathbb{R}, |t|^{n-1}\,dt; \mathbb{C}^2)} \\ &\le C' \left\| \begin{bmatrix} b \\ a \end{bmatrix} \right\|_{L^2(\mathbb{R}, |t|^{n-1}\,dt; \mathbb{C}^2)}, \end{aligned} \tag{XII.68}$$

with some $C,\ C' > 0$.

We will proceed by induction. Let us assume that, more generally,

$$\begin{aligned} &\| |t|^\alpha v\|_{L^\infty([-1,1])} + \| |t|^\alpha u\|_{L^\infty([-1,1])} \\ &\le C \left(\left\| \begin{bmatrix} b \\ a \end{bmatrix} \right\|_{L^2(\mathbb{R}, |t|^{n-1}\,dt; \mathbb{C}^2)} + \left\| \begin{bmatrix} b \\ a \end{bmatrix} \right\|_{L^\infty(\mathbb{R}; \mathbb{C}^2)} \right), \end{aligned} \tag{XII.69}$$

with $C > 0$ independent of $(b(t), a(t))$ and with some $\alpha \ge 0$. We note that the bound (XII.69) with $\alpha = (n-1)/2$ holds due to (XII.68); therefore, it also holds with $\alpha = n - 1/2$, with the same constant $C > 0$ (we will use the induction in half-integer values of α). The first equation from (XII.67) can be rewritten as

$$\partial_t \big(t^{n-1} u \big) = t^{n-1} \Big((1+2\kappa) \hat{V}(t)^{2\kappa} - \frac{1}{m+\omega} \Big) v(t) - t^{n-1} b(t). \tag{XII.70}$$

Since $u \in H^1(\mathbb{R}, |t|^{n-1}\,dt) \subset C(\mathbb{R} \setminus \{0\})$, $|t|^{\frac{n-1}{2}} u \in L^\infty([-1,1])$, and $n \ge 2$, one has $t^{n-1} u(t) \to 0$ as $t \to 0$; therefore, integrating the relation (XII.70), we arrive at

$$t^{n-1} u(t) = \int_0^t \Big(s^{n-1} \Big((1+2\kappa) \hat{V}(s)^{2\kappa} - \frac{1}{m+\omega} \Big) v(s) - s^{n-1} b(s) \Big)\, ds,$$

which yields

$$|t|^{n-1} |u(t)| \le \Big(\frac{C}{n-\alpha} |t|^{n-\alpha} \| |t|^\alpha v\|_{L^\infty([-1,1])} + \frac{1}{n} |t|^n \|b\|_{L^\infty([-1,1])} \Big), \tag{XII.71}$$

$t \in [-1,1]$, with $C > 0$ dependent on κ and $\hat{V}$ only; hence, taking into account that $\alpha \le n - 1/2$,

(XII.72)

$$|t|^{\alpha-1} |u(t)| \le \Big(2C \| |t|^\alpha v\|_{L^\infty([-1,1])} + \frac{|t|^\alpha \|b\|_{L^\infty([-1,1])}}{n} \Big), \quad t \in [-1,1] \setminus \{0\}.$$

Similarly, from the second equation in (XII.67) we deduce that

(XII.73)

$$\| |t|^\alpha \partial_t v\|_{L^\infty([-1,1])} \le |m+\omega| \| |t|^\alpha u\|_{L^\infty([-1,1])} + \| |t|^\alpha a\|_{L^\infty([-1,1])} =: C_*,$$

so that one has $|v'(t)| \le C_* |t|^{-\alpha}$; therefore,

$$|v(t)| \le |v(1)| + |v(-1)| + C_* |t|^{-\alpha+1} / |\alpha - 1|, \qquad t \in [-1,1] \setminus \{0\}.$$

It follows that

$$|t|^{\alpha-1} |v(t)| \le |t|^{\alpha-1} \big(|v(1)| + |v(-1)| \big) + \frac{C_*}{|\alpha - 1|}, \qquad t \in [-1,1] \setminus \{0\}. \tag{XII.74}$$

The inequalities (XII.72) and (XII.74) show that if (XII.69) holds with a half-integer $\alpha \geq 1/2$, $\alpha \leq n - 1/2$, then it also holds with $\alpha - 1$; this allows us to reduce the value of α in (XII.69) to $\alpha = -1/2$ (in at most n steps).

But then (XII.69) also holds with $\alpha = 0$. We use (XII.71), (XII.72) one more time (now with $\alpha = 0$), obtaining the bound

$$|u(t)/t| \leq \Big(\frac{C}{n}\|v\|_{L^\infty([-1,1])} + \frac{1}{n}\|b\|_{L^\infty([-1,1])}\Big), \qquad t \in [-1,1] \setminus \{0\}. \tag{XII.75}$$

Using the resulting bounds on $\|v\|_{L^\infty([-1,1])}$ and $\|u/t\|_{L^\infty([-1,1])}$ in the system (XII.67) yields the desired bounds on $\|\partial_t v\|_{L^\infty([-1,1])}$ and on $\|\partial_t u\|_{L^\infty([-1,1])}$. The continuity of the mapping (XII.65) is proved. □

The assumption $(\tilde V, \tilde U) \in \mathbf{Y}_{e,o} \subset \mathbf{X}_{e,o}$ leads to $\big(G_1(\tilde V,\tilde U), G_2(\tilde V,\tilde U)\big) \in \mathbf{X}_{e,o}$, with G_1, G_2 defined in (XII.59) and (XII.60). Due to invertibility of $A(\epsilon) : \mathbf{Y}_{e,o} \to \mathbf{X}_{e,o}$ (Lemma XII.7), the relation (XII.63) leads to

$$\tilde W = A(\epsilon)^{-1} G(\epsilon, \tilde W), \qquad \tilde W = \tilde W(t,\epsilon). \tag{XII.76}$$

Remark XII.8 The continuity of f is not enough to conclude that the map

$$\mu : \mathbf{X}_{e,o} \to \mathbf{Y}_{e,o} \subset \mathbf{X}_{e,o}, \qquad \mu : \tilde W \mapsto A(\epsilon)^{-1} G(\epsilon, \tilde W)$$

is a contraction, so we can not apply the contraction mapping principle to claim a unique fixed point of μ; we will use the Schauder fixed point theorem instead, proving the existence of a fixed point but missing its uniqueness. In the case $f \in C^1$, indeed the mapping μ can be shown to be a contraction on a particular subspace (see Lemma XII.30 below); this will allow us to prove uniqueness of a fixed point.

To be able to consider non-integer values of $\kappa > 0$, we need the following result.

Lemma XII.9 *For any $\kappa > 0$ and any a, $b \in \mathbb{R}$; one has:*

$$\big||a+b|^\kappa - |a|^\kappa\big| \leq 3^\kappa \left(|a|^{\kappa-\min(1,\kappa)} + |b|^{\kappa-\min(1,\kappa)}\right) |b|^{\min(1,\kappa)}, \tag{XII.77}$$

$$\big||a+b|^\kappa - |a|^\kappa - \kappa|a|^{\kappa-1} b \operatorname{sgn} a\big| \leq 3^\kappa \left(|a|^{\kappa-\min(2,\kappa)} + |b|^{\kappa-\min(2,\kappa)}\right) |b|^{\min(2,\kappa)}. \tag{XII.78}$$

PROOF. Since the inequalities (XII.77) and (XII.78) are homogeneous of degree κ in a and b, it is enough to give a proof for $a = 1$, $b \in \mathbb{R}$.

If $|b| \geq 1/2$, then $\big||1+b|^\kappa - 1\big| \leq \max(|1+b|^\kappa, 1) \leq 3^\kappa |b|^\kappa$. If $|b| < 1/2$, then, by the mean value theorem,

$$\big||1+b|^\kappa - 1\big| \leq \max_{c \in [1/2, 3/2]} \kappa |c|^{\kappa-1} |b|. \tag{XII.79}$$

If $\kappa \geq 1$, the right-hand side is bounded by $\kappa(3/2)^{\kappa-1}|b| \leq 3^\kappa |b|$ (since $\kappa(3/2)^{\kappa-1} < 3^\kappa$, $\forall \kappa \in \mathbb{R}$). If $\kappa \in (0,1)$, the right-hand side of (XII.79) is bounded by $\kappa 2^{1-\kappa}|b| = \kappa |2b|^{1-\kappa} |b|^\kappa \leq \kappa |b|^\kappa \leq 3^\kappa |b|^\kappa$. This completes the proof of (XII.77).

Now let us prove (XII.78); again, we only need to consider the case $a = 1$. For $b \geq 1/2$, one has

$$\big||1+b|^\kappa - 1 - \kappa b\big| \leq \max((3b)^\kappa, 1+\kappa b) \leq \max(3^\kappa,\ 2^\kappa + 2^{\kappa-1}\kappa) b^\kappa \leq 3^\kappa (b^{\kappa-\min(2,\kappa)} + b^\kappa).$$

In the last inequality, we took into account that, with $b \geq 1/2$,

$$1 \leq 2^{\min(2,\kappa)} b^{\min(2,\kappa)} \leq 3^\kappa b^{\min(2,\kappa)}, \qquad \kappa b \leq \kappa 2^{\kappa-1} b^\kappa \leq 3^\kappa b^\kappa.$$

For $b \le -1/2$, one similarly obtains

$$||1+b|^\kappa - 1 - \kappa b| \le \max(|b|^\kappa + \kappa|b|,\, 1) \le \max(1+2^{\kappa-1}\kappa,\, 2^\kappa)|b|^\kappa \le 3^\kappa |b|^\kappa,$$

since $1+2^{\kappa-1}\kappa < 3^\kappa$ for $\kappa > 0$.

Finally, for $|b| < 1/2$, by the mean value theorem,

$$||1+b|^\kappa - 1 - \kappa b| \le \max_{c\in[1/2,3/2]} \frac{\kappa|\kappa-1|}{2}|c|^{\kappa-2}|b|^2. \tag{XII.80}$$

If $\kappa \ge 2$, the right-hand side is bounded by

$$\frac{1}{2}\kappa(\kappa-1)(3/2)^{\kappa-2}|b|^2 \le 3^\kappa |b|^2,$$

since $\frac{1}{2}\kappa|\kappa-1|(3/2)^{\kappa-2} < 3^\kappa$, $\forall \kappa > 0$. If $\kappa \in (0,2)$, the right-hand side of (XII.80) is bounded by

$$\kappa|\kappa-1|2^{2-\kappa}|b|^2 = \kappa|\kappa-1||2b|^{2-\kappa}|b|^\kappa \le \kappa|\kappa-1||b|^\kappa \le 3^\kappa|b|^\kappa. \qquad \square$$

Recall that $\Lambda_\kappa > 0$ was defined in (XII.43).

Lemma XII.10 *There is $C > 0$ such that for any numbers*

$$\hat V,\, \hat U,\, \tilde V,\, \tilde U \in [-\Lambda_\kappa, \Lambda_\kappa], \qquad V = \hat V + \tilde V, \qquad U = \hat U + \tilde U,$$

one has

$$|G_1(\epsilon,\tilde V,\tilde U)| \le C\Big(h(\epsilon)(|V|+|U|)^{1+2\kappa} + \hat V^{1+2\kappa-\min(2,1+2\kappa)}|\tilde V|^{\min(2,1+2\kappa)} + |\tilde V|^{1+2\kappa} + \epsilon^2\hat V\Big),$$

$$|G_2(\epsilon,\tilde V,\tilde U)| \le C\left(\epsilon^2|V^2-\epsilon^2U^2|^\kappa|U| + \epsilon^2|\hat U|\right),$$

for all $\epsilon \in (0,\epsilon_0)$, with $\epsilon_0 > 0$ from Theorem XII.1.

Proof. Although most terms in the definition of G (cf. (XII.61)) are small, we have to be careful when we consider the general case $\kappa > 0$ when $f'(\tau)$ may not be uniformly bounded near $\tau = 0$. To bound $G_1(\epsilon,\tilde V,\tilde U)$ (cf. (XII.59)), we proceed as follows:

$$\begin{aligned}
|G_1| &\le \Big|\epsilon^{-2}f(\epsilon^{2/\kappa}(V^2-\epsilon^2U^2))V - \hat V^{2\kappa}\hat V - (1+2\kappa)\hat V^{2\kappa}\tilde V\Big| + \Big|\frac{\hat V}{m+\omega} - \frac{\hat V}{2m}\Big| \\
&\le \Big|\epsilon^{-2}f(\epsilon^{2/\kappa}(V^2-\epsilon^2U^2)) - |V^2-\epsilon^2U^2|^\kappa\Big|\,|V| \\
&\quad + \big||V^2-\epsilon^2U^2|^\kappa - |V|^{2\kappa}\big|\,|V| \\
&\quad + \Big||V|^{2\kappa}V - \hat V^{2\kappa}\hat V - (1+2\kappa)\hat V^{2\kappa}\tilde V\Big| + \Big(\frac{1}{m+\omega} - \frac{1}{2m}\Big)\hat V.
\end{aligned} \tag{XII.81}$$

We use (XII.50) to estimate the first term in the right-hand side by $h(\epsilon)|V^2 - \epsilon^2U^2|^\kappa|V|$. Other terms are dealt with by Lemma XII.9: we apply (XII.77) to the second term and (XII.78) (with $1+2\kappa$ instead of κ) to the third term, getting

$$\begin{aligned}
|G_1| &\le Ch(\epsilon)|V^2-\epsilon^2U^2|^\kappa|V| + 3^\kappa\left(|V|^{2\kappa-2\min(1,\kappa)}|\epsilon U|^{2\min(1,\kappa)} + |\epsilon U|^{2\kappa}\right)|V| \\
&\quad + 3^{1+2\kappa}\left(\hat V^{1+2\kappa-\min(2,1+2\kappa)}|\tilde V|^{\min(2,1+2\kappa)} + |\tilde V|^{1+2\kappa}\right) + \Big(\frac{1}{m+\omega} - \frac{1}{2m}\Big)|\hat V|,
\end{aligned}$$

which yields the desired bound on $|G_1|$. We took into account the definition of $h(\epsilon)$ in (XII.48).

The estimate on $|G_2(\epsilon,\tilde V,\tilde U)|$ immediately follows from (XII.60) and (XII.50). □

To apply the fixed point theorem, we will use the exponential weights, introducing compactness into (XII.76). We fix

$$\text{(XII.82)}\qquad \gamma\in(0,\gamma_0),\qquad \text{where}\quad \gamma_0:=\frac{1}{1+2\kappa}\inf_{\epsilon\in[0,\epsilon_0]}\frac{1}{1+\|A(\epsilon)^{-1}\|_{\mathbf{X}_{e,o}\to\mathbf{Y}_{e,o}}};$$

we note that, by Lemma XII.7, one has $\gamma_0>0$. Due to the exponential decay of $\hat V(t)$, $\hat U(t)$ (defined in (XII.11)), since $\gamma<1/(2\kappa+1)<1$, there are the following inclusions:

$$\text{(XII.83)}\qquad e^{(1+2\kappa)\gamma\langle t\rangle}\hat U\in\mathbf{Y},\qquad e^{(1+2\kappa)\gamma\langle t\rangle}\hat V\in\mathbf{Y},$$

with $\mathbf{Y}$ from (XII.14). We define

$$\text{(XII.84)}\qquad A_\gamma(\epsilon):=e^{(1+2\kappa)\gamma\langle t\rangle}\circ A(\epsilon)\circ e^{-(1+2\kappa)\gamma\langle t\rangle}=A(\epsilon)-(1+2\kappa)\gamma\frac{t}{\langle t\rangle}\begin{bmatrix}0&1\\1&0\end{bmatrix}.$$

Due to Lemma XII.7 and the choice of γ_0 in (XII.82), for any $\epsilon\in[0,\epsilon_0]$ the operator (XII.84) is closed and invertible, so that the mapping

$$\text{(XII.85)}\qquad A_\gamma(\epsilon)^{-1}=e^{(1+2\kappa)\gamma\langle t\rangle}\circ A(\epsilon)^{-1}\circ e^{-(1+2\kappa)\gamma\langle t\rangle},$$

$$A_\gamma(\epsilon)^{-1}:\ \mathbf{X}_{e,o}\to\mathbf{Y}_{e,o}:=H^1_{e,o}(\mathbb{R},\langle t\rangle^{n-1}\,dt;\,\mathbb{C}^2)$$

is bounded uniformly in $\epsilon\in[0,\epsilon_0]$. We multiply the fixed point problem (XII.76) by $e^{\gamma\langle t\rangle}$, rewriting it in the form

$$\text{(XII.86)}\qquad e^{\gamma\langle t\rangle}\tilde W=e^{-2\kappa\gamma\langle t\rangle}A_\gamma(\epsilon)^{-1}e^{(1+2\kappa)\gamma\langle t\rangle}G\big(\epsilon,e^{-\gamma\langle t\rangle}e^{\gamma\langle t\rangle}\tilde W\big).$$

Lemma XII.11 *There is $C>0$ such that if $\tilde W(\cdot,\epsilon)=\begin{bmatrix}\tilde V(\cdot,\epsilon)\\\tilde U(\cdot,\epsilon)\end{bmatrix}\in\mathbf{X}$, $\epsilon\in(0,\epsilon_0)$, satisfies the bound*

$$\text{(XII.87)}\qquad \|e^{\gamma\langle t\rangle}\tilde W(\cdot,\epsilon)\|_{\mathbf{X}}\le\left\|e^{\gamma\langle t\rangle}\begin{bmatrix}\hat V\\\hat U\end{bmatrix}\right\|_{\mathbf{X}},\qquad\forall\epsilon\in(0,\epsilon_0),$$

where $\langle t\rangle$ is considered as the operator of multiplication by $(1+t^2)^{1/2}$ and γ is from (XII.82), then one has the following bound on $G(\epsilon,\tilde V(\cdot,\epsilon),\tilde U(\cdot,\epsilon))$:

$$\text{(XII.88)}\qquad \|e^{(1+2\kappa)\gamma\langle t\rangle}G(\epsilon,\tilde W(\cdot,\epsilon))\|_{\mathbf{X}}\le C\left(h(\epsilon)+\|e^{\gamma\langle t\rangle}\tilde W(\cdot,\epsilon)\|_{\mathbf{X}}^{1+\min(1,2\kappa)}\right),$$

for all $\epsilon\in(0,\epsilon_0)$, with $h(\epsilon)$ from (XII.48).

PROOF. We use the pointwise estimates on G_1, G_2 from Lemma XII.10. There, the first term in the right-hand side of the bound on G_1 has a factor $h(\epsilon)$. Multiplying this term by $e^{(1+2\kappa)\gamma\langle t\rangle}$ and using (XII.83) and (XII.87), and also the fact that the space $\mathbf{X}$ defined in (XII.13) is closed under multiplication, we bound the resulting $\mathbf{X}$-norm by $Ch(\epsilon)$, with some $C>0$.

The terms

$$\hat V^{1+2\kappa-\min(2,1+2\kappa)}|\tilde V|^{\min(2,1+2\kappa)}+|\tilde V|^{1+2\kappa}$$

in the right-hand side of the bound on G_1 in Lemma XII.10, having no ϵ-factor, are of order higher than one in $\tilde V$, benefiting us when $|\tilde V|$ is small. Multiplying them by the factor $e^{(1+2\kappa)\gamma\langle t\rangle}$, which is absorbed by the terms which are homogeneous of order $(1+2\kappa)$ in $\hat V$ and $\tilde V$, we bound the $\mathbf{X}$-norm of the result by $C\|e^{\gamma\langle t\rangle}\tilde V\|_{\mathbf{X}}^{1+\min(1,2\kappa)}$.

We note that $\|e^{\gamma\langle t\rangle}\hat{V}\|_{\mathbf{X}}^{1+\min(1,2\kappa)}$ and $\|e^{\gamma\langle t\rangle}\hat{U}\|_{\mathbf{X}}^{1+\min(1,2\kappa)}$ are finite due to $\gamma < 1$ (see (XII.82)) and due to the exponential decay of $\hat{V}$ and $\hat{U}$.

For the last term in the right-hand side of the bound on G_1 from Lemma XII.10 multiplied by $e^{(1+2\kappa)\gamma\langle t\rangle}$, its $\mathbf{X}$-norm is bounded by $C\epsilon^2$ with the aid of (XII.83). We conclude that there is a constant $C > 0$ such that there is the desired bound

$$\|e^{(1+2\kappa)\gamma\langle t\rangle}G_1(\epsilon,\tilde{V},\tilde{U})\|_{\mathbf{X}} \le C\left(h(\epsilon) + \|e^{\gamma\langle t\rangle}\tilde{V}\|_{\mathbf{X}}^{1+\min(1,2\kappa)}\right), \qquad \forall\epsilon \in (0,\epsilon_0).$$

We now consider G_2. Due to the factor ϵ^2 in the right-hand side of the bound on G_2 in Lemma XII.10 and due to the exponential decay of $\hat{V}$, $\hat{U}$ (together with the bound (XII.83)), as well as due to the assumption (XII.87) about the exponential decay of $\tilde{V}$ and $\tilde{U}$, one has

$$\|e^{(1+2\kappa)\gamma\langle t\rangle}G_2(\epsilon,\tilde{V},\tilde{U})\|_{\mathbf{X}} \le C\epsilon^2, \qquad \forall\epsilon \in (0,\epsilon_0).$$

Lemma XII.11 is proved. □

Let us now complete the proof of Theorem XII.1. We consider the mapping

$$\text{(XII.89)} \qquad \mu_\gamma(\epsilon,\cdot):\ \mathbf{X}_{e,o} \to \mathbf{X}_{e,o} \to \mathbf{Y}_{e,o} \subset \mathbf{X}_{e,o},$$

$$\mu_\gamma(\epsilon,\cdot):\ Z \mapsto e^{(1+2\kappa)\gamma\langle t\rangle}G(\epsilon,e^{-\gamma\langle t\rangle}Z) \mapsto e^{-2\kappa\gamma\langle t\rangle}A_\gamma(\epsilon)^{-1}e^{(1+2\kappa)\gamma\langle t\rangle}G(\epsilon,e^{-\gamma\langle t\rangle}Z).$$

Note that $\tilde{W}$ is a solution to (XII.63) if and only if $Z = e^{\gamma\langle t\rangle}\tilde{W}$ is a fixed point of this map.

Lemma XII.12 *One can take $\epsilon_0 > 0$ smaller if necessary so that there is $a_0 > 0$ such that*

$$\text{(XII.90)} \qquad \mu_\gamma\left(\epsilon,\overline{\mathbb{B}_\rho(\mathbf{X}_{e,o})}\right) \subset \overline{\mathbb{B}_\rho(\mathbf{Y}_{e,o})}, \quad \rho = a_0 h(\epsilon), \qquad \forall\epsilon \in (0,\epsilon_0),$$

with $h(\epsilon)$ from (XII.48).

PROOF. If Z belongs to a closed ball $\overline{\mathbb{B}_\rho(\mathbf{X}_{e,o})} = \{\psi \in \mathbf{X}_{e,o}\colon\ \|\psi\|_{\mathbf{X}} \le \rho\}$, with

$$0 < \rho \le \left\|e^{\gamma\langle t\rangle}\begin{bmatrix}\hat{V}(t)\\ \hat{U}(t)\end{bmatrix}\right\|_{\mathbf{X}},$$

then Lemma XII.11 applies to $\tilde{W}(t) = e^{-\gamma\langle t\rangle}Z(t)$, giving us

$$\text{(XII.91)} \qquad \|e^{(1+2\kappa)\gamma\langle t\rangle}G(\epsilon,e^{-\gamma\langle t\rangle}Z)\|_{\mathbf{X}} \le C\left\{h(\epsilon) + \|Z\|_{\mathbf{X}}^{1+\min(1,2\kappa)}\right\}.$$

Therefore, to find the sufficient condition for (XII.90) to be satisfied, we use the definition of μ_γ from (XII.89) and apply the estimate (XII.91), arriving at the requirement

$$\text{(XII.92)} \qquad \|e^{-2\kappa\gamma\langle t\rangle}\circ A_\gamma(\epsilon)^{-1}\|_{\mathbf{X}_{e,o}\to\mathbf{Y}_{e,o}}C\left\{h(\epsilon) + \rho^{1+\min(1,2\kappa)}\right\} \le \rho.$$

Due to the continuity of the mapping (XII.85), the first factor in the left-hand side is bounded; thus, one can satisfy (XII.92) taking $\rho = O(h(\epsilon))$. This finishes the proof. □

Since it is not clear that the mapping $\mu_\gamma(\epsilon,\cdot):\ \mathbf{X}_{e,o} \to \mathbf{Y}_{e,o} \subset \mathbf{X}_{e,o}$ defined in (XII.89) is a contraction without assuming that f is sufficiently regular we can not apply the Banach fixed point theorem to (XII.89). Instead, we use the Schauder fixed point theorem (see e.g. [**GT83**, Corollary 11.2]):

Let Q be a closed, convex, bounded subset of a Banach space $\mathscr{X}$.
Let $\mu: Q \to Q$ a continuous compact map.
Then μ has a fixed point in Q.

Clearly, the mapping $\mu_\gamma(\epsilon, \cdot): \mathbf{X}_{e,o} \to \mathbf{Y}_{e,o}$ is continuous; note that, in particular,

$$(V, U) \mapsto \epsilon^{-2} f\big(\epsilon^{2/\kappa}(V^2 - \epsilon^2 U^2)\big) V$$

is continuous in the norm of the space $\mathbf{X}$ since the map $(V, U) \mapsto \epsilon^{-2} f\big(\epsilon^{2/\kappa}(V^2 - \epsilon^2 U^2)\big)$ is continuous as a map from $L^\infty(\mathbb{R}, \mathbb{C}^2)$ to $L^\infty(\mathbb{R})$. Then the mapping

$$e^{-2\kappa\gamma\langle t\rangle} \circ A_\gamma(\epsilon)^{-1}: \mathbf{X}_{e,o} \to \mathbf{Y}_{e,o} \to \mathbf{X}_{e,o},$$

is compact, since the multiplication by the decaying exponential weight is a compact map from $\mathbf{Y}_{e,o}$ to $\mathbf{X}_{e,o}$. Therefore, so is the mapping $\mu_\gamma(\epsilon, \cdot)$ when considered as a map from $\mathbf{X}_{e,o}$ into itself. By Lemma XII.12, the Schauder fixed point theorem gives a fixed point of the map $\mu_\gamma(\epsilon, \cdot)$ which belongs to a closed ball $\overline{\mathbb{B}_\rho(\mathbf{Y}_{e,o})}$ of radius $\rho = a_0 h(\epsilon)$, with $a_0 > 0$ which does not depend on $\epsilon \in (0, \epsilon_0)$. It follows that $\tilde{W} = e^{-\kappa\gamma\langle t\rangle} Z$ satisfies

$$\|e^{\kappa\gamma\langle t\rangle} \tilde{W}\|_{\mathbf{Y}} = \|Z\|_{\mathbf{Y}} \le \rho \le a_0 h(\epsilon), \qquad \forall \epsilon \in (0, \epsilon_0). \tag{XII.93}$$

This yields (XII.26).

Remark XII.13 The map $\tilde{W}(\epsilon)$ is not a sufficiently well-defined function to make it continuous in ϵ since the solution provided by the Schauder fixed point theorem is not necessarily unique, due to the absence of the contraction. The uniqueness of the mapping $\epsilon \mapsto \tilde{W}(\epsilon)$, under stronger assumptions on f, will be addressed in Section XII.5.2.

We note that

$$\|\tilde{W}\|_{L^\infty} \le \|e^{\kappa\gamma\langle t\rangle} \tilde{W}\|_{L^\infty} \le \|e^{\kappa\gamma\langle t\rangle} \tilde{W}\|_{\mathbf{Y}} \le a_0 h(\epsilon), \qquad \forall \epsilon \in (0, \epsilon_0);$$

thus, we can impose the condition that $\epsilon_0 > 0$ is small enough so that

$$|\tilde{V}(t, \epsilon)| + m|\tilde{U}(t, \epsilon)| < \Lambda_\kappa, \qquad \forall t \in \mathbb{R}, \qquad \forall \epsilon \in (0, \epsilon_0),$$

to satisfy our assumption (XII.44).

Finally, let us prove that $V, U \in C^1(\mathbb{R})$. We note that, due to the continuity of $\tilde{V}$ and $\tilde{U}$ (which follows from the application of Lemma XII.7 to (XII.63)), we know that $V = \hat{V} + \tilde{V}$ and $U = \hat{U} + \tilde{U}$ are continuous on the whole real axis $t \in \mathbb{R}$.

Lemma XII.14 *Fix $\epsilon \in (0, \epsilon_0)$. If $f \in C(\mathbb{R})$ and if $V, U \in C(\mathbb{R})$, with V even and U odd, are solutions to* (XII.57)*, then one has $V, U \in C^1(\mathbb{R})$, and the function $Y(t) := U(t)/t$, $t \ne 0$ could be extended to a continuous function on $\mathbb{R}$.*

Moreover, if there is $C > 0$ such that

$$|V(t)| + |U(t)| \le C, \qquad \forall t \in \mathbb{R}, \tag{XII.94}$$

then there is $C' > 0$ such that

$$|\partial_t V(t)| + |\partial_t U(t)| \le C', \qquad \forall t \in \mathbb{R}.$$

PROOF. The second equation in (XII.57) immediately gives $V \in C^1(\mathbb{R})$. To prove that one also has $U \in C^1(\mathbb{R})$, we write the first equation in (XII.57) as

$$\partial_t U(t) + \frac{n-1}{t} U(t) = B(t), \qquad t \in \mathbb{R}, \tag{XII.95}$$

with $B \in C(\mathbb{R})$ given by

$$\text{(XII.96)} \qquad B(t) = \frac{f\big(\epsilon^{2/\kappa}(V(t)^2 - \epsilon^2 U(t)^2)\big)}{\epsilon^2} V(t) - \frac{1}{m+\omega} V(t).$$

Due to the assumption (XII.20), $f(\tau) = |\tau|^\kappa + o(|\tau|^\kappa)$ for $|\tau| \le 1$, one has

$$\text{(XII.97)} \quad B(t) = \Big((V^2 - \epsilon^2 U^2)^\kappa + \frac{o(\epsilon^2(V^2 - \epsilon^2 U^2)^\kappa)}{\epsilon^2} - \frac{1}{m+\omega}\Big)V.$$

We note that at a particular $t \in \mathbb{R}$ the above expression defines $B(t)$ as a function continuous at t if $\epsilon > 0$ is small enough so that $\epsilon^{2/\kappa}|V(t)^2 - \epsilon^2 U(t)^2| \le 1$ (cf. (XII.20)). It is enough to prove that $Y(t) = U(t)/t \in C(\mathbb{R} \setminus \{0\})$ could be extended to a continuous function on $\mathbb{R}$ (then the same is true for U'). Thus, we need to show that $Y(t)$ has a finite limit as $t \to 0$. From (XII.95) we arrive at

$$\partial_t(U(t)t^{n-1}) = B(t)t^{n-1}, \qquad t \in \mathbb{R},$$

hence, one has

$$\text{(XII.98)} \qquad Y(t) = \frac{U(t)}{t} = \frac{\int_0^t B(\tau)\tau^{n-1}\,d\tau}{t^n}, \qquad t > 0,$$

which has a well-defined limit at the origin:

$$\lim_{t\to 0} Y(t) = \lim_{t\to 0} \frac{\int_0^t B(\tau)\tau^{n-1}\,d\tau}{t^n} = \lim_{t\to 0} \frac{B(t)}{n} = \frac{B(0)}{n}.$$

Let us show the uniform boundedness of the derivatives of V and U in the case when the uniform bound (XII.94) is satisfied. From the system (XII.57), due to bounds (XII.45), we conclude that there is $C > 0$ such that

$$|\partial_t V(t)| \le C, \qquad \forall t \in \mathbb{R}, \qquad \forall \epsilon \in (0, \epsilon_0).$$

We conclude from (XII.98) that $|Y(t)| \le \|B\|_{L^\infty}$ (which is finite due to (XII.94) and (XII.97)) and then from (XII.57) that $|\partial_t U(t)| \le 2\|B\|_{L^\infty}$, for all $t \in \mathbb{R}$. □

The proof of Theorem XII.1 *(1)* is finished.

XII.3 Positivity of $\phi_\omega^*\beta\phi_\omega$ and improved estimates

XII.3.1 Positivity of $\phi_\omega^*\beta\phi_\omega$ in the nonrelativistic limit. To be able to consider the nonlinearity $f(\tau) = |\tau|^\kappa + \dots$ which is not differentiable at $\tau = 0$ unless $\kappa \ge 1$, we will show that the quantity $\phi_\omega^*\beta\phi_\omega$, which is the argument in the nonlinearity, remains positive if $\omega \lessapprox m$. This will allow us to treat the nonlinear Dirac equation with fractional power nonlinearity using the Taylor-style estimates on the remainders instead of weaker estimates from Lemma XII.9.

So we proceed to the proof of Theorem XII.1 *(2)*, showing that U is pointwise dominated by V.

Proposition XII.15 *There is $\epsilon_1 \in (0, \epsilon_0)$ such that for all $\epsilon \in (0, \epsilon_1)$ one has*

$$\epsilon_1|U(t,\epsilon)| \le \frac{1}{2}|V(t,\epsilon)|, \qquad \forall t \in \mathbb{R}, \qquad \forall \epsilon \in (0, \epsilon_1).$$

Above, $\epsilon_0 > 0$ is from Theorem XII.1 *(1)*.

PROOF. We rewrite (XII.57) as follows:

(XII.99)
$$\begin{cases} \partial_t U = -\frac{1}{m+\omega}V - \frac{n-1}{t}U + |V|^{2\kappa}V + \Big(\epsilon^{-2}f\big(\epsilon^{2/\kappa}(V^2-\epsilon^2U^2)\big) - |V|^{2\kappa}\Big)V, \\ \partial_t V = -(m+\omega)U + \epsilon^2|V|^{2\kappa}U + \Big(f\big(\epsilon^{2/\kappa}(V^2-\epsilon^2U^2)\big) - \epsilon^2|V|^{2\kappa}\Big)U. \end{cases}$$

For any $\delta > 0$ and any $\nu \in (0, \nu_0)$, $\nu_0 = \min(\delta/8, m\delta/8)$, define the following closed sets (see Figure XII.1):

$$\mathcal{K}^+_{\delta,\nu} = \Big\{(V,U) \in \overline{\mathbb{B}^2_\delta} \subset \mathbb{R}^2:\quad U \geq \max\Big(0,\ \tfrac{V+\nu}{m},\ \tfrac{2V}{m}\Big)\Big\}, \qquad \text{(XII.100)}$$

$$\mathcal{K}^0_\delta = \Big\{(V,U) \in \overline{\mathbb{B}^2_\delta} \subset \mathbb{R}^2:\quad V \geq 0,\quad \tfrac{V}{4m} \leq U \leq \tfrac{2V}{m}\Big\}, \qquad \text{(XII.101)}$$

$$\mathcal{K}^-_{\delta,\nu} = \Big\{(V,U) \in \overline{\mathbb{B}^2_\delta} \subset \mathbb{R}^2:\quad V \geq 0,\ U \leq \min\big(\tfrac{V-\nu}{2m},\ \tfrac{V}{4m}\big)\Big\}. \qquad \text{(XII.102)}$$

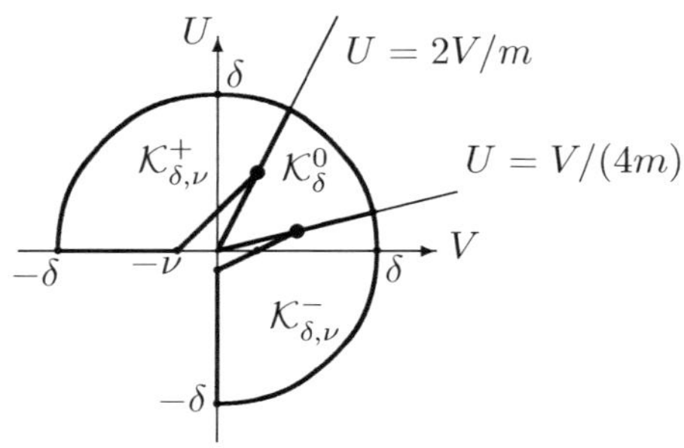

FIGURE XII.1. The regions $\mathcal{K}^+_{\delta,\nu}$, $\mathcal{K}^0_\delta$, $\mathcal{K}^-_{\delta,\nu}$ inside $\overline{\mathbb{B}^2_\delta}$.

The value of ν_0 is chosen so that for $\nu \in (0, \nu_0)$ the corner points of both $\mathcal{K}^+_{\delta,\nu}$ and $\mathcal{K}^-_{\delta,\nu}$ inside the first quadrant, $(\nu, 2\nu/m)$ and $(2\nu, \nu/(2m))$ (marked by big black dots on Figure XII.1), belong to $\mathbb{B}^2_{\delta/2}$:

$$\Big(\nu, \frac{2\nu}{m}\Big) \in \mathbb{B}^2_{\delta/2}, \qquad \Big(2\nu, \frac{\nu}{2m}\Big) \in \mathbb{B}^2_{\delta/2}. \qquad \text{(XII.103)}$$

Lemma XII.16 *If $\delta > 0$ is sufficiently small, then any C^1-solution to* (XII.99) *with $\epsilon \in (0, \epsilon_0)$ (with $\epsilon_0 > 0$ from Theorem* XII.1*) which satisfies*

$$(V(T), U(T)) \in \mathcal{K}^+_{\delta,\nu}$$

at some $T \geq 2n$, can only leave the region $\mathcal{K}^+_{\delta,\nu}$ through the boundary of the δ-disc: either

$$(V(t), U(t)) \in \mathcal{K}^+_{\delta,\nu} \qquad \forall t \geq T,$$

or else there is $T_* \in (T, +\infty)$ such that

$$(V(t), U(t)) \in \mathcal{K}^+_{\delta,\nu} \quad \forall t \in [T, T_*], \qquad (V(T_*), U(T_*)) \in \mathbb{S}^1_\delta.$$

PROOF. It suffices to check that at all pieces of $\partial\mathcal{K}^+_{\delta,\nu} \setminus \mathbb{S}^1_\delta$ the integral curves of (XII.99) are directed strictly inside $\mathcal{K}^+_{\delta,\nu}$ (defined in (XII.100)); that is, at all the points

$$(V,U) \quad \text{with} \quad U = \max\Big(0, \frac{V+\nu}{m}, \frac{2V}{m}\Big), \quad -\delta < V < \delta, \quad (V,U) \in \mathbb{B}^2_\delta,$$

one has $\xi \cdot (\dot V, \dot U) > 0$, with $\xi \in \mathbb{R}^2$ the inner normal to $\partial\mathcal{K}^+_{\delta,\nu}$ (as long as $t \geq 2n$).

On the piece $\{(V,0)\colon\quad -\delta \le V \le -\nu\} \subset \partial\mathcal{K}^+_{\delta,\nu}$, we compute:

$$(0,1)\cdot(\dot V,\dot U) = \dot U = -\frac{V}{m+\omega} + o(V) > 0,$$

as long as $\delta > 0$ is sufficiently small.

On the piece $\big\{(V,(V+\nu)/m)\colon\quad -\nu \le V \le \nu\big\} \subset \partial\mathcal{K}^+_{\delta,\nu}$, since $T \ge 2n$, one has:

$$\begin{aligned}(-1,m)\cdot(\dot V,\dot U) &= (m+\omega)U - m\Big(\frac{V}{m+\omega} + \frac{(n-1)U}{t}\Big) + o(|U|+|V|)\\ &\ge \Big(\frac{m}{2}+\omega\Big)U - \frac{mV}{m+\omega} + o(|U|+|V|)\\ &= \Big(\frac{1}{2}+\frac{\omega}{m}\Big)(V+\nu) - \frac{mV}{m+\omega} + o(|V+\nu|+|V|).\end{aligned}$$

When $-\nu \le V < 0$, the first two terms in the right-hand side are positive, dominating the last term if δ is sufficiently small. For $0 \le V \le \nu$, due to $\omega > m/2$ (cf. (XII.21)), the positive first term in the right-hand side dominates both the second term and the last term since

$$\frac{mV}{m+\omega} \le \frac{2}{3}V \le \frac{1}{3}(V+\nu).$$

On the piece of the boundary $\{(V,2V/m)\colon\quad V \ge \nu\} \cap \partial\mathcal{K}^+_{\delta,\nu}$, we get

$$(-2,m)\cdot(\dot V,\dot U) = -2\dot V + m\dot U = 2(m+\omega)U - m\Big(\frac{V}{m+\omega} + \frac{n-1}{t}U\Big) + o(V)$$

$$\ge 2(m+\omega)\frac{2V}{m} - \Big(V + \frac{n-1}{t}2V\Big) + o(V) \ge 4V + o(V) > 0.$$

We took into account that $\omega > m/2$ and that $t \ge 2n$. □

Lemma XII.17 *If $\delta > 0$ is sufficiently small, then any C^1-solution to* (XII.99) *with $0 < \epsilon \le \frac{m}{4}$ which satisfies*

$$(V(T),U(T)) \in \mathcal{K}^-_{\delta,\nu}$$

at some $T \ge 2n$ can only exit the region $\mathcal{K}^-_{\delta,\nu}$ through the boundary of the δ-disc: either

$$(V(t),\,U(t)) \in \mathcal{K}^-_{\delta,\nu} \qquad \forall t \ge T,$$

or else there is $T_* \in (T,+\infty)$ such that

$$(V(t),\,U(t)) \in \mathcal{K}^-_{\delta,\nu} \quad \forall t \in [T,T_*], \qquad (V(T_*),\,U(T_*)) \in \mathbb{S}^1_\delta.$$

We remind that $\mathcal{K}^-_{\delta,\nu}$ is defined in (XII.102).

Proof. The proof is similar to that of Lemma XII.17; we keep checking the positivity of the dot products of the inner normals to the boundary with $(\dot V,\dot U)$. For the pieces of the boundary given by $V = 0$, the proof is immediate (from (XII.99), one can see that $\dot V > 0$, as long as $\delta > 0$ is small enough so that the nonlinear terms are dominated by the linear part). On the piece given by $U = (V-\nu)/(2m)$, $0 \le V \le 2\nu$,

(XII.104)
$$(1,-2m)\cdot(\dot V,\dot U) = \dot V - 2m\dot U = -(m+\omega)U + \frac{2mV}{m+\omega} + \frac{2m(n-1)U}{t} + o(|V|+|U|).$$

At $V=0$, $U=-\nu/(2m)$, the linear part of the right-hand side of (XII.104) equals $\frac{m+\omega}{2m}\nu - \frac{n-1}{t}\nu$, which is positive for $\nu>0$, $\omega\in(0,m)$, $t\ge 2n$. At the other end of the interval, at $V=2\nu$, $U=\nu/(2m)$, the linear part of (XII.104) equals

$$-\frac{m+\omega}{2m}\nu + \frac{4m}{m+\omega}\nu + \frac{n-1}{t}\nu,$$

which is strictly positive for $t\ge 2n$, $\nu>0$, $\omega\in(m/2,m)$. Since the linear part is strictly positive, it dominates the error term $o(|V|+|U|)$ in (XII.104) as long as $\delta>0$ is sufficiently small.

On the piece of the boundary of $\mathcal{K}^-_{\delta,\nu}$ given by $U=V/(4m)$, $2\nu\le V\le\delta$, one has

$$\begin{aligned}(1,-4m)\cdot(\dot V,\dot U) &= \dot V - 4m\dot U\\ &= -(m+\omega)U + \frac{4mV}{m+\omega} + \frac{4m(n-1)U}{t} + o(|V|+|U|)\\ &= \Big(-\frac{m+\omega}{4m} + \frac{4m}{m+\omega} + \frac{n-1}{t}\Big)V + o(|V|).\end{aligned}$$

Since $\omega\in(m/2,m)$ (cf. (XII.21)), the linear part in the right-hand side is strictly positive, dominating the nonlinear part as long as $\delta>0$ is sufficiently small. □

Back to the proof of the proposition, we choose $\delta>0$ small enough so that both Lemma XII.16 and Lemma XII.17 are satisfied. By **[BL83a]**, $\hat V>0$ and $\hat U\ge 0$ are exponentially decaying, hence we can choose $T_1\ge 2n$ large enough and take $\delta>0$ smaller if necessary so that

(XII.105) $$\big(\hat V(T_1),\hat U(T_1)\big)\in Q_\delta := (\mathbb{B}^2_{3\delta/4}\setminus\mathbb{B}^2_{2\delta/3})\cap\{(V,U):\quad V\ge 0,\, U\ge 0\},$$

and so that

(XII.106) $$\big(\hat V(t),\hat U(t)\big)\in\mathbb{B}^2_{3\delta/4},\qquad \forall t\ge T_1.$$

By (XII.93),

(XII.107) $$\|\tilde V(\cdot,\epsilon)\|_{L^\infty} + \|\tilde U(\cdot,\epsilon)\|_{L^\infty} = O(h(\epsilon)).$$

Since Q_δ is strictly inside $\mathcal{K}^+_{\delta,\nu}\cup\mathcal{K}^0_\delta\cup\mathcal{K}^-_{\delta,\nu}$ (this is due to choosing $\nu_0>0$ such that (XII.103) is satisfied for $\nu\in(0,\nu_0)$), we use (XII.105) and (XII.107) to conclude that there is $\epsilon_1\in(0,\epsilon_0)$ such that for any $\epsilon\in(0,\epsilon_1)$,

(XII.108) $$\begin{aligned}&\big(V(T_1,\epsilon),U(T_1,\epsilon)\big)\\ &= \big(\hat V(T_1)+\tilde V(T_1,\epsilon),\ \hat U(T_1)+\tilde U(T_1,\epsilon)\big)\in\mathcal{K}^+_{\delta,\nu}\cup\mathcal{K}^0_\delta\cup\mathcal{K}^-_{\delta,\nu}.\end{aligned}$$

Moreover, by (XII.106) and (XII.107), we could take

$$\epsilon_1\in(0,\epsilon_0)$$

smaller if necessary so that for all $t \geq T_1$ and $\epsilon \in (0, \epsilon_1)$, one has

$$\text{(XII.109)} \qquad \big(V(t,\epsilon), U(t,\epsilon)\big) = \big(\hat{V}(t) + \tilde{V}(t,\epsilon),\ \hat{U}(t) + \tilde{U}(t,\epsilon)\big) \in \mathbb{B}^2_\delta.$$

Lemma XII.18 *For all $t \geq T_1$ and $\epsilon \in (0, \epsilon_1)$, one has*

$$V(t,\epsilon) > 0, \qquad U(t,\epsilon) > 0, \qquad \frac{V(t,\epsilon)}{4m} < U(t,\epsilon) < \frac{2V(t,\epsilon)}{m}.$$

PROOF. We claim that the solution $(V(t,\epsilon), U(t,\epsilon))$ remains in the region $\mathcal{K}^0_\delta$ for all $t \geq T_1$. First, we notice that if $(V(T_1,\epsilon), U(T_1,\epsilon)) \in \mathcal{K}^0_\delta$, then for $t \geq T_1$ the trajectory $(V(t), U(t))$ could not leave $\mathcal{K}^0_\delta$ through the arc of the δ-circle in the first quadrant (due to (XII.109)). At the same time, it can not leave $\mathcal{K}^0_\delta$ through $(V, U) = (0,0) \in \mathcal{K}^0_\delta$ because of the uniqueness of the solution passing through $(0,0)$ (for $t \geq T_1 \geq 2n$, the right-hand side of the system (XII.99) is Lipschitz in $(V, U) \in \mathcal{K}^0_\delta$); this unique solution is $V(t) \equiv U(t) \equiv 0$, $t \geq T_1$.

The solution also could not leave $\mathcal{K}^0_\delta$ through the side $U = 2V/m$ (with $V > 0$). Indeed, the assumption that $U(T_*,\epsilon) = 2V(T_*,\epsilon)/m > 0$ at some $T_* \geq T_1$ leads to a contradiction: we choose $\nu > 0$ small enough (one can take $\nu = \min(\nu_0, V(T_*,\epsilon)) > 0$) so that $(V(T_*,\epsilon), U(T_*,\epsilon)) \in \mathcal{K}^+_{\delta,\nu}$, and then Lemma XII.16 together with the bound (XII.109) show that the solution would be trapped in $\mathcal{K}^+_{\delta,\nu}$ for all $t \geq T_1$, hence would not be able to converge to zero as $t \to \infty$. For the same reason, the solution can not start in this region initially, at $t = T_1$: one should have $(V(T_1,\epsilon), U(T_1,\epsilon)) \notin \mathcal{K}^+_{\delta,\nu}$ for any $\nu \in (0, \nu_0]$.

The same argument (now with the aid of Lemma XII.17) shows that one can not have $U = V/(4m)$, $V > 0$ at some $T_* \geq T_1$, neither can the solution start at $t = T_1$ in $\mathcal{K}^-_{\delta,\nu}$ for any $\nu \in (0, \nu_0]$: the solution $(V(t,\epsilon), U(t,\epsilon))$ would be trapped in $\mathcal{K}^-_{\delta,\nu}$ for all $t \geq T_1$ and thus could not converge to zero.

Thus, by (XII.108), the trajectory $(V(t,\epsilon), U(t,\epsilon))$ starts strictly inside $\mathcal{K}^0_\delta$ at $t = T_1$ and stays there for all $t \geq T_1$. The statement of the lemma follows. □

Due to V being even and U being odd in t, Lemma XII.18 also yields the inequality

$$\text{(XII.110)} \qquad |U(t,\epsilon)| < \frac{2}{m} V(t,\epsilon), \qquad |t| \geq T_1, \qquad \epsilon \in (0, \epsilon_1).$$

Let us now consider the case $|t| \leq T_1$. By (XII.93), there is $C > 0$ such that

$$\text{(XII.111)} \qquad \sup_{|t| \leq T_1} |U(t,\epsilon)| \leq \sup_{|t| \leq T_1} \hat{U}(t) + \|\tilde{U}(\cdot,\epsilon)\|_{L^\infty} \leq \sup_{|t| \leq T_1} \hat{U}(t) + Ch(\epsilon);$$

on the other hand, again using (XII.93), we have, for all $\epsilon \in (0, \epsilon_1)$:

$$\begin{aligned} \inf_{|t| \leq T_1} V(t,\epsilon) &\geq \inf_{|t| \leq T_1} \hat{V}(t) - \|\tilde{V}(\cdot,\epsilon)\|_{L^\infty} \\ \text{(XII.112)} \qquad &\geq \inf_{|t| \leq T_1} \hat{V}(t) - Ch(\epsilon) \geq \inf_{|t| \leq T_1} \hat{V}(t)/2 > 0 \end{aligned}$$

if we choose $\epsilon_1 > 0$ is so small that $Ch(\epsilon_1) < \inf_{|t| \leq T_1} \hat{V}(t)/2$. It follows from (XII.111) and (XII.112) that for some $C' > 0$ we could write

$$\text{(XII.113)} \qquad |U(t,\epsilon)| < C' V(t,\epsilon), \qquad |t| \leq T_1, \qquad \epsilon \in (0, \epsilon_1).$$

We require that $\epsilon_1 > 0$ be small enough, satisfying $\epsilon_1 \leq \min(m/2, 1/(2C'))$; then the inequalities (XII.110) and (XII.113) yield (XII.27), finishing the proof of Proposition XII.15. □

Using the inequality (XII.27), one derives the bound (XII.28):

$$\phi_\omega^*\beta\phi_\omega = v^2 - u^2 = \epsilon^{\frac{2}{\kappa}}(V^2 - \epsilon^2 U^2) \ge \epsilon^{\frac{2}{\kappa}}\frac{3V^2}{4} \ge \epsilon^{\frac{2}{\kappa}}\frac{2V^2 + 2\epsilon^2 U^2}{4} = \frac{\phi_\omega^*\phi_\omega}{2}, \qquad \omega \in (\omega_1, m),$$

with $\omega_1 = \sqrt{m^2 - \epsilon_1^2}$. This completes the proof of Theorem XII.1 *(2)*.

XII.3.2 Sharp decay asymptotics and optimal estimates. Let us now prove Theorem XII.1 *(3)*. We will derive the sharp exponential decay of each of V, $\hat V$, U, $\hat U$ and then prove that, as the matter of fact, $\tilde V$ and $\tilde U$ are pointwise dominated by V. We recall that $\hat V$ and $\hat U$ are obtained from NLS solitary waves and that

$$V(t,\epsilon) = \hat V(t) + \tilde V(t,\epsilon), \qquad U(t,\epsilon) = \hat U(t) + \tilde U(t,\epsilon);$$

cf. (XII.11), (XII.25).

Lemma XII.19 *There are $C_1 > c_1 > 0$ such that for all $\epsilon \in (0, \epsilon_1)$ and all $t \ge T_1$ one has*

(XII.114) $$|V(t,\epsilon)| \ge c_1 t^{-(n-1)/2} e^{-t}, \qquad |V(t,\epsilon)| + |U(t,\epsilon)| \le C_1 t^{-(n-1)/2} e^{-t};$$

(XII.115) $$\hat V(t) \ge c_1 t^{-(n-1)/2} e^{-t}, \qquad \hat V(t) + |\hat U(t)| \le C_1 t^{-(n-1)/2} e^{-t};$$

(XII.116) $$|\tilde V(t,\epsilon)| + |\tilde U(t,\epsilon)| \le C_1 t^{-(n-1)/2} e^{-t}.$$

Above, $\epsilon_1 > 0$ is from Theorem XII.1 *(2)* and $T_1 \in (0,\infty)$ is from (XII.105).

PROOF. The inequality (XII.116) follows from (XII.114) and (XII.115). The inequalities (XII.115) have already been proved; see Lemma V.24. Now we will focus on inequalities (XII.114) which are more involved.

We introduce $\mathscr{V}(t,\epsilon)$ and $\mathscr{U}(t,\epsilon)$ such that

(XII.117) $$V(t,\epsilon) = t^{-(n-1)/2}\mathscr{V}(t,\epsilon),$$

(XII.118) $$U(t,\epsilon) = t^{-(n-1)/2}\Big(\mathscr{U}(t,\epsilon) + \frac{n-1}{2\mu t}\mathscr{V}(t,\epsilon)\Big),$$

where we use the notation

$$\mu = m + \omega, \qquad \omega = \sqrt{m^2 - \epsilon^2}.$$

Below, we will omit the dependence of V, U, $\mathscr{V}$, $\mathscr{U}$, ω, and μ on ϵ. By Lemma XII.18, for $t \ge T_1$, one has $\mathscr{V}(t) > 0$ (since so is $V(t)$). Then, applying inequalities from Lemma XII.18 to the relation

$$\mathscr{U}(t,\epsilon) = t^{(n-1)/2}\Big(U(t,\epsilon) - \frac{n-1}{2\mu t}V(t,\epsilon)\Big)$$

and using $\omega > m/2$, $t \ge T_1 \ge 2n$ (cf. (XII.21) and (XII.105)), we obtain:

(XII.119) $$\mathscr{U} \ge t^{(n-1)/2}\Big(\frac{V}{4m} - \frac{n-1}{2(3m/2)2n}V\Big) \ge t^{(n-1)/2}\frac{V}{12m} = \frac{\mathscr{V}}{12m} > 0, \qquad \forall t \ge T_1, \quad \forall \epsilon \in (0,\epsilon_1).$$

Substituting the expressions (XII.117), (XII.118) into the system (XII.57), we obtain the equation

$$\partial_t\Big(\mathscr{U} + \frac{n-1}{2\mu t}\mathscr{V}\Big) - \frac{n-1}{2t}\Big(\mathscr{U} + \frac{n-1}{2\mu t}\mathscr{V}\Big) + \frac{n-1}{t}\Big(\mathscr{U} + \frac{n-1}{2\mu t}\mathscr{V}\Big) + \frac{\mathscr{V}}{\mu} = \epsilon^{-2} f \mathscr{V},$$

which takes the form

$$\partial_t\mathscr{U} + \frac{n-1}{2\mu t}\partial_t\mathscr{V} + \frac{n-1}{2t}\mathscr{U} + \frac{(n-1)^2\mathscr{V}}{4\mu t^2} - \frac{(n-1)\mathscr{V}}{2\mu t^2} + \frac{\mathscr{V}}{\mu} = \frac{f}{\epsilon^2}\mathscr{V}, \tag{XII.120}$$

and the equation

$$\partial_t\mathscr{V} - \frac{n-1}{2t}\mathscr{V} + \mu\Big(\mathscr{U} + \frac{n-1}{2\mu t}\mathscr{V}\Big) = \partial_t\mathscr{V} + \mu\mathscr{U} = \Big(\mathscr{U} + \frac{n-1}{2\mu t}\mathscr{V}\Big)f. \tag{XII.121}$$

Above, f is from (XII.56):

$$f = f\big(\epsilon^{2/\kappa}V(t,\epsilon)^2 - \epsilon^{2+2/\kappa}U(t,\epsilon)^2\big).$$

Multiplying (XII.120) by μ and adding (XII.121), we get:

$$\begin{aligned}\partial_t(\mathscr{V}+\mu\mathscr{U}) + (\mathscr{V}+\mu\mathscr{U}) + \frac{n-1}{2t}\partial_t\mathscr{V} + \frac{\mu(n-1)}{2t}\mathscr{U} + \frac{(n-1)(n-3)\mathscr{V}}{4t^2}&\\ = \mu\frac{f}{\epsilon^2}\mathscr{V} + \Big(\mathscr{U} + \frac{n-1}{2\mu t}\mathscr{V}\Big)f.&\end{aligned} \tag{XII.122}$$

Using (XII.121) to simplify the two terms in the left-hand side which contain a factor $\frac{n-1}{2t}$, we get

$$\begin{aligned}\partial_t(\mathscr{V}+\mu\mathscr{U}) + (\mathscr{V}+\mu\mathscr{U}) + \frac{n-1}{2t}\Big(\mathscr{U} + \frac{n-1}{2\mu t}\mathscr{V}\Big)f + \frac{(n-1)(n-3)\mathscr{V}}{4t^2}&\\ = \mu\frac{f}{\epsilon^2}\mathscr{V} + \Big(\mathscr{U} + \frac{n-1}{2\mu t}\mathscr{V}\Big)f,&\end{aligned}$$

which yields the inequality

$$|\partial_t(\mathscr{V}+\mu\mathscr{U}) + (\mathscr{V}+\mu\mathscr{U})| \le \frac{C}{t^2}(\mathscr{V}+\mathscr{U}) + C\frac{|f|}{\epsilon^2}(\mathscr{V}+\mathscr{U}), \qquad \forall t \ge T_1, \qquad \forall \epsilon \in (0,\epsilon_1), \tag{XII.123}$$

with some $C > 0$; we took into account that both $\mathscr{V}$ and $\mathscr{U}$ are positive (cf. (XII.119)). Since one has $0 < \frac{\mathscr{V}}{\mathscr{V}+\mu\mathscr{U}} \le 1$ and $0 < \frac{\mathscr{U}}{\mathscr{V}+\mu\mathscr{U}} \le \frac{1}{\mu} \le \frac{1}{m}$, it follows from (XII.123) that there is $C' > 0$ such that

$$-1 - \frac{c}{t^2} - C'\frac{|f|}{\epsilon^2} \le \frac{\partial_t(\mathscr{V}+\mu\mathscr{U})}{\mathscr{V}+\mu\mathscr{U}} \le -1 + \frac{c}{t^2} + C'\frac{|f|}{\epsilon^2}, \qquad \forall t \ge T_1, \qquad \forall \epsilon \in (0,\epsilon_1). \tag{XII.124}$$

We note that, by (XII.46),

$$\frac{|f\big(\epsilon^{2/\kappa}(V(t,\epsilon)^2 - \epsilon^2U(t,\epsilon)^2)\big)|}{\epsilon^2} \le 2|V(t,\epsilon)^2 - \epsilon^2U(t,\epsilon)^2|^\kappa,$$

which is bounded and exponentially decreasing as $t \to +\infty$ (uniformly in $\epsilon \in (0,\epsilon_1)$) due to the exponential decay of $V(t,\epsilon) = \hat{V}(t) + \tilde{V}(t,\epsilon)$ and $U(t,\epsilon) = \hat{U}(t) + \tilde{U}(t,\epsilon)$ in t, which we proved in Theorem XII.1. Thus, there is $C'' > 0$ which does not depend on $\epsilon \in (0,\epsilon_1)$ such that

$$\int_{T_1}^{\infty}\Big(\frac{c}{t^2} + C'\frac{|f|}{\epsilon^2}\Big)\,dt \le C''.$$

This allows us to integrate (XII.124) from T_1 to an arbitrary value $t \ge T_1$; we get

$$-(t-T_1) - C'' \le \ln(\mathscr{V}(t) + \mu\mathscr{U}(t)) - \ln(\mathscr{V}(T_1) + \mu\mathscr{U}(T_1)) \le -(t-T_1) + C'',$$

which yields the desired inequalities (XII.114). □

The next result is a consequence of the inequality (XII.115) in Lemma XII.19 due to $\inf_{|t|\le T_1}\hat V>0$.

Corollary XII.20 *There are $C_1^*>c_1^*>0$ such that*

$$\hat V(t)\ge c_1^*\langle t\rangle^{-(n-1)/2}e^{-|t|},\qquad \hat V(t)+|\hat U(t)|\le C_1^*\langle t\rangle^{-(n-1)/2}e^{-|t|},\qquad \forall t\in\mathbb{R}.$$

We claim that the bound (XII.116) from Lemma XII.19 could be improved as follows.

Lemma XII.21 *There is $C_2>0$ and $T_2\in(T_1,+\infty)$ such that*

$$|\tilde V(t,\epsilon)|+|\tilde U(t,\epsilon)|\le C_2h(\epsilon)t^{-(n-1)/2}e^{-t},\qquad \forall t\ge T_2,\quad \forall\epsilon\in(0,\epsilon_1),$$

with $h(\epsilon)$ from (XII.48).

Above, $\epsilon_1>0$ is from Theorem XII.1 *(2)* and $T_1\in(0,\infty)$ is as in Lemma XII.19.

PROOF. We define $\tilde{\mathscr V}(t,\epsilon)$, $\tilde{\mathscr U}(t,\epsilon)$ by the relations similar to (XII.117), (XII.118):

$$\tilde V(t,\epsilon)=t^{-\frac{n-1}{2}}\tilde{\mathscr V}(t,\epsilon), \tag{XII.125}$$

$$\tilde U(t,\epsilon)=t^{-\frac{n-1}{2}}\Big(\tilde{\mathscr U}(t)+\frac{n-1}{2\mu t}\tilde{\mathscr V}(t,\epsilon)\Big), \tag{XII.126}$$

where

$$\mu=m+\omega,\qquad \omega=\sqrt{m^2-\epsilon^2},\qquad \epsilon\in(0,\epsilon_1).$$

By (XII.58), the functions $\tilde{\mathscr V}$, $\tilde{\mathscr U}$ satisfy

$$\partial_t\Big(\tilde{\mathscr U}+\frac{n-1}{2\mu t}\tilde{\mathscr V}\Big)+\frac{n-1}{2t}\Big(\tilde{\mathscr U}+\frac{n-1}{2\mu t}\tilde{\mathscr V}\Big)+\frac{\tilde{\mathscr V}}{\mu}=(1+2\kappa)\hat V^{2\kappa}\tilde{\mathscr V}-t^{\frac{n-1}{2}}G_1$$

and

$$\partial_t\tilde{\mathscr V}-\frac{n-1}{2t}\tilde{\mathscr V}+\mu\big(\tilde{\mathscr U}+\frac{n-1}{2\mu t}\tilde{\mathscr V}\big)=t^{\frac{n-1}{2}}G_2,$$

which we rewrite as

$$\partial_t\tilde{\mathscr U}+\frac{n-1}{2\mu t}\partial_t\tilde{\mathscr V}+\frac{n-1}{2t}\tilde{\mathscr U}+\frac{(n-1)(n-3)}{4\mu t^2}\tilde{\mathscr V}+\frac{\tilde{\mathscr V}}{\mu}=(1+2\kappa)\hat V^{2\kappa}\tilde{\mathscr V}-t^{\frac{n-1}{2}}G_1, \tag{XII.127}$$

$$\partial_t\tilde{\mathscr V}+\mu\tilde{\mathscr U}=t^{\frac{n-1}{2}}G_2. \tag{XII.128}$$

We multiply (XII.127) by μ; adding and subtracting (XII.128), we obtain, respectively,

$$\partial_t(\mu\tilde{\mathscr U}+\tilde{\mathscr V})+(\mu\tilde{\mathscr U}+\tilde{\mathscr V})=(1+2\kappa)\mu\hat V^{2\kappa}\tilde{\mathscr V}-\tfrac{(n-1)(n-3)}{4t^2}\tilde{\mathscr V}+t^{\frac{n-1}{2}}\Big(G_2-\mu G_1-\tfrac{n-1}{2t}G_2\Big),$$

$$\partial_t(\mu\tilde{\mathscr U}-\tilde{\mathscr V})-(\mu\tilde{\mathscr U}-\tilde{\mathscr V})=(1+2\kappa)\mu\hat V^{2\kappa}\tilde{\mathscr V}-\tfrac{(n-1)(n-3)}{4t^2}\tilde{\mathscr V}-t^{\frac{n-1}{2}}\Big(\mu G_1+\tfrac{n-1}{2t}G_2+G_2\Big).$$

Multiplying the above relations by e^t and e^{-t}, respectively, we rewrite them as
(XII.129)
$$\partial_t(e^t(\upmu\tilde{\mathscr{U}}+\tilde{\mathscr{V}})) = e^t\Big((1+2\kappa)\upmu\hat{V}^{2\kappa}\tilde{\mathscr{V}} - \tfrac{(n-1)(n-3)}{4t^2}\tilde{\mathscr{V}} + t^{\frac{n-1}{2}}\big(G_2-\upmu G_1-\tfrac{n-1}{2t}G_2\big)\Big),$$

(XII.130)
$$\partial_t(e^{-t}(\upmu\tilde{\mathscr{U}}-\tilde{\mathscr{V}})) = e^{-t}\Big((1+2\kappa)\upmu\hat{V}^{2\kappa}\tilde{\mathscr{V}} - \tfrac{(n-1)(n-3)}{4t^2}\tilde{\mathscr{V}} - t^{\frac{n-1}{2}}\big(\upmu G_1+\tfrac{n-1}{2t}G_2+G_2\big)\Big).$$

We are to integrate the above relations in t; before we do this, we need a special treatment for the last term in the right-hand side of (XII.129).

Lemma XII.22 *There is $C>0$ such that the functions $G_1(\epsilon,\tilde{V}(t,\epsilon),\tilde{U}(t,\epsilon))$ and $G_2(\epsilon,\tilde{V}(t,\epsilon),\tilde{U}(t,\epsilon))$ defined in* (XII.59) *and* (XII.60) *satisfy*
$$\left|\int_T^{T'} t^{\frac{n-1}{2}}e^t\Big(G_2-\upmu G_1-\frac{n-1}{2t}G_2\Big)\,dt\right| \le Ch(\epsilon),\quad \forall T'\ge T\ge T_1,\quad \forall\epsilon\in(0,\epsilon_1),$$
with $h(\epsilon)$ from (XII.48).

PROOF. Applying the bounds on G_1 and G_2 from Lemma XII.10, we can treat all the terms (obtaining the desired bound $O(h(\epsilon))$) except for the ones linear in $\hat{V}$ and $\hat{U}$; the worry comes from e.g. $\langle t\rangle^{(n-1)/2}e^t\hat{V}(t)\ge c_1^*>0$ (cf. Corollary XII.20), whose contribution to the integral considered in the lemma would not be bounded uniformly in T, T'. Collecting from the expression $G_2-\upmu G_1-\frac{n-1}{2t}G_2$ all the terms which are linear in $\hat{V}$ and $\hat{U}$, we have:
$$(m-\omega)\hat{U}-(m+\omega)\frac{m-\omega}{2m(m+\omega)}\hat{V}-\frac{n-1}{2t}(m-\omega)\hat{U} = (m-\omega)\Big(\hat{U}-\frac{\hat{V}}{2m}-\frac{n-1}{2t}\hat{U}\Big).$$
Using (XII.12), we rewrite the right-hand side as $(m-\omega)\Big(\hat{U}+\partial_t\hat{U}+\frac{n-1}{2t}\hat{U}-|\hat{V}|^{2\kappa}\hat{V}\Big)$. Since
$$\int_T^{T'} t^{\frac{n-1}{2}}e^t\Big(\hat{U}+\partial_t\hat{U}+\frac{n-1}{2t}\hat{U}-|\hat{V}|^{2\kappa}\hat{V}\Big)\,dt = \int_T^{T'}\partial_t\big(t^{\frac{n-1}{2}}e^t\hat{U}\big)\,dt - \int_T^{T'} t^{\frac{n-1}{2}}e^t|\hat{V}|^{2\kappa}\hat{V}\,dt,$$
with both integrals in the right-hand side being bounded uniformly in $T'\ge T\ge T_1$ (due to the bounds on $\hat{U}$ and $\hat{V}$ from Lemma XII.19), while $m-\omega=O(\epsilon^2)$, the conclusion follows. □

For some fixed $T_2\ge T_1$ (to be specified later), we denote
$$M(\epsilon)=\sup_{t\ge T_2} e^t\Big(|\tilde{\mathscr{V}}(t,\epsilon)|+|\tilde{\mathscr{U}}(t,\epsilon)|\Big),\qquad \epsilon\in(0,\epsilon_1). \tag{XII.131}$$

Note that due to the bounds (XII.116) from Lemma XII.19 and the definitions (XII.125) and (XII.126) one has
$$\sup_{\epsilon\in(0,\epsilon_1)} M(\epsilon)<\infty.$$

Integrating (XII.129) from T_2 to some $t \ge T_2$ and using Lemma XII.22, one gets:

$$\begin{aligned}&\Big|e^t|\mu\tilde{\mathscr{U}}(t,\epsilon)+\tilde{\mathscr{V}}(t,\epsilon)| - e^{T_2}|\mu\tilde{\mathscr{U}}(T_2,\epsilon)+\tilde{\mathscr{V}}(T_2,\epsilon)|\Big| \\ &\le C\int_{T_2}^t \Big(\hat V(s)^{2\kappa}+\frac{1}{s^2}\Big)e^s\tilde{\mathscr{V}}(s,\epsilon)\,ds + Ch(\epsilon). \end{aligned} \tag{XII.132}$$

Taking into account that, due to Theorem XII.1 and (XII.125) and (XII.126), one has

$$|\tilde{\mathscr{V}}(t,\epsilon)|+|\tilde{\mathscr{U}}(t,\epsilon)| = O(h(\epsilon)), \qquad \forall t\ge T_1, \quad \forall \epsilon\in(0,\epsilon_1), \tag{XII.133}$$

and using (XII.131), we rewrite (XII.132) as

$$e^t|\mu\tilde{\mathscr{U}}(t,\epsilon)+\tilde{\mathscr{V}}(t,\epsilon)| \le M(\epsilon)C\int_{T_2}^t \Big(\hat V(s)^{2\kappa}+\frac{1}{s^2}\Big)\,ds + Ch(\epsilon), \tag{XII.134}$$

with some $C>0$ (which does not depend on $\epsilon\in(0,\epsilon_1)$, $T_2\ge T_1$, and $t\ge T_2$).

We now integrate (XII.130) from $t\ge T_2$ to $+\infty$. Due to the presence of the factor e^{-t} in the right-hand side, the last term does not need a special treatment such as in Lemma XII.22: the bounds on G_1 and G_2 from Lemma XII.10 together with the exponential decay of V, U, $\hat V$, $\hat U$ from Lemma XII.19 are sufficient. The integration yields

$$e^{-t}|\mu\tilde{\mathscr{U}}(t,\epsilon)-\tilde{\mathscr{V}}(t,\epsilon)| \le C\int_t^\infty \Big(\hat V^{2\kappa}(s)+\frac{1}{s^2}\Big)e^{-s}\tilde{\mathscr{V}}(s,\epsilon)\,ds + Ch(\epsilon)e^{-2t},$$

again with some $C>0$ which does not depend on $\epsilon\in(0,\epsilon_1)$, T_2, and t. We took into account that in the left-hand side the boundary term at $t=\infty$ disappears due to (XII.133). Using (XII.131), we rewrite the above relation as

$$e^t|\mu\tilde{\mathscr{U}}(t,\epsilon)-\tilde{\mathscr{V}}(t,\epsilon)| \le M(\epsilon)C\int_t^\infty e^{2t-2s}\Big(\hat V^{2\kappa}(s)+\frac{1}{s^2}\Big)\,ds + Ch(\epsilon). \tag{XII.135}$$

Since

$$|\tilde{\mathscr{V}}|+|\tilde{\mathscr{U}}| \le \frac{|\mu\tilde{\mathscr{U}}+\tilde{\mathscr{V}}|+|\mu\tilde{\mathscr{U}}-\tilde{\mathscr{V}}|}{2} + \frac{|\mu\tilde{\mathscr{U}}+\tilde{\mathscr{V}}|+|\mu\tilde{\mathscr{U}}-\tilde{\mathscr{V}}|}{2\mu},$$

the inequalities (XII.134) and (XII.135) lead to the bound

$$M(\epsilon)\le M(\epsilon)C\int_{T_2}^\infty \Big(\hat V(s)^{2\kappa}+\frac{1}{s^2}\Big)\,ds + Ch(\epsilon), \qquad \forall\epsilon\in(0,\epsilon_1), \tag{XII.136}$$

with some constant $C>0$ (which does not depend on $\epsilon\in(0,\epsilon_1)$ and T_2). Now we can choose T_2: we set $T_2\ge T_1$ to be sufficiently large so that the coefficient at $M(\epsilon)$ in the right-hand side is smaller than $1/2$ (due to the exponential decay of $\hat V$ and $\tilde{\mathscr{V}}$, such a value of T_2 could be chosen independent of $\epsilon\in(0,\epsilon_1)$). Now (XII.136) turns into the inequality

$$M(\epsilon)\le 2Ch(\epsilon) \qquad \forall\epsilon\in(0,\epsilon_1),$$

and (XII.131) gives

$$|\tilde{\mathscr{V}}(t)|+|\tilde{\mathscr{U}}(t)| \le 2Ch(\epsilon)e^{-t}, \qquad \forall t\ge T_2, \quad \forall\epsilon\in(0,\epsilon_1),$$

yielding the bounds stated in the lemma. □

Lemma XII.23 *There is $C_3 > 0$ such that*

$$|\tilde{V}(t,\epsilon)| + |\tilde{U}(t,\epsilon)| \le C_3 h(\epsilon)\hat{V}(t), \qquad \forall t \in \mathbb{R}, \quad \forall \epsilon \in (0,\epsilon_1), \tag{XII.137}$$

with $h(\epsilon)$ from (XII.48).

Above, $\epsilon_1 > 0$ is from Theorem XII.1 *(2)*.

PROOF. Using the bound from below on $\hat{V}$ from Lemma XII.19 and bound from above on $\tilde{V}$ and $\tilde{U}$ from Lemma XII.21, we conclude that the inequality (XII.137) takes place for $t \ge T_2$ (and also for $t \le -T_2$) and for all $\epsilon \in (0,\epsilon_1)$. Let us now consider the case $|t| \le T_2$. By the inequality (XII.93), there is $C > 0$ such that

$$\|\tilde{V}(\cdot,\epsilon)\|_{L^\infty} + \|\tilde{U}(\cdot,\epsilon)\|_{L^\infty} \le Ch(\epsilon) \le \frac{C}{\hat{V}(T_2)} h(\epsilon)\hat{V}(t), \quad \forall |t| \le T_2, \quad \forall \epsilon \in (0,\epsilon_0);$$

in the last inequality, we used the fact that $\hat{V}(t)$ is positive and monotonically decreasing for $t > 0$. This proves the desired inequality for $|t| \le T_2$. □

Lemma XII.23 proves (XII.29).

The pointwise bound (XII.30) follows from the inequality

$$\hat{V}(t) \le C_1^* \langle t\rangle^{-(n-1)/2} e^{-|t|}, \qquad \forall t \in \mathbb{R}, \qquad \forall \epsilon \in (0,\epsilon_1)$$

(see Corollary XII.20) and also from (XII.27) and (XII.29) which show that $\tilde{V}$, $\hat{U}$, and $\tilde{U}$ are all pointwise dominated by $\hat{V}$.

This completes the proof of Theorem XII.1 *(3)*.

For our convenience, we take ϵ_1 small enough so that $C_3 h(\epsilon_1) < 1/2$; then, for the later use, we have

$$|\tilde{V}(t,\epsilon)| + |\tilde{U}(t,\epsilon)| < \frac{1}{2}\hat{V}(t), \qquad \forall t \in \mathbb{R}, \qquad \forall \epsilon \in (0,\epsilon_1). \tag{XII.138}$$

XII.4 Improved error estimates

Now we prove Theorem XII.1 *(5)*. The assumption (XII.32), together with the bounds on the amplitude of solitary waves (XII.45), allows us to assume that there is $c > 0$ such that

$$|f(\tau) - |\tau|^\kappa| \le c|\tau|^K, \qquad |f(\tau)| \le (c+1)|\tau|^\kappa, \qquad \tau \in \mathbb{R}. \tag{XII.139}$$

The improvement of the estimates stated in Theorem XII.1 *(1)* and *(3)* comes from having better bounds on the second and third terms from the right-hand side of (XII.81): when estimating e.g. $|V^2 - \epsilon^2 U^2|^\kappa - |V^2|^\kappa$, we no longer have to rely on Lemma XII.9, being able to use the Taylor expansions instead.

We recall that $\Lambda_\kappa > 0$ was defined in (XII.43).

Lemma XII.24 *There is $C_4 > 0$ such that for any real numbers*

$$\hat{V},\ \hat{U},\ \tilde{V},\ \tilde{U} \in [-\Lambda_\kappa, \Lambda_\kappa], \qquad V = \hat{V} + \tilde{V}, \qquad U = \hat{U} + \tilde{U}$$

which satisfy

$$\epsilon_1 |U| \le \frac{1}{2} V, \qquad |\tilde{V}| \le \frac{1}{2}\hat{V}, \tag{XII.140}$$

one has

$$\left| f\big(\epsilon^{2/\kappa}(V^2 - \epsilon^2 U^2)\big) - \epsilon^2 \hat{V}^{2\kappa} \right| \le C_4 \left(\epsilon^{2+2\varkappa}\hat{V}^{2\kappa} + \epsilon^2 \hat{V}^{2\kappa-1}|\tilde{V}| \right), \qquad \forall \epsilon \in (0,\epsilon_1).$$

Above,

$$\varkappa := \min\left(1, \frac{K}{\kappa} - 1\right)$$

was defined in (XII.35).

PROOF. We proceed as follows:

$$\begin{aligned}
&|f(\epsilon^{2/\kappa}(V^2 - \epsilon^2 U^2)) - \epsilon^2 \hat{V}^{2\kappa}| \\
&\le |f(\epsilon^{2/\kappa}(V^2 - \epsilon^2 U^2)) - \epsilon^2 (V^2 - \epsilon^2 U^2)^\kappa| \\
&\qquad + \epsilon^2 |(V^2 - \epsilon^2 U^2)^\kappa - V^{2\kappa}| + \epsilon^2 |V^{2\kappa} - \hat{V}^{2\kappa}| \\
&\le c\epsilon^{2K/\kappa}(V^2 - \epsilon^2 U^2)^K + O(\epsilon^2 V^{2(\kappa-1)}\epsilon^2 U^2) + O(\epsilon^2 \hat{V}^{2\kappa-1}\tilde{V}),
\end{aligned}$$

where the three terms from the second line were estimated using (XII.139) and (XII.140). The conclusion follows. □

Here is an improvement of Lemma XII.10.

Lemma XII.25 *There is $C > 0$ such that for any numbers*

$$\hat{V}, \hat{U}, \tilde{V}, \tilde{U} \in [-\Lambda_\kappa, \Lambda_\kappa], \qquad V = \hat{V} + \tilde{V}, \qquad U = \hat{U} + \tilde{U}$$

which satisfy (XII.140) *and additionally*

$$|\hat{U}| \le \frac{C_1}{c_1}\hat{V}, \tag{XII.141}$$

with c_1 and C_1 from Lemma XII.19, *one has:*

$$\left| G_1 - \left(\frac{1}{m+\omega} - \frac{1}{2m}\right)\hat{V} - \kappa\epsilon^2 V^{2\kappa-1}U^2 \right| \le C\left((\epsilon^{2\frac{K}{\kappa}-2} + \epsilon^4)\hat{V}^{2\kappa+1} + \hat{V}^{2\kappa-1}\tilde{V}^2\right),$$

$$|G_2 - \epsilon^2 \hat{V}^{2\kappa}\hat{U} - (m-\omega)\hat{U}| \le C\,(\epsilon^{2+2\varkappa}\hat{V}^{2\kappa} + \epsilon^2 \hat{V}^{2\kappa-1}\tilde{V})|U| + \epsilon^2 \hat{V}^{2\kappa}|\tilde{U}|,$$

$$|G_1| + |G_2| \le C\epsilon^{2\varkappa}\hat{V} + C\hat{V}^{2\kappa-1}\tilde{V}^2,$$

for all $\epsilon \in (0, \epsilon_1)$. Above, $G_1 = G_1(\epsilon, \tilde{V}, \tilde{U})$ and $G_2 = G_2(\epsilon, \tilde{V}, \tilde{U})$.

PROOF. We start with the definition (XII.59) of $G_1(\epsilon, \tilde{V}, \tilde{U})$ and apply the inequalities (XII.140):

$$\begin{aligned}
&G_1(\epsilon, \tilde{V}, \tilde{U}) \\
&= -\epsilon^{-2} f(\epsilon^{2/\kappa}(V^2 - \epsilon^2 U^2))V + \hat{V}^{2\kappa}\hat{V} + (1+2\kappa)\hat{V}^{2\kappa}\tilde{V} + \frac{\hat{V}}{m+\omega} - \frac{\hat{V}}{2m} \\
&= -\left(\epsilon^{-2} f(\epsilon^{2/\kappa}(V^2 - \epsilon^2 U^2)) - |V^2 - \epsilon^2 U^2|^\kappa\right)V - \left(|V^2 - \epsilon^2 U^2|^\kappa - V^{2\kappa}\right)V \\
&\qquad - \left(V^{2\kappa+1} - \hat{V}^{2\kappa+1} - (1+2\kappa)\hat{V}^{2\kappa}\tilde{V}\right) + \left(\frac{1}{m+\omega} - \frac{1}{2m}\right)\hat{V} \\
&= O(\epsilon^{2\frac{K}{\kappa}-2}V^{2K+1}) + \kappa\epsilon^2 V^{2\kappa-1}U^2 + O(\epsilon^4 V^{2\kappa-3}U^4) + O(\hat{V}^{2\kappa-1}\tilde{V}^2) \\
&\qquad + \left(\frac{1}{m+\omega} - \frac{1}{2m}\right)\hat{V}.
\end{aligned}$$

Let us point out that the third term in the right hand side in the line above has the factor of ϵ^4, which contributes ϵ^4 into the first conclusion of the lemma.

For $G_2(\epsilon, \tilde{V}, \tilde{U})$ from (XII.60), we have:

$$\begin{aligned} G_2 - \epsilon^2 \hat{V}^{2\kappa}\hat{U} - (m-\omega)\hat{U} &= f\big(\epsilon^{2/\kappa}(V^2 - \epsilon^2 U^2)\big)U - \epsilon^2 \hat{V}^{2\kappa}\hat{U} \\ &= \big(f\big(\epsilon^{2/\kappa}(V^2 - \epsilon^2 U^2)\big) - \epsilon^2 \hat{V}^{2\kappa}\big)U + \epsilon^2 \hat{V}^{2\kappa}\tilde{U}. \end{aligned}$$

Applying Lemma XII.24 to the right-hand side, we have:

$$\left|G_2 - \epsilon^2 \hat{V}^{2\kappa}\hat{U} - (m-\omega)\hat{U}\right| \le C_4 \left(\epsilon^{2+2\varkappa}\hat{V}^{2\kappa} + \epsilon^2 \hat{V}^{2\kappa-1}|\tilde{V}|\right)|U| + \epsilon^2 \hat{V}^{2\kappa}|\tilde{U}|.$$

The second conclusion of the lemma follows.

Taking into account (XII.140), the bound on $|G_1| + |G_2|$ also follows; we need to mention that, due to (XII.140) and (XII.141), both $|\hat{U}|$ and $|\tilde{U}|$ are estimated by $\hat{V}$. □

We notice that, due to (XII.27) and (XII.138), the functions $\hat{V}(t)$, $\hat{U}(t)$, $\tilde{V}(t,\epsilon)$, and $\tilde{U}(t,\epsilon)$ satisfy inequalities (XII.140) for all $t \in \mathbb{R}$ and $\epsilon \in (0, \epsilon_1)$. Also, $\hat{U}(t)$ and $\hat{V}(t)$ satisfy the inequality (XII.141) due to (XII.115) from Lemma XII.19. Using Lemma XII.25 in place of Lemma XII.10, we can rewrite the proof of Lemma XII.11 as follows.

Lemma XII.26 *There is $C > 0$ such that if $\tilde{W}(\cdot,\epsilon) = \begin{bmatrix}\tilde{V}(\cdot,\epsilon)\\ \tilde{U}(\cdot,\epsilon)\end{bmatrix} \in \mathbf{X}_{e,o}$, $\epsilon \in (0, \epsilon_1)$, satisfies the bound*

$$\text{(XII.142)} \qquad \|e^{\gamma\langle t\rangle}\tilde{W}(\cdot,\epsilon)\|_{\mathbf{X}} \le \left\|e^{\gamma\langle t\rangle}\begin{bmatrix}\hat{V}\\ \hat{U}\end{bmatrix}\right\|_{\mathbf{X}}, \qquad \forall \epsilon \in (0, \epsilon_1),$$

where $\langle t\rangle$ is considered as the operator of multiplication by $(1+t^2)^{1/2}$ and γ is from (XII.82), *and also the functions*

$$V(t,\epsilon) = \hat{V}(t) + \tilde{V}(t,\epsilon), \qquad U(t,\epsilon) = \hat{U}(t) + \tilde{U}(t,\epsilon), \qquad t \in \mathbb{R}, \quad \epsilon \in (0, \epsilon_1),$$

satisfy the relations

(XII.143)

$$\epsilon_1 |U(t,\epsilon)| \le \frac{1}{2}V(t,\epsilon), \qquad |\tilde{V}(t,\epsilon)| \le \frac{1}{2}\hat{V}(t), \qquad \forall t \in \mathbb{R}, \qquad \forall \epsilon \in (0, \epsilon_1),$$

then there is the following bound on $G(\epsilon, \tilde{W}(\cdot,\epsilon))$:

(XII.144)

$$\|e^{(1+2\kappa)\gamma\langle t\rangle}G(\epsilon, \tilde{W}(\cdot,\epsilon))\|_{\mathbf{X}} \le C\left(\epsilon^{2\varkappa} + \|e^{\gamma\langle t\rangle}\tilde{W}(\cdot,\epsilon)\|_{\mathbf{X}}^2\right), \qquad \forall \epsilon \in (0, \epsilon_1),$$

with $\varkappa$ from (XII.35).

Proof. For $\tilde{V}(t,\epsilon)$ and $\tilde{U}(t,\epsilon)$ which satisfy the the assumptions of the lemma, due to Lemma XII.25, one has

$$\text{(XII.145)} \qquad |G_1(\epsilon, \tilde{V}, \tilde{U})| + |G_2(\epsilon, \tilde{V}, \tilde{U})| \le C\epsilon^{2\varkappa}\hat{V}(t) + C\hat{V}(t)^{2\kappa-1}\tilde{V}(t,\epsilon)^2, \qquad \forall \epsilon \in (0, \epsilon_1), \qquad \forall t \in \mathbb{R}.$$

Multiplying the first term in the right-hand side by $e^{(1+2\kappa)\gamma\langle t\rangle}$ and using (XII.83) and (XII.142), we bound the resulting $\mathbf{X}$-norm by $C\epsilon^{2\varkappa}$, with some $C > 0$. The second term in the right-hand side of (XII.145) is homogeneous of order $1 + 2\kappa$ in $\tilde{V}$ and $\hat{V}$; we multiply it by the factor $e^{(1+2\kappa)\gamma\langle t\rangle}$, absorbing $e^{\gamma\langle t\rangle}$ into each power of $\hat{V}$ and $\tilde{V}$ and bounding the $\mathbf{X}$-norm of the result by $C\|e^{\gamma\langle t\rangle}\hat{V}\|_{\mathbf{X}}^{2\kappa-1}\|e^{\gamma\langle t\rangle}\tilde{V}\|_{\mathbf{X}}^2 \le C'\|e^{\gamma\langle t\rangle}\tilde{V}\|_{\mathbf{X}}^2$. □

Now we use Lemma XII.26 to improve the estimates on $\tilde{W}$.

Lemma XII.27 *One can take $\epsilon_1 > 0$ smaller if necessary so that, for some $b_1 > 0$,*

$$\|e^{\gamma\langle t\rangle}\tilde{W}(\cdot,\epsilon)\|_{\mathbf{Y}} \le b_1\epsilon^{2\varkappa}, \qquad \forall \epsilon \in (0,\epsilon_1).$$

PROOF. We recall the relation (XII.86) satisfied by $\tilde{W}(t,\epsilon)$:

$$e^{\gamma\langle t\rangle}\tilde{W} = e^{-2\kappa\gamma\langle t\rangle}A_\gamma(\epsilon)^{-1}e^{(1+2\kappa)\gamma\langle t\rangle}G\big(\epsilon, e^{-\gamma\langle t\rangle}e^{\gamma\langle t\rangle}\tilde{W}\big).$$

Using the continuity of the mapping (XII.85) and estimating $G(\epsilon,\tilde{W})$ by Lemma XII.26, we obtain:

$$\begin{aligned}\|e^{\gamma\langle t\rangle}\tilde{W}\|_{\mathbf{Y}} &= \|e^{-2\kappa\gamma\langle t\rangle}A_\gamma(\epsilon)^{-1}\|_{\mathbf{X}\to\mathbf{Y}}\|e^{(1+2\kappa)\gamma\langle t\rangle}G(\epsilon,\tilde{W})\|_{\mathbf{X}}\\ &\le C(e^{2\varkappa} + \|e^{\gamma\langle t\rangle}\tilde{W}\|_{\mathbf{X}}^2).\end{aligned}$$

Since $\|e^{\gamma\langle t\rangle}\tilde{W}\|_{\mathbf{X}} \le \|e^{\gamma\langle t\rangle}\tilde{W}\|_{\mathbf{Y}}$ (cf. (XII.15)), the above relation yields the bound stated in the lemma as long as $\|e^{\gamma\langle t\rangle}\tilde{W}\|_{\mathbf{Y}}$ is sufficiently small (which holds due to (XII.93) as long as $\epsilon \in (0,\epsilon_1)$ with $\epsilon_1 > 0$ small enough). □

Lemma XII.27 improves the estimates from Theorem XII.1 *(1)* on the error terms $\tilde{V}$, $\tilde{U}$, proving (XII.33).

We also do the second pass over the proof of Theorem XII.1 *(3)*, improving in the inequality (XII.29) the factor $h(\epsilon)$ to $\epsilon^{2\varkappa}$. For this, we rewrite the proof of Lemma XII.21, where the bounds on G_1, G_2 come from Lemma XII.25 instead of Lemma XII.10. We also rewrite the proof of Lemma XII.23 with $\epsilon^{2\varkappa}$ instead of $h(\epsilon)$ (we use (XII.33) in place of (XII.93)). This brings us at $|\tilde{V}(t,\epsilon)| + |\tilde{U}(t,\epsilon)| \le C\epsilon^{2\varkappa}\hat{V}(t)$, with some $C > 0$, valid for all $t \in \mathbb{R}$ and all $\epsilon \in (0,\epsilon_1)$, thus proving (XII.34).

This completes the proof of Theorem XII.1.

XII.5 Solitary waves in the nonrelativistic limit: the case $f \in C^1$

We now turn to the case when $f \in C^1(\mathbb{R}\setminus\{0\}) \cap C(\mathbb{R})$ satisfies both the assumption (XII.36) and (XII.37). Just like the former assumption leads to (XII.139), the assumption (XII.37) allows us to accept that there is $C > 0$ such that

$$\text{(XII.146)} \qquad |\tau f'(\tau) - \kappa|\tau|^\kappa| \le C|\tau|^K, \qquad |\tau f'(\tau)| \le (C+\kappa)|\tau|^\kappa, \qquad \tau \in \mathbb{R},$$

where $\kappa \in (0, 2/(n-2))$ (any $\kappa > 0$ if $n \le 2$) and $K > \kappa$. Now we will be able to prove uniqueness and regularity of the family of solitary waves bifurcating from the nonrelativistic limit. This amounts to noticing that in (XII.89), taking into account Theorem XII.1 *(2)*, we actually recover some features of the implicit function theorem. A careful analysis shows that the main obstacle to its application is the lack of regularity of the mapping f in (XII.59), (XII.60). This closer look shows that the unique obstacle are the terms $f\big(\epsilon^{2/\kappa}(V^2 - \epsilon^2U^2)\big)V$ and $f\big(\epsilon^{2/\kappa}(V^2 - \epsilon^2U^2)\big)U$, which with (XII.146) can now be treated.

XII.5.1 Improved regularity of the groundstate. Let us prove Theorem XII.3 *(1)*. By Theorem XII.1, we already have $\phi_\omega \in H^1(\mathbb{R}^n, \mathbb{C}^N)$, $\omega \in (\omega_1, m)$, with $\omega_1 = \sqrt{m^2 - \epsilon_1^2}$, with $\epsilon_1 > 0$ from Theorem XII.1 *(2)*; we need to show how to get the improvement in the regularity of ϕ_ω under better regularity of f.

We start with the improvement of regularity of V, U proved in Lemma XII.14.

Lemma XII.28 *Let $\omega \in (\omega_1, m)$. If in* (XII.57) *one has $f \in C^1(\mathbb{R}\setminus\{0\}) \cap C(\mathbb{R})$ which satisfies the assumptions* (XII.146)*, and $V, U \in C^1(\mathbb{R})$, with V even and U odd, are solutions to* (XII.57) *from Lemma* XII.14*, then one has $V, U \in C^2(\mathbb{R})$, and the function $Y(t) := U(t)/t$, $t \neq 0$, could be extended to a function $Y \in C^1(\mathbb{R})$.*

PROOF. We start with the case $f \in C^1(\mathbb{R})$. We proceed similarly to Lemma XII.14. The inclusion $V \in C^2(\mathbb{R})$ immediately follows from the second equation in (XII.57). Let us prove that $U \in C^2(\mathbb{R})$. Equation (XII.57) takes the form (XII.95) with

$$B(t) = \frac{f}{\epsilon^2} V(t) - \frac{1}{m+\omega} V(t), \qquad f = f\big(\epsilon^{\frac{2}{\kappa}} V(t)^2 - \epsilon^{2+\frac{2}{\kappa}} U(t)^2\big).$$

We note that now $B \in C^1(\mathbb{R})$ and is even. It is enough to prove that $Y(t) = U(t)/t$ could be extended to a C^1 function on $\mathbb{R}$. Since $Y(t)$ is even, it is enough to prove that $\lim_{t\to 0} \partial_t Y(t) = 0$. Taking the derivative of (XII.98) at $t > 0$, we arrive at

$$\partial_t Y(t) = \frac{B(t) t^n - n \int_0^t B(\tau) \tau^{n-1}\, d\tau}{t^{n+1}} = \frac{\int_0^t B'(\tau) \tau^n \, d\tau}{t^{n+1}},$$

therefore

$$\lim_{t\to 0} \partial_t Y(t) = \lim_{t \to 0} \frac{\int_0^t B'(\tau)\tau^n\, d\tau}{t^{n+1}} = \lim_{t\to 0} \frac{B'(t)}{n+1} = \frac{B'(0)}{n+1} = 0,$$

where we took into account that $B \in C^1(\mathbb{R})$ is even.

The above argument still applies if we only require that $f \in C^1(\mathbb{R}\setminus\{0\}) \cap C(\mathbb{R})$: due to Theorem XII.1, the argument of f, given by

$$\tau(t) = \epsilon^{\frac{2}{\kappa}} V(t)^2 - \epsilon^{2+\frac{2}{\kappa}} U(t)^2,$$

always belongs to $\mathbb{R}_+ = (0, +\infty)$, hence in (XII.57) one has $f(\tau(t))$ which is a C^1 function of $t \in \mathbb{R}$. Moreover, one can deduce from (XII.146) that $\epsilon^{-2} V(t) \frac{d}{dt} f(\tau(t))$ remains bounded pointwise by $C\hat{V}(t)^\kappa (|\partial_t V(t)| + |\partial_t U(t)|)$ uniformly in $t \in \mathbb{R}$ and in $\epsilon \in (0, \epsilon_1)$:

$$\begin{aligned}
\left| \epsilon^{-2} V(t) \frac{d}{dt} f(\tau(t)) \right| &= |\epsilon^{-2} V(t) f'(\tau(t)) (2\epsilon^{\frac{2}{\kappa}} V(t) \partial_t V(t) - 2\epsilon^{2+\frac{2}{\kappa}} U(t) \partial_t U(t))| \\
&\le \epsilon^{-2} |V(t)| \frac{\kappa|\tau|^\kappa + C|\tau|^K}{|\tau|} |2\epsilon^{\frac{2}{\kappa}} V(t)\partial_t V(t) - 2\epsilon^{2+\frac{2}{\kappa}} U(t) \partial_t U(t)| \\
&\le \frac{C}{\epsilon^2} (\kappa|\tau|^\kappa + C|\tau|^K)(|\partial_t V| + |\partial_t U|) \le C\hat{V}(t)^\kappa (|\partial_t V(t)| + |\partial_t U(t)|).
\end{aligned}$$

Above, we used (XII.146) to deal with f' (note that $\tau > 0$ by Theorem XII.1 *(2)* and *(3)*), and then Theorem XII.1 *(3)* to estimate $|V(t)|$ and $|U(t)|$ with the aid of $\hat{V}(t)$. So, we again have $B \in C^1(\mathbb{R})$ and proceed as in the first part of the argument. □

Now we can show that $\phi_\omega \in H^2(\mathbb{R}^n, \mathbb{C}^N)$ for $\omega \in (\omega_1, m)$. From the Ansatz (XII.22), taking into account that $Y(t) = U(t)/t$ belongs to $C^1(\mathbb{R})$ (as we proved in Lemma XII.28), we conclude that $\phi_\omega \in C^1(\mathbb{R}^n, \mathbb{C}^N)$. Therefore, the nonlinear term $f(\phi_\omega^*\beta\phi_\omega)\beta\phi_\omega$ is in $C^1(\mathbb{R}^n, \mathbb{C}^N)$ as a function of $x \in \mathbb{R}^n$, and one has:

$$\begin{aligned} |\nabla (f(\phi_\omega^*\beta\phi_\omega)\beta\phi_\omega)| &\le |f'(\phi_\omega^*\beta\phi_\omega)||\operatorname{Re}(\phi_\omega^*\beta\nabla\phi_\omega)||\phi_\omega| + |f(\phi_\omega^*\beta\phi_\omega)||\nabla\phi_\omega| \\ &\le C\left(|f'(\phi_\omega^*\beta\phi_\omega)||\phi_\omega|^2 + |f(\phi_\omega^*\beta\phi_\omega)|\right)|\nabla\phi_\omega|. \end{aligned} \tag{XII.147}$$

By Theorem XII.1, $\phi_\omega \in L^\infty(\mathbb{R}^n, \mathbb{C}^N)$ and $\nabla\phi_\omega \in L^2(\mathbb{R}^n, \mathbb{C}^{n\times N})$; using the bounds (XII.139), (XII.146), we conclude that the right-hand side of (XII.147) is in $L^2(\mathbb{R}^n)$. Then (XII.147) shows that $f(\phi_\omega^*\beta\phi_\omega)\beta\phi_\omega$ is in $H^1(\mathbb{R}^n, \mathbb{C}^N)$, and then from

$$\phi_\omega = -(D_m - \omega)^{-1} f(\phi_\omega^*\beta\phi_\omega)\beta\phi_\omega,$$

with some $\omega \in (\omega_1, m)$, we deduce the inclusion $\phi_\omega \in H^2(\mathbb{R}^n, \mathbb{C}^N)$.

XII.5.2 Uniqueness, continuity, and differentiability of the mapping $\omega \mapsto \phi_\omega$. We start with the following technical result. Recall that $\Lambda_\kappa > 0$ was defined in (XII.43).

Lemma XII.29 *There is $C > 0$ such that for all $\epsilon \in (0, \epsilon_1)$ and for all numbers*

$$\hat{V}, \hat{U}, \tilde{V}, \tilde{U} \in [-\Lambda_\kappa, \Lambda_\kappa], \qquad V = \hat{V} + \tilde{V}, \qquad U = \hat{U} + \tilde{U}$$

which satisfy

$$\epsilon_1|U| \le \frac{1}{2}V, \qquad |\tilde{V}| \le b_2\epsilon^{2\varkappa}\hat{V}, \tag{XII.148}$$

one has

$$\left\| \frac{\partial G(\epsilon, \tilde{V}, \tilde{U})}{\partial(\tilde{V}, \tilde{U})} \right\|_{\operatorname{End}(\mathbb{C}^2)} \le C\epsilon^{2\varkappa},$$

where $G(\epsilon, \tilde{V}, \tilde{U}) = \begin{bmatrix} G_1(\epsilon, \tilde{V}, \tilde{U}) \\ G_2(\epsilon, \tilde{V}, \tilde{U}) \end{bmatrix}$ *(cf.* (XII.59), (XII.60)*).*

Above, $\epsilon_1 > 0$ is from Theorem XII.1 *(2)* and $b_2 > 0$ is from Theorem XII.1 *(5)*.

PROOF. Let us consider

$$\begin{aligned} \frac{\partial G(\epsilon, \tilde{V}, \tilde{U})}{\partial(\tilde{V}, \tilde{U})} &= \begin{bmatrix} \partial_{\tilde{V}} G_1 & \partial_{\tilde{U}} G_1 \\ \partial_{\tilde{V}} G_2 & \partial_{\tilde{U}} G_2 \end{bmatrix} \\ &= \begin{bmatrix} -2f'\epsilon^{\frac{2}{\kappa}-2}V^2 - \epsilon^{-2}f + (1+2\kappa)\hat{V}^{2\kappa} & 2f'\epsilon^{\frac{2}{\kappa}}VU \\ 2f'\epsilon^{\frac{2}{\kappa}}VU & f - 2f'\epsilon^{2+\frac{2}{\kappa}}U^2 \end{bmatrix}. \end{aligned}$$

Above, f and f' are evaluated at $\tau = \epsilon^{2/\kappa}(V^2 - \epsilon^2U^2)$. All the terms except for $\partial_{\tilde{V}}G_1$ are immediately $O(\epsilon^2)$; we now focus on $\partial_{\tilde{V}}G_1$. Denoting $\tau = \epsilon^{2/\kappa}(V^2 - \epsilon^2U^2) = O(\epsilon^{2/\kappa})$, one has:

$$|\epsilon^{-2}f(\tau) - \hat{V}^{2\kappa}| \le \epsilon^{-2}|f(\tau) - \tau^\kappa| + |(V^2 - \epsilon^2U^2)^\kappa - V^{2\kappa}| + |V^{2\kappa} - \hat{V}^{2\kappa}| \le C\epsilon^{2\varkappa}.$$

We estimated the three terms in the middle using (XII.139) and (XII.148). Similarly,

$$\begin{aligned} &|f'(\tau)\epsilon^{\frac{2}{\kappa}-2}2V^2 - 2\kappa\hat{V}^{2\kappa}| \\ &\le \frac{2V^2\epsilon^{\frac{2}{\kappa}}}{\epsilon^2}|f'(\tau) - \kappa\tau^{\kappa-1}| + \frac{2\kappa V^2\epsilon^{\frac{2}{\kappa}}}{\epsilon^2}|\tau^{\kappa-1} - (\epsilon^{\frac{2}{\kappa}}V^2)^{\kappa-1}| + 2\kappa|V^{2\kappa} - \hat{V}^{2\kappa}| \le C\epsilon^{2\varkappa}; \end{aligned}$$

we used (XII.146) and (XII.148). So,

$$\begin{aligned}|\partial_{\tilde{V}} G_1| &= |-2f'\epsilon^{\frac{2}{\kappa}-2}V^2 - \epsilon^{-2}f + (1+2\kappa)\hat{V}^{2\kappa}| \\ &\le |\epsilon^{-2}f - \hat{V}^{2\kappa}| + |2f'\epsilon^{\frac{2}{\kappa}-2}V^2 - 2\kappa\hat{V}^{2\kappa}| = O(\epsilon^{2\varkappa}).\end{aligned}$$

□

We claim that the mapping

$$\mu : \mathbf{X}_{e,o} \to \mathbf{Y}_{e,o} \subset \mathbf{X}_{e,o}, \qquad \mu : \tilde{W} \mapsto A(\epsilon)^{-1}G(\epsilon, \tilde{W})$$

is a contraction when considered on a certain subset of a sufficiently small ball.

Lemma XII.30 *Let $f \in C^1(\mathbb{R} \setminus \{0\}) \cap C(\mathbb{R})$ satisfy* (XII.146). *Then we can take $\epsilon_1 > 0$ smaller if necessary so that for any $\epsilon \in (0, \epsilon_1)$ and any*

$$\tilde{W}_0 = \begin{bmatrix}\tilde{V}_0 \\ \tilde{U}_0\end{bmatrix} \in \overline{\mathbb{B}_\rho(\mathbf{X}_{e,o})}, \qquad \tilde{W}_1 = \begin{bmatrix}\tilde{V}_1 \\ \tilde{U}_1\end{bmatrix} \in \overline{\mathbb{B}_\rho(\mathbf{X}_{e,o})}, \qquad \textit{with} \quad \rho = b_1\epsilon^{2\varkappa},$$

with $b_1 > 0$ from Lemma XII.27, *which satisfy the inequalities*

$$\text{(XII.149)} \qquad \epsilon_1|\hat{U}(t) + \tilde{U}_s(t)| \le \frac{1}{2}(\hat{V}(t) + \tilde{V}_s(t)), \qquad \forall t \in \mathbb{R}, \qquad \forall s = 0,\, 1,$$

$$\text{(XII.150)} \qquad |\tilde{V}_s(t)| \le b_2\epsilon_1^{2\varkappa}\hat{V}(t), \qquad \forall t \in \mathbb{R}, \qquad \forall s = 0,\, 1,$$

with $b_2 > 0$ from Lemma XII.29, *one has:*

$$\text{(XII.151)} \qquad \|\mu(\epsilon, \tilde{W}_1) - \mu(\epsilon, \tilde{W}_0)\|_{\mathbf{Y}} \le \frac{1}{2}\|\tilde{W}_1 - \tilde{W}_0\|_{\mathbf{X}}.$$

Proof. We consider the linear interpolations

$$\text{(XII.152)} \qquad \tilde{W}_s(t) = \begin{bmatrix}\tilde{V}_s(t) \\ \tilde{U}_s(t)\end{bmatrix} = (1-s)\begin{bmatrix}\tilde{V}_0(t) \\ \tilde{U}_0(t)\end{bmatrix} + s\begin{bmatrix}\tilde{V}_1(t) \\ \tilde{U}_1(t)\end{bmatrix}, \qquad s \in [0,1],$$

with $\tilde{W}_s \in \mathbf{X}_{e,o}$ for all $s \in [0,1]$, and we also set

$$\text{(XII.153)} \qquad W_s(t) = \begin{bmatrix}V_s(t) \\ U_s(t)\end{bmatrix} = \begin{bmatrix}\hat{V}(t) \\ \hat{U}(t)\end{bmatrix} + \begin{bmatrix}\tilde{V}_s(t) \\ \tilde{U}_s(t)\end{bmatrix}, \qquad s \in [0,1].$$

with $W_s \in \mathbf{X}_{e,o}$ for all $s \in [0,1]$. We notice that, due to (XII.149) and (XII.150), the interpolation (XII.152) is such that the equivalents of (XII.149) and (XII.150) are satisfied for all $s \in [0,1]$:

(XII.154)

$$\epsilon_1|U_s(t)| \le \frac{1}{2}V_s(t), \qquad |\tilde{V}_s(t)| \le b_2\epsilon_1^{2\varkappa}\hat{V}(t), \qquad \forall t \in \mathbb{R}, \quad \forall s \in [0,1].$$

Let us pick $\epsilon \in (0, \epsilon_1)$ and consider the relation

$$\text{(XII.155)} \qquad \mu(\epsilon, \tilde{W}_1) - \mu(\epsilon, \tilde{W}_0) = A(\epsilon)^{-1}\big(G(\epsilon, \tilde{W}_1) - G(\epsilon, \tilde{W}_0)\big).$$

To estimate the right-hand side, we express

$$\begin{aligned} G(\epsilon, \tilde{W}_1(t)) - G(\epsilon, \tilde{W}_0(t)) &= \int_0^1 ds\, \frac{d}{ds} G\big(\epsilon, \tilde{W}_s(t)\big) \\ \text{(XII.156)} \qquad &= \big(\tilde{W}_1(t) - \tilde{W}_0(t)\big)\int_0^1 \partial_{\tilde{W}} G\big(\epsilon, \tilde{W}_s(t)\big)\, ds, \qquad t \in \mathbb{R}.\end{aligned}$$

Since the inequalities (XII.154) are satisfied, we can use Lemma XII.29 to estimate $\partial_{\tilde{W}} G$ in the right-hand side of (XII.156), getting

$$\|G(\epsilon,\tilde{W}_1)-G(\epsilon,\tilde{W}_0)\|_{\mathbf{X}} \le C\epsilon^{2\varkappa}\|\tilde{W}_1-\tilde{W}_0\|_{\mathbf{X}},$$

with some $C>0$, and then, by (XII.155),

$$\|\mu(\epsilon,\tilde{W}_1)-\mu(\epsilon,\tilde{W}_0)\|_{\mathbf{Y}} \le \|A(\epsilon)^{-1}\|_{\mathbf{X}_{e,o}\to\mathbf{Y}_{e,o}} C\epsilon^{2\varkappa}\|\tilde{W}_1-\tilde{W}_0\|_{\mathbf{X}}.$$

We can take $\epsilon_1>0$ smaller if necessary so that

$$\sup_{\epsilon\in[0,\epsilon_1]}\|A(\epsilon)^{-1}\|_{\mathbf{X}_{e,o}\to\mathbf{Y}_{e,o}} C\epsilon_1^{2\varkappa}\le 1/2.$$

The lemma follows. □

For each $\epsilon\in(0,\epsilon_1)$, Lemma XII.30 proves the uniqueness of the fixed point $\tilde{W}\in\mathbf{X}_{e,o}$ of $\mu(\epsilon,\cdot)$ which we constructed in Theorem XII.1. Indeed, $\tilde{W}$ satisfies the assumptions (XII.149) (by Theorem XII.1 *(2)*) and (XII.150) (by Theorem XII.1 *(3)*). Since $\|\cdot\|_{\mathbf{X}}\le\|\cdot\|_{\mathbf{Y}}$ (see (XII.15)), the inequality (XII.151) proves uniqueness of the fixed point $\tilde{W}$. Thus, we have a well-defined map

$$\begin{aligned}(0,\epsilon_1) &\to \mathbb{B}_\rho(\mathbf{Y}_{e,o}), &\quad \rho=b_1\epsilon_1^{2\varkappa};\\ \epsilon &\mapsto \tilde{W}(t,\epsilon), &\quad \|e^{\gamma\langle t\rangle}\tilde{W}(\cdot,\epsilon)\|_{H^1(\mathbb{R},\mathbb{C}^2)}\le b_1\epsilon^{2\varkappa}.\end{aligned} \tag{XII.157}$$

The above argument also implies the continuity of the fixed point $\tilde{W}(\epsilon)$ as a function of ϵ, since for any ϵ, $\epsilon'\in(0,\epsilon_1)$ one has

$$\begin{aligned}\tilde{W}(\epsilon')-\tilde{W}(\epsilon) &= A(\epsilon')^{-1}\big(G(\epsilon',\tilde{W}(\epsilon'))-G(\epsilon',\tilde{W}(\epsilon))\big)\\ &\quad +A(\epsilon')^{-1}G(\epsilon',\tilde{W}(\epsilon))-A(\epsilon)^{-1}G(\epsilon,\tilde{W}(\epsilon)).\end{aligned}$$

We evaluate $\mathbf{Y}$-norm of the above relation, applying Lemma XII.30 to the first term in the right-hand side to bound its norm with one half of the $\mathbf{Y}$-norm of the left-hand side; this yields

$$\|\tilde{W}(\epsilon')-\tilde{W}(\epsilon)\|_{\mathbf{Y}}\le 2\big\|A(\epsilon')^{-1}G(\epsilon',\tilde{W}(\epsilon))-A(\epsilon)^{-1}G(\epsilon,\tilde{W}(\epsilon))\big\|_{\mathbf{Y}},\quad \forall\epsilon,\ \epsilon'\in(0,\epsilon_1).$$

Due to the continuous dependence of A and G on $\epsilon>0$, the above relation shows the continuity of the map (XII.157) in $\epsilon\in(0,\epsilon_1)$.

We now turn to the differentiability of $\tilde{W}$ with respect to ϵ. Let us take α, $\beta\in(0,\epsilon_1)$. Without loss of generality, we may assume that $\alpha<\beta$. For both α and β, we denote the unique fixed points of $\mu(\alpha,\cdot)$ and $\mu(\beta,\cdot)$ (the images of α, $\beta\in(0,\epsilon_1)$ under the mapping (XII.157)) by $\tilde{W}(t,\alpha)=\begin{bmatrix}\tilde{V}(t,\alpha)\\ \tilde{U}(t,\alpha)\end{bmatrix}$ and $\tilde{W}(t,\beta)=\begin{bmatrix}\tilde{V}(t,\beta)\\ \tilde{U}(t,\beta)\end{bmatrix}$. By Theorem XII.1 *(2)* and *(3)*, these fixed points satisfy

$$\epsilon_1|U(t,\alpha)|\le\frac{1}{2}V(t,\alpha),\qquad |\tilde{V}(t,\alpha)|\le b_2\alpha^{2\varkappa}\hat{V}(t),\qquad \forall t\in\mathbb{R},$$

$$\epsilon_1|U(t,\beta)|\le\frac{1}{2}V(t,\beta),\qquad |\tilde{V}(t,\beta)|\le b_2\beta^{2\varkappa}\hat{V}(t),\qquad \forall t\in\mathbb{R},$$

therefore the linear interpolation

$$\tilde{W}_s(t)=\begin{bmatrix}\tilde{V}_s(t)\\ \tilde{U}_s(t)\end{bmatrix}=(1-s)\tilde{W}(t,\alpha)+s\tilde{W}(t,\beta),\qquad s\in[0,1],$$

satisfies

(XII.158)
$$\epsilon_1 |U_s(t)| \le \frac{1}{2} V_s(t), \qquad |\tilde{V}_s(t)| \le b_2 \beta^{2\varkappa} \hat{V}(t), \qquad \forall t \in \mathbb{R}, \quad \forall s \in [0,1],$$

where $V_s(t) = \hat{V}(t) + \tilde{V}_s(t)$ and $U_s(t) = \hat{U}(t) + \tilde{U}_s(t)$ (cf. (XII.153)); in the second inequality in (XII.158), we took into account that $\alpha < \beta$. We have:

$$\frac{\tilde{W}(\beta) - \tilde{W}(\alpha)}{\beta - \alpha} = \frac{\mu(\beta, \tilde{W}(\beta)) - \mu(\beta, \tilde{W}(\alpha))}{\beta - \alpha} + \frac{\mu(\beta, \tilde{W}(\alpha) - \mu(\alpha, \tilde{W}(\alpha))}{\beta - \alpha}$$
$$= A(\beta)^{-1} \left(\int_0^1 \partial_{\tilde{W}} G\big(\beta, (1-s)\tilde{W}(\alpha) + s\tilde{W}(\beta)\big)\, ds \right) \frac{\tilde{W}(\beta) - \tilde{W}(\alpha)}{\beta - \alpha}$$
$$+ \frac{\mu(\beta, \tilde{W}(\alpha)) - \mu(\alpha, \tilde{W}(\alpha))}{\beta - \alpha}.$$

The above relation takes place at each $t \in \mathbb{R}$; we omitted the dependence of $\tilde{W}$ on t. By Lemma XII.29, which is applicable due to (XII.158), we can choose $\epsilon_1 > 0$ smaller if necessary so that the operator $B(t, \alpha, \beta) \in \operatorname{End}(\mathbb{C}^2)$ defined by

$$B(t, \alpha, \beta) = I_2 - A(\beta)^{-1} \int_0^1 \partial_{\tilde{W}} G\big(\beta, (1-s)\tilde{W}(t, \alpha) + s\tilde{W}(t, \beta)\big)\, ds$$

is invertible, with the inverse bounded uniformly in $t \in \mathbb{R}$ and α, $\beta \in (0, \epsilon_1)$; we then have:

$$\frac{\tilde{W}(\beta) - \tilde{W}(\alpha)}{\beta - \alpha} = B(\alpha, \beta)^{-1} \frac{\mu(\beta, \tilde{W}(\alpha)) - \mu(\alpha, \tilde{W}(\alpha))}{\beta - \alpha}.$$

Since B is continuous in α and β while $\mu(\epsilon, \tilde{W}) = A(\epsilon)^{-1} G(\epsilon, \tilde{W})$, with both $A(\epsilon)^{-1}$ and $G(\epsilon, \tilde{W})$ differentiable in ϵ, we deduce that $\big(\tilde{W}(t,\beta) - \tilde{W}(t,\alpha)\big)/(\beta - \alpha)$ has a limit as $\beta \to \alpha$; setting $\alpha = \epsilon$, we have:

$$\partial_\epsilon \tilde{W} = B^{-1} \frac{\partial}{\partial \epsilon}\big(A^{-1} G(\epsilon, \tilde{W})\big) = B^{-1} A^{-1}\big(- \partial_\epsilon A\, A^{-1} G(\epsilon, \tilde{W}) + \partial_\epsilon G(\epsilon, \tilde{W})\big)$$

(XII.159)
$$= B^{-1} A^{-1}\big(- \partial_\epsilon A\, \tilde{W} + \partial_\epsilon G(\epsilon, \tilde{W})\big),$$

where $\tilde{W} = \tilde{W}(t, \epsilon)$, $A = A(\epsilon)$, and

(XII.160)
$$B = B(t, \epsilon) := B(t, \epsilon, \epsilon) = I_2 - A(\epsilon)^{-1} \partial_{\tilde{W}} G\big(\epsilon, \tilde{W}(t, \epsilon)\big).$$

In the last equality in (XII.159), we took into account that $\tilde{W}(t, \epsilon) = A(\epsilon)^{-1} G(\epsilon, \tilde{W})$ (cf. (XII.63)).

Lemma XII.31 *For $\epsilon \in (0, \epsilon_1)$, one has:*

$$\left\| e^{\gamma\langle t\rangle} \frac{\partial G}{\partial \epsilon}\big(\epsilon, \tilde{W}(t, \epsilon)\big) \right\|_{\mathbf{X}} = O(\epsilon^{2\varkappa - 1}),$$

$$\left\| e^{\gamma\langle t\rangle} \left(\frac{\partial G}{\partial \epsilon}\big(\epsilon, \tilde{W}(t, \epsilon)\big) - \epsilon \begin{bmatrix} 2\kappa \hat{U}^2 \hat{V}^{2\kappa - 1} + \frac{\hat{V}}{4m^3} \\ 2\hat{U}\hat{V}^{2\kappa} + \frac{\hat{U}}{m} \end{bmatrix} \right) \right\|_{\mathbf{X}} = O\big(\epsilon^{\frac{2K}{\kappa} - 3}\big) + o(\epsilon).$$

PROOF. Due to $2\varkappa - 1 \le 1$ and due to the exponential decay of $\hat{V}$ and $\hat{U}$, the first estimate stated in the lemma follows from the second one. By (XII.59) and (XII.60), $\partial_\epsilon G$ is given by

(XII.161)

$$\frac{\partial G(\epsilon, \tilde{W})}{\partial \epsilon} = \begin{bmatrix} \Big(2\epsilon^{-3} f - \epsilon^{-2} \frac{2}{\kappa} \epsilon^{\frac{2}{\kappa}-1} (V^2 - \epsilon^2 U^2) f' + 2\epsilon^{-2} \epsilon^{1+\frac{2}{\kappa}} U^2 f'\Big) V + \frac{\hat{V}}{(m+\omega)^2} \frac{\epsilon}{\omega} \\ (\frac{2}{\kappa} \epsilon^{\frac{2}{\kappa}-1} V^2 - \frac{2\kappa+2}{\kappa} \epsilon^{1+\frac{2}{\kappa}} U^2) U f' + \hat{U} \frac{\epsilon}{\omega} \end{bmatrix},$$

with f, f' evaluated at $\tau = \epsilon^{2/\kappa}(V^2 - \epsilon^2 U^2)$. We recall that $\tilde{W} = \begin{bmatrix} \tilde{V} \\ \tilde{U} \end{bmatrix}$, $V = \hat{V} + \tilde{V}$, $U = \hat{U} + \tilde{U}$; cf. (XII.11), (XII.25). By (XII.146), taking into account the exponential decay of $\hat{V}$ and $\hat{U}$, and also $\|e^{\gamma\langle t\rangle} \tilde{W}\|_{H^1(\mathbb{R},\mathbb{C}^2)} = O(\epsilon^{2\varkappa})$ (cf. Theorem XII.1 *(5)*), one has:

$$\|e^{\gamma\langle t\rangle} \big(f(\tau) - \frac{\tau}{\kappa} f'(\tau)\big)\|_{\mathbf{X}} = \|e^{\gamma\langle t\rangle} O(|\tau|^K)\|_{\mathbf{X}} = O(\epsilon^{2K/\kappa}),$$

$$\begin{aligned} \|e^{\gamma\langle t\rangle} \epsilon^{2/\kappa} U^2 f'(\tau)\|_{\mathbf{X}} &\le C \|e^{\gamma\langle t\rangle} \epsilon^{2/\kappa} V^2 f'(\tau)\|_{\mathbf{X}} \\ &\le C \|e^{\gamma\langle t\rangle} \tau f'(\tau)\|_{\mathbf{X}} = \|O(|\tau|^\kappa)\|_{\mathbf{X}} = O(\epsilon^2), \end{aligned}$$

where $\tau(t) = \epsilon^{2/\kappa}(V(t)^2 - \epsilon^2 U(t)^2)$. Applying these bounds to terms in (XII.161), one arrives at the second estimate stated in the lemma. □

Multiplying (XII.159) by $e^{\gamma\langle t\rangle}$, we have:

(XII.162) $$e^{\gamma\langle t\rangle} \partial_\epsilon \tilde{W} = B^{-1} \circ e^{\gamma\langle t\rangle} \circ A^{-1} \circ e^{-\gamma\langle t\rangle} \circ \big(\partial_\epsilon A \circ e^{\gamma\langle t\rangle} \circ \tilde{W} + e^{\gamma\langle t\rangle} \partial_\epsilon G\big),$$

with $G = G(\epsilon, \tilde{W})$. Above, $e^{\pm\gamma\langle t\rangle}$ are understood as the multiplication operators; we note that they commute with

$$\partial_\epsilon A(\epsilon) = \frac{\epsilon}{\omega} \begin{bmatrix} -\frac{1}{(m+\omega)^2} & 0 \\ 0 & -1 \end{bmatrix}.$$

The operator $B(t, \epsilon)$ (see (XII.160)) defines a mapping

(XII.163) $$B(t, \epsilon)^{-1} : \mathbf{Y} \to \mathbf{Y}$$

which is continuous since both $\|B(t,\epsilon)^{-1}\|_{\mathrm{End}(\mathbb{C}^2)}$ and $\|\partial_t B(t,\epsilon)\|_{\mathrm{End}(\mathbb{C}^2)}$ are bounded uniformly in $t \in \mathbb{R}$ and $\epsilon \in (0, \epsilon_1)$, as long as $\epsilon_1 > 0$ is sufficiently small; we took into account that $\|\partial_{\tilde{W}} G(\epsilon, \tilde{W})\|_{\mathrm{End}(\mathbb{C}^2)} = O(\epsilon^{2\varkappa})$ by Lemma XII.29, while the derivatives $\partial_t V(t,\epsilon)$ and $\partial_t U(t,\epsilon)$ are bounded pointwise, uniformly in $t \in \mathbb{R}$ and $\epsilon \in (0, \epsilon_1)$, due to Lemma XII.14, and hence so is $\partial_t \tilde{W}(t, \epsilon)$.

Since $\|e^{\gamma\langle t\rangle} \tilde{W}\|_{\mathbf{Y}} = O(\epsilon^{2\varkappa})$ (cf. Lemma XII.27) and the mapping

$$e^{\gamma\langle t\rangle} \circ A(\epsilon)^{-1} \circ e^{-\gamma\langle t\rangle} : \mathbf{X} \to \mathbf{Y}$$

is continuous (just like the mapping

$$e^{(1+2\kappa)\gamma\langle t\rangle} \circ A(\epsilon)^{-1} \circ e^{-(1+2\kappa)\gamma\langle t\rangle} : \mathbf{X} \to \mathbf{Y}$$

in (XII.85)), while (XII.163) is continuous in $\mathbf{Y}$, it follows that the $\mathbf{Y}$-norm of the right-hand side of (XII.162) is bounded by

$$C \Big(\epsilon \|e^{\gamma\langle t\rangle} \tilde{W}(t,\epsilon)\|_{\mathbf{Y}} + \|e^{\gamma\langle t\rangle} \partial_\epsilon G(\epsilon, \tilde{W}(t,\epsilon))\|_{\mathbf{X}}\Big) = O(\epsilon^{2\varkappa-1}), \qquad \forall \epsilon \in (0, \epsilon_1);$$

we estimated the second term in the left-hand side with the aid of Lemma XII.31. Thus, the relation (XII.162) gives

$$\text{(XII.164)} \qquad \partial_\epsilon \tilde{W} \in \mathbf{Y}_{e,o}, \qquad \|e^{\gamma\langle t\rangle}\partial_\epsilon \tilde{W}\|_{\mathbf{Y}} = O(\epsilon^{2\varkappa-1}), \qquad \epsilon\in(0,\epsilon_1),$$

proving (XII.38).

We can now estimate $\|\partial_\omega\phi_\omega\|_{L^2}^2$. We have:

$$\begin{aligned}\|\partial_\omega\phi_\omega\|_{L^2}^2 &= \frac{\epsilon^2}{\omega^2}\left\|\frac{d}{d\epsilon}\phi_\omega\right\|_{L^2}^2\\ &= \frac{\epsilon^2\operatorname{vol}(\mathbb{S}^{n-1})}{\omega^2}\int_0^\infty \Big((\partial_\epsilon(\epsilon^{\frac{1}{\kappa}}V(\epsilon r,\epsilon)))^2 + (\partial_\epsilon(\epsilon^{1+\frac{1}{\kappa}}U(\epsilon r,\epsilon)))^2\Big)\, r^{n-1}\,dr.\end{aligned}$$

Let us estimate the above integral. Since

$$\partial_\epsilon(\epsilon^{\frac{1}{\kappa}}V(\epsilon r,\epsilon)) = \frac{1}{\kappa}\epsilon^{\frac{1}{\kappa}-1}V(\epsilon r,\epsilon) + \epsilon^{\frac{1}{\kappa}}r\partial_t V(\epsilon r,\epsilon) + \epsilon^{\frac{1}{\kappa}}\partial_\epsilon V(\epsilon r,\epsilon),$$

we have:

$$\begin{aligned}&\int_0^\infty (\partial_\epsilon(\epsilon^{\frac{1}{\kappa}}V(\epsilon r,\epsilon)))^2 r^{n-1}\,dr\\ &= \epsilon^{-n}\int_0^\infty \Big(\frac{\epsilon^{\frac{1}{\kappa}-1}}{\kappa}V(t,\epsilon) + \epsilon^{\frac{1}{\kappa}-1}t\partial_t V(t,\epsilon) + \epsilon^{\frac{1}{\kappa}}\partial_\epsilon V(t,\epsilon)\Big)^2 t^{n-1}\,dt\\ &= \epsilon^{-n+\frac{2}{\kappa}-2}\int_0^\infty \left(\frac{V(t,\epsilon)}{\kappa} + t\partial_t V(t,\epsilon) + \epsilon\partial_\epsilon V(t,\epsilon)\right)^2 t^{n-1}\,dt\\ &= \epsilon^{-n+\frac{2}{\kappa}-2}\left(C + O(\epsilon^{2\varkappa})\right),\end{aligned}$$

with

$$C = \int_0^\infty \Big(\frac{\hat{V}(t)}{\kappa} + t\partial_t\hat{V}(t)\Big)^2 t^{n-1}\,dt > 0.$$

We used Theorem XII.1 *(5)* for the L^2-norm of $t\partial_t V(t,\epsilon)$ and (XII.38) for the L^2-norm of $\partial_\epsilon V(t,\epsilon) = \partial_\epsilon\tilde{V}(t,\epsilon)$. We omit the computations for the part containing U since its contribution will be of the order $O(\epsilon^2)$ smaller, which is dominated by the $O(\epsilon^{2\varkappa})$ error term. It follows that

$$\|\partial_\omega\phi_\omega\|_{L^2}^2 = \frac{\epsilon^2}{\omega^2}\|\partial_\epsilon\phi_\omega\|_{L^2}^2 = \epsilon^{-n+\frac{2}{\kappa}}\frac{\operatorname{vol}(\mathbb{S}^{n-1})}{\omega^2}(C + O(\epsilon^{2\varkappa})),$$

proving (XII.39).

This completes the proof of Theorem XII.3 *(1)*.

XII.6 The Kolokolov condition for the nonlinear Dirac equation

Let us prove Theorem XII.3 *(2)*. Given a positive solution u_κ to the stationary Schrödinger equation (XII.4),

$$-\frac{1}{2m}u_\kappa = -\frac{1}{2m}\Delta u_\kappa - u_\kappa^{1+2\kappa},$$

with $n\in\mathbb{N}$ and

$$0<\kappa<\frac{2}{n-2} \qquad (\text{any } \kappa>0 \text{ if } n\le 2),$$

one can use u_κ to construct the solitary wave solutions to (XII.3) for any $\omega<0$:

$$\varphi_\omega(x) = (2m|\omega|)^{1/(2\kappa)}u_\kappa\big(\sqrt{2m|\omega|}x\big).$$

When $\kappa = 2/n$, it follows that the L^2-norm of φ_ω does not depend on ω; $\frac{d}{d\omega}\|\varphi_\omega\|^2 = 0$.

We are going to show that in the case of the nonlinear Dirac equation in n spatial dimensions with the "critical" value $\kappa = 2/n$ (and absent or sufficiently small higher order terms), the charge is no longer constant; instead, $\partial_\omega Q(\phi_\omega) < 0$ for $\omega \lessapprox m$. This reduces the degeneracy of the zero eigenvalue of the linearization at the corresponding solitary wave; see Section IX.6.2

Lemma XII.32 *Let $n \in \mathbb{N}$. Let us assume that $f \in C^1(\mathbb{R}\setminus\{0\}) \cap C(\mathbb{R})$ satisfies the assumption* (XII.146) *with some $K > \kappa > 0$. One has:*

$$\langle \hat{V}, \partial_\epsilon \tilde{V}\rangle = \epsilon q_1 + \epsilon\Big(\frac{1}{\kappa} - \frac{n}{2}\Big) q_2 + O(\epsilon^{\frac{2K}{\kappa}-3} + \epsilon^{4\varkappa-1}) + o(\epsilon), \qquad \epsilon \in (0, \epsilon_1),$$

with

$$q_1 = \int_{\mathbb{R}^n} (4m\hat{V}^{2\kappa}\hat{U}^2 + \hat{U}^2)\,dy > 0, \qquad q_2 = \int_{\mathbb{R}^n} \Big(\frac{\hat{V}^2}{4m^2} + 2m\hat{V}^{2\kappa}\hat{U}^2 + \hat{U}^2\Big)\,dy > 0.$$

Proof. By (XII.9), $\mathfrak{l}_+\big(-\frac{m}{\kappa}\hat{V} - mx\cdot\nabla\hat{V}\big) = \hat{V}$; hence,

$$A(0)\begin{bmatrix} \frac{\hat{V}}{\kappa} + x\cdot\nabla\hat{V} \\ -\frac{1}{2m}\partial_r(\frac{\hat{V}}{\kappa} + x\cdot\nabla\hat{V})\end{bmatrix} = \frac{1}{m}\begin{bmatrix}\hat{V}\\0\end{bmatrix}.$$

Therefore,

$$\begin{aligned}
\frac{\langle \hat{V}, \partial_\epsilon\tilde{V}\rangle}{m} &= \Big\langle \frac{1}{m}\begin{bmatrix}\hat{V}\\0\end{bmatrix}, \partial_\epsilon\begin{bmatrix}\tilde{V}\\\tilde{U}\end{bmatrix}\Big\rangle = \Big\langle A(0)\begin{bmatrix} \frac{\hat{V}}{\kappa} + y\cdot\nabla\hat{V} \\ -\frac{1}{2m}\partial_r(\frac{\hat{V}}{\kappa} + y\cdot\nabla\hat{V})\end{bmatrix}, \partial_\epsilon\begin{bmatrix}\tilde{V}\\\tilde{U}\end{bmatrix}\Big\rangle \\
&= \Big\langle \begin{bmatrix} \frac{\hat{V}}{\kappa} + y\cdot\nabla\hat{V} \\ -\frac{1}{2m}\partial_r(\frac{\hat{V}}{\kappa} + y\cdot\nabla\hat{V})\end{bmatrix}, A(0)\partial_\epsilon\begin{bmatrix}\tilde{V}\\\tilde{U}\end{bmatrix}\Big\rangle \\
&= \Big\langle \begin{bmatrix} \frac{\hat{V}}{\kappa} + y\cdot\nabla\hat{V} \\ -\frac{1}{2m}\partial_r(\frac{\hat{V}}{\kappa} + y\cdot\nabla\hat{V})\end{bmatrix}, A(\epsilon)\partial_\epsilon\begin{bmatrix}\tilde{V}\\\tilde{U}\end{bmatrix}\Big\rangle + O(\epsilon^2)\|\partial_\epsilon\tilde{W}\|_{L^\infty}.
\end{aligned}$$

We took into account that the operator $A(\epsilon)$ defined in (XII.62) is self-adjoint on $\mathbf{Y}_{e,o}$ and that $\|A(\epsilon) - A(0)\|_{L^\infty(\mathbb{R},\mathrm{End}(\mathbb{C}^2))} = O(\epsilon^2)$. Taking the derivative of (XII.63) with respect to ϵ, we derive:

$$\begin{aligned}
&\frac{\langle \hat{V}, \partial_\epsilon\tilde{V}\rangle}{m} \\
&= \Big\langle \begin{bmatrix} \frac{\hat{V}}{\kappa} + y\cdot\nabla\hat{V} \\ -\frac{1}{2m}\partial_r(\frac{\hat{V}}{\kappa} + y\cdot\nabla\hat{V})\end{bmatrix}, \partial_{\tilde{W}}G\partial_\epsilon\tilde{W} + \partial_\epsilon G - \partial_\epsilon A(\epsilon)\tilde{W}\Big\rangle + O\big(\epsilon^2\big)\|\partial_\epsilon\tilde{W}\|_{L^\infty} \\
&= \Big\langle \begin{bmatrix} \frac{\hat{V}}{\kappa} + y\cdot\nabla\hat{V} \\ -\frac{1}{2m}\partial_r(\frac{\hat{V}}{\kappa} + y\cdot\nabla\hat{V})\end{bmatrix}, \partial_\epsilon G\Big\rangle + O\big(\epsilon^{4\varkappa-1} + \epsilon^{2\varkappa+1}\big).
\end{aligned}$$

We used the estimates $\|\partial_{\tilde{W}}G\|_{L^\infty(\mathbb{R},\mathrm{End}(\mathbb{C}^2))} = O(\epsilon^{2\varkappa})$ (cf. Lemma XII.29), $\|\tilde{W}\|_{L^\infty} = O(\epsilon^{2\varkappa})$ (cf. Theorem XII.1 *(5)*), and $\|\partial_\epsilon\tilde{W}\|_{L^\infty} = O(\epsilon^{2\varkappa-1})$ (cf. Theorem XII.3 *(1)*).

Taking into account Lemma XII.31 to express $\partial_\epsilon G(\epsilon, \tilde{W})$, we continue:

$$\frac{\langle \hat{V}, \partial_\epsilon \tilde{V} \rangle}{m}$$

$$= \epsilon \Big\langle \begin{bmatrix} \frac{\hat{V}}{\kappa} + y \cdot \nabla \hat{V} \\ -\frac{1}{2m}\partial_r(\frac{\hat{V}}{\kappa} + y \cdot \nabla \hat{V}) \end{bmatrix}, \begin{bmatrix} 2\kappa \hat{U}^2 \hat{V}^{2\kappa-1} + \frac{\hat{V}}{4m^3} \\ 2\hat{U}\hat{V}^{2\kappa} + \frac{\hat{U}}{m} \end{bmatrix} \Big\rangle + O(\epsilon^{\frac{2K}{\kappa}-3} + \epsilon^{4\varkappa-1}) + o(\epsilon)$$

$$= \epsilon \int\limits_{\mathbb{R}^n} \Big[\Big(\frac{\hat{V}}{\kappa} + y \cdot \nabla \hat{V} \Big) \Big(2\kappa \hat{V}^{2\kappa-1} \hat{U}^2 + \frac{\hat{V}}{4m^3} \Big)$$

$$+ \Big(\frac{1+\kappa}{\kappa} \hat{U} + y \cdot \nabla \hat{U} \Big) \Big(2\hat{V}^{2\kappa} \hat{U} + \frac{\hat{U}}{m} \Big) \Big] \, dy + O(\epsilon^{\frac{2K}{\kappa}-3} + \epsilon^{4\varkappa-1}) + o(\epsilon);$$

we took into account that $-\frac{1}{2m}\partial_r\Big(\frac{1}{\kappa}\hat{V} + y \cdot \nabla \hat{V}\Big) = \frac{1}{\kappa}\hat{U} + y \cdot \nabla \hat{U} + \hat{U}$. The integral $\int_{\mathbb{R}^n} [\dots] \, dy$ is evaluated by parts as follows:

$$\int\limits_{\mathbb{R}^n} \Big[\tfrac{\hat{V}^2}{4m^3\kappa} + 2\hat{V}^{2\kappa}\hat{U}^2 + \tfrac{y \cdot \nabla \hat{V}^2}{8m^3} + \tfrac{(1+\kappa)\hat{U}^2}{m\kappa} + \tfrac{2(1+\kappa)\hat{V}^{2\kappa}\hat{U}^2}{\kappa} + y \cdot \nabla(\hat{V}^{2\kappa}\hat{U}^2) + \tfrac{y \cdot \nabla \, \hat{U}^2}{2m} \Big] dy$$

$$= \int\limits_{\mathbb{R}^n} \Big[\tfrac{\hat{V}^2}{4m^3\kappa} + 2\hat{V}^{2\kappa}\hat{U}^2 - \tfrac{n\hat{V}^2}{8m^3} + \tfrac{(1+\kappa)\hat{U}^2}{m\kappa} + \tfrac{2(1+\kappa)\hat{V}^{2\kappa}\hat{U}^2}{\kappa} - n\hat{V}^{2\kappa}\hat{U}^2 - \tfrac{n\hat{U}^2}{2m} \Big] \, dy$$

$$= \int_{\mathbb{R}^n} \left[\frac{\hat{V}^2}{4m^3}\Big(\frac{1}{\kappa} - \frac{n}{2}\Big) + 2\Big(2 + \frac{1}{\kappa} - \frac{n}{2}\Big)\hat{V}^{2\kappa}\hat{U}^2 + \Big(1 + \frac{1}{\kappa} - \frac{n}{2}\Big)\frac{\hat{U}^2}{m} \right] \, dy.$$

□

Lemma XII.33 *Let $n \in \mathbb{N}$. Let $f \in C^1(\mathbb{R} \setminus \{0\})$ satisfy*

$$f(\tau) = |\tau|^\kappa + O(|\tau|^K), \qquad \tau \in \mathbb{R}.$$

(1) Assume that in the assumption (XII.146) *one has either $\kappa \in (0, 2/n)$, $K > \kappa$, or $\kappa = 2/n$, $K > 4/n$. Then there is $\omega_* \in (\omega_1, m)$ such that $\partial_\omega Q(\phi_\omega) < 0$ for $\omega \in (\omega_*, m)$.*

(2) If in the assumption (XII.146) *one has $\kappa \in (2/n,\, 2/(n-2))$ (any $\kappa > 2/n$ if $n \leq 2$), then there is $\omega_* \in (\omega_1, m)$ such that $\partial_\omega Q(\phi_\omega) > 0$ for $\omega \in (\omega_*, m)$.*

Above, $\omega_1 = \sqrt{m^2 - \epsilon_1^2}$, with $\epsilon_1 > 0$ from Theorem XII.3 *(1)*.

Proof. We recall that, by Theorem XII.1,

$$v(x,\omega) = \epsilon^{\frac{1}{\kappa}}\big(\hat{V}(\epsilon x) + \tilde{V}(\epsilon x, \epsilon)\big), \qquad u(x,\omega) = \epsilon^{\frac{1}{\kappa}+1}\big(\hat{U}(\epsilon x) + \tilde{U}(\epsilon x, \epsilon)\big);$$

$$Q(\phi_\omega) = \int_{\mathbb{R}^n} |\phi_\omega(x)|^2 \, dx = \epsilon^{\frac{2}{\kappa}-n} \int_{\mathbb{R}^n} \big(V(|y|,\epsilon)^2 + \epsilon^2 U(|y|,\epsilon)^2\big) \, dy.$$

Let us evaluate the contribution to the derivative of $Q(\phi_\omega)$, $\omega = \sqrt{m^2 - \epsilon^2}$, with respect to ϵ:

$$\begin{aligned}\partial_\epsilon Q &= \Big(\frac{2}{\kappa} - n\Big)\epsilon^{\frac{2}{\kappa}-n-1}(\langle V, V\rangle + \epsilon^2\langle U, U\rangle) + \epsilon^{\frac{2}{\kappa}-n}\partial_\epsilon\left(\langle V, V\rangle + \epsilon^2\langle U, U\rangle\right)\\ &= \Big(\frac{2}{\kappa} - n\Big)\epsilon^{\frac{2}{\kappa}-n-1}(\langle V, V\rangle + \epsilon^2\langle U, U\rangle)\\ &\quad + \epsilon^{\frac{2}{\kappa}-n}\left(2\langle\hat{V}, \partial_\epsilon\tilde{V}\rangle + 2\epsilon\langle\hat{U}, \hat{U}\rangle + O(\epsilon^{4\varkappa-1})\right).\end{aligned}$$

The estimate on the error terms in the right-hand side, such as $\langle\tilde{V}, \partial_\epsilon\tilde{V}\rangle = O(\epsilon^{4\varkappa-1})$, follows from (XII.24), (XII.25), while the **Y**-bounds on $\tilde{W}$ and $\partial_\epsilon\tilde{W}$ follow from Theorem XII.1 *(5)* and Theorem XII.3 *(1)*, respectively. By Lemma XII.32, in the non-critical case, when $\kappa \neq 2/n$ and $K > \kappa$, one has

$$\begin{aligned}\partial_\epsilon Q &= \Big(\frac{2}{\kappa} - n\Big)\epsilon^{\frac{2}{\kappa}-n-1}\langle\hat{V}, \hat{V}\rangle + O(\epsilon^{\frac{2}{\kappa}-n}\epsilon^{\frac{2K}{\kappa}-3}) + o(\epsilon^{\frac{2}{\kappa}-n-1})\\ &= \Big(\frac{2}{\kappa} - n\Big)\epsilon^{\frac{2}{\kappa}-n-1}\langle\hat{V}, \hat{V}\rangle + o(\epsilon^{\frac{2}{\kappa}-n-1});\end{aligned}$$

hence, for $\epsilon > 0$ sufficiently small, the sign of $\partial_\epsilon Q$ is determined by the sign of $\frac{2}{\kappa} - n$. Thus, if $\kappa \in (0, 2/n)$, one has $\partial_\omega Q = -\frac{\epsilon}{\omega}\partial_\epsilon Q < 0$ as long as $\omega < m$ is sufficiently close to m. In the critical case $\kappa = 2/n$, again by Lemma XII.32,

$$\partial_\epsilon Q(\omega) = 2\langle\hat{V}, \partial_\epsilon\tilde{V}\rangle + 2\epsilon\langle\hat{U}, \hat{U}\rangle + O(\epsilon^{4\varkappa-1})$$

$$= 2\epsilon\int_{\mathbb{R}^n}(4m\hat{V}^{2\kappa}\hat{U}^2 + \hat{U}^2)\,dy + 2\epsilon\int_{\mathbb{R}^n}\hat{U}^2\,dy + O(\epsilon^{\frac{2K}{\kappa}-3} + \epsilon^{4\varkappa-1} + \epsilon^{2\varkappa+1}) + o(\epsilon).$$

If $K/\kappa > 2$, $\varkappa = \min\left(1, \frac{K}{\kappa} - 1\right) = 1$, then, for $\epsilon > 0$ sufficiently small, the above is dominated by the terms of order one in ϵ, hence is strictly positive. Thus, in this case, $\partial_\omega Q = -\frac{\epsilon}{\omega}\partial_\epsilon Q < 0$ as long as $\omega < m$ is sufficiently close to m. This finishes the proof of Lemma XII.33. □

This concludes the proof of Theorem XII.3 *(2)*.

CHAPTER XIII

Spectral stability in the nonrelativistic limit

We now undertake a detailed study of the bifurcations of nonzero-real-part eigenvalues "from the nonrelativistic limit", as mentioned in Chapter XI. It is of no surprise that the behaviour of eigenvalues of the linearized operator near $\lambda = 0$, in the nonrelativistic limit $\omega \lessapprox m$ ($\omega \in (m - \varepsilon, m)$ for some $\varepsilon > 0$ small enough), follows closely the pattern which one observes in the nonlinear Schrödinger equation with the same nonlinearity. In other words, if the linearizations of the nonlinear Dirac equation at solitary waves with $\omega \lessapprox m$ admit a family of eigenvalues Λ_ω which continuously depends on ω, such that $\Lambda_\omega \to 0$ as $\omega \to m$, then this family is merely a deformation of an eigenvalue family $\Lambda_\omega^{\mathrm{NLS}}$ of the linearization of the nonlinear Schrödinger equation with the same nonlinearity (linearized at corresponding solitary waves). In formal agreement with the Kolokolov stability criterion **[Kol73]**, one expects that the solitary wave solutions to the "charge-subcritical" nonlinear Dirac equation in the nonrelativistic limit $\omega \lessapprox m$ are spectrally stable. This has been verified numerically in one- and in two-dimensional cases **[BC12a, CMKS$^+$16, Lak18]**. For a rigorous proof of spectral stability, one considers the spectral problem for the linearization at a solitary wave with $\omega \lessapprox m$, applies the rescaling with respect to $m - \omega \ll 1$, and uses the reduction based on the Schur complement method, recovering in the nonrelativistic limit $\omega \to m$ the linearization of the nonlinear Schrödinger equation, and then applying the Rayleigh–Schrödinger perturbation theory; in **[CGG14]**, this approach was developed to prove the linear instability of small amplitude solitary waves $\phi_\omega(x)e^{-\mathrm{i}\omega t}$ in the "charge-supercritical" NLD, in the nonrelativistic limit $\omega \lessapprox m$.

In the nonrelativistic limit $\omega \lessapprox m$, there could be eigenvalue families of the linearization of the nonlinear Dirac equation bifurcating not only from the origin, but also from the embedded thresholds (that is, such that $\lim_{\omega \to m} \Lambda_i(\omega) = \pm 2m\mathrm{i}$). Rescaling and using the Schur complement approach shows that there could be at most $N/2$ such families bifurcating from each of $\pm 2m\mathrm{i}$, with N the number of components of a spinor field. (More precisely, to study the bifurcations off the embedded thresholds, we consider the "nonlinear" eigenvalue problem, following the theory of characteristic roots developed by M. Keldysh.) Could these eigenvalues go off the imaginary axis into the complex plane? While for the nonlinear Dirac equations with a general nonlinearity the answer to this question is unknown, in the Soler model we can exclude this scenario. One can show that there are exact eigenvalues $\lambda_\pm(\omega) = \pm 2\omega\mathrm{i}$, each being of multiplicity $N/2$; thus, we know precisely what happens to the eigenvalues which bifurcate from $\pm 2m\mathrm{i}$, excluding the possibility of bifurcations of eigenvalues off the imaginary axis.

Let us mention that the presence of eigenvalues $\pm 2\omega\mathrm{i}$ of the linearization at a solitary wave in the Soler model (IX.1) is a consequence of the existence of bi-frequency solitary wave solutions, in any dimension and for any nonlinearity; see Chapter X.

We consider the nonlinear Dirac equation (IX.1),

$$\text{(XIII.1)}\qquad \mathrm{i}\partial_t\psi = D_m\psi - f(\psi^*\beta\psi)\beta\psi, \qquad \psi(t,x)\in\mathbb{C}^N, \quad x\in\mathbb{R}^n, \quad n\in\mathbb{N},$$

with the Dirac operator given by (III.46):

$$D_m = -\mathrm{i}\boldsymbol{\alpha}\cdot\nabla + \beta m, \qquad m>0.$$

We assume that the nonlinearity is represented by some function $f(\tau)\sim|\tau|^\kappa$, $\kappa>0$:

Assumption XIII.1 $f\in C^1(\mathbb{R}\setminus\{0\})\cap C(\mathbb{R})$, and there are $\kappa>0$ and $K>\kappa$ such that

$$|f(\tau)-|\tau|^\kappa| = O(|\tau|^K), \qquad |\tau f'(\tau)-\kappa|\tau|^\kappa| = O(|\tau|^K); \qquad |\tau|\le 1.$$

If $n\ge 3$, we additionally assume that $\kappa<2/(n-2)$.

We will consider the solitary wave solutions $\phi_\omega e^{-\mathrm{i}\omega t}$ to (XIII.1) in the nonrelativistic limit $\omega\lessapprox m$; these solitary waves were obtained in Theorems XII.1 and XII.3 as bifurcations from the solitary wave solutions to the nonlinear Schrödinger equation (XII.3). (It is for the application of these results that we needed the restriction $\kappa<2/(n-2)$ in Assumption XIII.1.) Since the L^∞-norm of the solitary waves from Theorem XII.1 goes to zero as $\omega\to m$, we could then take $\omega_1\lessapprox m$ sufficiently close to m so that $\|\phi_\omega\|_{L^\infty}$ remains smaller than one for $\omega\in(\omega_1,m)$. Therefore, if f satisfies Assumption XIII.1, then, modifying $f(\tau)$ for $|\tau|\ge 1$, we may assume that there are c, $C>0$ such that

$$\text{(XIII.2)}\qquad |f(\tau)-|\tau|^\kappa|\le c|\tau|^K, \qquad |f(\tau)|\le(c+1)|\tau|^\kappa, \qquad \forall\tau\in\mathbb{R};$$

$$\text{(XIII.3)}\qquad |\tau f'(\tau)-\kappa|\tau|^\kappa|\le C|\tau|^K, \qquad |\tau f'(\tau)|\le(C+\kappa)|\tau|^\kappa, \qquad \forall\tau\in\mathbb{R}.$$

By Theorem XI.1, the eigenvalues of the linearization at solitary waves $\phi_\omega e^{-\mathrm{i}\omega t}$ with $\omega=\omega_j$ and with $\omega_j\to m$ ("the nonrelativistic limit") can only accumulate to $\lambda=\pm 2m\mathrm{i}$ and $\lambda=0$. We are going to relate the families of eigenvalues of the linearized nonlinear Dirac equation bifurcating from $\lambda=0$ and from $\lambda=\pm 2m\mathrm{i}$ to the eigenvalues of the linearization of the nonlinear Schrödinger equation (XII.3) at a solitary wave, proving the spectral stability of Dirac solitary waves in $\mathbb{R}^n$, $n\in\mathbb{N}$, for certain values of $\kappa\le 2/n$.

XIII.1 Main results

We need to recall some notations from the stability theory for nonlinear Schrödinger equation. With $u_\kappa(x)$ from (XII.5), a strictly positive spherically symmetric exponentially decaying solution to the stationary nonlinear Schrödinger equation (XII.4), the solitary wave $\psi(t,x)=u_\kappa(x)e^{-\mathrm{i}\omega t}$ with $\omega=-\frac{1}{2m}$ is a solution to the nonlinear Schrödinger equation (XII.3). The linearization at this solitary wave is given by (see (XII.7) and (XII.8))

$$\text{(XIII.4)}\qquad \partial_t\begin{bmatrix}\operatorname{Re}\rho\\ \operatorname{Im}\rho\end{bmatrix} = \begin{bmatrix}0 & \mathfrak{l}_-\\ -\mathfrak{l}_+ & 0\end{bmatrix}\begin{bmatrix}\operatorname{Re}\rho\\ \operatorname{Im}\rho\end{bmatrix},$$

with $\mathfrak{l}_\pm$ defined by

$$\mathfrak{l}_- = \frac{1}{2m} - \frac{\Delta}{2m} - u_\kappa^{2\kappa}, \qquad \mathfrak{l}_+ = \frac{1}{2m} - \frac{\Delta}{2m} - (1+2\kappa)u_\kappa^{2\kappa}, \tag{XIII.5}$$

with domain $\mathfrak{D}(\mathfrak{l}_\pm) = H^2(\mathbb{R}^n)$; we recall that, by Lemma V.24 *(3)*, the multiplication by $u_\kappa^{2\kappa}$ is a continuous linear operator in $L^2(\mathbb{R}^n)$.

Theorem XIII.2 (Bifurcations of eigenvalues from the origin) *Let $n \in \mathbb{N}$. Let*

$$f \in C^1(\mathbb{R}\setminus\{0\}) \cap C(\mathbb{R})$$

satisfy Assumption XIII.1 *with some values of κ, K. Let $\phi_\omega e^{-\mathrm{i}\omega t}$, $\omega \in (\omega_1, m)$, be the family of solitary wave solutions to* (XIII.1) *constructed in Theorems* XII.1 *and* XII.3.

Let $(\omega_j)_{j\in\mathbb{N}}$, $\omega_j \in (\omega_1, m)$, be a sequence such that $\omega_j \to m$, and assume that λ_j is an eigenvalue of the linearization of (XIII.1) *at the solitary wave $\phi_{\omega_j}(x)e^{-\mathrm{i}\omega_j t}$ (see Section* XIII.2*) and that $\lambda_j \to 0$. Denote*

$$\Lambda_j := \frac{\lambda_j}{\epsilon_j^2} \in \mathbb{C}, \qquad \epsilon_j := (m^2 - \omega_j^2)^{1/2} > 0, \qquad j \in \mathbb{N},$$

and let $\Lambda_0 \in \mathbb{C} \cup \{\infty\}$ be an accumulation point of the sequence $(\Lambda_j)_{j\in\mathbb{N}}$. Then:

(1)
$$\Lambda_0 \in \sigma\Big(\begin{bmatrix} 0 & \mathfrak{l}_- \\ -\mathfrak{l}_+ & 0 \end{bmatrix}\Big) \cup \sigma(\mathrm{i}\mathfrak{l}_-) \cup \sigma(-\mathrm{i}\mathfrak{l}_-);$$
in particular, $\Lambda_0 \neq \infty$. If moreover $N = 2$, then $\Lambda_0 \in \sigma\Big(\begin{bmatrix} 0 & \mathfrak{l}_- \\ -\mathfrak{l}_+ & 0 \end{bmatrix}\Big)$.

(2) If $\operatorname{Re}\lambda_j \neq 0$ for all $j \in \mathbb{N}$, then
$$\Lambda_0 \in \sigma_{\mathrm{p}}\Big(\begin{bmatrix} 0 & \mathfrak{l}_- \\ -\mathfrak{l}_+ & 0 \end{bmatrix}\Big) \cap \mathbb{R}.$$

(3) If $\operatorname{Re}\lambda_j \neq 0$ for all $j \in \mathbb{N}$, then $\Lambda_0 = 0$ could be an accumulation point of $(\Lambda_j)_{j\in\mathbb{N}}$ only when $\kappa = 2/n$ and $\partial_\omega Q(\phi_\omega) > 0$ for $\omega \in (\omega_*, m)$, with some $\omega_* < m$. If, moreover, $\Lambda_j \to \Lambda_0 = 0$, then $\lambda_j \in \mathbb{R}$ for all but finitely many $j \in \mathbb{N}$.

In other words, as long as $\partial_\omega Q(\phi_\omega) < 0$ for $\omega \lessapprox m$, there can be no linear instability due to bifurcations from the origin: there would be no eigenvalues λ_j of the linearization at solitary waves with $\omega_j \to m$ such that $\operatorname{Re}\lambda_j \neq 0$, $\lambda_j \to 0$; let us mention that this is in formal agreement with the Kolokolov stability criterion **[Kol73]**. In the proof of Theorem XIII.2 in Section XIII.3, we will make this rigorous by applying the rescaling and the Schur complement to the linearization of the nonlinear Dirac equation and recovering in the nonrelativistic limit $\omega \to m$ the linearization of the nonlinear Schrödinger equation. Consequently, the absence of eigenvalues with nonzero real part in the vicinity of $\lambda = 0$ is controlled by the Kolokolov condition $\partial_\omega Q(\phi_\omega) < 0$, $\omega \lessapprox m$.

We note that quintic nonlinear Schrödinger equation in one spatial dimension and the cubic one in two spatial dimensions are "charge critical", providing the scaling such that all groundstate solitary waves have the same charge. As a consequence, by **[Kol73]**, the linearization at any solitary wave has a 4×4 Jordan block at $\lambda = 0$. On the contrary, by Theorem XII.3 *(2)*, for the nonlinear Dirac with the “Schrödinger charge-critical” power $f(\tau) = |\tau|^\kappa$ with $\kappa = 2/n$ (in any dimension $n \in \mathbb{N}$) the charge of solitary waves is no longer the same, satisfying $\partial_\omega Q(\phi_\omega) < 0$ for $\omega \lessapprox m$, where $Q(\phi_\omega) = \int_{\mathbb{R}^n} \phi_\omega(x)^* \phi_\omega(x)\,dx$ is the corresponding charge. This reduces the degeneracy of the zero eigenvalue of the linearization at the corresponding solitary wave (see Section IX.6.2).

Remark XIII.3 The set $\sigma(\mathrm{i}\mathfrak{l}_-)$ appears in Theorem XIII.2 *(1)* since the linearized operator has a certain degeneracy if $N \geq 4$; see Lemma XIII.17 below.

In order to state the next theorem, we need to recall the concept of virtual levels (also known in this context as *threshold resonances* and *zero-energy resonances*) for Schrödinger operators $-\Delta + V$, where V is some multiplication operator. For definitions and more details see e.g. [**JK79**] (for the case $n = 3$), [**JN01**, Sections 5, 6] (for $n = 1, 2$), and [**Yaf10**, Sections 5.2 and 7.4] (for $n = 1$ and $n \geq 3$). For practical purposes, in the case $V \not\equiv 0$, $n \leq 2$, the threshold point is a *virtual level* if it corresponds to a *virtual state*, which in turn can be characterized as an L^∞-solution to $(-\Delta + V)\Psi = 0$ which does not belong to L^2 (see e.g. [**JN01**, Theorem 5.2 and Theorem 6.2]). According to [**Yaf10**, Proposition 7.4.8] for $n = 3$, the virtual levels can be characterized as $H^2_{-1/2-s}$-solutions with an arbitrarily small $s > 0$ (see also [**JK79**, Lemma 3.2]); for $n = 4$, the virtual levels can be characterized as H^2_{-s}-solutions with an arbitrarily small $s > 0$; there are no virtual levels of Schrödinger operators in dimension larger than four, although there could be L^2-eigenstates at the threshold.

Theorem XIII.4 (Bifurcations of eigenvalues from $\pm 2m\mathrm{i}$) *Let $n \in \mathbb{N}$. Let*

$$f \in C^1(\mathbb{R} \setminus \{0\}) \cap C(\mathbb{R})$$

satisfy Assumption XIII.1 *with some values of κ, K. Let $\phi_\omega(x)e^{-\mathrm{i}\omega t}$, $\omega \in (\omega_1, m)$, be the family of solitary wave solutions to* (XIII.1) *constructed in Theorems* XII.1 *and* XII.3.

Let $(\omega_j)_{j\in\mathbb{N}}$, $\omega_j \in (\omega_1, m)$, be a sequence such that $\omega_j \to m$ and assume that λ_j is an eigenvalue of the linearization of (XIII.1) *at the solitary wave $\phi_{\omega_j}(x)e^{-\mathrm{i}\omega_j t}$ (see Section* XIII.2*) and that $\lambda_j \to 2m\mathrm{i}$. Denote*

$$\text{(XIII.6)} \qquad Z_j = -\frac{2\omega_j + \mathrm{i}\lambda_j}{\epsilon_j^2} \in \mathbb{C}, \qquad \epsilon_j := (m^2 - \omega_j^2)^{1/2}, \qquad j \in \mathbb{N},$$

and let $Z_0 \in \mathbb{C} \cup \{\infty\}$ be an accumulation point of the sequence $(Z_j)_{j\in\mathbb{N}}$. Then:

(1) $Z_0 \in \{\frac{1}{2m}\} \cup \sigma_{\mathrm{d}}(\mathfrak{l}_-)$, *with $\mathfrak{l}_-$ from* (XIII.5). *In particular, $Z_0 \neq \infty$.*
(2) If the edge of the essential spectrum of $\mathfrak{l}_-$ at $1/(2m)$ is a regular point of the essential spectrum of $\mathfrak{l}_-$ (neither a virtual level nor an eigenvalue), then $Z_0 \neq 1/(2m)$.
(3) If $Z_j \to Z_0$ and $Z_0 = 0$, then $\lambda_j = 2\omega_j\mathrm{i}$ for all but finitely many $j \in \mathbb{N}$.

In other words, as long as $\mathfrak{l}_-$ has no nonzero point spectrum and the threshold $z = 1/(2m)$ is a regular point of its essential spectrum, the linearization at the Dirac solitary waves with $\omega \lessapprox m$ can have no nonzero-real-part eigenvalues near $\pm 2m\mathrm{i}$. We prove Theorem XIII.4 in Section XIII.4.

Remark XIII.5 We do not need to study the case $\lambda_j \to -2m\mathrm{i}$ since the eigenvalues of the linearization at solitary waves are symmetric with respect to real and imaginary axes; see Lemma IX.25.

While the behaviour of eigenvalues of the linearized operator near $\lambda = 0$, in the nonrelativistic limit $\omega \lessapprox m$, closely follows the pattern that one finds in the nonlinear Schrödinger equation with the same nonlinearity (this is the content of Theorem XIII.2), there could be eigenvalue families of the which satisfy $\lim_{\omega\to m}\lambda_a(\omega) = \pm 2m\mathrm{i}$, as allowed by Theorem XI.1. Could these eigenvalues be located in the complex plane? Theorem XIII.4 states that in the Soler model, under certain spectral assumptions, this scenario could be excluded. Rescaling and the Schur complement approach will show that there could be at most $N/2$ families of eigenvalues with nonnegative real part (with N being the number of spinor components) bifurcating from each of $\pm 2m\mathrm{i}$ when $\omega = m$; this essentially follows from Section XIII.4.2 below (see Lemma XIII.40). At the same time, the linearization at a solitary wave has eigenvalues $\lambda = \pm 2\omega\mathrm{i}$, each of multiplicity (at least) $N/2$; this follows from the existence of bi-frequency solitary waves in the Soler model (Chapter X). Namely, if there is a solitary wave solution (XII.22) to the nonlinear Dirac equation (XIII.1), then there are also bi-frequency solitary wave solutions of the form

$$\text{(XIII.7)}\qquad \psi(t,x) = \phi_{\omega,\xi}(x)e^{-\mathrm{i}\omega t} + \chi_{\omega,\eta}(x)e^{\mathrm{i}\omega t}, \quad \xi,\eta\in\mathbb{C}^{N/2}, \quad |\xi|^2 - |\eta|^2 = 1,$$

where

$$\text{(XIII.8)}\qquad \phi_{\omega,\xi}(x) = \begin{bmatrix} v(r,\omega)\xi \\ \mathrm{i}\frac{x}{r}\cdot\boldsymbol{\sigma}\, u(r,\omega)\xi \end{bmatrix}, \qquad \chi_{\omega,\eta}(x) = \begin{bmatrix} -\mathrm{i}\frac{x}{r}\cdot\boldsymbol{\sigma}^*\, u(r,\omega)\eta \\ v(r,\omega)\eta \end{bmatrix},$$

with $r = |x|$ and $v(r,\omega)$ and $u(r,\omega)$ from (XII.22). The form of these bi-frequency solitary waves allows us to conclude that $\pm 2\omega\mathrm{i}$ are eigenvalues of the linearization at a solitary wave of multiplicities $N/2$ (see Lemma XIII.10 below). We emphasize that this is of utmost importance to us: it follows from Lemma XIII.40 that there could be *at most* $N/2$ bifurcations from embedded thresholds at $\pm 2m\mathrm{i}$, and we are now able to conclude that there could be no other eigenvalue families starting from $\pm 2m\mathrm{i}$ except for $\pm 2\omega\mathrm{i}$; in particular, no families of eigenvalues with nonzero real part.

As the matter of fact, since the points $\pm 2m\mathrm{i}$ belong to the essential spectrum, the perturbation theory can not be applied immediately for the analysis of families of eigenvalues which bifurcate from $\pm 2m\mathrm{i}$. We use the limiting absorption principle to rewrite the eigenvalue problem in such a way that the eigenvalue no longer appears as embedded. When doing so, we find out that the eigenvalues $\pm 2\omega\mathrm{i}$ become isolated solutions to the *nonlinear eigenvalue problem*, which are known as the *characteristic roots* (or, informally, *nonlinear eigenvalues*). To make sure that we end up with *isolated* nonlinear eigenvalues, we need to be able to vary the spectral parameter to both sides of the imaginary axis. To avoid the jump of the resolvent at the essential spectrum, we use the analytic continuation of the resolvent in the exponentially weighted spaces. Finally, we show that under the circumstances of the problem the *isolated nonlinear eigenvalues* can not bifurcate off the imaginary axis. This part is based on the theory of the characteristic roots of holomorphic operator-valued functions [**Kel51**, **Kel71**, **MS70**, **GS71**]; more recent references are [**Mar88**] and [**MM03**, Chapter I]. We give the necessary details of this theory in Section III.9.

We use Theorems XIII.2 and XIII.4 to prove the spectral stability of small amplitude solitary waves.

Theorem XIII.6 (Spectral stability of solitary waves of the nonlinear Dirac equation) *Let $n \in \mathbb{N}$. Let*

$$f \in C^1(\mathbb{R} \setminus \{0\}) \cap C(\mathbb{R})$$

satisfy Assumption XIII.1, *with κ, K such that either $0 < \kappa < 2/n$, $K > \kappa$ (this is a* charge-subcritical case*), or $\kappa = 2/n$ and $K > 4/n$ (*charge-critical case*). Further, assume that $\sigma_{\rm d}(\mathfrak{l}_-) = \{0\}$, and that the threshold $z = 1/(2m)$ of the operator $\mathfrak{l}_-$ is a regular point of its essential spectrum. Let $\phi_\omega(x)e^{-\mathrm{i}\omega t}$, $\phi_\omega \in H^2(\mathbb{R}^n, \mathbb{C}^N)$, $\omega \lessapprox m$, be the family of solitary waves constructed in Theorems* XII.1 *and* XII.3. *Then there is $\omega_* \in (0, m)$ such that for each $\omega \in (\omega_*, m)$ the corresponding solitary wave is spectrally stable.*

Remark XIII.7 We note that, if either $\kappa < 2/n$, $K > \kappa$, or $\kappa = 2/n$, $K > 4/n$, then, by Theorem XII.3, for $\omega \lessapprox m$ one has $\partial_\omega Q(\phi_\omega) < 0$, which is formally in agreement with the Kolokolov stability criterion [**Kol73**].

Proof. We consider the family of solitary wave solutions $\phi_\omega e^{-\mathrm{i}\omega t}$, $\omega \lessapprox m$ constructed in Theorems XII.1 and XII.3. Let us assume that there is a sequence $\omega_j \to m$ and a family of eigenvalues λ_j of the linearization at solitary waves $\phi_{\omega_j} e^{-\mathrm{i}\omega_j t}$ such that $\operatorname{Re}\lambda_j \neq 0$.

By Theorem XI.1, the only possible accumulation points of the sequence $\big(\lambda_j\big)_{j\in\mathbb{N}}$ are $\lambda = \pm 2m\mathrm{i}$ and $\lambda = 0$. By Theorem XIII.4, as long as $\sigma_{\rm d}(\mathfrak{l}_-) = \{0\}$ and the threshold of $\mathfrak{l}_-$ is a regular point of the essential spectrum, $\lambda = \pm 2m\mathrm{i}$ can not be an accumulation point of nonzero-real-part eigenvalues; it remains to consider the case $\lambda_j \to \lambda = 0$. By Theorem XIII.2 *(2)*, if $\operatorname{Re}\lambda_j \neq 0$ and Λ_0 is an accumulation point of the sequence $\Lambda_j := \lambda_j/(m^2 - \omega_j^2)$, then

$$\Lambda_0 \in \sigma_{\rm p}\Big(\begin{bmatrix} 0 & \mathfrak{l}_- \\ -\mathfrak{l}_+ & 0 \end{bmatrix}\Big) \cap \mathbb{R}. \tag{XIII.9}$$

Above, $\mathfrak{l}_\pm$ (see (XIII.5)) correspond to the linearization at a solitary wave $u_\kappa(x)e^{-\mathrm{i}\omega t}$ with $\omega = -1/(2m)$ of the nonlinear Schrödinger equation (XII.3). For $\kappa \le 2/n$, the spectrum of the linearization of the corresponding NLS at a solitary wave is purely imaginary:

$$\sigma_{\rm p}\Big(\begin{bmatrix} 0 & \mathfrak{l}_- \\ -\mathfrak{l}_+ & 0 \end{bmatrix}\Big) \subset \mathrm{i}\mathbb{R}.$$

We conclude from (XIII.9) that one could only have $\Lambda_0 = 0$; by Theorem XIII.2 *(3)*, this would require that $\kappa = 2/n$ and $\partial_\omega Q(\phi_\omega) > 0$ for $\omega \lessapprox m$. On the other hand, as long as $\kappa = 2/n$ and $K > 4/n$, Theorem XII.3 yields $\partial_\omega Q(\phi_\omega) < 0$ for $\omega \lessapprox m$, hence $\Lambda_0 = 0$ would not be possible. We conclude that there is no family of eigenvalues $(\lambda_j)_{j\in\mathbb{N}}$ with $\operatorname{Re}\lambda_j \neq 0$. □

Remark XIII.8 We can not claim the spectral stability for all subcritical values $\kappa \in (0, 2/n)$: the nonlinear Schrödinger equation with nonlinearity of order $1 + 2\kappa$ linearized at a solitary wave has a rich discrete spectrum for small values $\kappa \gtrapprox 0$, and potentially any of its points could become a source of nonzero-real-part eigenvalues of linearization of the nonlinear Dirac. Such cases would require a more detailed analysis. (In particular, in one spatial dimension, we only prove the spectral stability for $1 < \kappa \le 2$; the critical, quintic case ($\kappa = 2$) is included, but our proof formally does not cover the cubic case $\kappa = 1$ because of the virtual level at the threshold in the spectrum of the linearization operator corresponding to the one-dimensional cubic NLS.) Our numerics show that $\sigma_{\rm p}(\mathfrak{l}_-) = \{0\}$ and the threshold $1/(2m)$ is a regular point of the essential spectrum of $\mathfrak{l}_-$, with $\mathfrak{l}_-$ corresponding to the nonlinear Schrödinger equation in $\mathbb{R}^n$ (thus the spectral hypotheses of Theorem XIII.2 and Theorem XIII.4 are satisfied) as long as

$$\kappa > \kappa_n, \qquad \text{where} \quad \kappa_1 = 1, \quad \kappa_2 \approx 0.62, \quad \kappa_3 \approx 0.46\,.$$

Our numerical values of κ_2 and κ_3 seem to be in agreement with the plots in **[CGNT08]**. In three dimensions, according to **[DS06**, Fig. 3], $\mathfrak{l}_-$ does not have nonzero eigenvalues for $\kappa > \kappa_3$ with some $\kappa_3 < 2/3$.

XIII.2 The linearization operator

We assume that $f \in C^1(\mathbb{R} \setminus \{0\}) \cap C(\mathbb{R})$ satisfies Assumption XIII.1 (recall Remark XIII). Let $\phi_\omega(x)e^{-\mathrm{i}\omega t}$ be a solitary wave solution to equation (XIII.1) of the form (XII.22), with $\omega \in (\omega_1, m)$, where $\omega_1 \in (m/2, m)$ is from Theorem XII.3. Consider the solution to (XIII.1) in the form of the Ansatz

$$\psi(t,x) = (\phi_\omega(x) + \rho(t,x))e^{-\mathrm{i}\omega t},$$

so that $\rho(t,x) \in \mathbb{C}^N$ is a small perturbation of the solitary wave. By (IX.83) and (IX.84), the linearization at the solitary wave $\phi_\omega(x)e^{-\mathrm{i}\omega t}$ (the linearized equation on ρ) is given by

$$\text{(XIII.10)} \qquad \mathrm{i}\partial_t \rho = \mathcal{L}(\omega)\rho,$$

$$\mathcal{L}(\omega) = D_m - \omega - f(\phi_\omega^*\beta\phi_\omega)\beta - 2f'(\phi_\omega^*\beta\phi_\omega)\,\mathrm{Re}(\phi_\omega^*\beta\ \cdot)\beta\phi_\omega,$$

with the domain $\mathfrak{D}(\mathcal{L}(\omega)) = H^1(\mathbb{R}^n, \mathbb{C}^N)$.

Remark XIII.9 Even if $f'(\tau)$ is not continuous at $\tau = 0$, there are no singularities in (XIII.10) for solitary waves with $\omega \lessapprox m$ constructed in Theorems XII.1 and XII.3: in view of the bound $f'(\tau) = O(|\tau|^{\kappa-1})$ (see (XIII.3)) and the bound from below $\phi_\omega^*\beta\phi_\omega \ge |\phi_\omega|^2/2$ (see Theorem XII.3), the last term in the expression for $\mathcal{L}(\omega)$ in (XIII.10) could be estimated by $O(|\phi_\omega|^{2\kappa})$.

Lemma XIII.10 *The operator $\mathcal{L}(\omega)$ from* (XIII.10) *corresponding to the linearization at a (one-frequency) solitary wave has the eigenvalue -2ω of geometric multiplicity (at least) $N/2$, with the eigenspace containing the subspace* $\mathrm{Span}\{\chi_{\omega,\eta}\colon\ \eta \in \mathbb{C}^{N/2}\}$, *with $\chi_{\omega,\eta}$ defined in* (XIII.8). *The operator $\mathbf{J}\mathbf{L}(\omega)$ of the linearization at the solitary wave (see* (IX.92)*) has eigenvalues $\pm 2\omega\mathrm{i}$ of geometric multiplicity (at least) $N/2$.*

Proof. This could be concluded from the expressions for the bi-frequency solitary waves (XIII.7). Since

$$\psi(t,x) = \big((1+\varepsilon^2)^{1/2}\phi_{\omega,\xi}(x) + \varepsilon\chi_{\omega,\eta}(x)e^{2\mathrm{i}\omega t}\big)e^{-\mathrm{i}\omega t},$$

for any $0 < \varepsilon \ll 1$ and ξ, $\eta \in \mathbb{C}^{N/2}$, satisfies the nonlinear Dirac equation (XIII.1), one concludes that $r(t,x) = \chi_{\omega,\eta}(x)e^{2\mathrm{i}\omega t}$ is a solution to (XIII.1) linearized at $\phi_{\omega,\xi}e^{-\mathrm{i}\omega t}$. This shows that $2\omega\mathrm{i}$ is an eigenvalue of the linearization. Due to the symmetry of the spectrum with respect to $\mathrm{Re}\,\lambda = 0$ and $\mathrm{Im}\,\lambda = 0$ (see Lemma IX.25), so is $-2\omega\mathrm{i}$.

The existence could also be verified directly. One has

$$-2\omega\chi_{\omega,\eta} = (-\mathrm{i}\boldsymbol{\alpha}\cdot\nabla + (m-f)\beta - \omega)\chi_{\omega,\eta},$$

and then one takes into account that $\phi_{\omega,\xi}(x)^*\beta\chi_{\omega,\eta}(x) = 0$, so that the last term in the expression for the operator $\mathcal{L}(\omega)$ from (XIII.10) vanishes when applied to $\chi_{\omega,\eta}$. □

Remark XIII.11 The presence of the eigenvalues $\pm 2\omega\mathrm{i}$ in the spectrum of the linearization of the Soler model at a solitary wave was noticed in [**BC12a**] (initially in the one-dimensional case) and eventually led to the conclusion that there exist bi-frequency solitary waves. We note that the existence of such bi-frequency solutions could have already been deduced applying the Bogoliubov transformation (X.3) from [**Gal77**] to one-frequency solitary waves $\phi(x)e^{-\mathrm{i}\omega t}$ which were constructed in [**Sol70**].

XIII.2.1 Scaling, projections, and estimates on the effective potential. Let

$$\pi_P = (1+\boldsymbol{\beta})/2, \qquad \pi_A = (1-\boldsymbol{\beta})/2, \qquad \pi^{\pm} = (1 \mp \mathrm{i}\mathbf{J})/2 \tag{XIII.11}$$

be the projectors corresponding to $\pm 1 \in \sigma(\boldsymbol{\beta})$ ("particle" and "antiparticle" components) and to $\pm\mathrm{i} \in \sigma(\mathbf{J})$ ($\mathbb{C}$-antilinear and $\mathbb{C}$-linear). These projectors commute; we denote their compositions by

$$\pi_P^{\pm} = \pi^{\pm} \circ \pi_P, \qquad \pi_A^{\pm} = \pi^{\pm} \circ \pi_A. \tag{XIII.12}$$

With $\xi \in \mathbb{C}^{N/2}$, $|\xi| = 1$, from Theorem XII.3 (see (XII.22)), we denote

$$\Xi = \begin{bmatrix} \operatorname{Re}\xi \\ 0 \\ \operatorname{Im}\xi \\ 0 \end{bmatrix} \in \mathbb{R}^{2N}, \qquad |\Xi| = 1. \tag{XIII.13}$$

For the future use, we introduce the orthogonal projection onto Ξ:

$$\Pi_\Xi = \Xi\langle \Xi, \cdot \rangle_{\mathbb{C}^{2N}} \in \operatorname{End}(\mathbb{C}^{2N}). \tag{XIII.14}$$

We note that, since $\boldsymbol{\beta}\Xi = \Xi$, one has

$$\Pi_\Xi \circ \pi_P = \pi_P \circ \Pi_\Xi = \Pi_\Xi, \qquad \Pi_\Xi \circ \pi_A = \pi_A \circ \Pi_\Xi = 0. \tag{XIII.15}$$

Denote by $\boldsymbol{\psi}_j \in H^1(\mathbb{R}^n, \mathbb{C}^{2N})$ eigenfunctions which correspond to the eigenvalues $\lambda_j \in \sigma_{\mathrm{p}}(\mathbf{J}\mathbf{L}(\omega_j))$; thus, one has

$$\mathbf{L}(\omega_j)\boldsymbol{\psi}_j = \big(\mathbf{J}\boldsymbol{\alpha}\cdot\nabla_x + \boldsymbol{\beta}m - \omega_j + \mathbf{v}(\omega_j)\big)\boldsymbol{\psi}_j = -\mathbf{J}\lambda_j\boldsymbol{\psi}_j, \qquad j \in \mathbb{N}, \tag{XIII.16}$$

where the potential $\mathbf{v}(\omega)$ is defined by (cf. (IX.90))

$$\mathbf{v}(\omega)\boldsymbol{\psi} = -f(\boldsymbol{\phi}_\omega^*\boldsymbol{\beta}\boldsymbol{\phi}_\omega)\boldsymbol{\beta}\boldsymbol{\psi} - 2\boldsymbol{\phi}_\omega^*\boldsymbol{\beta}\boldsymbol{\psi}\, f'(\boldsymbol{\phi}_\omega^*\boldsymbol{\beta}\boldsymbol{\phi}_\omega)\boldsymbol{\beta}\boldsymbol{\phi}_\omega. \tag{XIII.17}$$

We will use the notations

$$y = \epsilon_j x \in \mathbb{R}^n,$$

where $\epsilon_j = \sqrt{m^2 - \omega_j^2}$, so that $\mathbf{J}\boldsymbol{\alpha}\cdot\nabla_x = \epsilon_j\mathbf{J}\boldsymbol{\alpha}\cdot\nabla_y =: \epsilon_j\mathbf{D}_0$ and $\Delta = \epsilon_j^2\Delta_y$. For $\mathbf{v}$ from (XIII.17), define the potential $\mathbf{V}(y,\epsilon) \in \operatorname{End}(\mathbb{C}^{2N})$ by

$$\mathbf{V}(y,\epsilon) = \epsilon^{-2}\mathbf{v}(\epsilon^{-1}y, \omega), \qquad \omega = \sqrt{m^2 - \epsilon^2}, \quad y \in \mathbb{R}^n, \quad \epsilon \in (0, \epsilon_1). \tag{XIII.18}$$

We define $\boldsymbol{\Psi}_j(y) = \epsilon_j^{-n/2}\boldsymbol{\psi}_j(\epsilon_j^{-1}y)$. With $\mathbf{V}$ from (XIII.18), $\mathbf{L}(\omega_j)$ is given by

$$\mathbf{L}(\omega_j) = \epsilon_j\mathbf{D}_0 + \boldsymbol{\beta}m - \omega_j + \epsilon_j^2\mathbf{V}(\omega_j), \tag{XIII.19}$$

and the relation (XIII.16) takes the form

$$\big(\epsilon_j \mathbf{D}_0 + \boldsymbol{\beta} m - \omega_j + \mathbf{J}\lambda_j + \epsilon_j^2 \mathbf{V}(\omega_j)\big)\boldsymbol{\Psi}_j = 0. \tag{XIII.20}$$

We project (XIII.20) onto "particle" and "antiparticle" components and onto the $\mp\mathrm{i}$ spectral subspaces of $\mathbf{J}$ with the aid of projectors (XIII.11):

$$\epsilon_j \mathbf{D}_0 \pi_A^- \boldsymbol{\Psi}_j + (m - \omega_j - \mathrm{i}\lambda_j)\pi_P^- \boldsymbol{\Psi}_j + \epsilon_j^2 \pi_P^- \mathbf{V}\boldsymbol{\Psi}_j = 0, \tag{XIII.21}$$

$$\epsilon_j \mathbf{D}_0 \pi_P^- \boldsymbol{\Psi}_j - (m + \omega_j + \mathrm{i}\lambda_j)\pi_A^- \boldsymbol{\Psi}_j + \epsilon_j^2 \pi_A^- \mathbf{V}\boldsymbol{\Psi}_j = 0, \tag{XIII.22}$$

$$\epsilon_j \mathbf{D}_0 \pi_A^+ \boldsymbol{\Psi}_j + (m - \omega_j + \mathrm{i}\lambda_j)\pi_P^+ \boldsymbol{\Psi}_j + \epsilon_j^2 \pi_P^+ \mathbf{V}\boldsymbol{\Psi}_j = 0, \tag{XIII.23}$$

$$\epsilon_j \mathbf{D}_0 \pi_P^+ \boldsymbol{\Psi}_j - (m + \omega_j - \mathrm{i}\lambda_j)\pi_A^+ \boldsymbol{\Psi}_j + \epsilon_j^2 \pi_A^+ \mathbf{V}\boldsymbol{\Psi}_j = 0. \tag{XIII.24}$$

We need several estimates on the potential $\mathbf{V}$.

Lemma XIII.12 *There is $C > 0$ such that for all $y \in \mathbb{R}^n$ and all $\epsilon \in (0, \epsilon_1)$ one has the following pointwise bounds:*

$$\|\mathbf{V}(y,\epsilon)\|_{\mathrm{End}(\mathbb{C}^{2N})} \le C|u_\kappa(y)|^{2\kappa},$$

$$\|\pi_P \circ \mathbf{V}(y,\epsilon) \circ \pi_A\|_{\mathrm{End}(\mathbb{C}^{2N})} + \|\pi_A \circ \mathbf{V}(y,\epsilon) \circ \pi_P\|_{\mathrm{End}(\mathbb{C}^{2N})} \le C\epsilon |u_\kappa(y)|^{2\kappa},$$

$$\|\pi_A \circ \big(\mathbf{V}(y,\epsilon) + |u_\kappa(y)|^{2\kappa}(1 + 2\kappa\Pi_{\Xi})\boldsymbol{\beta}\big) \circ \pi_A\|_{\mathrm{End}(\mathbb{C}^{2N})} \le C\epsilon^{2\varkappa}|u_\kappa(y)|^{2\kappa},$$

$$\|\pi_P \circ \big(\mathbf{V}(y,\epsilon) + |u_\kappa(y)|^{2\kappa}(1 + 2\kappa\Pi_{\Xi})\boldsymbol{\beta}\big) \circ \pi_P\|_{\mathrm{End}(\mathbb{C}^{2N})} \le C\epsilon^{2\varkappa}|u_\kappa(y)|^{2\kappa}.$$

Above, u_κ is the positive radially symmetric ground state of the nonlinear Schrödinger equation (XII.4); $\varkappa = \min\big(1, K/\kappa - 1\big) > 0$ was defined in (XII.35).

PROOF. The bound on $\|\mathbf{V}(y,\epsilon)\|_{\mathrm{End}(\mathbb{C}^{2N})}$ follows from the estimates (XIII.2) and (XIII.3):

$$\|\mathbf{V}(y,\epsilon)\|_{\mathrm{End}(\mathbb{C}^{2N})} \le C\epsilon^{-2}\big(|f(v^2 - u^2)| + v^2|f'(v^2 - u^2)|\big) \le C\epsilon^{-2}v^{2\kappa} \le C|u_\kappa(y)|^{2\kappa},$$

where $\omega = \sqrt{m^2 - \epsilon^2}$, $\epsilon \in (0, \epsilon_1)$, and

$$|v(\epsilon^{-1}|y|,\omega)| \le C\hat{V}(|y|)\epsilon^{\frac{1}{\kappa}}, \qquad |u(\epsilon^{-1}|y|,\omega)| \le C\hat{V}(|y|)\epsilon^{1+\frac{1}{\kappa}},$$

in the notations from Theorem XII.1, with $\hat{V}(|y|) = u_\kappa(y)$ (see (XII.11)).

The bounds on $\|\pi_P \circ \mathbf{V}(y,\epsilon) \circ \pi_A\|_{\mathrm{End}(\mathbb{C}^{2N})}$ and $\|\pi_A \circ \mathbf{V}(y,\epsilon) \circ \pi_P\|_{\mathrm{End}(\mathbb{C}^{2N})}$ follow from

$$\|\pi_P \circ \mathbf{V} \circ \pi_A\|_{\mathrm{End}(\mathbb{C}^{2N})} + \|\pi_A \circ \mathbf{V} \circ \pi_P\|_{\mathrm{End}(\mathbb{C}^{2N})} \le C|\epsilon^{-2} f'(\phi_\omega^* \beta \phi_\omega) v u|,$$

where $|f'(\phi_\omega^* \beta \phi_\omega)| = |f'(v^2 - u^2)| \le C|v|^{2\kappa - 2}$ by (XII.28) and (XIII.3), with v, u bounded as above.

Let us prove the bound on $\|\pi_A \circ \big(\mathbf{V}(y,\epsilon) + |u_\kappa(y)|^{2\kappa}(1 + 2\kappa\Pi_{\Xi})\boldsymbol{\beta}\big) \circ \pi_A\|_{\mathrm{End}(\mathbb{C}^{2N})}$. For any numbers $\hat{V} > 0$ and $\hat{U}$, $\tilde{V}$, $\tilde{U} \in \mathbb{R}$ which satisfy

$$\epsilon_1|U| \le \frac{V}{2}, \qquad |\tilde{V}| \le \min\Big(\frac{1}{2}, C\epsilon^{2\varkappa}\Big)\hat{V}, \tag{XIII.25}$$

with $V = \hat{V} + \tilde{V}$ and $U = \hat{U} + \tilde{U}$, there are the following bounds:

(XIII.26)
$$
\begin{aligned}
&|f\big(\epsilon^{\frac{2}{\kappa}}(V^2 - \epsilon^2 U^2)\big) - \epsilon^2 \hat{V}^{2\kappa}| \\
&\le |f\big(\epsilon^{\frac{2}{\kappa}}(V^2 - \epsilon^2 U^2)\big) - \epsilon^2 (V^2 - \epsilon^2 U^2)^\kappa| + \epsilon^2 |(V^2 - \epsilon^2 U^2)^\kappa - V^{2\kappa}| + \epsilon^2 |V^{2\kappa} - \hat{V}^{2\kappa}| \\
&\le c\epsilon^{2K/\kappa}(V^2 - \epsilon^2 U^2)^K + O(\epsilon^2 V^{2(\kappa-1)}\epsilon^2 U^2) + O(\epsilon^2 \hat{V}^{2\kappa-1}\tilde{V}) \\
&\le C\epsilon^{2+2\varkappa}\hat{V}^{2\kappa},
\end{aligned}
$$

where we used (XIII.25) and also applied (XIII.2); similarly, using (XIII.3),

(XIII.27)
$$
\begin{aligned}
&|f'\big(\epsilon^{\frac{2}{\kappa}}(V^2 - \epsilon^2 U^2)\big)\epsilon^{\frac{2}{\kappa}} V^2 - \kappa\epsilon^2 \hat{V}^{2\kappa}| \\
&\le \Big| f'\big(\epsilon^{\frac{2}{\kappa}}(V^2 - \epsilon^2 U^2)\big) - \kappa\big(\epsilon^{\frac{2}{\kappa}}(V^2 - \epsilon^2 U^2)\big)^{\kappa-1}\Big|\epsilon^{\frac{2}{\kappa}} V^2 + \kappa\epsilon^2 |(V^2 - \epsilon^2 U^2)^{\kappa-1} V^2 - \hat{V}^{2\kappa}| \\
&\le C|\epsilon^{\frac{2}{\kappa}}(V^2 - \epsilon^2 U^2)|^{K-1}\epsilon^{\frac{2}{\kappa}} V^2 + \kappa\epsilon^2 |(V^2 - \epsilon^2 U^2)^{\kappa-1} V^2 - V^{2\kappa}| + \kappa\epsilon^2 |V^{2\kappa} - \hat{V}^{2\kappa}| \\
&\le C\epsilon^{2+2\varkappa}\hat{V}^{2\kappa};
\end{aligned}
$$

(XIII.28)
$$
|f'\big(\epsilon^{\frac{2}{\kappa}}(V^2 - \epsilon^2 U^2)\big)\epsilon^{1+\frac{2}{\kappa}} UV| \le Cn|\epsilon^{\frac{2}{\kappa}}(V^2 - \epsilon^2 U^2)|^{\kappa-1}\epsilon^{1+\frac{2}{\kappa}}|UV| \le C\epsilon^3 \hat{V}^{2\kappa}.
$$

By (XII.6), (XII.33), and (XII.30), we may assume that $\epsilon_1 > 0$ in Theorem XII.3 is sufficiently small so that for $\epsilon \in (0, \epsilon_1)$ the functions $V(t,\epsilon)$, $U(t,\epsilon)$, $\hat{V}(t)$, $\tilde{V}(t,\epsilon)$ satisfy (XIII.25), pointwise in $t \in \mathbb{R}$. Then, by (XIII.26), one has $|\epsilon^{-2} f - \hat{V}^{2\kappa}| \le C\epsilon^{2\varkappa}\hat{V}^{2\kappa}$; so,

$$
\|\pi_A \circ \big(\mathbf{V} + \hat{V}^{2\kappa}(1 + 2\kappa\Pi_{\boldsymbol{\Xi}})\boldsymbol{\beta}\big) \circ \pi_A\|_{\mathrm{End}(\mathbb{C}^{2N})} \le \|\pi_A \circ \big(\mathbf{V} - \hat{V}^{2\kappa}\big) \circ \pi_A\|_{\mathrm{End}(\mathbb{C}^{2N})}
$$
$$
\le C|\epsilon^{-2} f - \hat{V}^{2\kappa}| + C|\epsilon^{-2} f' \epsilon^{2+\frac{2}{\kappa}} U^2| \le C\epsilon^{2\varkappa}\hat{V}^{2\kappa};
$$

in the first inequality, we also took into account (XIII.15). The above is understood pointwise in $y \in \mathbb{R}^n$; $\hat{V}$ and U are evaluated at $t = |y|$, f and f' are evaluated at $\phi_\omega^* \beta \phi_\omega = V^2 - \epsilon^2 U^2$ and are estimated with the aid of (XIII.2) and (XIII.3).

The bound on $\|\pi_P \circ \big(\mathbf{V}(y,\epsilon) + |u_\kappa(y)|^{2\kappa}(1 + 2\kappa\Pi_{\boldsymbol{\Xi}})\boldsymbol{\beta}\big) \circ \pi_P\|_{\mathrm{End}(\mathbb{C}^{2N})}$ is derived similarly. We have:

$$
\begin{aligned}
&\|\pi_P \circ \big(\mathbf{V} + \hat{V}^{2\kappa}(1 + 2\kappa\Pi_{\boldsymbol{\Xi}})\boldsymbol{\beta}\big) \circ \pi_P\|_{\mathrm{End}(\mathbb{C}^{2N})} \\
&\le \| - \epsilon^{-2} f - \epsilon^{-2} 2(\boldsymbol{\phi}_\omega^* \pi_P \,\cdot\,) f' \pi_P \boldsymbol{\phi}_\omega + \hat{V}^{2\kappa}(1 + 2\kappa\Pi_{\boldsymbol{\Xi}})\|_{\mathrm{End}(\mathbb{C}^{2N})} \\
&\le |\epsilon^{-2} f - \hat{V}^{2\kappa}| + |\epsilon^{-2} 2 f' v^2 - \hat{V}^{2\kappa} 2\kappa| \le C|\epsilon^{-2} f - \hat{V}^{2\kappa}| + C|\epsilon^{-2} f' u^2| \le C\epsilon^{2\varkappa}\hat{V}^{2\kappa}.
\end{aligned}
$$

Above, we used the inequalities (XIII.26), (XIII.27), and (XIII.28), and also took into account that

$$
(\boldsymbol{\phi}_\omega^* \boldsymbol{\beta} \pi_P \,\cdot\,)\pi_P \boldsymbol{\phi}_\omega = ((\pi_P \boldsymbol{\beta} \boldsymbol{\phi}_\omega)^* \,\cdot\,)\pi_P \boldsymbol{\phi}_\omega = ((\pi_P \boldsymbol{\phi}_\omega)^* \,\cdot\,)\pi_P \boldsymbol{\phi}_\omega = v^2 \Pi_{\boldsymbol{\Xi}},
$$

with $\Pi_{\boldsymbol{\Xi}}$ from (XIII.14) and $v = v(|x|,\omega)$ from (XII.22), since

$$
\frac{1}{2}(1 + \beta)\phi_\omega(x) = v(|x|,\omega)\begin{bmatrix}\xi \\ 0\end{bmatrix}, \qquad \text{hence} \quad \pi_P \boldsymbol{\phi}_\omega(x) = v(|x|,\omega)\boldsymbol{\Xi}. \qquad \square
$$

XIII.2.2 The limiting absorption principle for $\mathfrak{l}_-$. We will need the estimates on the resolvent of $\mathfrak{l}_- = \frac{1}{2m} - \frac{\Delta}{2m} - u_\kappa^{2\kappa}$ from (XIII.5) in the spaces with exponential weights.

Lemma XIII.13 *Let $\Omega \subset \mathbb{C}$ be a compact set. Assume that*

$$\Omega \cap \sigma_{\mathrm{d}}(\mathfrak{l}_-) = \emptyset.$$

If $z = 1/(2m)$ is either an eigenvalue or a virtual level of $\mathfrak{l}_-$, assume further that

$$1/(2m) \notin \Omega.$$

Then there is $C = C(\Omega) > 0$ such that the map

$$u_\kappa^\kappa \circ (\mathfrak{l}_- - z)^{-1} \circ u_\kappa^\kappa : \; L^2(\mathbb{R}^n) \to H^2(\mathbb{R}^n), \qquad z \in \mathbb{C} \setminus \sigma(\mathfrak{l}_-)$$

satisfies

$$\|u_\kappa^\kappa \circ (\mathfrak{l}_- - z)^{-1} \circ u_\kappa^\kappa\|_{L^2 \to H^2} \le C, \qquad \forall z \in \Omega \setminus [1/(2m), +\infty).$$

Remark XIII.14 The $L^2_s \to L^2_{-s}$ estimate on the resolvent improves as the spectral parameter goes to infinity, while $L^2_s \to H^2_{-s}$ estimate does not necessarily so (see e.g. Lemma VI.11 with different values of ν); this is why in the above lemma we need to consider the spectral parameter restricted to a compact set $\Omega \subset \mathbb{C}$.

PROOF. The first part (bounds away from an open neighborhood of the threshold $z = 1/(2m)$) follows from [**Agm75**, Appendix A].

If the threshold $z = 1/(2m)$ is a regular point of the essential spectrum of $\mathfrak{l}_-$ (neither an eigenvalue nor a virtual level), then the result follows from [**JK79**] for $n = 3$ and from [**Yaf10**, Proposition 7.4.6] for $n \ge 3$.

The case $n \le 2$, which is left to prove, follows from [**JN01**]. We recall the terminology from that article. Given the operators $H_0 = -\Delta$ and $H = H_0 + V$ in $L^2(\mathbb{R}^n)$, we denote

$$U = \begin{cases} 1, & V \ge 0; \\ -1, & V < 0 \end{cases}; \qquad v = |V|^{1/2}, \qquad w = Uv,$$

$$M(\varsigma) = U + v(H_0 + \varsigma^2)^{-1} v : \; L^2(\mathbb{R}^n) \to L^2(\mathbb{R}^n), \qquad \operatorname{Re} \varsigma > 0.$$

There is the identity $(1 - w(H + \varsigma^2)^{-1} v)(1 + w(H_0 + \varsigma^2)^{-1} v) = 1$; hence, if $M(\varsigma)$ is invertible,

$$1 - w(H+\varsigma^2)^{-1} v = (1 + w(H_0+\varsigma^2)^{-1} v)^{-1} = (U + v(H_0+\varsigma^2)^{-1} v)^{-1} U = M(\varsigma)^{-1} U,$$

$$U - w(H + \varsigma^2)^{-1} w = M(\varsigma)^{-1}, \qquad w(H + \varsigma^2)^{-1} w = U - M(\varsigma)^{-1}$$

(cf. [**JN01**, Equation (4.8)]). In the case at hand, $V = -u_\kappa^{2\kappa}$, $U = -1$, $v = -w = u_\kappa^\kappa$ (understood as operators of multiplication); thus,

$$M(\varsigma) = -1 + u_\kappa^\kappa \circ \Big(-\frac{\Delta}{2m} + \varsigma^2 \Big)^{-1} \circ u_\kappa^\kappa, \tag{XIII.29}$$

and, when $M(\varsigma)$ is invertible,

$$u_\kappa^\kappa \circ (\mathfrak{l}_- - z)^{-1} \circ u_\kappa^\kappa = u_\kappa^\kappa \circ (-\Delta - u_\kappa^{2\kappa} + \varsigma^2)^{-1} \circ u_\kappa^\kappa = -1 - M(\varsigma)^{-1}, \tag{XIII.30}$$

with z and ς related by

$$\frac{1}{2m} - z = \varsigma^2.$$

Using the expression for the integral kernels of $(H_0 + \varsigma^2)^{-1}$ in dimensions $n \le 2$ (see [**JN01**, Section 3, (3.13) and (3.14)]), the operator $M(\varsigma)$ extends by continuity to the region

$$\{\varsigma \in \mathbb{C} \setminus \{0\} \colon \quad \operatorname{Re} \varsigma \ge 0\},$$

with a singularity $-\frac{1}{2\pi}\log(\varsigma)$ at $\varsigma = 0$. If $-\Delta - u_\kappa^{2\kappa}$ has no eigenvalue or virtual level at $\lambda = 0$, then $M(\varsigma)$ is invertible in the orthogonal complement of v (with the inverse bounded uniformly as $\varsigma \to 0$), while $M(\varsigma)/\log(\varsigma)$ is always invertible (with the inverse bounded uniformly as $\varsigma \to 0$) in the span of v. We deduce that, as long as $|\varsigma|$ is small enough and $\varsigma \ne 0$, $M(\varsigma)$ is invertible, with a uniform bound on $\|M(\varsigma)^{-1}\|$ in an open neighborhood of 0 in the half-plane $\operatorname{Re}\varsigma \ge 0$. Now the conclusion of the lemma for the case $n = 2$ follows from (XIII.30). The one-dimensional case is dealt with similarly. □

XIII.3 Bifurcations from the origin

Here we prove Theorem XIII.2.

Lemma XIII.15 *Let $\omega_j \in (0, m)$, $j \in \mathbb{N}$; $\omega_j \to m$. If there are eigenvalues $\lambda_j \in \sigma_{\mathrm{p}}(\mathbf{JL}(\omega_j))$, $\operatorname{Re}\lambda_j \ne 0$, $j \in \mathbb{N}$, such that $\lim\limits_{j\to\infty} \lambda_j = 0$, then the sequence*

$$\Lambda_j := \frac{\lambda_j}{\epsilon_j^2}, \qquad j \in \mathbb{N}$$

does not have the accumulation point at infinity.

Proof. Due to the exponential decay of solitary waves stated in Theorem XII.1, there is $C > 0$ and $s > 1/2$ such that

$$\|\langle r\rangle^{2s}\mathbf{V}(\cdot,\omega)\|_{L^\infty(\mathbb{R}^n,\operatorname{End}(\mathbb{C}^{2N}))} \le C, \qquad \forall \omega \in (0, m). \tag{XIII.31}$$

Let $\mathbf{\Psi}_j \in L^2(\mathbb{R}^n, \mathbb{C}^{2N})$, $j \in \mathbb{N}$ be the eigenfunctions of $\mathbf{JL}(\omega_j)$ corresponding to λ_j; we then have $(\epsilon_j \mathbf{D}_0 + \boldsymbol{\beta} m - \omega_j + \mathbf{J}\lambda_j)\mathbf{\Psi}_j = -\epsilon_j^2 \mathbf{V}(\omega_j)\mathbf{\Psi}_j$ (see (XIII.20)). Applying to this relation $\pi^\pm = (1 \mp \mathrm{i}\mathbf{J})/2$ and denoting $\mathbf{\Psi}_j^\pm = \pi^\pm \mathbf{\Psi}_j$, we arrive at the system

$$\begin{aligned} (\epsilon_j \mathbf{D}_0 + \boldsymbol{\beta} m - \omega_j + \mathrm{i}\lambda_j)\mathbf{\Psi}_j^+ &= -\epsilon_j^2 \pi^+ \mathbf{V}(\omega_j)\mathbf{\Psi}_j, \\ (\epsilon_j \mathbf{D}_0 + \boldsymbol{\beta} m - \omega_j - \mathrm{i}\lambda_j)\mathbf{\Psi}_j^- &= -\epsilon_j^2 \pi^- \mathbf{V}(\omega_j)\mathbf{\Psi}_j. \end{aligned} \tag{XIII.32}$$

Due to $\omega_j \to m$, without loss of generality, we can assume that $\omega_j > m/2$ for all $j \in \mathbb{N}$. Since the spectrum $\sigma(\mathbf{JL})$ is symmetric with respect to real and imaginary axes, we may assume, without loss of generality, that $\operatorname{Im}\lambda_j \ge 0$ for all $j \in \mathbb{N}$, so that $\operatorname{Re}(\mathrm{i}\lambda_j) \le 0$ (see Figure XIII.1). Since $\lambda_j \to 0$, we can also assume that $|\lambda_j| \le m/2$ for all $j \in \mathbb{N}$. With $\epsilon_j \mathbf{D}_0 + \boldsymbol{\beta} m - \omega_j = D_m - \omega_j$ (considered in $L^2(\mathbb{R}^n, \mathbb{C} \otimes_{\mathbb{R}} \mathbb{R}^{2N})$) being self-adjoint, one has

$$\|(D_m - \omega_j - \mathrm{i}\lambda_j)^{-1}\| = \frac{1}{\operatorname{dist}(\mathrm{i}\lambda_j, \sigma(D_m - \omega_j))} = \frac{1}{|m - \omega_j - \mathrm{i}\lambda_j|}, \tag{XIII.33}$$

with $j \in \mathbb{N}$.

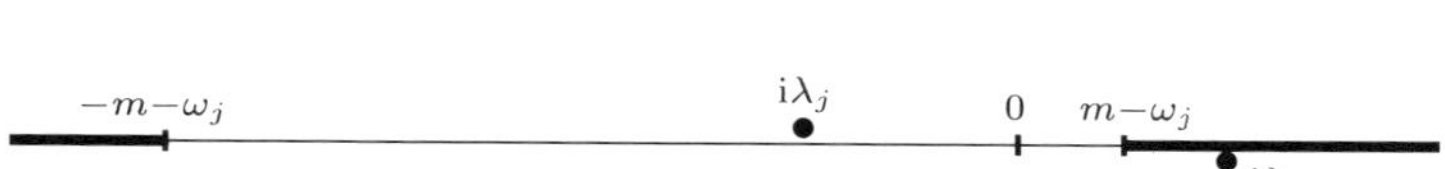

Figure XIII.1. $\sigma(D_m - \omega_j)$ and $\pm\mathrm{i}\lambda_j$.

From (XIII.32) and (XIII.33), using the bound (XIII.31) on $\mathbf{V}$, we obtain (XIII.34)

$$\|\boldsymbol{\Psi}_j^-\|_{L^2} \le \frac{\epsilon_j^2\|\pi^-\mathbf{V}\boldsymbol{\Psi}_j\|}{|m-\omega_j-\mathrm{i}\lambda_j|} \le C\frac{\epsilon_j^2}{|m-\omega_j-\mathrm{i}\lambda_j|}\|\langle r\rangle^{-2s}\boldsymbol{\Psi}_j\|, \quad \forall j\in\mathbb{N}.$$

From (XIII.32) we have:

(XIII.35) $$\begin{bmatrix} m-\omega_j+\mathrm{i}\lambda_j & \epsilon_j\mathbf{D}_0 \\ \epsilon_j\mathbf{D}_0 & -m-\omega_j+\mathrm{i}\lambda_j\end{bmatrix}\begin{bmatrix}\pi_P^+\boldsymbol{\Psi}_j\\ \pi_A^+\boldsymbol{\Psi}_j\end{bmatrix} = -\epsilon_j^2\begin{bmatrix}\pi_P^+\mathbf{V}\boldsymbol{\Psi}_j\\ \pi_A^+\mathbf{V}\boldsymbol{\Psi}_j\end{bmatrix},$$

hence

$$\Big(\frac{(-m-\omega_j+\mathrm{i}\lambda_j)(m-\omega_j+\mathrm{i}\lambda_j)}{\epsilon_j^2}+\Delta\Big)\begin{bmatrix}\pi_P^+\boldsymbol{\Psi}_j\\ \pi_A^+\boldsymbol{\Psi}_j\end{bmatrix}$$
$$=\begin{bmatrix} m+\omega_j-\mathrm{i}\lambda_j & \epsilon_j\mathbf{D}_0 \\ \epsilon_j\mathbf{D}_0 & -(m-\omega_j+\mathrm{i}\lambda_j)\end{bmatrix}\begin{bmatrix}\pi_P^+\mathbf{V}\boldsymbol{\Psi}_j\\ \pi_A^+\mathbf{V}\boldsymbol{\Psi}_j\end{bmatrix}.$$

Denote

(XIII.36) $$\mu_j = \frac{(-m-\omega_j+\mathrm{i}\lambda_j)(m-\omega_j+\mathrm{i}\lambda_j)}{\epsilon_j^2}, \qquad j\in\mathbb{N}.$$

We note that $\operatorname{Im}\mu_j = -2\epsilon_j^{-2}(\omega_j+\operatorname{Im}\lambda_j)\operatorname{Re}\lambda_j \ne 0$ for all $j\in\mathbb{N}$, so that $\mu_j+\Delta$ is invertible. We may assume that $\inf_{j\in\mathbb{N}}|\mu_j|>0$ (or else there would be nothing to prove: if $\mu_j\to 0$, we would have $\lambda_j-\mathrm{i}(m-\omega_j)|=o(\epsilon_j^2)$, hence $|\lambda_j|=O(\epsilon_j^2)$). Then, by the limiting absorption principle (cf. Lemma VI.11),

$$\|\langle r\rangle^{-s}\pi^+\boldsymbol{\Psi}_j\| \le C|\mu_j|^{-1/2}\|\langle r\rangle^{s}\pi^+\mathbf{V}\boldsymbol{\Psi}_j\| + C\epsilon_j\|\langle r\rangle^{s}\pi^+\mathbf{V}\boldsymbol{\Psi}_j\|.$$

The above, together with (XIII.34) and the bound (XIII.31) on $\mathbf{V}$, leads to

$$\|\langle r\rangle^{-s}\boldsymbol{\Psi}_j\| \le \|\langle r\rangle^{-s}\boldsymbol{\Psi}_j^-\|+\|\langle r\rangle^{-s}\boldsymbol{\Psi}_j^+\| \le C\Big(\frac{\epsilon_j^2}{|m-\omega_j-\mathrm{i}\lambda_j|}+\frac{1}{|\mu_j|^{\frac12}}+\epsilon_j\Big)\|\langle r\rangle^{-s}\boldsymbol{\Psi}_j\|.$$

If we had $|\lambda_j|/\epsilon_j^2\to\infty$, then $|m-\omega_j-\mathrm{i}\lambda_j|\ge|\lambda_j|/2$ for j large enough, hence

$$|\mu_j|\ge m|m-\omega_j-\mathrm{i}\lambda_j|/\epsilon_j^2 \ge m|\lambda_j|/(2\epsilon_j^2)$$

for j large enough (since $\omega_j\to m$ and $\lambda_j\to 0$ in (XIII.36)),

$$\|\langle r\rangle^{-s}\boldsymbol{\Psi}_j\| \le C\Big(\frac{\epsilon_j^2}{|\lambda_j|}+\frac{\epsilon_j}{|\lambda_j|^{\frac12}}+\epsilon_j\Big)\|\langle r\rangle^{-s}\boldsymbol{\Psi}_j\|.$$

Due to $|\lambda_j|/\epsilon_j^2\to\infty$, the above relation would lead to a contradiction since $\boldsymbol{\Psi}_j\ne 0$ for all $j\in\mathbb{N}$. We conclude that the sequence $\Lambda_j=\lambda_j/\epsilon_j^2$ can not have an accumulation point at infinity. □

Lemma XIII.16 *For any $\eta\in\mathbb{C}\setminus\overline{\mathbb{R}_+}$ there is $s_0(\eta)\in(0,1)$, lower semicontinuous in η, such that the resolvent $(-\Delta-\eta)^{-1}$ defines a continuous mapping*

$$(-\Delta-\eta)^{-1}:\ L^2_s(\mathbb{R}^n)\to H^2_s(\mathbb{R}^n), \qquad 0\le s<s_0(\eta).$$

PROOF. Let $f\in L^2_s(\mathbb{R}^n)$; define $u=(-\Delta-\eta)^{-1}f\in H^2(\mathbb{R}^n)$. There is the identity

(XIII.37) $$(-\Delta-\eta)(\langle r\rangle^s u)+[\langle r\rangle^s,-\Delta]u=\langle r\rangle^s(-\Delta-\eta)u,$$

which holds in the sense of distributions. Taking into account that

$$\|[\langle r\rangle^s,-\Delta]u\| \le C\|u\|_{H^1}O(s)\le C\|f\|_{L^2_s}O(s),$$

one concludes from (XIII.37) that $(-\Delta-\eta)(\langle r\rangle^s u)\in L^2(\mathbb{R}^n)$ and hence $\langle r\rangle^s u\in L^2(\mathbb{R}^n)$, both being bounded by $C\|f\|_{L^2_s}$, with some $C=C(\eta)>0$, thus so is $\|u\|_{H^2_s}$. □

It is convenient to introduce the following operator acting in $L^2(\mathbb{R}^n,\mathbb{C}^{2N})$:

$$\text{(XIII.38)}\qquad \mathbf{K}=\Big(\frac{1}{2m}-\frac{\Delta}{2m}-u_\kappa^{2\kappa}(1+2\kappa\Pi_{\Xi})\Big)I_{2N},\qquad \mathfrak{D}(\mathbf{K})=H^2(\mathbb{R}^n,\mathbb{C}^{2N}).$$

Above, $\Pi_\Xi\in\operatorname{End}(\mathbb{C}^{2N})$ is the orthogonal projector onto $\Xi\in\mathbb{R}^{2N}$; see (XIII.13), (XIII.14).

Lemma XIII.17 *(1) If $N=2$, then*

$$\sigma(\mathbf{JK}|_{\mathfrak{R}(\pi_P)})=\sigma\Big(\begin{bmatrix}0&\mathfrak{l}_-\\-\mathfrak{l}_+&0\end{bmatrix}\Big).$$

The equality also holds for the point spectra.

If $N\ge 4$, then

$$\sigma(\mathbf{JK}|_{\mathfrak{R}(\pi_P)})=\sigma\Big(\begin{bmatrix}0&\mathfrak{l}_-\\-\mathfrak{l}_+&0\end{bmatrix}\Big)\cup\sigma(\mathrm{i}\mathfrak{l}_-)\cup\sigma(-\mathrm{i}\mathfrak{l}_-).$$

The equality also holds for the point spectra.

(2) One has:

$$\text{(XIII.39)}\qquad \dim\mathfrak{L}(\mathbf{JK}|_{\mathfrak{R}(\pi_P)})=\begin{cases}2n+N, & \kappa\ne 2/n;\\ 2n+N+2, & \kappa=2/n.\end{cases}$$

Above, $\mathbf{K}$ is from (XIII.38) and $\mathfrak{l}_\pm$ were introduced in (XIII.5).

Proof. We decompose $L^2(\mathbb{R}^n,\mathfrak{R}(\pi_P))$ into the direct sum $\mathscr{X}_1\oplus\mathscr{X}_2$, where

$$\text{(XIII.40)}\qquad \begin{aligned}\mathscr{X}_1&=L^2\big(\mathbb{R}^n,\operatorname{Span}\{\Xi,\mathbf{J}\Xi)\big\},\\ \mathscr{X}_2&=L^2\big(\mathbb{R}^n,\big(\operatorname{Span}\{\Xi,\mathbf{J}\Xi\}\big)^\perp\cap\mathfrak{R}(\pi_P)\big).\end{aligned}$$

Note that both Ξ and $\mathbf{J}\Xi$ belong to $\mathfrak{R}(\pi_P)$. The proof of *Part (1)* follows once we notice that $\mathbf{JK}$ is invariant in the spaces $\mathscr{X}_1$ and $\mathscr{X}_2$, and that $\mathbf{JK}|_{\mathscr{X}_1}$ is represented in $L^2\big(\mathbb{R}^n,\operatorname{Span}\{\Xi,\mathbf{J}\Xi\}\big)$ by

$$\begin{bmatrix}0&\mathfrak{l}_-\\-\mathfrak{l}_+&0\end{bmatrix},$$

while $\mathbf{JK}|_{\mathscr{X}_2}$ is represented in $L^2\big(\mathbb{R}^n,\big(\operatorname{Span}\{\Xi,\mathbf{J}\Xi\}\big)^\perp\cap\mathfrak{R}(\pi_P)\big)$ by

$$I_{N/2-1}\otimes_{\mathbb{C}}\begin{bmatrix}0&\mathfrak{l}_-\\-\mathfrak{l}_-&0\end{bmatrix}.$$

We also notice that if $N=2$, then $\mathscr{X}_2=\{0\}$.

The proof of *Part (2)* also follows from the above decomposition and the relations

$$\dim\mathfrak{L}(\mathbf{JK}|_{\mathscr{X}_1})=\dim\mathfrak{L}\Big(\begin{bmatrix}0&\mathfrak{l}_-\\-\mathfrak{l}_+&0\end{bmatrix}\Big)=\begin{cases}2n+2, & \kappa\ne 2/n;\\ 2n+4, & \kappa=2/n\end{cases}$$

(cf. Lemma V.27) and

$$\dim\mathfrak{L}(\mathbf{JK}|_{\mathscr{X}_2})=(N-2)\dim\mathfrak{L}(\mathfrak{l}_-)=(N-2)\dim\mathbf{ker}(\mathfrak{l}_-)=N-2.\qquad\square$$

Remark XIII.18 $\mathbf{JK}|_{\mathfrak{R}(\pi_A)}$ is represented in $L^2\big(\mathbb{R}^n,\mathfrak{R}(\pi_A)\big)$ by $I_{N/2}\otimes_{\mathbb{C}}\begin{bmatrix}0&\mathfrak{l}_-\\-\mathfrak{l}_-&0\end{bmatrix}$.

Since $\sigma(\mathbf{JL})$ is symmetric with respect to real and imaginary axes, we assume without loss of generality that λ_j satisfies

$$\operatorname{Im} \lambda_j \geq 0, \qquad \forall j \in \mathbb{N}. \tag{XIII.41}$$

Passing to a subsequence, we assume that

$$\Lambda_j = \frac{\lambda_j}{\epsilon_j^2} \to \Lambda_0 \in \mathbb{C}. \tag{XIII.42}$$

Lemma XIII.19 *(1) If $\Lambda_0 \notin \sigma(\mathbf{JK})$, then there is $C > 0$ such that*

$$\|\pi_P \mathbf{\Psi}_j\| + \epsilon_j^{-1} \|\pi_A \mathbf{\Psi}_j\| \leq C\epsilon_j^{2\varkappa} \|u_\kappa^\kappa \mathbf{\Psi}_j\|, \qquad \forall j \in \mathbb{N}.$$

(2) For $s > 0$ sufficiently small, there is $C > 0$ such that

$$\|\pi_P^- \mathbf{\Psi}_j\|_{H_s^1} + \epsilon_j^{-1} \|\pi_A^- \mathbf{\Psi}_j\|_{H_s^1} \leq C\|u_\kappa^\kappa \mathbf{\Psi}_j\|, \quad \forall j \in \mathbb{N}.$$

(3) If $\Lambda_0 \notin \mathrm{i}[1/(2m), +\infty)$, then there is $C > 0$ such that

$$\|\pi_P^+ \mathbf{\Psi}_j\|_{H_s^1} + \epsilon_j^{-1} \|\pi_A^+ \mathbf{\Psi}_j\|_{H_s^1} \leq C\|u_\kappa^\kappa \mathbf{\Psi}_j\|, \quad \forall j \in \mathbb{N}.$$

(4) If $\Lambda_0 \in \mathrm{i}[1/(2m), +\infty)$, then there is $C > 0$ such that

$$\|u_\kappa^\kappa \pi_P^+ \mathbf{\Psi}_j\| + \epsilon_j^{-1} \|u_\kappa^\kappa \pi_A^+ \mathbf{\Psi}_j\| \leq C\|u_\kappa^\kappa \mathbf{\Psi}_j\|, \quad \forall j \in \mathbb{N}.$$

PROOF. Let us prove *Part (1)*. We divide (XIII.21), (XIII.23) by ϵ_j^2 and (XIII.22), (XIII.24) by ϵ_j, arriving at

$$\begin{bmatrix} \frac{1}{m+\omega_j} + \mathbf{J}\Lambda_j & \mathbf{D}_0 \\ \mathbf{D}_0 & -m - \omega_j + \epsilon_j^2 \mathbf{J}\Lambda_j \end{bmatrix} \begin{bmatrix} \pi_P \mathbf{\Psi}_j \\ \epsilon_j^{-1} \pi_A \mathbf{\Psi}_j \end{bmatrix} = - \begin{bmatrix} \pi_P \mathbf{V}(y, \epsilon_j) \mathbf{\Psi}_j \\ \epsilon_j \pi_A \mathbf{V}(y, \epsilon_j) \mathbf{\Psi}_j \end{bmatrix},$$

which we rewrite as

$$\begin{bmatrix} \frac{1}{m+\omega_j} - u_\kappa^{2\kappa}(1 + 2\kappa\Pi_\Xi) + \mathbf{J}\Lambda_j & \mathbf{D}_0 \\ \mathbf{D}_0 & -m - \omega_j \end{bmatrix} \begin{bmatrix} \pi_P \mathbf{\Psi}_j \\ \epsilon_j^{-1} \pi_A \mathbf{\Psi}_j \end{bmatrix}$$

$$= - \begin{bmatrix} \pi_P(\mathbf{V}(y, \epsilon_j) + u_\kappa^{2\kappa}(1 + 2\kappa\Pi_\Xi))\mathbf{\Psi}_j \\ \epsilon_j \pi_A(\mathbf{V}(y, \epsilon_j) + \mathbf{J}\Lambda_j)\mathbf{\Psi}_j \end{bmatrix}. \tag{XIII.43}$$

We denote the operator in the left-hand side by A. The Schur complement of the entry $A_{22} = -m - \omega_j$ (see (III.39)) is given by

$$T_j = \frac{1}{m + \omega_j} + \Lambda_j \mathbf{J} - u_\kappa^{2\kappa}(1 + 2\kappa\Pi_\Xi) - \frac{\Delta}{m + \omega_j}.$$

If $\Lambda_j \to \Lambda_0 \notin \sigma(\mathbf{JK})$, then T_j has a bounded inverse for j sufficiently large, and then the operator in the left-hand side of (XIII.43) has a bounded inverse (see (III.39)). Now the proof of *Part (1)* follows from (XIII.43) once we take into account the bounds from Lemma XIII.12.

Let us prove *Part (2)*. We apply $\pi^\pm$ to (XIII.43) and rewrite the result as

$$\begin{bmatrix} \frac{1}{m+\omega_j} \pm \mathrm{i}\Lambda_j & \mathbf{D}_0 \\ \mathbf{D}_0 & -m - \omega_j \end{bmatrix} \begin{bmatrix} \pi_P^\pm \mathbf{\Psi}_j \\ \epsilon_j^{-1} \pi_A^\pm \mathbf{\Psi}_j \end{bmatrix} = - \begin{bmatrix} \pi_P^\pm \mathbf{V}(y, \epsilon_j) \mathbf{\Psi}_j \\ \epsilon_j \pi_A^\pm (\mathbf{V}(y, \epsilon_j) \pm \mathrm{i}\Lambda_j) \mathbf{\Psi}_j \end{bmatrix}. \tag{XIII.44}$$

Denote the matrix-valued operator in the left-hand side of (XIII.44) by $A^\pm$. The Schur complement of the entry $A_{22}^\pm = -m - \omega_j$ (cf. (III.39)) is given by

$$T_j^\pm = A_{11}^\pm - A_{12}^\pm (A_{22}^\pm)^{-1} A_{21}^\pm = \frac{1}{m + \omega_j} \pm \mathrm{i}\Lambda_j - \frac{\Delta}{m + \omega_j}. \tag{XIII.45}$$

Since $\operatorname{Im}\Lambda_0 \geq 0$ (cf. (XIII.41)), T_j^- is invertible in L^2 (except perhaps at finitely many values of j which we disregard); writing the inverse of A^- in terms of T_j^-, we conclude from (XIII.44) that $\|\pi_P^-\mathbf{\Psi}_j\|_{H^1} + \epsilon_j^{-1}\|\pi_A^-\mathbf{\Psi}_j\|_{H^1} \leq C\|\mathbf{V}\mathbf{\Psi}_j\|$. Moreover, by Lemma XIII.16, for sufficiently small $s > 0$,

$$\|\pi_P^-\mathbf{\Psi}_j\|_{H^1_s} + \epsilon_j^{-1}\|\pi_A^-\mathbf{\Psi}_j\|_{H^1_s} \leq C\|\mathbf{V}\mathbf{\Psi}_j\|_{L^2_s} \leq C\|u_\kappa^\kappa\mathbf{\Psi}_j\|, \qquad j \in \mathbb{N}.$$

This proves *Part (2)*. As long as $\Lambda_0 \notin \mathrm{i}[1/(2m), +\infty)$, *Part (3)* is proved in the same way as *Part (2)*.

To prove *Part (4)*, we write

$$\text{(XIII.46)} \quad \begin{bmatrix} \frac{1}{m+\omega_j} + \mathrm{i}\Lambda_j + \mu u_\kappa^{2\kappa} & \mathbf{D}_0 \\ \mathbf{D}_0 & -m-\omega_j \end{bmatrix} \begin{bmatrix} \pi_P^+\mathbf{\Psi}_j \\ \epsilon_j^{-1}\pi_A^+\mathbf{\Psi}_j \end{bmatrix} = - \begin{bmatrix} \pi_P^+(\mathbf{V} - \mu u_\kappa^{2\kappa})\mathbf{\Psi}_j \\ \epsilon_j\pi_A^+(\mathbf{V} + \mathrm{i}\Lambda_j)\mathbf{\Psi}_j \end{bmatrix}.$$

Above, the value of $\mu \geq 0$ is to be specified. Like in (XIII.45), the Schur complement of $-m - \omega_j$ is given by

$$T_j = \frac{1}{m+\omega_j} + \mathrm{i}\Lambda_j + \mu u_\kappa^{2\kappa} - \frac{\Delta}{m+\omega_j}.$$

We pick $\mu \geq 0$ such that the threshold $z = 1/(2m)$ is a regular point of the essential spectrum of the operator $\frac{1}{2m} + \mu u_\kappa^{2\kappa} - \frac{\Delta}{2m}$, which is by **[JN01]** a generic situation; indeed, by (XIII.29), the virtual levels correspond to the case when $M = -1 + \mu\mathcal{K}$ is not invertible, with $\mathcal{K}$ a compact operator. (For $n \geq 3$, enough to take $\mu = 0$ since $-\Delta$ has no virtual level at $z = 0$; for $n \leq 2$, enough to take $\mu > 0$ small **[Sim76]**.) Then, by Lemma XIII.13, $u_\kappa^\kappa \circ T_j^{-1} \circ u_\kappa^\kappa$ (for j large enough) is bounded in L^2 and the conclusion follows from (XIII.46). □

The inclusion $\Lambda_0 \in \sigma(\mathbf{JK})$ immediately follows from Lemma XIII.19 *(1)* which shows that if the sequence Λ_j were to converge to a point away from $\sigma(\mathbf{JK})$, then at most finitely many of $\mathbf{\Psi}_j$ could be different from zero. Together with the results on the spectrum of $\mathbf{JK}$ (cf. Lemma XIII.17), this proves Theorem XIII.2 *(1)*.

Proposition XIII.20 *If $\Lambda_0 = \lim_{j\to\infty} \lambda_j/\epsilon_j^2$ and $\operatorname{Re}\lambda_j \neq 0$ for all $j \in \mathbb{N}$, then*

$$\Lambda_0 \in \sigma_{\mathrm{p}}(\mathbf{JK}) \cap \mathbb{R}.$$

PROOF. From now on, we assume that the corresponding eigenfunctions $\mathbf{\Psi}_j$ (cf. (XIII.20)) are normalized:

$$\text{(XIII.47)} \qquad \|\mathbf{\Psi}_j\|^2 = 1, \qquad j \in \mathbb{N}.$$

As one can readily deduce from (XIII.43), $\epsilon_j^{-1}\mathbf{D}_0\pi_A\mathbf{\Psi}_j$ is bounded in L^2 uniformly in $j \in \mathbb{N}$. By Lemma XIII.19 *(2)*, *(3)* and *(4)*, one can see that so is $u_\kappa^\kappa\epsilon_j^{-1}\pi_A\mathbf{\Psi}_j$. Again by (XIII.43), $u_\kappa^\kappa\mathbf{D}_0\pi_P\mathbf{\Psi}_j$ is uniformly bounded in L^2. It follows that both $\epsilon_j^{-1}\pi_P\mathbf{\Psi}_j$ and $\epsilon_j^{-1}\pi_A\mathbf{\Psi}_j$ belong to $H^1_{\mathrm{loc}}(\mathbb{R}^n, \mathbb{C}^{2N})$ and contain weakly convergent subsequences; we denote their limits by

$$\text{(XIII.48)} \qquad \hat{\mathbf{P}} \in H^1_{\mathrm{loc}}(\mathbb{R}^n, \mathbb{C}^{2N}), \qquad \hat{\mathbf{A}} \in H^1_{\mathrm{loc}}(\mathbb{R}^n, \mathbb{C}^{2N}).$$

Passing to the limit in (XIII.43) and using the bounds from Lemma XIII.12, we arrive at the following system (valid in the sense of distributions):

$$\text{(XIII.49)} \qquad \begin{bmatrix} \frac{1}{2m} - u_\kappa^{2\kappa}(1 + 2\kappa\Pi_{\mathbf{\Xi}}) + \mathbf{J}\Lambda_0 & \mathbf{D}_0 \\ \mathbf{D}_0 & -2m \end{bmatrix} \begin{bmatrix} \hat{\mathbf{P}} \\ \hat{\mathbf{A}} \end{bmatrix} = 0.$$

Let us argue that if $\operatorname{Re}\lambda_j \neq 0$ for all $j \in \mathbb{N}$, then $\begin{bmatrix}\hat{\mathbf{P}}\\ \hat{\mathbf{A}}\end{bmatrix}$ is not identically zero. By Lemma XIII.19 *(2)*, using the compactness of the Sobolev embedding $H^1_s \subset L^2$, we conclude that there is an infinite subsequence (which we again enumerate by $j \in \mathbb{N}$) such that

$$\text{(XIII.50)} \quad \pi_P^-\mathbf{\Psi}_j \to \pi^-\hat{\mathbf{P}} \in H^1(\mathbb{R}^n, \mathbb{C}^{2N}), \qquad \epsilon_j^{-1}\pi_A^-\mathbf{\Psi}_j \to \pi^-\hat{\mathbf{A}} \in H^1(\mathbb{R}^n, \mathbb{C}^{2N})$$

as $j \to \infty$, with the strong convergence in L^2.

Remark XIII.21 If additionally $\Lambda_0 \notin \mathrm{i}[1/(2m), +\infty)$, then T_j^+ from (XIII.45) is also invertible; just like above, one concludes that there is an infinite subsequence (which we again enumerate by $j \in \mathbb{N}$) such that

$$\text{(XIII.51)} \quad \pi_P^+\mathbf{\Psi}_j \to \pi^+\hat{\mathbf{P}} \in H^1(\mathbb{R}^n, \mathbb{C}^{2N}), \qquad \epsilon_j^{-1}\pi_A^+\mathbf{\Psi}_j \to \pi^+\hat{\mathbf{A}} \in H^1(\mathbb{R}^n, \mathbb{C}^{2N})$$

as $j \to \infty$, with the strong convergence in L^2.

Lemma XIII.22 *Let J be skew-symmetric and invertible (as a map onto its image) and L be symmetric linear operators on a Hilbert space $\mathbf{H}$, with $\mathfrak{R}(L) \subset \mathfrak{D}(J)$. That is,*

$$\langle \psi, J\phi\rangle = -\langle J\psi, \phi\rangle \quad \forall \psi,\, \phi \in \mathfrak{D}(J); \qquad \langle \psi, L\phi\rangle = \langle L\psi, \phi\rangle \quad \forall \psi,\, \phi \in \mathfrak{D}(L)$$

If $\mathbf{\Psi} \in \mathbf{H}$ is an eigenvector of JL corresponding to an eigenvalue λ with $\operatorname{Re}\lambda \neq 0$, then $\langle \mathbf{\Psi}, J^{-1}\mathbf{\Psi}\rangle = 0$ and $\langle \mathbf{\Psi}, L\mathbf{\Psi}\rangle = 0$.

Proof. If $\operatorname{Re}\lambda \neq 0$, then both sides of the identity $\langle \mathbf{\Psi}, L\mathbf{\Psi}\rangle = \lambda\langle \mathbf{\Psi}, J^{-1}\mathbf{\Psi}\rangle$ equal zero since $\langle \mathbf{\Psi}, L\mathbf{\Psi}\rangle \in \mathbb{R}$, $\langle \mathbf{\Psi}, J^{-1}\mathbf{\Psi}\rangle \in \mathrm{i}\mathbb{R}$. □

Lemma XIII.23 *If* $\operatorname{Re}\lambda_j \neq 0$ *for all* $j \in \mathbb{N}$, *then* $\hat{\mathbf{P}}^- \neq 0$.

Proof. Lemma XIII.22 yields

$$0 = \langle \mathbf{\Psi}_j, \mathbf{J}\mathbf{\Psi}_j\rangle = \mathrm{i}\|\mathbf{\Psi}_j^+\|^2 - \mathrm{i}\|\mathbf{\Psi}_j^-\|^2, \qquad \forall j \in \mathbb{N};$$

thus, by (XIII.47),

$$\text{(XIII.52)} \qquad \|\mathbf{\Psi}_j^+\|^2 = \|\mathbf{\Psi}_j^-\|^2 = \frac{1}{2}\|\mathbf{\Psi}_j\|^2 = \frac{1}{2}, \qquad \forall j \in \mathbb{N}.$$

Therefore,

$$\text{(XIII.53)} \quad \|\hat{\mathbf{P}}^-\|^2 = \lim_{j\to\infty}\|\pi_P^-\mathbf{\Psi}_j\|^2 = \lim_{j\to\infty}\left(\|\pi_P^-\mathbf{\Psi}_j\|^2 + \|\pi_A^-\mathbf{\Psi}_j\|^2\right) = \frac{1}{2}, \quad \forall j \in \mathbb{N}.$$

Above, in the first two relations, we took into account (XIII.50). □

Thus, $\begin{bmatrix}\hat{\mathbf{P}}\\ \hat{\mathbf{A}}\end{bmatrix} \in L^2(\mathbb{R}^n, \mathbb{C}^{2N} \times \mathbb{C}^{2N})$ is not identically zero, hence $\Lambda_0 \in \sigma_{\mathrm{p}}(\mathbf{JK})$. It remains to prove that

$$\text{(XIII.54)} \qquad \Lambda_0 \in \mathbb{R}.$$

Let us assume that, on the contrary, $\Lambda_0 \in \sigma_{\mathrm{p}}(\mathbf{JK}) \cap (\mathrm{i}\mathbb{R} \setminus \{0\})$. By (XIII.41), it is enough to consider

$$\text{(XIII.55)} \qquad \Lambda_0 = \mathrm{i}a, \qquad a > 0.$$

By (XIII.53), $\|\hat{\mathbf{P}}^-\|^2 = 1/2$. Since

$$\|\hat{\mathbf{P}}^+\|^2 \le \lim_{j\to\infty}\|\pi_P^+\mathbf{\Psi}_j\|^2 \le \lim_{j\to\infty}\left(\|\pi_P^+\mathbf{\Psi}_j\|^2 + \|\pi_A^+\mathbf{\Psi}_j\|^2\right) = 1/2,$$

we arrive at the inequality

$$\|\hat{\mathbf{P}}^+\|^2 - \|\hat{\mathbf{P}}^-\|^2 \leq 0. \tag{XIII.56}$$

From the above and from $\mathbf{JK}\hat{\mathbf{P}} = \mathrm{i}a\hat{\mathbf{P}}$ it follows that

$$\langle \hat{\mathbf{P}}, \mathbf{K}\hat{\mathbf{P}}\rangle = \langle \hat{\mathbf{P}}, -\mathrm{i}a\mathbf{J}\hat{\mathbf{P}}\rangle = a\langle \hat{\mathbf{P}}^+, \hat{\mathbf{P}}^+\rangle - a\langle \hat{\mathbf{P}}^-, \hat{\mathbf{P}}^-\rangle \leq 0. \tag{XIII.57}$$

Remark XIII.24 If Λ_0 belongs to the spectral gap of $\mathbf{JK}$ ($\Lambda_0 \in \mathrm{i}\mathbb{R}$, $|\Lambda_0| < 1/(2m)$), then both $\pi_P^\pm \mathbf{\Psi}_j$ and $\epsilon_j^{-1}\pi_A^\pm\mathbf{\Psi}_j$ (up to choosing a subsequence) converge to

$$\pi^\pm\hat{\mathbf{P}} \in H^1(\mathbb{R}^n, \mathbb{C}^{2N}) \qquad \text{and} \qquad \pi^\pm\hat{\mathbf{A}} \in H^1(\mathbb{R}^n, \mathbb{C}^{2N}),$$

strongly in L^2 (cf. Remark XIII.21). Then, by the above arguments, $\|\hat{\mathbf{P}}^\pm\|^2 = 1/2$ and hence $\langle \hat{\mathbf{P}}, \mathbf{J}\hat{\mathbf{P}}\rangle = 0 = \langle \hat{\mathbf{P}}, \mathbf{K}\hat{\mathbf{P}}\rangle$.

Lemma XIII.25 *If $\Lambda_0 \in \mathrm{i}\mathbb{R}\setminus\{0\}$, $\Lambda_0 \in \sigma_{\mathrm{p}}\Big(\begin{bmatrix} 0 & \mathfrak{l}_- \\ -\mathfrak{l}_+ & 0\end{bmatrix}\Big)$, and $\boldsymbol{z} \in L^2(\mathbb{R}^n, \mathbb{C}^2)$ is a corresponding eigenfunction, then*

$$\Big\langle \boldsymbol{z}, \begin{bmatrix} \mathfrak{l}_+ & 0 \\ 0 & \mathfrak{l}_-\end{bmatrix}\boldsymbol{z}\Big\rangle > 0.$$

Proof. Let $\boldsymbol{z}$ be an eigenfunction which corresponds to the eigenvalue

$$\Lambda_0 \in \sigma_{\mathrm{p}}\Big(\begin{bmatrix} 0 & \mathfrak{l}_- \\ -\mathfrak{l}_+ & 0\end{bmatrix}\Big) \cap \mathrm{i}\mathbb{R}, \qquad \Lambda_0 \neq 0.$$

Let $p,\, q \in L^2(\mathbb{R}^n, \mathbb{C})$ be such that $\boldsymbol{z} = \begin{bmatrix} p \\ \mathrm{i}q\end{bmatrix}$ and let $\Lambda_0 = \mathrm{i}a$ with $a \in \mathbb{R}\setminus\{0\}$. Then $\mathrm{i}a\begin{bmatrix} p \\ \mathrm{i}q\end{bmatrix} = \begin{bmatrix} 0 & \mathfrak{l}_- \\ -\mathfrak{l}_+ & 0\end{bmatrix}\begin{bmatrix} p \\ \mathrm{i}q\end{bmatrix}$ results in $ap = \mathfrak{l}_- q$ and $aq = \mathfrak{l}_+ p$ (note that $q \notin \mathbf{ker}(\mathfrak{l}_-)$; otherwise one would conclude that $p \equiv 0$ and then also $q \equiv 0$, so that $\boldsymbol{z} \equiv 0$, hence not an eigenvector). These relations lead to

$$\langle p, \mathfrak{l}_+ p\rangle = a\langle p, q\rangle = a\overline{\langle q, p\rangle} = \overline{\langle q, ap\rangle} = \overline{\langle q, \mathfrak{l}_- q\rangle} = \langle q, \mathfrak{l}_- q\rangle,$$

hence

$$\Big\langle \begin{bmatrix} p \\ \mathrm{i}q\end{bmatrix}, \begin{bmatrix} \mathfrak{l}_+ & 0 \\ 0 & \mathfrak{l}_-\end{bmatrix}\begin{bmatrix} p \\ \mathrm{i}q\end{bmatrix}\Big\rangle = \langle p, \mathfrak{l}_+ p\rangle + \langle q, \mathfrak{l}_- q\rangle = 2\langle q, \mathfrak{l}_- q\rangle > 0,$$

where we took into account that $\mathfrak{l}_-$ is semi-positive-definite and that $q \notin \mathbf{ker}(\mathfrak{l}_-)$. □

Since $\mathbf{K}$ is invariant in the subspaces $\mathscr{X}_1$ and $\mathscr{X}_2$ defined in (XIII.40), where it is represented by $\begin{bmatrix} \mathfrak{l}_+ & 0 \\ 0 & \mathfrak{l}_-\end{bmatrix}$ and by a positive-definite operator $I_{\frac{N}{2}-1} \otimes_{\mathbb{C}} \begin{bmatrix} \mathfrak{l}_- & 0 \\ 0 & \mathfrak{l}_-\end{bmatrix}$, respectively (see the proof of Lemma XIII.17), it follows from Lemma XIII.25 that the quadratic form

$$\langle\,\cdot\,, \mathbf{K}\,\cdot\,\rangle = \langle\,\cdot\,, \mathbf{K}\,\cdot\,\rangle|_{\mathscr{X}_1} + \langle\,\cdot\,, \mathbf{K}\,\cdot\,\rangle|_{\mathscr{X}_2} = \langle\,\cdot\,, \mathbf{K}\,\cdot\,\rangle|_{\mathscr{X}_1} + \langle\,\cdot\,, (I_{N-2}\otimes \mathfrak{l}_-)\,\cdot\,\rangle|_{\mathscr{X}_2}$$

is strictly positive-definite on any eigenspace of $\mathbf{JK}$ corresponding to $\Lambda_0 = \mathrm{i}a \in \sigma_{\mathrm{p}}(\mathbf{JK})$, $a > 0$. Therefore,

$$\langle \hat{\mathbf{P}}, \mathbf{K}\hat{\mathbf{P}}\rangle > 0. \tag{XIII.58}$$

The relations (XIII.57) and (XIII.58) lead to a contradiction; we conclude that (XIII.54) is satisfied. □

Proposition XIII.20 concludes the proof of Theorem XIII.2 *(2)*.

The case $\Lambda_0 = 0$. Now we turn to Theorem XIII.2 *(3)*, which treats the case $\Lambda_0 = 0$. Let us find the dimension of the spectral subspace of $\mathbf{JL}(\omega)$ corresponding to all eigenvalues which satisfy $|\lambda| = o(\epsilon^2)$.

Proposition XIII.26 *There is $\delta > 0$ sufficiently small and $\epsilon_2 > 0$ such that $\partial\mathbb{D}_{\delta\epsilon^2} \subset \rho(\mathbf{JL})$ for all $\epsilon \in (0, \epsilon_2)$, and for the Riesz projector*

$$P_{\delta,\epsilon} = -\frac{1}{2\pi \mathrm{i}} \oint_{|\eta|=\delta\epsilon^2} \big(\mathbf{JL}(\omega) - \eta\big)^{-1}\, d\eta, \qquad \omega = \sqrt{m^2 - \epsilon^2} \tag{XIII.59}$$

one has $\operatorname{rank} P_{\delta,\epsilon} = 2n + N$ *if* $\kappa \neq 2/n$, *and* $2n + N + 2$ *otherwise. One also has* $\dim \mathbf{ker}(\mathbf{JL}(\omega)) = n + N - 1$.

Remark XIII.27 Let us first give an informal calculation of rank $P_{\delta,\epsilon}$, which is the dimension of the generalized null space of $\mathbf{JL}$. By Lemma IX.26, due to the unitary and translational invariance, the null space is of dimension (at least) $n + 1$, and there is (at least) a 2×2 Jordan block corresponding to each of these null vectors, resulting in $\dim \mathfrak{L}(\mathbf{JL}(\omega)) \geq 2n + 2$. Moreover, the ground states of the nonlinear Dirac equation from Theorems XII.1 and XII.3 have additional degeneracy due to the choice of the direction $\xi \in \mathbb{C}^{N/2}$, $|\xi| = 1$ (cf. (XII.22)). The tangent space to the sphere on which ξ lives is of complex dimension $N/2 - 1$. (Let us point out that the real dimension is $N - 2$, as it should be; we did not expect to have the real dimension $N - 1$ since we have already factored out the action of the unitary group.) Thus,

$$\dim \mathfrak{L}(\mathbf{JL}(\omega)) \geq 2(n + 1) + 2\big(N/2 - 1\big) = 2n + N, \qquad \omega \lessapprox m. \tag{XIII.60}$$

Whether this is a strict inequality, depends on the Kolokolov condition $\partial_\omega Q(\phi_\omega) = 0$ which indicates the jump by 2 in size of the Jordan block corresponding to the unitary symmetry, and on the energy vanishing $E(\phi_\omega) = 0$, which indicates jumps in size of Jordan blocks corresponding to the translational symmetry (see Section IX.6.3).

PROOF OF PROPOSITION XIII.26. Let $\delta > 0$ be such that

$$\overline{\mathbb{D}}_\delta \cap \sigma\left(\begin{bmatrix} 0 & \mathrm{l}_- \\ -\mathrm{l}_+ & 0 \end{bmatrix}\right) = \{0\}.$$

Let us define the operator

$$\mathscr{L}(\omega) = \epsilon^{-2}\mathbf{L}(\omega) = \epsilon^{-1}\mathbf{D}_0 + \epsilon^{-2}(\boldsymbol{\beta} m - \omega) + \mathbf{V}(y, \omega) \tag{XIII.61}$$

(cf. (XIII.19)), where $y = \epsilon x$, $\epsilon = \sqrt{m^2 - \omega^2}$, and $\mathbf{D}_0$ is the Dirac operator in the variables $y = \epsilon x$ (we recall that $\epsilon\mathbf{D}_0 = \epsilon\mathbf{J}\boldsymbol{\alpha} \cdot \nabla_y = \mathbf{J}\boldsymbol{\alpha} \cdot \nabla_x$). We rewrite (XIII.59) as follows:

$$P_{\delta,\epsilon} = -\frac{1}{2\pi \mathrm{i}} \oint_{|\eta|=\delta} (\mathbf{J}\mathscr{L}(\omega) - \eta)^{-1}\, d\eta, \qquad \omega = \sqrt{m^2 - \epsilon^2}.$$

Lemma XIII.28 *Let*

$$p_\delta = -\frac{1}{2\pi \mathrm{i}} \oint_{|\eta|=\delta} (\mathbf{JK} - \eta)^{-1} \pi_P\, d\eta$$

be the Riesz projector onto the generalized null space of $\mathbf{JK}|_{\mathfrak{R}(\pi_P)}$. *Then:*

(1) $$\left\| \begin{bmatrix} \pi_P P_{\delta,\epsilon} \pi_P & \pi_P P_{\delta,\epsilon} \pi_A \\ \pi_A P_{\delta,\epsilon} \pi_P & \pi_A P_{\delta,\epsilon} \pi_A \end{bmatrix} - \begin{bmatrix} p_\delta & 0 \\ 0 & 0 \end{bmatrix} \right\|_{L^2(\mathbb{R}^n, \mathbb{C}^{4N}) \to L^2(\mathbb{R}^n, \mathbb{C}^{4N})} \to 0 \text{ as } \epsilon \to 0;$$

(2) There is $\epsilon_2 > 0$ *such that, for any* $\epsilon \in (0, \epsilon_2)$, *one has* $\operatorname{rank} P_{\delta,\epsilon} = \operatorname{rank} p_\delta$.

PROOF. By Lemma XIII.17,

$$\sigma(\mathbf{JK}) \subset \sigma\Big(\begin{bmatrix} 0 & \mathrm{l}_- \\ -\mathrm{l}_+ & 0 \end{bmatrix}\Big) \cup \sigma(\mathrm{i}\mathrm{l}_-) \cup \sigma(-\mathrm{i}\mathrm{l}_-),$$

hence $(\mathbf{JK} - \eta)|_{\mathfrak{R}(\pi_P)}$ has a bounded inverse

$$(\mathbf{JK} - \eta)^{-1} : H^{-1}(\mathbb{R}^n, \mathfrak{R}(\pi_P)) \to H^1(\mathbb{R}^n, \mathfrak{R}(\pi_P))$$

on the circle $|\eta| = \delta$, $\eta \in \mathbb{C}$, with $\delta > 0$ sufficiently small (cf. Lemma V.28).

On the direct sum $(\mathfrak{R}(\pi_P)) \oplus (\mathfrak{R}(\pi_A))$, the operator $\mathbf{J}\mathscr{L}(\omega) - \eta$ is represented by the matrix

$$\begin{bmatrix} A_{11}(\epsilon,\eta) & A_{12}(\epsilon) \\ A_{21}(\epsilon) & A_{22}(\epsilon,\eta) \end{bmatrix} := \begin{bmatrix} \pi_P \mathbf{J}\mathscr{L}\pi_P - \eta & \pi_P \mathbf{J}\mathscr{L}\pi_A \\ \pi_A \mathbf{J}\mathscr{L}\pi_P & \pi_A \mathbf{J}\mathscr{L}\pi_A - \eta \end{bmatrix}.$$

According to (XIII.61),

$$\|A_{12}(\epsilon)\|_{H^1\to L^2} + \|A_{12}(\epsilon)\|_{L^2\to H^{-1}} + \|A_{21}(\epsilon)\|_{H^1\to L^2} = O(\epsilon^{-1}),$$

$$A_{22}(\epsilon,\eta)^{-1}|_{\mathfrak{R}(\pi_A)} = -\tfrac{\epsilon^2}{2m}\mathbf{J}^{-1} + O_{L^2\to L^2}(\epsilon^4). \tag{XIII.62}$$

In the last equality, we used the following relation (cf. (XIII.61)):

$$A_{22}(\epsilon,\eta) = \pi_A \mathbf{J}\mathscr{L}\pi_A - \eta = -\epsilon^{-2}(m+\omega)\mathbf{J} + \pi_A \mathbf{J}\mathbf{V}(y,\omega)\pi_A - \eta.$$

The Schur complement of $A_{22}(\epsilon,\eta)$ (see (III.39)) is given by

$$\begin{aligned} T(\epsilon,\eta) &= A_{11}(\epsilon,\eta) - A_{12}(\epsilon)A_{22}(\epsilon,\eta)^{-1}A_{21}(\epsilon) \\ &= \pi_P\Big(\frac{\mathbf{J}}{m+\omega} + \mathbf{J}\mathbf{V} - \eta\Big)\pi_P \\ &\quad - \pi_P(\epsilon^{-1}\mathbf{J}\mathbf{D}_0 + \mathbf{J}\mathbf{V})\pi_A A_{22}(\epsilon,\eta)^{-1}\pi_A(\epsilon^{-1}\mathbf{J}\mathbf{D}_0 + \mathbf{J}\mathbf{V})\pi_P, \end{aligned} \tag{XIII.63}$$

which we consider as an operator $T(\epsilon,\eta) : H^1(\mathbb{R}^n, \mathbb{C}^{2N}) \to H^{-1}(\mathbb{R}^n, \mathbb{C}^{2N})$. With the expression (XIII.62) for $A_{22}(\epsilon,\eta)^{-1}$, the Schur complement (XIII.63) takes the form

$$T(\epsilon,\eta) = \pi_P\Big(\frac{\mathbf{J}}{m+\omega} + \mathbf{J}\mathbf{V} - \eta - \frac{\mathbf{J}\Delta}{2m} + O_{H^1\to H^{-1}}(\epsilon^2)\Big)\pi_P. \tag{XIII.64}$$

Using the expression (XIII.64), we can write the inverse of $\mathbf{J}\mathscr{L}(\omega) - \eta$, considered as a map

$$(\mathbf{J}\mathscr{L}(\omega) - \eta)^{-1} : L^2(\mathbb{R}^n, \mathfrak{R}(\pi_P) \oplus \mathfrak{R}(\pi_A)) \to L^2(\mathbb{R}^n, \mathfrak{R}(\pi_P) \oplus \mathfrak{R}(\pi_A)),$$

as follows (see (III.37)):

$$(\mathbf{J}\mathscr{L} - \eta)^{-1} = \begin{bmatrix} T^{-1} & T^{-1}A_{12}A_{22}^{-1} \\ -A_{22}^{-1}A_{21}T^{-1} & A_{22}^{-1} + A_{22}^{-1}A_{21}T^{-1}A_{12}A_{22}^{-1} \end{bmatrix}. \tag{XIII.65}$$

Above, $T = T(\epsilon,\eta)$, $A_{12} = A_{12}(\epsilon)$, $A_{21} = A_{21}(\epsilon)$, $A_{22} = A_{22}(\epsilon,\eta)$. Since

$$\|\big(T(\epsilon,\eta) - (\mathbf{JK} - \eta)\big)|_{\mathfrak{R}(\pi_P)}\|_{H^1\to H^{-1}} = O(\epsilon), \tag{XIII.66}$$

uniformly in $|\eta| = \delta$, while $\mathbf{JK} - \eta : H^1(\mathbb{R}^n, \mathbb{C}^{2N}) \to H^{-1}(\mathbb{R}^n, \mathbb{C}^{2N})$ has a bounded inverse for $|\eta| = \delta$, the operator $T(\epsilon,\eta)|_{\mathfrak{R}(\pi_P)}$ is also invertible for $|\eta| = \delta$ as long as $\epsilon > 0$ is sufficiently small, with its inverse being a bounded map from $H^{-1}(\mathbb{R}^n, \mathfrak{R}(\pi_P))$ to $H^1(\mathbb{R}^n, \mathfrak{R}(\pi_P))$. Using (XIII.62), we conclude that the matrix (XIII.65) has all its entries, except the top left one, of order $O(\epsilon)$ (when considered in the $L^2 \to L^2$ operator norm). Hence, it follows from (XIII.65) and (XIII.66) that, considering $P_{\delta,\epsilon}$ as an operator on $\mathfrak{R}(\pi_P) \oplus \mathfrak{R}(\pi_A)$,

$$\left\| P_{\delta,\epsilon} - \begin{bmatrix} p_\delta & 0 \\ 0 & 0 \end{bmatrix} \right\| = \left\| \frac{1}{2\pi \mathrm{i}} \oint_{|\eta|=\delta} \begin{bmatrix} T(\epsilon,\eta)^{-1} - (\mathbf{JK} - \eta)^{-1} & 0 \\ 0 & 0 \end{bmatrix} d\eta \right\| + O(\epsilon) = O(\epsilon),$$

where the norms refer to $L^2 \to L^2$ operator norm. This proves Lemma XIII.28 *(1)*. The statement *(2)* follows since both $P_{\delta,\epsilon}$ and p_δ are projectors. □

The statement of Proposition XIII.26 on the rank of $P_{\delta,\epsilon}$ follows from Lemma XIII.28 and Lemma XIII.17 *(2)*. The dimension of the kernel of $\mathbf{JL}(\omega)$ follows from considering the rank of the projection onto the neighborhood of the eigenvalue $\lambda = 0$ of the self-adjoint operator $\mathscr{L}$:

$$\hat{P}_{\delta,\epsilon} = -\frac{1}{2\pi \mathrm{i}} \oint_{|\eta|=\delta} (\mathscr{L}(\omega) - \eta)^{-1} \, d\eta, \qquad \omega = \sqrt{m^2 - \epsilon^2},$$

similarly to how it was done for $P_{\delta,\epsilon}$, and from the relation

$$\mathbf{ker}(\mathbf{J}\mathscr{L}(\omega)) = \mathbf{ker}(\mathscr{L}(\omega)) = \mathfrak{R}(\hat{P}_{\delta,\epsilon}), \qquad \epsilon \in (0, \epsilon_2).$$

Above, $\delta > 0$ is small enough so that

$$\overline{\mathbb{D}}_\delta \cap \sigma\Big(\begin{bmatrix} \mathfrak{l}_+ & 0 \\ 0 & \mathfrak{l}_- \end{bmatrix} \Big) = \{0\}.$$

This finishes the proof of Proposition XIII.26. □

Now we return to the proof of Theorem XIII.2 *(3)*. If there is an eigenvalue family $(\lambda_j)_{j\in\mathbb{N}}$, $\lambda_j \in \sigma_{\mathrm{p}}(\mathbf{JL}(\omega_j))$, such that $\Lambda_j \neq 0$ and $\Lambda_j = \frac{\lambda_j}{m^2-\omega_j^2} \to 0$ as $\omega_j \to m$, then the dimension of the generalized kernel of the nonrelativistic limit of the rescaled system jumps up, so that $\dim \mathfrak{L}(\mathbf{JL}(\omega))|_{\omega<m} + 1 \geq 2n + N + 1$, or, taking into account the symmetry of $\sigma(\mathbf{JL}(\omega))$ with respect to reflections relative to the axes $\mathbb{R}$ and $\mathrm{i}\mathbb{R}$, we see that there is at least one more eigenvalue family, hence the dimension of the generalized kernel of the nonrelativistic limit jumps up by at least two:

$$\dim \mathfrak{L}(\mathbf{JL}(\omega))|_{\omega<m} + 2 \geq 2n + N + 2.$$

Comparing this inequality to Lemma XIII.17 *(2)* shows that the assumption $\Lambda_j \neq 0$ for $j \in \mathbb{N}$, $\Lambda_j \to 0$ leads to $\dim \mathfrak{L}\Big(\begin{bmatrix} 0 & \mathfrak{l}_- \\ -\mathfrak{l}_+ & 0 \end{bmatrix} \Big) \geq 2n + 4$. By Lemma V.27, this is only possible in the charge-critical case $\kappa = 2/n$.

Thus, we know that $\kappa = 2/n$. The remaining part of the argument further develops the approach from **[CP03]** to show that there could be no subsequence $\Lambda_j \to 0$ with $\operatorname{Re}\Lambda_j \neq 0$ in the case when $\partial_\omega Q(\phi_\omega) < 0$ for $\omega \lessapprox m$, in a formal agreement with the Kolokolov stability condition **[Kol73]**. We define

$$\text{(XIII.67)} \qquad \begin{aligned} \mathbf{\Phi}(y,\omega) &= \epsilon^{-\frac{1}{\kappa}} \boldsymbol{\phi}_\omega(\epsilon^{-1}y), \\ \boldsymbol{e}_1(y,\omega) &= \epsilon^{-\frac{1}{\kappa}} \mathbf{J}\boldsymbol{\phi}_\omega(\epsilon^{-1}y), \\ \boldsymbol{e}_2(y,\omega) &= \epsilon^{2-\frac{1}{\kappa}} (\partial_\omega \boldsymbol{\phi}_\omega)(\epsilon^{-1}y); \end{aligned}$$

here and below, $\epsilon = \sqrt{m^2 - \omega^2}$. Noting the factor ϵ^{-2} in the definition of $\mathscr{L}$ in (XIII.61), we deduce from (IX.93) and (IX.94) the relations

$$\text{(XIII.68)} \qquad \mathbf{J}\mathscr{L}(\omega)\boldsymbol{e}_1(\omega) = 0, \qquad \mathbf{J}\mathscr{L}(\omega)\boldsymbol{e}_2(\omega) = \boldsymbol{e}_1(\omega), \qquad \omega \in (\omega_1, m).$$

With

$$\theta(y) = -\frac{m}{\kappa}u_\kappa(y) - m\,y\cdot\nabla u_\kappa(y) \qquad \theta \in H^1(\mathbb{R}^n) \tag{XIII.69}$$

and real-valued α, $\beta \in H^2(\mathbb{R}^n)$ such that

$$\mathfrak{l}_+\theta(y) = u_\kappa(y), \qquad \mathfrak{l}_-\alpha(y) = \theta(y), \qquad \mathfrak{l}_+\beta(y) = \alpha(y) \tag{XIII.70}$$

(see (XII.9) and (XII.10)), we define

$$E_3(y) = -\mathbf{J}\Xi\alpha(y), \qquad E_4(y) = -\Xi\beta(y),$$

with $\Xi \in \mathbb{R}^{2N}$ from (XIII.13), so that E_3, $E_4 \in H^2(\mathbb{R}^n, \mathbb{R}^{2N})$ satisfy

$$\mathbf{JK}E_3(y) = \Xi\theta(y), \qquad \mathbf{JK}E_4(y) = E_3(y). \tag{XIII.71}$$

Lemma XIII.29 *Let $\omega_2 = \sqrt{m^2 - \epsilon_2^2}$, with ϵ_2 from Proposition* XIII.26. *The functions*

$$\boldsymbol{e}_a(\omega),\ \ 1 \le a \le 2, \qquad \boldsymbol{e}_4(\omega) = P_{\delta,\epsilon}E_4, \qquad \boldsymbol{e}_3(\omega) = \mathbf{JL}(\omega)\boldsymbol{e}_4(\omega), \qquad \omega \in (\omega_2, m),$$

can be extended to continuous maps $\boldsymbol{e}_a : (\omega_2, m] \to L^2(\mathbb{R}^n, \mathbb{R}^{2N})$, $1 \le a \le 4$, with $\boldsymbol{e}_1(m) = \mathbf{J}\Xi u_\kappa$, $\boldsymbol{e}_2(m) = \Xi\theta$, and $\boldsymbol{e}_a(m) = \lim_{\omega\to m}\boldsymbol{e}_a(\omega) = E_a$, $3 \le a \le 4$, so that

$$\mathbf{JK}\boldsymbol{e}_1(\omega) = 0, \quad \mathbf{JK}\boldsymbol{e}_2(\omega) = \boldsymbol{e}_1(\omega), \quad \omega \in (\omega_2, m];$$

$$\mathbf{JK}\boldsymbol{e}_3(m) = \boldsymbol{e}_2(m), \quad \mathbf{JK}\boldsymbol{e}_4(m) = \boldsymbol{e}_3(m). \tag{XIII.72}$$

PROOF. By Theorem XII.1,

$$\boldsymbol{e}_1(y,\omega) = \epsilon^{-\frac{1}{\kappa}}\mathbf{J}\boldsymbol{\phi}_\omega(\epsilon^{-1}y) = \mathbf{J}\Xi u_\kappa(y) + O_{H^1(\mathbb{R}^n,\mathbb{C}^{2N})}(\epsilon^{2\varkappa}),$$

so $\lim\limits_{\omega\to m}\boldsymbol{e}_1(y,\omega)$ is defined in $H^1(\mathbb{R}^n, \mathbb{C}^{2N})$. Since

$$v(r,\omega) = \epsilon^{1/\kappa}(\hat V(\epsilon r) + \tilde V(\epsilon r, \epsilon)) \quad \text{and} \quad u(r,\omega) = \epsilon^{1+1/\kappa}(\hat U(\epsilon r) + \tilde U(\epsilon r, \epsilon)),$$

with $\hat V$, $\hat U$ from (XII.11), one has $\partial_\omega v(x,\omega) = \frac{\partial\epsilon}{\partial\omega}\partial_\epsilon\big(\epsilon^{\frac{1}{\kappa}}\hat V(\epsilon x) + \epsilon^{\frac{1}{\kappa}}\tilde V(\epsilon x, \epsilon)\big)$, so that

$$\begin{aligned}\epsilon^{2-\frac{1}{\kappa}}&\partial_\omega v(\epsilon^{-1}y,\omega) \\ &= -\omega\Big(\frac{\hat V(y)}{\kappa} + y\cdot\nabla\hat V(y) + \frac{\tilde V(y,\epsilon)}{\kappa} + y\cdot\nabla\tilde V(y,\epsilon) + \epsilon\partial_\epsilon\tilde V(y,\epsilon)\Big).\end{aligned} \tag{XIII.73}$$

Using (XII.33) to bound the $y\cdot\nabla\tilde V$-term, one has

$$\||y|\nabla_y\tilde V(|y|,\epsilon)\|_{L^2(\mathbb{R}^n)} = O(\epsilon^{2\varkappa});$$

due to (XII.38) from Theorem XII.3, $\|\partial_\epsilon\tilde V(\cdot,\epsilon)\|_{H^1(\mathbb{R}^n,\mathbb{R}^2)} = O(\epsilon^{2\varkappa-1})$. Taking into account these estimates in (XIII.73), we arrive at

$$\epsilon^{2-\frac{1}{\kappa}}(\partial_\omega v)(\epsilon^{-1}y,\epsilon) = -\omega\Big(\frac{1}{\kappa}\hat V(y) + y\cdot\nabla\hat V(y)\Big) + O_{L^2(\mathbb{R}^n)}(\epsilon^{2\varkappa}),$$

with a similar expression for $\epsilon^{2-\frac{1}{\kappa}}\partial_\omega u$. This leads to

$$\epsilon^{2-\frac{1}{\kappa}}(\partial_\omega\boldsymbol{\phi}_\omega)(\epsilon^{-1}y) = -\omega\Big(\frac{1}{\kappa}\hat V(y) + y\cdot\nabla\hat V(y)\Big)\Xi + O_{L^2(\mathbb{R}^n)}(\epsilon^{2\varkappa}). \tag{XIII.74}$$

Taking into account that $\boldsymbol{e}_2(y,\omega) = \epsilon^{2-\frac{1}{\kappa}}(\partial_\omega\boldsymbol{\phi}_\omega)(\epsilon^{-1}y)$ (cf. (XIII.67)), the relation (XIII.74) allows us to define

$$\boldsymbol{e}_2(m) := \lim_{\omega\to m}\boldsymbol{e}_2(\omega) = \lim_{\omega\to m}\epsilon^{2-\frac{1}{\kappa}}(\partial_\omega\boldsymbol{\phi}_\omega)(\epsilon^{-1}\,\cdot\,) = \Xi\theta, \tag{XIII.75}$$

with θ from (XIII.69). By (XIII.74), the convergence in (XIII.75) is in $L^2(\mathbb{R}^n, \mathbb{C}^{2N})$.

For $a = 4$, one has:

$$\lim_{\omega\to m} \boldsymbol{e}_4(\omega) = \lim_{\omega\to m} P_{\delta,\epsilon}E_4 = E_4 + \lim_{\omega\to m}(P_{\delta,\epsilon} - p_\delta)E_4 = E_4,$$

with the limit holding in L^2 norm. In the last relation, we used the relation $p_\delta E_a = E_a$, $1 \le a \le 4$, and Lemma XIII.28.

For $a = 3$, the result follows from $\boldsymbol{e}_3(\omega) = \mathbf{JL}(\omega)\boldsymbol{e}_4(\omega) = \mathbf{JL}(\omega)P_{\delta,\omega}\boldsymbol{e}_4(\omega)$ since $\mathbf{JL}(\omega)P_{\delta,\omega}$ is a bounded operator.

We also point out that not only $\boldsymbol{e}_1(\omega)$ and $\boldsymbol{e}_2(\omega)$, but also $\boldsymbol{e}_3(\omega)$ and $\boldsymbol{e}_4(\omega)$ are real-valued; this follows from the observation that $E_4 \in L^2(\mathbb{R}^n, \mathbb{R}^{2N})$ is real-valued, while $P_{\delta,\omega}$ commutes with the operator $\boldsymbol{K} : \mathbb{C}^{2N} \to \mathbb{C}^{2N}$ of complex conjugation since $\mathbf{JL}$ has real coefficients. □

In the vector space $\mathfrak{R}(P_{\delta,\epsilon})$ we may choose the basis (cf. Lemma IX.26)

$$\text{(XIII.76)}\qquad \big\{\boldsymbol{e}_a(\omega),\ 1 \le a \le 4;\quad \partial_{y^i}\boldsymbol{\Phi},\ \ \boldsymbol{\alpha}^i\boldsymbol{\Phi} - 2\omega y^i\mathbf{J}\boldsymbol{\Phi},\ \ 1 \le i \le n;\\ \boldsymbol{\Theta}_k,\ 1 \le k \le N-2\big\},$$

where $\boldsymbol{\Phi}(\omega)$ is from (XIII.67) and $\boldsymbol{\Theta}_k(\omega)$ are certain vectors from $\mathbf{ker}(\mathbf{J}\mathscr{L}(\omega))$, with $1 \le k \le N-2$ due to Proposition XIII.26 which states that

$$\operatorname{rank} P_{\delta,\epsilon} = 2n + N + 2, \qquad \dim \mathbf{ker}(\mathbf{J}\mathscr{L}(\omega)|_{P_{\delta,\epsilon}}) = n + N - 1.$$

Remark XIII.30 When $n = 3$ and $N = 4$, there are three vectors $\boldsymbol{\Theta}_k(\omega)$ corresponding to infinitesimal rotations around three coordinate axes, but, as it was mentioned in Section IX.3.6, the span of these vectors, $\operatorname{span}\{\boldsymbol{\Theta}_k;\ 1 \le k \le 3\}$, turns out to contain the null eigenvector $\boldsymbol{e}_1(\omega)$.

In the basis (XIII.76) of the space $\mathfrak{R}(P_{\delta,\epsilon})$, the operator $(\mathbf{J}\mathscr{L}(\omega) - \lambda I_N)|_{\mathfrak{R}(P_{\delta,\epsilon})}$ is represented by

$$\text{(XIII.77)}\quad M_\omega - \lambda I_N = \begin{bmatrix} -\lambda & 1 & \sigma_1(\omega) & 0 & 0 & 0 & 0 \\ 0 & -\lambda & \sigma_2(\omega) & 0 & 0 & 0 & 0 \\ 0 & 0 & \sigma_3(\omega) - \lambda & 1 & 0 & 0 & 0 \\ 0 & 0 & \sigma_4(\omega) & -\lambda & 0 & 0 & 0 \\ \vdots & \vdots & \vdots & \vdots & -\lambda I_n & I_n & 0 \\ \vdots & \vdots & \vdots & \vdots & 0 & -\lambda I_n & 0 \\ \vdots & \vdots & \vdots & \vdots & 0 & 0 & -\lambda I_{N-2} \end{bmatrix},$$

where vertical dots denote columns of irrelevant coefficients and

$$\sigma_a(\omega), \qquad 1 \le a \le 4,$$

are some continuous functions of ω. We used (XIII.72). Considering (XIII.77) at $\lambda = 0$ and $\epsilon = 0$, one concludes from (XIII.72) that

$$\text{(XIII.78)}\qquad \sigma_1(m) = \sigma_3(m) = \sigma_4(m) = 0, \qquad \sigma_2(m) = 1.$$

From (XIII.77), we also have

$$\text{(XIII.79)}\qquad \det(M_\omega - \lambda) = (-\lambda)^{2n+N}(\lambda^2 - \lambda\sigma_3(\omega) - \sigma_4(\omega)).$$

Expanding $\mathbf{J}\mathscr{L}\boldsymbol{e}_3(\omega)$ over the basis in $\mathfrak{R}(P_{\delta,\epsilon})$ (see (XIII.76)), we conclude that for some continuous functions $\gamma_i(\omega)$ and $\rho_i(\omega)$, $1 \le i \le n$, and $\tau_k(\omega)$, $1 \le k \le N-2$, there is a relation

$$\text{(XIII.80)}\quad \mathbf{J}\mathscr{L}\boldsymbol{e}_3=\sum_{a=1}^{4}\sigma_a\boldsymbol{e}_a+\sum_{i=1}^{n}\Big(\gamma_i\partial_{y^i}\boldsymbol{\Phi}+\rho_i(\boldsymbol{\alpha}^i\boldsymbol{\Phi}-2\omega y^i\mathbf{J}\boldsymbol{\Phi})\Big)+\sum_{k=1}^{N-2}\tau_k\boldsymbol{\Theta}_k,$$

valid for $\omega\in(\omega_2,m]$. Above, $\sigma_a(\omega)$, $1\le a\le 4$, are continuous functions from (XIII.77). Pairing (XIII.80) with $\boldsymbol{\Phi}=\mathbf{J}^{-1}\boldsymbol{e}_1(\omega)$, we get:

$$\text{(XIII.81)}\ \ 0=\sigma_2(\omega)\left\langle\mathbf{J}^{-1}\boldsymbol{e}_1(\omega),\boldsymbol{e}_2(\omega)\right\rangle+\sigma_4(\omega)\left\langle\mathbf{J}^{-1}\boldsymbol{e}_1(\omega),\boldsymbol{e}_4(\omega)\right\rangle,\quad \omega\in(\omega_2,m].$$

We took into account that one has $\langle\boldsymbol{\Phi},\boldsymbol{v}\rangle=\langle\mathscr{L}\boldsymbol{e}_2,\boldsymbol{v}\rangle=\langle\boldsymbol{e}_2,\mathscr{L}\boldsymbol{v}\rangle=0$ for any $\boldsymbol{v}\in\mathbf{ker}(\mathbf{J}\mathscr{L})$, the identities

$$\left\langle\mathbf{J}^{-1}\boldsymbol{e}_1,\mathbf{J}\mathscr{L}\boldsymbol{e}_3(\omega)\right\rangle=-\left\langle\mathscr{L}\boldsymbol{e}_1,\boldsymbol{e}_3(\omega)\right\rangle=0,$$

$$\langle\mathbf{J}^{-1}\boldsymbol{e}_1,\boldsymbol{e}_3\rangle=\langle\mathbf{J}^{-1}\boldsymbol{e}_1,\mathbf{J}\mathscr{L}\boldsymbol{e}_4\rangle=-\langle\mathscr{L}\boldsymbol{e}_1,\boldsymbol{e}_4\rangle=0,$$

and also the identity $\langle\boldsymbol{\Phi},\boldsymbol{\alpha}^i\boldsymbol{\Phi}-2\omega y^i\mathbf{J}\boldsymbol{\Phi}\rangle=0$ which holds due to Lemma IX.14 and due to $\boldsymbol{\Phi}^*\mathbf{J}\boldsymbol{\Phi}\equiv 0$ (the left-hand side is skew-adjoint while all the quantities are real-valued). Since

$$\begin{aligned}\langle\mathbf{J}^{-1}\boldsymbol{e}_1(\omega),\boldsymbol{e}_2(\omega)\rangle &= \epsilon^{-\frac{2}{\kappa}}\langle\boldsymbol{\phi}_\omega(\epsilon^{-1}\cdot),(\partial_\omega\boldsymbol{\phi}_\omega)(\epsilon^{-1}\cdot)\rangle\\ &= \epsilon^{n-\frac{2}{\kappa}}\langle\boldsymbol{\phi}_\omega,\partial_\omega\boldsymbol{\phi}_\omega\rangle=\partial_\omega Q(\phi_\omega)/2\end{aligned}\qquad\text{(XIII.82)}$$

(we took into account that $n-\frac{2}{\kappa}=0$), the relation (XIII.81) takes the form

$$\text{(XIII.83)}\qquad \sigma_2(\omega)\partial_\omega Q(\phi_\omega)/2=\mu(\omega)\sigma_4(\omega),\qquad \omega\in(\omega_2,m],$$

where $\mu(\omega):=-\langle\mathbf{J}^{-1}\boldsymbol{e}_1(\omega),\boldsymbol{e}_4(\omega)\rangle$ is a continuous function of $\omega\in(\omega_2,m]$.

Remark XIII.31 By (XIII.82) and Lemma XIII.29, $\partial_\omega Q(\phi_\omega)$ is a continuous function of $\omega\in(\omega_2,m]$.

Lemma XIII.32 *There is $\omega_3\in(\omega_2,m)$ such that $\mu(\omega)>0$ for $\omega_3<\omega\le m$.*

PROOF. We have

$$\mu(\omega)=-\langle\boldsymbol{\Phi},\boldsymbol{e}_4\rangle=-\langle\boldsymbol{\Phi},P_{\delta,\epsilon}(\boldsymbol{e}_4)\rangle=-\langle\mathbf{J}^{-1}\boldsymbol{e}_1(m),\boldsymbol{e}_4(m)\rangle+O(\epsilon),$$

while (XIII.72) yields

$$-\langle\mathbf{J}^{-1}\boldsymbol{e}_1,\boldsymbol{e}_4\rangle|_{\omega=m}=-\langle\mathbf{K}\boldsymbol{e}_2,\boldsymbol{e}_4\rangle|_{\omega=m}=-\langle\mathbf{J}\mathbf{K}\boldsymbol{e}_3,\mathbf{J}^{-1}\boldsymbol{e}_3\rangle|_{\omega=m}=\left\langle\Xi\alpha,\mathbf{K}\Xi\alpha\right\rangle>0.$$

Above, we used (XIII.71) and the explicit form of E_2 and E_3. □

Lemma XIII.33 *There is $\omega_4\in[\omega_3,m)$ such that $\sigma_3(\omega)\equiv 0$ for $\omega\in[\omega_4,m]$.*

PROOF. Applying $(\mathbf{J}\mathscr{L}(\omega))^2$ to (XIII.80), we get

$$(\mathbf{J}\mathscr{L})^3\boldsymbol{e}_3(\omega)=\sigma_3(\omega)(\mathbf{J}\mathscr{L})^2\boldsymbol{e}_3(\omega)+\sigma_4(\omega)(\mathbf{J}\mathscr{L})^2\boldsymbol{e}_4(\omega).$$

Coupling this relation with $\mathbf{J}^{-1}\boldsymbol{e}_4$ and using the identities

$$\langle\mathbf{J}^{-1}\boldsymbol{e}_4,(\mathbf{J}\mathscr{L})^3\boldsymbol{e}_3\rangle=\langle\boldsymbol{e}_3,\mathscr{L}\mathbf{J}\mathscr{L}\boldsymbol{e}_3\rangle=0$$

and

$$\langle\mathbf{J}^{-1}\boldsymbol{e}_4,(\mathbf{J}\mathscr{L})^2\boldsymbol{e}_4\rangle=-\langle\boldsymbol{e}_4,\mathscr{L}\mathbf{J}\mathscr{L}\boldsymbol{e}_4\rangle=0$$

(both of these due to skew-adjointness of $\mathscr{L}\mathbf{J}\mathscr{L}$, taking into account that $\boldsymbol{e}_a(\omega)$, $1\le a\le 4$, are real-valued by Lemma XIII.29, while $\mathbf{J}$ and $\mathscr{L}$ have real coefficients), we have

$$\text{(XIII.84)}\qquad \sigma_3(\omega)\langle\mathbf{J}^{-1}\boldsymbol{e}_4,(\mathbf{J}\mathscr{L})^2\boldsymbol{e}_3\rangle=0.$$

The factor at $\sigma_3(\omega)$ is nonzero for $\omega < m$ sufficiently close to m. Indeed, using (XIII.78),

$$\langle \mathbf{J}^{-1}\boldsymbol{e}_4, (\mathbf{J}\mathscr{L})^2\boldsymbol{e}_3\rangle\big|_{\omega=m} = \langle \mathbf{J}^{-1}\boldsymbol{e}_4, \sigma_2\boldsymbol{e}_1 + \sigma_3\mathbf{J}\mathscr{L}\boldsymbol{e}_3 + \sigma_4\boldsymbol{e}_3\rangle\big|_{\omega=m}$$
$$= \langle \mathbf{J}^{-1}\boldsymbol{e}_4, \boldsymbol{e}_1\rangle\big|_{\omega=m} = -\langle \boldsymbol{e}_4, \boldsymbol{\phi}_\omega\rangle\big|_{\omega=m},$$

which is positive due to Lemma XIII.32. Due to continuity in ω of the coefficient at $\sigma_3(\omega)$ in (XIII.84), we conclude that $\sigma_3(\omega)$ is identically zero for $\omega \in [\omega_4, m]$, with some $\omega_4 < m$. □

Since $\sigma_3(\omega)$ is identically zero for $\omega \in [\omega_4, m]$, we conclude from (XIII.79) that the nonzero eigenvalues of $\mathbf{J}\mathscr{L}(\omega)$ satisfy $\lambda^2 - \sigma_4(\omega) = 0$, $\omega \in [\omega_4, m]$. By (XIII.78) and Lemma XIII.32, the relation (XIII.83) shows that $\sigma_4(\omega)$ is of the same sign as $\partial_\omega Q(\phi_\omega)$. Thus, if $\partial_\omega Q(\phi_\omega) > 0$ for $\omega \lessapprox m$, then for these values of ω there are two nonzero real eigenvalues of $\mathbf{J}\mathscr{L}(\omega)$, one positive (indicating the linear instability) and one negative, both of magnitude $\sim \sqrt{\partial_\omega Q(\phi_\omega)}$ for $\omega \lessapprox m$; hence, there are two real eigenvalues of $\mathbf{JL}$, of magnitude $\sim \epsilon^2\sqrt{\partial_\omega Q(\phi_\omega)}$. This completes the proof of Theorem XIII.2.

XIII.4 Bifurcations from embedded thresholds

XIII.4.1 Absence of bifurcations from the essential spectrum. Now we proceed to the proof of Theorem XIII.4 *(1)* and *(2)*: we need to prove that the sequence (XIII.6),

$$Z_j = -\frac{2\omega_j + \mathrm{i}\lambda_j}{\epsilon_j^2} \in \mathbb{C}, \qquad j \in \mathbb{N}, \tag{XIII.85}$$

can only accumulate to either the discrete spectrum or the threshold of the operator $\mathfrak{l}_-$ from (XIII.5). Moreover, we will see that the accumulation to the threshold is not possible when it is a regular point of the essential spectrum of $\mathfrak{l}_-$ (neither an L^2-eigenvalue nor a virtual level).

Lemma XIII.34 *There is $C > 0$ such that*

$$\|u_\kappa^\kappa \pi^+ \boldsymbol{\Psi}_j\|_{L^2} \le C\epsilon_j \|u_\kappa^\kappa \boldsymbol{\Psi}_j\|_{L^2}, \qquad \forall j \in \mathbb{N}.$$

Proof. The relations (XIII.23) and (XIII.24) yield

$$\begin{bmatrix} m - \omega_j + \mathrm{i}\lambda_j & \epsilon_j \mathbf{D}_0 \\ \epsilon_j \mathbf{D}_0 & -(m + \omega_j - \mathrm{i}\lambda_j)\end{bmatrix}\begin{bmatrix}\pi_P^+ \boldsymbol{\Psi}_j \\ \pi_A^+ \boldsymbol{\Psi}_j\end{bmatrix} = -\epsilon_j^2 \begin{bmatrix}\pi_P^+ \mathbf{V}\boldsymbol{\Psi}_j \\ \pi_A^+ \mathbf{V}\boldsymbol{\Psi}_j\end{bmatrix},$$

hence

$$\begin{bmatrix}\pi_P^+ \boldsymbol{\Psi}_j \\ \pi_A^+ \boldsymbol{\Psi}_j\end{bmatrix} = \begin{bmatrix} m + \omega_j - \mathrm{i}\lambda_j & \epsilon_j \mathbf{D}_0 \\ \epsilon_j \mathbf{D}_0 & -(m - \omega_j + \mathrm{i}\lambda_j)\end{bmatrix}\left(\Delta_y + \zeta_j^2\right)^{-1}\begin{bmatrix}\pi_P^+ \mathbf{V}\boldsymbol{\Psi}_j \\ \pi_A^+ \mathbf{V}\boldsymbol{\Psi}_j\end{bmatrix}, \tag{XIII.86}$$

with

$$\zeta_j^2 := \frac{(\omega_j - \mathrm{i}\lambda_j)^2 - m^2}{\epsilon_j^2} = \frac{8m^2 + o(1)}{\epsilon_j^2}, \qquad \operatorname{Re}\zeta_j \ge 0; \qquad j \in \mathbb{N}. \tag{XIII.87}$$

(Note that since $\omega_j \to m$, $\operatorname{Re}\lambda_j \ne 0$, and $\lambda_j \to 2m\mathrm{i}$, one has $\operatorname{Im}\zeta_j^2 \ne 0$ for all but finitely many $j \in \mathbb{N}$ which we discard; then ζ_j^2 for all $j \in \mathbb{N}$ is in the resolvent set of

$-\Delta$.) The limiting absorption principle for the Laplace operator from Lemma VI.11 with $\nu = 0$, 1 and with $z = \zeta_j^2$ shows that there is $C > 0$ such that

$$\|u_\kappa^\kappa \circ (\Delta + \zeta_j^2)^{-1} \circ u_\kappa^\kappa\| \le C|\zeta_j|^{-1}, \qquad \forall j \in \mathbb{N}, \tag{XIII.88}$$

$$\|u_\kappa^\kappa \circ \epsilon_j \mathbf{D}_0 (\Delta + \zeta_j^2)^{-1} \circ u_\kappa^\kappa\| \le C\epsilon_j, \qquad \forall j \in \mathbb{N}, \tag{XIII.89}$$

where $u_\kappa^\kappa \circ \ldots \circ u_\kappa^\kappa$ denotes the compositions with the operators of multiplication by u_κ^κ. Applying (XIII.88) and (XIII.89) to (XIII.86) leads to

$$\begin{aligned}\|u_\kappa^\kappa \pi^+ \mathbf{\Psi}_j\| &\le C\big(\|u_\kappa^\kappa \epsilon_j \mathbf{D}_0 (\Delta_y + \zeta_j^2)^{-1} \pi^+ \mathbf{V}\mathbf{\Psi}_j\| + \|u_\kappa^\kappa (\Delta_y + \zeta_j^2)^{-1} \pi^+ \mathbf{V}\mathbf{\Psi}_j\|\big) \\ &\le C\big(\|u_\kappa^\kappa \circ \epsilon_j \mathbf{D}_0 (\Delta_y + \zeta_j^2)^{-1} \circ u_\kappa^\kappa\| + \|u_\kappa^\kappa \circ (\Delta_y + \zeta_j^2)^{-1} \circ u_\kappa^\kappa\|\big) \|u_\kappa^\kappa \mathbf{\Psi}_j\| \\ &\le C\epsilon_j \|u_\kappa^\kappa \mathbf{\Psi}_j\|, \qquad \forall j \in \mathbb{N}.\end{aligned}$$

We used the bound $\|\mathbf{V}(y, \epsilon)\|_{\mathrm{End}(\mathbb{C}^{2N})} \le C|u_\kappa(y)|^{2\kappa}$ from Lemma XIII.12. □

Lemma XIII.35 *The values Z_j, $j \in \mathbb{N}$ from* (XIII.85) *are uniformly bounded: There is $C > 0$ such that*

$$|Z_j| \le C, \qquad \forall j \in \mathbb{N}.$$

In other words, the sequence $\big(Z_j\big)_{j \in \mathbb{N}}$ has no accumulation point at infinity.

PROOF. Let us consider the values of $j \in \mathbb{N}$ for which the following inequality is satisfied:

$$|(\omega_j + \mathrm{i}\lambda_j)^2 - m^2| < \epsilon_j^2. \tag{XIII.90}$$

This implies that $(\omega_j + \mathrm{i}\lambda_j)^2 = m^2 + O(\epsilon_j^2)$, hence

$$-\omega_j - \mathrm{i}\lambda_j = m + O(\epsilon_j^2)$$

(since $\lambda_j \to -2m\mathrm{i}$ as $\omega_j \to m$); taking into account the relation $\epsilon_j^2 Z_j = -2\omega_j - \mathrm{i}\lambda_j$ (see (XIII.85)), we arrive at

$$\epsilon_j^2 Z_j = m - \omega_j + O(\epsilon_j^2) = O(\epsilon_j^2),$$

which shows that $|Z_j|$ are uniformly bounded for the values of j for which (XIII.90) takes place. Now we discard these values of $j \in \mathbb{N}$; assuming that there are infinitely many left (or else there is nothing to prove), we have:

$$|(\omega_j + \mathrm{i}\lambda_j)^2 - m^2| \ge \epsilon_j^2, \qquad \forall j \in \mathbb{N}. \tag{XIII.91}$$

We write (XIII.21), (XIII.22) as the following system:

$$\begin{bmatrix} m - \omega_j - \mathrm{i}\lambda_j & \epsilon_j \mathbf{D}_0 \\ \epsilon_j \mathbf{D}_0 & -(m + \omega_j + \mathrm{i}\lambda_j) \end{bmatrix} \begin{bmatrix} \pi_P^- \mathbf{\Psi}_j \\ \pi_A^- \mathbf{\Psi}_j \end{bmatrix} = -\epsilon_j^2 \begin{bmatrix} \pi_P^- \mathbf{V}\mathbf{\Psi}_j \\ \pi_A^- \mathbf{V}\mathbf{\Psi}_j \end{bmatrix}, \tag{XIII.92}$$

for all $j \in \mathbb{N}$, which can then be rewritten as follows:

$$\begin{bmatrix} \pi_P^- \mathbf{\Psi}_j \\ \pi_A^- \mathbf{\Psi}_j \end{bmatrix} = \begin{bmatrix} m + \omega_j + \mathrm{i}\lambda_j & \epsilon_j \mathbf{D}_0 \\ \epsilon_j \mathbf{D}_0 & -(m - \omega_j - \mathrm{i}\lambda_j) \end{bmatrix} \Big(\Delta_y + \mathrm{v}_j\Big)^{-1} \begin{bmatrix} \pi_P^- \mathbf{V}\mathbf{\Psi}_j \\ \pi_A^- \mathbf{V}\mathbf{\Psi}_j \end{bmatrix}, \tag{XIII.93}$$

with

$$\mathrm{v}_j := \frac{(\omega_j + \mathrm{i}\lambda_j)^2 - m^2}{\epsilon_j^2}, \qquad j \in \mathbb{N}. \tag{XIII.94}$$

We notice that $|\nu_j| \geq 1$ by (XIII.91) and that $\operatorname{Im} \nu_j \neq 0$ except perhaps for finitely many values of j, which we discard. Applying (XIII.88) and (XIII.89) to (XIII.93), we derive:

$$\text{(XIII.95)} \qquad \|u_\kappa^\kappa \pi^- \boldsymbol{\Psi}_j\| \leq C\big(\epsilon_j + |\nu_j|^{-1/2}\big)\|u_\kappa^\kappa \boldsymbol{\Psi}_j\|, \qquad \forall j \in \mathbb{N}.$$

By Lemma XIII.34 and (XIII.95), there is $C > 0$ such that

$$\text{(XIII.96)} \qquad \|u_\kappa^\kappa \boldsymbol{\Psi}_j\| \leq \|u_\kappa^\kappa \pi^+ \boldsymbol{\Psi}_j\| + \|u_\kappa^\kappa \pi^- \boldsymbol{\Psi}_j\| \leq C\big(\epsilon_j + |\nu_j|^{-1/2}\big)\|u_\kappa^\kappa \boldsymbol{\Psi}_j\|,$$

for all $j \in \mathbb{N}$. Assume that $\limsup_{j\to\infty} |\nu_j| = +\infty$. Then the coefficient at $\|u_\kappa^\kappa \boldsymbol{\Psi}_j\|$ in the right-hand side of (XIII.96) would go to zero for an infinite subsequence of $j \to \infty$; since $\boldsymbol{\Psi}_j \not\equiv 0$, we arrive at the contradiction.

Thus, $|\nu_j|$, $j \in \mathbb{N}$, are uniformly bounded. From (XIII.94), we derive:

$$\omega_j + \mathrm{i}\lambda_j = -\sqrt{m^2 + \epsilon_j^2 \nu_j} = -m + O(\epsilon_j^2),$$

where we chose the "positive" branch of the square root (for all but finitely many $j \in \mathbb{N}$) since $\omega_j \to m$ and $\lambda_j \to 2m\mathrm{i}$ as $j \to \infty$. This yields

$$Z_j = -\frac{2\omega_j + \mathrm{i}\lambda_j}{\epsilon_j^2} = -\frac{\omega_j + (\omega_j + \mathrm{i}\lambda_j)}{\epsilon_j^2} = -\frac{\sqrt{m^2 - \epsilon_j^2} - \sqrt{m^2 + \epsilon_j^2 \nu_j}}{\epsilon_j^2}, \qquad j \in \mathbb{N};$$

therefore, $Z_j = O(1)$ are uniformly bounded and could not accumulate at infinity. □

Substituting

$$-\frac{m + \omega_j + \mathrm{i}\lambda_j}{\epsilon_j^2} = -\frac{2\omega_j + \mathrm{i}\lambda_j}{\epsilon_j^2} - \frac{m - \omega_j}{\epsilon_j^2} = Z_j - \frac{1}{m + \omega_j},$$

we rewrite (XIII.92) as

(XIII.97)

$$\begin{bmatrix} m - \omega_j - \mathrm{i}\lambda_j & \mathbf{D}_0 \\ \mathbf{D}_0 & Z_j - \frac{1}{m+\omega_j} + u_\kappa^{2\kappa} \end{bmatrix} \begin{bmatrix} \pi_P^- \boldsymbol{\Psi}_j \\ \epsilon_j \pi_A^- \boldsymbol{\Psi}_j \end{bmatrix} = -\begin{bmatrix} \epsilon_j^2 \pi_P^- \mathbf{V} \boldsymbol{\Psi}_j \\ \epsilon_j \pi_A^- \circ \big(\mathbf{V} - u_\kappa^{2\kappa}\big) \boldsymbol{\Psi}_j \end{bmatrix}.$$

We denote the matrix-valued operator in the left-hand side of (XIII.97) by

$$\text{(XIII.98)} \qquad A := \begin{bmatrix} m - \omega_j - \mathrm{i}\lambda_j & \mathbf{D}_0 \\ \mathbf{D}_0 & Z_j - \frac{1}{m+\omega_j} + u_\kappa^{2\kappa} \end{bmatrix},$$

$$A: \ L^2(\mathbb{R}^n, \mathbb{C}^{4N}) \to L^2(\mathbb{R}^n, \mathbb{C}^{4N}), \qquad \mathfrak{D}(A) = H^1(\mathbb{R}^n, \mathbb{C}^{4N}).$$

We note that $A_{11} \to 2m$ as $j \to \infty$, hence, for j large enough, A_{11} is invertible. By (III.37), its Schur complement is given by

$$\text{(XIII.99)} \quad S_j = A_{22} - A_{21} A_{11}^{-1} A_{12} = \Big(Z_j - \frac{1}{m + \omega_j} + u_\kappa^{2\kappa} + \frac{\Delta}{m - \omega_j - \mathrm{i}\lambda_j}\Big) I_{2N},$$

$j \in \mathbb{N}$, with $S_j : L^2(\mathbb{R}^n, \mathbb{C}^{2N}) \to L^2(\mathbb{R}^n, \mathbb{C}^{2N})$, $\mathfrak{D}(S_j) = H^2(\mathbb{R}^n, \mathbb{C}^{2N})$ for $j \in \mathbb{N}$.

Let Z_0 be the limit point of the sequence $\big(Z_j\big)_{j\in\mathbb{N}}$; by Lemma XIII.35, $Z_0 \neq \infty$.

Lemma XIII.36 *$Z_0 \in \sigma_{\mathrm{d}}(\mathfrak{l}_-) \cup \{1/(2m)\}$. If, moreover, the threshold $z = 1/(2m)$ is a regular point of the essential spectrum of $\mathfrak{l}_-$, then $Z_0 \in \sigma_{\mathrm{d}}(\mathfrak{l}_-)$.*

PROOF. We write the inverse of the matrix-valued operator in the left-hand side of (XIII.97) using (III.37):

$$\begin{bmatrix}\pi_P^-\boldsymbol{\Psi}_j\\ \epsilon_j\pi_A^-\boldsymbol{\Psi}_j\end{bmatrix}=-\frac{1}{a_j}\begin{bmatrix}1+\dfrac{\mathbf{D}_0 S_j^{-1}\mathbf{D}_0}{a_j} & -\mathbf{D}_0 S_j^{-1}\\ -S_j^{-1}\mathbf{D}_0 & a_j S_j^{-1}\end{bmatrix}\begin{bmatrix}\epsilon_j^2\pi_P^-\mathbf{V}\boldsymbol{\Psi}_j\\ \epsilon_j\pi_A^-(\mathbf{V}-u_\kappa^{2\kappa})\boldsymbol{\Psi}_j\end{bmatrix}. \tag{XIII.100}$$

Above, the Schur complement of $a_j := A_{11} = m-\omega_j-\mathrm{i}\lambda_j$ (so that $a_j\to 2m$ as $j\to\infty$) is given by the expression (XIII.99):

$$S_j = h_j I_{2N},\qquad j\in\mathbb{N},$$

with $h_j : L^2(\mathbb{R}^n)\to L^2(\mathbb{R}^n)$ given by

$$\begin{aligned}h_j &= Z_j-\frac{1}{m+\omega_j}+u_\kappa^{2\kappa}+\frac{\Delta}{m-\omega_j-\mathrm{i}\lambda_j}\\ &= Z_j+\frac{1}{m-\omega_j-\mathrm{i}\lambda_j}-\frac{1}{m+\omega_j}-\frac{1}{m-\omega_j-\mathrm{i}\lambda_j}\\ &\quad+\Big(1-\frac{2m}{m-\omega_j-\mathrm{i}\lambda_j}\Big)u_\kappa^{2\kappa}+\frac{2mu_\kappa^{2\kappa}}{m-\omega_j-\mathrm{i}\lambda_j}+\frac{\Delta}{m-\omega_j-\mathrm{i}\lambda_j}\\ &= -\frac{2m}{m-\omega_j-\mathrm{i}\lambda_j}(\mathfrak{l}_- - \hat{Z}_j)+\Big(1-\frac{2m}{m-\omega_j-\mathrm{i}\lambda_j}\Big)u_\kappa^{2\kappa},\end{aligned}$$

$\mathfrak{D}(h_j)=H^2(\mathbb{R}^n)$, where the sequence

$$\hat{Z}_j=\frac{m-\omega_j-\mathrm{i}\lambda_j}{2m}\Big(Z_j+\frac{1}{m-\omega_j-\mathrm{i}\lambda_j}-\frac{1}{m+\omega_j}\Big),\qquad j\in\mathbb{N},$$

has the same limit as the sequence $\big(Z_j\big)_{j\in\mathbb{N}}$.

Let $\Omega\subset\mathbb{C}$ be a compact set as in Lemma XIII.13: $\Omega\cap\sigma_{\mathrm{d}}(\mathfrak{l}_-)=\emptyset$, and if $z=1/(2m)$ is an eigenvalue or a virtual level of $\mathfrak{l}_-$, then we also assume that $1/(2m)\notin\Omega$. Then, by Lemma XIII.13, there is $C>0$ such that

$$\|u_\kappa^\kappa\circ(\mathfrak{l}_- - z)^{-1}\circ u_\kappa^\kappa\|_{L^2\to H^2}\le C,\qquad \forall z\in\Omega\setminus[1/(2m),+\infty).$$

Since $Z_0\in\Omega$, one has $\hat{Z}_j\in\Omega$' and also $\hat{Z}_j\notin\sigma(\mathfrak{l}_-)$ for $j\in\mathbb{N}$ large enough. We note that

$$\hat{Z}_j=\frac{m-\omega_j-\mathrm{i}\lambda_j}{2m}\Big(\frac{2\omega_j+\mathrm{i}\lambda_j}{m^2-\omega_j^2}+\frac{1}{m-\omega-\mathrm{i}\lambda_j}-\frac{1}{m-\omega_j}\Big)=\frac{1}{2m}-\frac{(m-\omega_j-\mathrm{i}\lambda_j)^2}{2m(m^2-\omega_j^2)};$$

since $\operatorname{Im}\lambda_j\to 2m$ and $\operatorname{Re}\lambda_j>0$, we conclude that $\operatorname{Im}\hat{Z}_j=\frac{(m-\omega_j+\operatorname{Im}\lambda_j)\operatorname{Re}\lambda_j}{2m(m^2-\omega_j^2)}\ne 0$ (hence $\hat{Z}_j\notin\sigma(\mathfrak{l}_-)$) for all $j\in\mathbb{N}$ except at most for finitely many values of j which we discard. Then the mapping

$$u_\kappa^\kappa\circ S_j^{-1}\circ u_\kappa^\kappa :\ L^2(\mathbb{R}^n)\to H^2(\mathbb{R}^n),\qquad j\in\mathbb{N},$$

is bounded uniformly in $j\in\mathbb{N}$. By duality, so is the mapping $H^{-2}(\mathbb{R}^n)\to L^2(\mathbb{R}^n)$, and, by complex interpolation, so is

$$u_\kappa^\kappa\circ S_j^{-1}\circ u_\kappa^\kappa :\ H^{-1}(\mathbb{R}^n)\to H^1(\mathbb{R}^n).$$

It follows that the following maps are also continuous:

$$u_\kappa^\kappa \circ S_j^{-1} \circ \mathbf{D}_0 \circ u_\kappa^\kappa : \quad L^2(\mathbb{R}^n) \to H^1(\mathbb{R}^n),$$
$$u_\kappa^\kappa \circ \mathbf{D}_0 \circ S_j^{-1} \circ u_\kappa^\kappa : \quad H^{-1}(\mathbb{R}^n) \to L^2(\mathbb{R}^n),$$
$$u_\kappa^\kappa \circ \mathbf{D}_0 \circ S_j^{-1} \circ \mathbf{D}_0 \circ u_\kappa^\kappa : \quad L^2(\mathbb{R}^n) \to L^2(\mathbb{R}^n),$$

with the bounds which are uniform in $j \in \mathbb{N}$. We note that, due to Lemma V.24 *(3)*, the multiplication by u_κ^κ extends to a continuous linear operator in $L^2(\mathbb{R}^n)$, $H^1(\mathbb{R}^n)$, and by duality in $H^{-1}(\mathbb{R}^n)$. Similarly, the matrix multiplication by $[\mathbf{D}_0, u_\kappa^\kappa] = \kappa u_\kappa^{\kappa-1}\mathbf{J}\boldsymbol{\alpha}^i\partial_{x^i}u_\kappa$ extends to a continuous linear operator in $L^2(\mathbb{R}^n, \mathbb{C}^{2N})$.

Therefore, if either $Z_j \to Z_0 \in \mathbb{C} \setminus \sigma(\mathfrak{l}_-)$ or $Z_j \to Z_0 \in [1/(2m), +\infty)$ (with the assumption that $Z_0 \neq 1/(2m)$ if $z = 1/(2m)$ is either an eigenvalue or a virtual level of $\mathfrak{l}_-$), with $\operatorname{Im} Z_j \neq 0$, then, taking into account that $\lambda_j \to 2m\mathrm{i}$ as $\omega_j \to m$, we conclude that the relation (XIII.100) yields the following inequality:

(XIII.101)

$$\begin{aligned}
&\|u_\kappa^\kappa \pi_P^- \boldsymbol{\Psi}_j\| + \epsilon_j \|u_\kappa^\kappa \pi_A^- \boldsymbol{\Psi}_j\| \\
&\quad \le C\epsilon_j^2 \Big\|\frac{1}{u_\kappa^\kappa}\pi_P^- \mathbf{V}\boldsymbol{\Psi}_j\Big\| + C\epsilon_j \Big\|\frac{1}{u_\kappa^\kappa}\big(\pi_A^- \mathbf{V}\boldsymbol{\Psi}_j - u_\kappa^{2\kappa}\pi_A^- \boldsymbol{\Psi}_j\big)\Big\| \\
&\quad \le C\epsilon_j^2 \Big\|\frac{1}{u_\kappa^\kappa}\pi_P^- \mathbf{V}\boldsymbol{\Psi}_j\Big\| + C\epsilon_j \Big\|\frac{1}{u_\kappa^\kappa}\pi_A^- \mathbf{V}\pi_P^- \boldsymbol{\Psi}_j + \frac{1}{u_\kappa^\kappa}\big(\pi_A^-(\mathbf{V} - u_\kappa^{2\kappa})\pi_A^- \boldsymbol{\Psi}_j\big)\Big\|,
\end{aligned}$$

with $C = C(Z_0) > 0$, and then, using the bounds

$$\|\mathbf{V}\|_{\operatorname{End}(\mathbb{C}^{2N})} \le C|u_\kappa(y)|^{2\kappa}, \qquad \|\pi_A \circ \mathbf{V} \circ \pi_P\|_{\operatorname{End}(\mathbb{C}^{2N})} \le C\epsilon|u_\kappa(y)|^{2\kappa},$$

and

$$\|\pi_A^-(\mathbf{V} + u_\kappa^{2\kappa}(1 + 2\kappa\Pi_{\boldsymbol{\Xi}})\boldsymbol{\beta})\pi_A^-\|_{\operatorname{End}(\mathbb{C}^{2N})} \le C\epsilon^{2\varkappa}|u_\kappa(y)|^{2\kappa}$$

from Lemma XIII.12 and taking into account the identity

$$\pi_A^-(\mathbf{V} - u_\kappa^{2\kappa})\pi_A^- = \pi_A^-(\mathbf{V} + u_\kappa^{2\kappa}(1 + 2\kappa\Pi_{\boldsymbol{\Xi}})\boldsymbol{\beta})\pi_A^-,$$

we obtain the estimate

$$\|u_\kappa^\kappa \pi^- \boldsymbol{\Psi}_j\| \le C\epsilon_j^{\min(1,2\varkappa)}\|u_\kappa^\kappa \boldsymbol{\Psi}_j\|, \qquad \forall j \in \mathbb{N}. \tag{XIII.102}$$

The inequality (XIII.102) and Lemma XIII.34 lead to $\|u_\kappa^\kappa \boldsymbol{\Psi}_j\| = O\big(\epsilon_j^{\min(1,2\varkappa)}\big)\|u_\kappa^\kappa \boldsymbol{\Psi}_j\|$, in contradiction to $\boldsymbol{\Psi}_j \not\equiv 0$, $j \in \mathbb{N}$. Thus, the assumption $Z_0 \in \mathbb{C} \setminus \sigma(\mathfrak{l}_-)$ leads to a contradiction, and so does the assumption $Z_0 \in (1/(2m), +\infty)$, and so does the assumption $Z_0 = 1/(2m)$ when the threshold $z = 1/(2m)$ is a regular point of the essential spectrum of $\mathfrak{l}_-$. □

This finishes the proof of Theorem XIII.4 *(1)* and *(2)*.

XIII.4.2 Characteristic roots of the nonlinear eigenvalue problem. Let us prove Theorem XIII.4 *(3)*, showing that $Z_0 = 0$ is only possible when $\lambda_j = 2\omega_j\mathrm{i}$ for all but finitely many $j \in \mathbb{N}$. First, we claim that the relations (XIII.23) and (XIII.24) allow one to express $\mathbf{Y} := \pi^+\boldsymbol{\Psi}_j$ in terms of $\mathbf{X} := \pi^-\boldsymbol{\Psi}_j$.

Lemma XIII.37 *There is* $\epsilon_2 \in (0, \epsilon_1)$ *such that for any* $\epsilon \in (0, \epsilon_2)$ *and any* $z \in \mathbb{D}_1$ *the relations*

$$\epsilon\mathbf{D}_0\pi_A\mathbf{Y} - (\omega - \mathrm{i}\lambda - m)\pi_P\mathbf{Y} + \epsilon^2\pi_P^+\mathbf{V}(\mathbf{X} + \mathbf{Y}) = 0,$$
$$\epsilon\mathbf{D}_0\pi_P\mathbf{Y} - (\omega - \mathrm{i}\lambda + m)\pi_A\mathbf{Y} + \epsilon^2\pi_A^+\mathbf{V}(\mathbf{X} + \mathbf{Y}) = 0,$$

where $\omega = \sqrt{m^2 - \epsilon^2}$ *and*

$$\lambda = \lambda(z) = (2\omega + \epsilon^2 z)\mathrm{i} \tag{XIII.103}$$

(cf. (XIII.6)*), define a linear map*

$$\vartheta(\cdot, \epsilon, z): \ L^{2,-\kappa}(\mathbb{R}^n, \mathfrak{R}(\pi^-)) \to L^{2,-\kappa}(\mathbb{R}^n, \mathfrak{R}(\pi^+)), \qquad \vartheta(\cdot, \epsilon, z): \ \mathbf{X} \mapsto \mathbf{Y},$$

which is analytic in z*, where for* $\mu \in \mathbb{R}$ *the exponentially weighted spaces are defined by*

$$L^{2,\mu}(\mathbb{R}^n) = \left\{u \in L^2_{\mathrm{loc}}(\mathbb{R}^n);\ e^{\mu\langle r\rangle} u \in L^2(\mathbb{R}^n)\right\}, \qquad \|u\|_{L^{2,\mu}} := \|e^{\mu\langle r\rangle} u\|_{L^2}.$$

Moreover, there is $C > 0$ *such that*

$$\|\vartheta(\cdot, \epsilon, z)\|_{L^{2,-\kappa}(\mathbb{R}^n,\mathbb{C}^{2N}) \to L^{2,-\kappa}(\mathbb{R}^n,\mathbb{C}^{2N})} \le C\epsilon, \qquad \epsilon \in (0, \epsilon_2), \quad z \in \mathbb{D}_1,$$
$$\|\partial_z\vartheta(\cdot, \epsilon, z)\|_{L^{2,-\kappa}(\mathbb{R}^n,\mathbb{C}^{2N}) \to L^{2,-\kappa}(\mathbb{R}^n,\mathbb{C}^{2N})} \le C\epsilon^3, \qquad \epsilon \in (0, \epsilon_2), \quad z \in \mathbb{D}_1.$$

PROOF. By (XIII.86),

$$\begin{bmatrix}\pi_P\mathbf{Y}\\ \pi_A\mathbf{Y}\end{bmatrix} = \begin{bmatrix}\omega - \mathrm{i}\lambda + m & \epsilon\mathbf{D}_0\\ \epsilon\mathbf{D}_0 & \omega - \mathrm{i}\lambda - m\end{bmatrix}\left(\Delta_y + \frac{(\omega - \mathrm{i}\lambda)^2 - m^2}{\epsilon^2} I\right)^{-1}\begin{bmatrix}\pi_P^+\mathbf{V}(\mathbf{X}+\mathbf{Y})\\ \pi_A^+\mathbf{V}(\mathbf{X}+\mathbf{Y})\end{bmatrix}.$$

This leads to

$$\mathbf{Y} = \pi^+\left\{(\omega - \mathrm{i}\lambda) + m(\pi_P - \pi_A) + \epsilon\mathbf{D}_0\right\}\left(\Delta + \zeta^2 I\right)^{-1}\mathbf{V}(\mathbf{X}+\mathbf{Y}), \tag{XIII.104}$$

where $\zeta \in \mathbb{C}$ is related to $\lambda \in \mathbb{C}$ by (cf. (XIII.87))

$$\zeta^2 = \frac{(\omega - \mathrm{i}\lambda)^2 - m^2}{\epsilon^2}, \qquad \operatorname{Re}\zeta \ge 0; \tag{XIII.105}$$

the values of λ with $\operatorname{Re}\lambda > 0$ correspond to the values of ζ with $\operatorname{Im}\zeta < 0$ (since $\lambda \in \mathbb{D}_{\epsilon_2}(2\omega\mathrm{i})$ and $\omega \in (\omega_2, m)$, where $\omega_2 = \sqrt{m^2 - \epsilon_2^2}$, with $\epsilon_2 \in (0, \epsilon_1)$ small enough).

For $\epsilon > 0$ and $z \in \mathbb{C}$, $\operatorname{Im} z < 0$ (so that $\operatorname{Re}\lambda(z) > 0$ by (XIII.103) and then $\operatorname{Im}\zeta < 0$), we define the linear map

$$\Phi(\cdot, \epsilon, z): \ L^{2,-\kappa}(\mathbb{R}^n, \mathbb{C}^{2N}) \to L^{2,-\kappa}(\mathbb{R}^n, \mathfrak{R}(\pi^+)), \tag{XIII.106}$$
$$\Phi(\mathbf{\Psi}, \epsilon, z) = \pi^+\left\{(\omega - \mathrm{i}\lambda) + m(\pi_P - \pi_A) + \epsilon\mathbf{D}_0\right\}\left(\Delta + \zeta^2 I\right)^{-1}\mathbf{V}\mathbf{\Psi},$$

where

$$\zeta^2 = \frac{(\omega - \mathrm{i}\lambda(z))^2 - m^2}{\epsilon^2}, \qquad \lambda(z) = (2\omega + \epsilon^2 z)\mathrm{i};$$

see (XIII.103) and (XIII.105).

Remark XIII.38 The inverse of $\Delta + \frac{(\omega - \mathrm{i}\lambda)^2 - m^2}{\epsilon^2}$ is not continuous in λ at $\operatorname{Re}\lambda = 0$; this discontinuity could result in two different families of eigenvalues bifurcating from an embedded eigenvalue even when its algebraic multiplicity is one. We will use the resolvent corresponding to $\operatorname{Re}\lambda > 0$ and then use its analytic continuation through $\operatorname{Re}\lambda = 0$.

Using the definition (XIII.106), the relation (XIII.104) takes the form

$$\mathbf{Y} = \Phi(\mathbf{X}+\mathbf{Y}, \epsilon, z). \tag{XIII.107}$$

Since the norm of

$$\Phi(\cdot, \epsilon, z) \in \mathscr{B}\big(L^{2,-\kappa}(\mathbb{R}^n, \mathbb{C}^{2N}), L^{2,-\kappa}(\mathbb{R}^n, \mathbb{C}^{2N})\big), \qquad z \in \mathbb{D}_1,$$

is small (as long as $\epsilon > 0$ is small enough), we will be able to use the above relation to express $\mathbf{Y} \in L^{2,-\kappa}(\mathbb{R}^n, \mathfrak{R}(\pi^+))$ as a function of $\mathbf{X} \in L^{2,-\kappa}(\mathbb{R}^n, \mathfrak{R}(\pi^-))$.

Due to Lemma XIII.12, one has

$$\|\mathbf{V}(y,\epsilon)\|_{\mathrm{End}(\mathbb{C}^{2N})} \le Ce^{-2\kappa|y|}, \qquad \forall y \in \mathbb{R}^n, \quad \forall \epsilon \in (0, \epsilon_1).$$

Using this exponential decay together with the analytic continuation of the resolvent from Proposition VI.28, the mapping (XIII.106) could be extended from $\{\operatorname{Im}\zeta < 0\}$ to $\{\zeta \in \mathbb{C};\ \operatorname{Im}\zeta < \kappa\} \setminus \overline{\mathrm{i}\mathbb{R}_+}$. For the uniformity, we require that

$$|\operatorname{Im}\zeta| < \kappa, \tag{XIII.108}$$

considering the resolvent $(-\Delta - \zeta^2)^{-1}$ for $\operatorname{Im}\zeta < 0$ (this corresponds to $\operatorname{Re}\lambda > 0$) and its analytic continuation into the strip $0 \le \operatorname{Im}\zeta < \kappa$, $\operatorname{Re}\zeta \ne 0$ (this corresponds to $\operatorname{Re}\lambda \le 0$). Due to our assumptions that $\omega \to m$ and $\lambda \to 2m\mathrm{i}$, one has

$$\operatorname{Re}((\omega - \mathrm{i}\lambda)^2 - m^2) = 8m^2 + O(\operatorname{Re}\lambda) + O(\operatorname{Im}\lambda - 2m) + O(\epsilon^2). \tag{XIII.109}$$

Therefore, by (XIII.105),

$$\operatorname{Re}\zeta = \epsilon^{-1}\operatorname{Re}\sqrt{(\omega - \mathrm{i}\lambda)^2 - m^2} = O(\epsilon^{-1}), \tag{XIII.110}$$

showing that for ϵ sufficiently small one has $\zeta \in \mathbb{C} \setminus \mathbb{D}_1$ (we take $\epsilon_2 > 0$ smaller if necessary). Since we only consider $z \in \mathbb{D}_1$, the relation (XIII.103) yields $|\operatorname{Re}\lambda| \le |z|\epsilon^2 \le \epsilon^2$, and then (XIII.105) and (XIII.110) lead to

$$\operatorname{Im}\zeta = \frac{1}{2\operatorname{Re}\zeta}\operatorname{Im}\zeta^2 = O(\epsilon)\frac{\operatorname{Im}((\omega - \mathrm{i}\lambda)^2 - m^2)}{\epsilon^2} = O(\epsilon)\frac{O(\operatorname{Re}\lambda)}{\epsilon^2} = O(\epsilon),$$

showing that the condition (XIII.108) holds true for $\epsilon \in (0, \epsilon_2)$ with $\epsilon_2 > 0$ sufficiently small, satisfying assumptions of Proposition VI.28. Then, by that proposition, there is $C > 0$ such that for any $\mathbf{\Psi} \in L^{2,-\kappa}(\mathbb{R}^n, \mathbb{C}^{2N})$ the map (XIII.106) satisfies

(XIII.111)
$$\|\Phi(\mathbf{\Psi}, \epsilon, z))\|_{L^{2,-\kappa}(\mathbb{R}^n,\mathbb{C}^{2N})} \le C\epsilon\|\mathbf{\Psi}\|_{L^{2,-\kappa}(\mathbb{R}^n,\mathbb{C}^{2N})}, \quad \forall \epsilon \in (0, \epsilon_2), \quad \forall z \in \mathbb{D}_1.$$

We take $\epsilon_2 > 0$ smaller if necessary so that $\epsilon_2 \le 1/(2C)$ (with $C > 0$ from (XIII.111)); then the linear map

$$I - \Phi(\cdot, \epsilon, z) :\ \mathbf{Y} \mapsto \mathbf{Y} - \Phi(\mathbf{Y}, \epsilon, z)$$

is invertible, with

$$\|(I - \Phi(\cdot, \epsilon, z))^{-1}\|_{L^{2,-\kappa} \to L^{2,-\kappa}} \le 2, \qquad \epsilon \in (0, \epsilon_2), \qquad z \in \mathbb{D}_1. \tag{XIII.112}$$

Since $\Phi(\cdot, \epsilon, z)$ is linear, writing (XIII.107) in the form $\mathbf{Y} - \Phi(\mathbf{Y}) = \Phi(\mathbf{X})$, we can express $\mathbf{Y} = (I - \Phi)^{-1}\Phi(\mathbf{X})$. Thus, for each $\epsilon \in (0, \epsilon_2)$ and $z \in \mathbb{D}_1$, we may define the mapping $(\mathbf{X}, \epsilon, z) \mapsto \mathbf{Y}$, which we denote by ϑ:

$$\vartheta(\cdot, \epsilon, z) :\ L^{2,-\kappa}(\mathbb{R}^n, \mathfrak{R}(\pi^-)) \to L^{2,-\kappa}(\mathbb{R}^n, \mathfrak{R}(\pi^+)),$$

$$\vartheta(\cdot, \epsilon, z) :\ \mathbf{X} \mapsto \mathbf{Y} = (I - \Phi(\cdot, \epsilon, z))^{-1}\Phi(\mathbf{X}, \epsilon, z). \tag{XIII.113}$$

By (XIII.111) and (XIII.112), one has $\|\vartheta(\cdot, \epsilon, z)\|_{L^{2,-\kappa} \to L^{2,-\kappa}} \le 2C\epsilon$, for $\epsilon \in (0, \epsilon_2)$ and $z \in \mathbb{D}_1$.

Finally, let us discuss the differentiability of ϑ with respect to z. The map Φ can be differentiated in the strong sense with respect to z. First, we notice that, by (XIII.105) and then by (XIII.103),

$$(\text{XIII.114})\qquad 2|\zeta\partial_z\zeta| = \left|\partial_z\frac{(\omega-\mathrm{i}\lambda)^2-m^2}{\epsilon^2}\right| = \left|\frac{2(\omega-\mathrm{i}\lambda)}{\epsilon^2}\partial_z(2\omega+\epsilon^2 z)\right| = 2|\omega-\mathrm{i}\lambda|,$$

with the right-hand side bounded uniformly in $\epsilon\in(0,\epsilon_2)$ and $z\in\mathbb{D}_1$. Therefore, using the bound for the derivative of the analytic continuation of the resolvent (cf. Proposition VI.28, which we apply with $\nu=0$ and also with $\nu=1$ to accommodate the operator $\epsilon\mathbf{D}_0$ from the definition of Φ in (XIII.106)), we conclude that there is $C>0$ such that

$$\|\partial_z\Phi(\cdot,\epsilon,z)\|_{L^{2,-\kappa}\to L^{2,-\kappa}} \le \|\partial_\zeta\Phi\|_{L^{2,-\kappa}\to L^{2,-\kappa}}|\partial_z\zeta| \le \frac{C}{\langle\zeta\rangle^2}\frac{|\omega-\mathrm{i}\lambda|}{|\zeta|} \le C\epsilon^3$$

for all $\epsilon\in(0,\epsilon_2)$ and $z\in\mathbb{D}_1$. Above, ∂_z is considered as a gradient in $\mathbb{R}^2\cong\mathbb{C}$; we used (XIII.114) and the estimate (XIII.110). Then it follows from (XIII.113) that there is $C>0$ such that one also has

$$\|\partial_z\vartheta(\cdot,\epsilon,z)\|_{L^{2,-\kappa}\to L^{2,-\kappa}} \le C\epsilon^3,$$

for all $\epsilon\in(0,\epsilon_2)$ and $z\in\mathbb{D}_1$. One can see from (XIII.106) that Φ is analytic in the complex parameter z, hence so is ϑ. □

As long as $j\in\mathbb{N}$ is sufficiently large so that $\epsilon_j\in(0,\epsilon_2)$ and $z_j\in\mathbb{D}_1$, then, by Lemma XIII.37, the relations (XIII.23) and (XIII.24) allow us to express

$$\pi^+\mathbf{\Psi}_j = \vartheta(\pi^-\mathbf{\Psi}_j,\epsilon_j,z_j),$$

with $\vartheta(\cdot,\epsilon,z)$ a linear map from $L^{2,-\kappa}(\mathbb{R}^n,\mathbb{C}^{2N})$ into itself, with

$$\|\vartheta(\cdot,\epsilon,z)\|_{L^{2,-\kappa}(\mathbb{R}^n,\mathbb{C}^{2N})\to L^{2,-\kappa}(\mathbb{R}^n,\mathbb{C}^{2N})} \le C\epsilon.$$

Now (XIII.21) and (XIII.22) can be written as

$$(\text{XIII.115})\qquad \epsilon_j\mathbf{D}_0\pi_A^-\mathbf{\Psi}_j + (m-\omega_j-\mathrm{i}\lambda_j)\pi_P^-\mathbf{\Psi}_j + \epsilon_j^2\pi_P^-\mathbf{V}^\vartheta\pi^-\mathbf{\Psi}_j = 0,$$

$$(\text{XIII.116})\qquad \epsilon_j\mathbf{D}_0\pi_P^-\mathbf{\Psi}_j - (m+\omega_j+\mathrm{i}\lambda_j)\pi_A^-\mathbf{\Psi}_j + \epsilon_j^2\pi_P^-\mathbf{V}^\vartheta\pi^-\mathbf{\Psi}_j = 0,$$

where $\mathbf{V}^\vartheta=\mathbf{V}^\vartheta(y,\epsilon_j,\lambda_j)$, with $\mathbf{V}^\vartheta(y,\epsilon,z):=\mathbf{V}(y,\epsilon)\circ(1+\vartheta(\cdot,\epsilon,z))$ satisfying

$$\|\mathbf{V}^\vartheta(\epsilon,\lambda)\|_{L^2\to L^2} \le \|\mathbf{V}(\epsilon)\|_{L^{2,-\kappa}\to L^2}\big(1+\|\vartheta(\cdot,\epsilon,z)\|_{L^{2,-\kappa}\to L^{2,-\kappa}}\big) \le C,$$

so that

$$\mathbf{V}(\epsilon_j)\mathbf{\Psi}_j = \mathbf{V}(\epsilon_j)(\pi^-\mathbf{\Psi}_j+\pi^+\mathbf{\Psi}_j) = \mathbf{V}(\epsilon_j)\big(\pi^-\mathbf{\Psi}_j+\vartheta(\pi^-\mathbf{\Psi}_j,\epsilon_j,z_j)\big) = \mathbf{V}^\vartheta\pi^-\mathbf{\Psi}_j.$$

We recall the definition $Z_j=-(2\omega_j+\mathrm{i}\lambda_j)/\epsilon_j^2$ (cf. (XIII.85)) and rewrite (XIII.115), (XIII.116), as the following system:

$$(\text{XIII.117})\qquad \begin{bmatrix}\pi_P^-(\frac{m+\omega_j}{\epsilon_j^2}+Z_j+\mathbf{V}^\vartheta)\pi_P^- & \pi_P^-(\epsilon_j^{-1}\mathbf{D}_0+\mathbf{V}^\vartheta)\pi_A^- \\ \pi_A^-(\epsilon_j^{-1}\mathbf{D}_0+\mathbf{V}^\vartheta)\pi_P^- & \pi_A^-(Z_j-\frac{1}{m+\omega_j}+\mathbf{V}^\vartheta)\pi_A^-\end{bmatrix}\begin{bmatrix}\pi_P^-\mathbf{\Psi}_j\\ \pi_A^-\mathbf{\Psi}_j\end{bmatrix} = 0,$$

with $j\in\mathbb{N}$. We rewrite the above as the *nonlinear eigenvalue problem*

$$A(\epsilon,z)\begin{bmatrix}\pi_P^-\mathbf{\Psi}\\ \pi_A^-\mathbf{\Psi}\end{bmatrix} = 0,$$

$$(\text{XIII.118})\qquad A(\epsilon,z) := \begin{bmatrix}\pi_P^-(\frac{m+\omega}{\epsilon^2}+z+\mathbf{V}^\vartheta)\pi_P^- & \pi_P^-(\epsilon^{-1}\mathbf{D}_0+\mathbf{V}^\vartheta)\pi_A^- \\ \pi_A^-(\epsilon^{-1}\mathbf{D}_0+\mathbf{V}^\vartheta)\pi_P^- & \pi_A^-(z-\frac{1}{m+\omega}+\mathbf{V}^\vartheta)\pi_A^-\end{bmatrix},$$

where the operator $A(\epsilon,z): L^2(\mathbb{R}^n,\mathfrak{R}(\pi^-)) \to L^2(\mathbb{R}^n,\mathfrak{R}(\pi^-))$ is considered as

$$A(\epsilon,z): \ L^2(\mathbb{R}^n,\mathfrak{R}(\pi_P^-)\times\mathfrak{R}(\pi_A^-)) \to L^2(\mathbb{R}^n,\mathfrak{R}(\pi_P^-)\times\mathfrak{R}(\pi_A^-)),$$

with domain $\mathfrak{D}(A(\epsilon,z)) = H^1(\mathbb{R}^n,\mathfrak{R}(\pi_P^-)\times\mathfrak{R}(\pi_A^-))$. Note that the operator $A(\epsilon,z)$ depends on z analytically via ϑ (cf. Lemma XIII.37). By Weyl's theorem,

$$\text{(XIII.119)} \qquad \sigma_{\mathrm{ess}}(A(\epsilon,z)) = \big(-\infty,-(m+\omega)^{-1}+z\big] \cup \big[(m-\omega)^{-1}+z,+\infty\big),$$

so that $0\notin\sigma_{\mathrm{ess}}(A(\epsilon,z))$. Thus, the values Z_j defined in (XIII.85) are such that the kernel of $A(\epsilon_j,z)$ is nontrivial at $z=Z_j$; such values of z are called the *characteristic roots* (or, informally, *nonlinear eigenvalues*) of $A(\epsilon,z)$.

To study the operator $A(\epsilon,z)$, we will apply the Keldysh theory of characteristic roots presented in Section III.9. First, we will do the reduction of A using the Schur complement of its invertible block. By (XIII.119), we may assume that there is a sufficiently small open neighborhood U of $z=0$ and that $\omega_*\in(0,m)$ is sufficiently large so that

$$\sigma_{\mathrm{ess}}(A(\epsilon,z))\cap\mathbb{D}_{1/(4m)}=\emptyset \qquad \forall\omega\in(\omega_*,m), \qquad \forall\epsilon\in(0,\epsilon_*), \qquad \forall z\in U,$$

where $\epsilon_*=\sqrt{m^2-\omega_*^2}$. Since $0\in\sigma_{\mathrm{d}}(\mathfrak{l}_-)$, with $\mathfrak{l}_-$ from (XIII.5), we may assume that the open neighborhood $U\ni\{0\}$ is small enough so that

$$\text{(XIII.120)} \qquad U\cap\sigma(\mathfrak{l}_-)=\{0\}.$$

Let $A_{ij}(\epsilon,z)$, $1\le i,\,j\le 2$, denote the operators which are the entries of $A(\epsilon,z)$ defined in (XIII.4.2), so that $A(\epsilon,z)=\begin{bmatrix}A_{11}(\epsilon,z) & A_{12}(\epsilon,z)\\ A_{21}(\epsilon,z) & A_{22}(\epsilon,z)\end{bmatrix}$. We then have:

$$\text{(XIII.121)} \qquad \begin{gathered}\|A_{12}\|_{H^1\to L^2}+\|A_{21}\|_{H^1\to L^2}+\|A_{21}\|_{L^2\to H^{-1}}=O(\epsilon^{-1}),\\ \|A_{11}\|_{L^2\to L^2}=O(\epsilon^{-2}),\end{gathered}$$

and, taking $\epsilon_*>0$ smaller if necessary, one has

$$\text{(XIII.122)} \qquad A_{11}^{-1}=\frac{\epsilon^2}{2m}+O_{L^2\to L^2}(\epsilon^4), \qquad \forall\epsilon\in(0,\epsilon_*), \qquad \forall z\in U,$$

with the estimate $O_{L^2\to L^2}(\epsilon^4)$ uniform in $z\in U$. This suggests that we study the invertibility of $A(\epsilon,z)$ in terms of the Schur complement of $A_{11}(\epsilon,z)$ (see (III.37)), which is defined by

$$\text{(XIII.123)} \quad S(\epsilon,z)=A_{22}-A_{21}A_{11}^{-1}A_{12}: \ H^1(\mathbb{R}^n,\mathfrak{R}(\pi_A^-))\to H^{-1}(\mathbb{R}^n,\mathfrak{R}(\pi_A^-)),$$

where A_{11}, A_{12}, A_{21}, and A_{22} are evaluated at $(\epsilon,z)\in(0,\epsilon_*)\times U$. Let us derive the explicit expression for $S(\epsilon,z)$:

$$\begin{aligned}S(\epsilon,z)&=\pi_A^-\big(z-\frac{1}{m+\omega}+\mathbf{V}^\vartheta\big)\pi_A^-\\ &\qquad-\pi_A^-(\mathbf{D}_0+\epsilon\mathbf{V}^\vartheta)\pi_P^-\big(m+\omega+\epsilon^2(\mathbf{V}^\vartheta+z)\big)^{-1}\pi_P^-(\mathbf{D}_0+\epsilon\mathbf{V}^\vartheta)\pi_A^-\\ &=\pi_A^-\big(z-\frac{1}{m+\omega}+u_\kappa^{2\kappa}+\frac{\Delta}{m+\omega}\big)\pi_A^-+\pi_A^-(\mathbf{V}^\vartheta-u_\kappa^{2\kappa})\pi_A^-\\ &\qquad+\pi_A^-\frac{\mathbf{D}_0^2}{m+\omega}\pi_A^--\pi_A^-(\mathbf{D}_0+\epsilon\mathbf{V}^\vartheta)\pi_P^-\frac{1}{m+\omega}\pi_P^-(\mathbf{D}_0+\epsilon\mathbf{V}^\vartheta)\pi_A^-\\ &\qquad-\pi_A^-(\mathbf{D}_0+\epsilon\mathbf{V}^\vartheta)\pi_P^-\frac{1}{m+\omega}\Big(\Big(1+\frac{\epsilon^2(\mathbf{V}^\vartheta+z)}{m+\omega}\Big)^{-1}-1\Big)\pi_P^-(\mathbf{D}_0+\epsilon\mathbf{V}^\vartheta)\pi_A^-.\end{aligned}$$

We recall that $u_\kappa = u_\kappa(x)$ is the ground state of the nonlinear Schrödinger equation (XII.4). We note that, by (XIII.17),

$$\pi_A^- \mathbf{V}^\vartheta \pi_A^- = \pi_A^- \mathbf{V} \circ (1+\vartheta)\pi_A^- = \pi_A^- u_\kappa^{2\kappa} + O_{L^2 \to L^2}(\epsilon);$$

we used the bounds $\|\pi_A \boldsymbol{\phi}_\omega\|_{L^\infty} = O(\epsilon^{1+\frac{1}{\kappa}})$ (cf. Theorem XII.1) and $\|\vartheta\|_{L^2_{-s} \to L^2_{-s}} = O(\epsilon)$ (cf. Lemma XIII.37) which yield, for any fixed $s > 1/2$,

$$\|\mathbf{V} \circ \vartheta\|_{L^2 \to L^2} \le \|\mathbf{V}\|_{L^2_{-s} \to L^2_s} \|\vartheta\|_{L^2_{-s} \to L^2_{-s}} = O(\epsilon).$$

Thus, taking into account Lemma V.28, the operator $S(\epsilon, z)$ defined in (XIII.123) takes the form

$$S(\epsilon, z) = \pi_A^- \Big(z - \frac{1}{2m} + u_\kappa^{2\kappa} + \frac{\Delta}{2m} + O_{H^1 \to H^{-1}}(\epsilon) \Big) \pi_A^-, \quad \epsilon \in [0, \epsilon_*), \tag{XIII.124}$$

with the estimate $O_{H^1 \to H^{-1}}(\epsilon)$ uniform in $z \in U$. Above, we extended $S(\epsilon, z)$ defined in (XIII.124) from $\epsilon \in (0, \epsilon_*)$ to $\epsilon \in [0, \epsilon_*)$ by continuity.

The following lemma allows us to reduce the problem of studying the characteristic roots of $A(\epsilon, z)$ (see (XIII.4.2)) to the characteristic roots of $S(\epsilon, z)$ (cf. (XIII.123)).

Lemma XIII.39 *If $\epsilon_* > 0$ is sufficiently small, then for all $\epsilon \in [0, \epsilon_*)$ the point $z_0 \in \mathbb{D}_{1/(2m)}$ is a characteristic root of $A(\epsilon, z)$ if and only if it is a characteristic root of $S(\epsilon, z)$.*

PROOF. Since the operator $A_{11}(\epsilon, z) : L^2(\mathbb{R}^n, \mathfrak{R}(\pi_P^-)) \to L^2(\mathbb{R}^n, \mathfrak{R}(\pi_P^-))$ is invertible with bounded inverse for $\epsilon > 0$ small enough (see (XIII.122)), the expression for the Schur complement (III.37) and the relation (III.35) show that the operator $A(\epsilon, z)$ from (XIII.4.2) is invertible if and only if so is $S(\epsilon, z)$. □

Lemma XIII.40 *Multiplicity of the characteristic root $z_0 = 0$ of $S(0, z)$ is $\alpha = N/2$.*

PROOF. By (XIII.124), $S(0, 0) = (z - \mathfrak{l}_-)\pi_A^-$. Since $\dim \mathbf{ker}(\mathfrak{l}_-) = 1$, one has $\dim \mathbf{ker}(S(0,0)) = \operatorname{rank} \pi_A^- = N/2$. Let $\boldsymbol{e}_i \in \mathbb{C}^{2N}$, $1 \le i \le N/2$, be the basis in $\mathfrak{R}(\pi_A^-)$; then $\{u_\kappa \boldsymbol{e}_i\}_{1 \le i \le N/2}$ is the basis in $\mathbf{ker}(S(0,0))$. We do not need to use the Riesz projectors since the operator $S(0, z)$ is invariant in this space, being represented by $\mathbf{S}(0, z) = z I_{N/2}$; thus, $\det \mathbf{S}(0, z) = z^{N/2}$. □

Lemma XIII.41 *There is no sequence of characteristic roots $Z_j \ne 0$ of $A(\epsilon_j, z)$ such that $Z_j \to 0$ as $j \to \infty$.*

PROOF. Let $\delta > 0$ be such that $\partial \mathbb{D}_\delta \in \rho(A(\epsilon, z_0))$, $z_0 = 0$. Due to the continuity of the resolvent in z and ϵ, there is $\epsilon_* \in (0, \epsilon_2)$ and an open neighborhood $U \subset \mathbb{D}_{1/(2m)}$, $z_0 \in U$, such that $\partial \mathbb{D}_\delta \subset \rho(A(\epsilon, z))$ for all $z \in U$ and $\epsilon \in [0, \epsilon_*)$.

By (XIII.120) and Lemma XIII.40, the sum of multiplicities of the characteristic roots of $S(0, z)$ in U equals $N/2$, and by Theorem III.147 the same is true for $S(\epsilon, z)$ for all $\epsilon \in [0, \epsilon_*)$. At the same time, by Lemma XIII.10 and Lemma III.145, $z = 0$ is a characteristic root of $S(\epsilon, z)$, $\epsilon \in [0, \epsilon_*)$, of multiplicity *at least* $\alpha = N/2$. Hence, there can be no other, nonzero characteristic roots $z \in U$ of $S(\epsilon, z)$ for any $\epsilon \in [0, \epsilon_*)$, and, in particular, given a sequence $\epsilon_j \to 0$, there is no sequence of characteristic roots Z_j of $S(\epsilon_j, z)$ such that $Z_j \ne 0$ for $j \in \mathbb{N}$, $Z_j \to 0$ as $j \to \infty$. By Lemma XIII.39, the same conclusion holds true for $A(\epsilon, z)$. □

By Lemma XIII.41, $Z_j = 0$ for all but finitely many $j \in \mathbb{N}$. By (XIII.6), this implies that $\lambda_j = 2\omega_j\mathrm{i}$ for all but finitely many $j \in \mathbb{N}$. This concludes the proof of Theorem XIII.4 *(3)*.

The proof of Theorem XIII.4 is now complete.

Bibliography

[Abe98] S. Abenda, *Solitary waves for Maxwell-Dirac and Coulomb-Dirac models* (English, with English and French summaries), Ann. Inst. H. Poincaré Phys. Théor. **68** (1998), no. 2, 229–244. MR1618672

[ABG82] W. Amrein, A.-M. Berthier, and V. Georgescu, *Estimations du type Hardy ou Carleman pour des opérateurs différentiels à coefficients opératoriels* (French, with English summary), C. R. Acad. Sci. Paris Sér. I Math. **295** (1982), no. 10, 575–578. MR685028

[AC81] A. Alvarez and B. Carreras, *Interaction dynamics for the solitary waves of a nonlinear Dirac model*, Phys. Lett. A **86** (1981), no. 6-7, 327–332, DOI 10.1016/0375-9601(81)90548-X. MR637021

[AD70] D. L. T. Anderson and G. H. Derrick, *Stability of time-dependent particlelike solutions in nonlinear field theories. I*, Journal of Mathematical Physics **11** (1970), pp. 1336–1346.

[AF03] R. A. Adams and J. J. F. Fournier, *Sobolev spaces*, 2nd ed., Pure and Applied Mathematics (Amsterdam), vol. 140, Elsevier/Academic Press, Amsterdam, 2003. MR2424078

[Agm75] S. Agmon, *Spectral properties of Schrödinger operators and scattering theory*, Ann. Scuola Norm. Sup. Pisa Cl. Sci. (4) **2** (1975), no. 2, 151–218. MR0397194

[And71] D. L. T. Anderson, *Stability of time-dependent particlelike solutions in nonlinear field theories. II*, Journal of Mathematical Physics **12** (1971), pp. 945–952.

[Apo76] C. Apostol, *On the closed range points in the spectrum of operators*, Rev. Roumaine Math. Pures Appl. **21** (1976), no. 8, 971–975. MR0425639

[BC81] H. Berestycki and T. Cazenave, *Instabilité des états stationnaires dans les équations de Schrödinger et de Klein-Gordon non linéaires* (French, with English summary), C. R. Acad. Sci. Paris Sér. I Math. **293** (1981), no. 9, 489–492. MR646873

[BC12a] G. Berkolaiko and A. Comech, *On spectral stability of solitary waves of nonlinear Dirac equation in 1D*, Math. Model. Nat. Phenom. **7** (2012), no. 2, 13–31, DOI 10.1051/mmnp/20127202. MR2892774

[BC12b] N. Boussaid and S. Cuccagna, *On stability of standing waves of nonlinear Dirac equations*, Comm. Partial Differential Equations **37** (2012), no. 6, 1001–1056, DOI 10.1080/03605302.2012.665973. MR2924465

[BC14] N. Bournaveas and T. Candy, *Global well-posedness for the massless cubic Dirac equation*, Int. Math. Res. Not. IMRN **22** (2016), 6735–6828. MR3632067

[BC16] N. Boussaïd and A. Comech, *On spectral stability of the nonlinear Dirac equation*, J. Funct. Anal. **271** (2016), no. 6, 1462–1524, DOI 10.1016/j.jfa.2016.04.013. MR3530581

[BC17] N. Boussaïd and A. Comech, *Nonrelativistic asymptotics of solitary waves in the Dirac equation with Soler-type nonlinearity*, SIAM J. Math. Anal. **49** (2017), no. 4, 2527–2572, DOI 10.1137/16M1081385. MR3670258

[BC18] N. Boussaïd and A. Comech, *Spectral stability of bi-frequency solitary waves in Soler and Dirac-Klein-Gordon models*, Commun. Pure Appl. Anal. **17** (2018), no. 4, 1331–1347, DOI 10.3934/cpaa.2018065. MR3842864

[BC19] N. Boussaïd and A. Comech, *Spectral stability of small amplitude solitary waves of the Dirac equation with the Soler-type nonlinearity*, J. Funct. Anal. **51** (2019), to appear, DOI 10.1016/j.jfa.2019.108289

[BCS15] G. Berkolaiko, A. Comech, and A. Sukhtayev, *Vakhitov-Kolokolov and energy vanishing conditions for linear instability of solitary waves in models of classical self-interacting spinor fields*, Nonlinearity **28** (2015), no. 3, 577–592, DOI 10.1088/0951-7715/28/3/577. MR3311594

[BD64] J. D. Bjorken and S. D. Drell, *Relativistic quantum mechanics*, McGraw-Hill Book Co., New York-Toronto-London, 1964. MR0187641

[BD65] J. D. Bjorken and S. D. Drell, *Relativistic quantum fields*, McGraw-Hill Book Co., New York-Toronto-London-Sydney, 1965. MR0187642

[BdMBMP93] A. Boutet de Monvel-Berthier, D. Manda, and R. Purice, *Limiting absorption principle for the Dirac operator* (English, with English and French summaries), Ann. Inst. H. Poincaré Phys. Théor. **58** (1993), no. 4, 413–431. MR1241704

[Bec11] M. Beceanu, *New estimates for a time-dependent Schrödinger equation*, Duke Math. J. **159** (2011), no. 3, 417–477, DOI 10.1215/00127094-1433394. MR2831875

[BES05] J.-M. Barbaroux, M. J. Esteban, and E. Séré, *Some connections between Dirac-Fock and electron-positron Hartree-Fock*, Ann. Henri Poincaré **6** (2005), no. 1, 85–102, DOI 10.1007/s00023-005-0199-7. MR2119356

[BFS98] V. Bach, J. Fröhlich, and I. M. Sigal, *Quantum electrodynamics of confined nonrelativistic particles*, Adv. Math. **137** (1998), no. 2, 299–395, DOI 10.1006/aima.1998.1734. MR1639713

[BG87] A. Berthier and V. Georgescu, *On the point spectrum of Dirac operators*, J. Funct. Anal. **71** (1987), no. 2, 309–338, DOI 10.1016/0022-1236(87)90007-3. MR880983

[BG10] N. Boussaid and S. Golénia, *Limiting absorption principle for some long range perturbations of Dirac systems at threshold energies*, Comm. Math. Phys. **299** (2010), no. 3, 677–708, DOI 10.1007/s00220-010-1099-3. MR2718928

[BGK83] H. Berestycki, T. Gallouët, and O. Kavian, *Équations de champs scalaires euclidiens non linéaires dans le plan* (French, with English summary), C. R. Acad. Sci. Paris Sér. I Math. **297** (1983), no. 5, 307–310. MR734575

[BH92] E. Balslev and B. Helffer, *Limiting absorption principle and resonances for the Dirac operator*, Adv. in Appl. Math. **13** (1992), no. 2, 186–215, DOI 10.1016/0196-8858(92)90009-L. MR1162140

[BH00] P. Binding and R. Hryniv, *Relative boundedness and relative compactness for linear operators in Banach spaces*, Proc. Amer. Math. Soc. **128** (2000), no. 8, 2287–2290, DOI 10.1090/S0002-9939-00-05729-4. MR1756088

[BH15] I. Bejenaru and S. Herr, *The cubic Dirac equation: small initial data in $H^1(\mathbb{R}^3)$*, Comm. Math. Phys. **335** (2015), no. 1, 43–82, DOI 10.1007/s00220-014-2164-0. MR3314499

[BH16] I. Bejenaru and S. Herr, *The cubic Dirac equation: small initial data in $H^{\frac{1}{2}}(\mathbb{R}^2)$*, Comm. Math. Phys. **343** (2016), pp. 515–562.

[BKS91] M. Sh. Birman, G. E. Karadzhov, and M. Z. Solomyak, *Boundedness conditions and spectrum estimates for the operators $b(X)a(D)$ and their analogs*, Estimates and asymptotics for discrete spectra of integral and differential equations (Leningrad, 1989), Adv. Soviet Math., vol. 7, Amer. Math. Soc., Providence, RI, 1991, pp. 85–106. MR1306510

[BL79] H. Berestycki and P.-L. Lions, *Existence d'ondes solitaires dans des problèmes nonlinéaires du type Klein-Gordon* (French, with English summary), C. R. Acad. Sci. Paris Sér. A-B **288** (1979), no. 7, A395–A398. MR552061

[BL83a] H. Berestycki and P.-L. Lions, *Nonlinear scalar field equations. I. Existence of a ground state*, Arch. Rational Mech. Anal. **82** (1983), no. 4, 313–345, DOI 10.1007/BF00250555. MR695535

[BL83b] H. Berestycki and P.-L. Lions, *Nonlinear scalar field equations. II. Existence of infinitely many solutions*, Arch. Rational Mech. Anal. **82** (1983), no. 4, 347–375, DOI 10.1007/BF00250556. MR695536

[BM18] H. Brezis and P. Mironescu, *Gagliardo-Nirenberg inequalities and non-inequalities: the full story*, Ann. Inst. H. Poincaré Anal. Non Linéaire **35** (2018), no. 5, 1355–1376, DOI 10.1016/j.anihpc.2017.11.007. MR3813967

[Bog58] N. N. Bogoljubov, *On a new method in the theory of superconductivity* (English, with Italian summary), Nuovo Cimento (10) **7** (1958), 794–805. MR0106740

[Boh13] N. Bohr, *On the constitution of atoms and molecules*, Phil. Mag. **26** (1913), pp. 1–25.

[Bou96] N. Bournaveas, *Local existence for the Maxwell-Dirac equations in three space dimensions*, Comm. Partial Differential Equations **21** (1996), no. 5-6, 693–720, DOI 10.1080/03605309608821204. MR1391520

[Bou04] N. Bournaveas, *Local and global solutions for a nonlinear Dirac system*, Adv. Differential Equations **9** (2004), no. 5-6, 677–698. MR2099976

[Bou06] N. Boussaid, *Stable directions for small nonlinear Dirac standing waves*, Comm. Math. Phys. **268** (2006), no. 3, 757–817, DOI 10.1007/s00220-006-0112-3. MR2259214

[Bou08a] N. Bournaveas, *Local well-posedness for a nonlinear Dirac equation in spaces of almost critical dimension*, Discrete Contin. Dyn. Syst. **20** (2008), no. 3, 605–616, DOI 10.3934/dcds.2008.20.605. MR2373206

[Bou08b] N. Boussaid, *On the asymptotic stability of small nonlinear Dirac standing waves in a resonant case*, SIAM J. Math. Anal. **40** (2008), no. 4, 1621–1670, DOI 10.1137/070684641. MR2466169

[BP92a] V. S. Buslaev and G. S. Perel′man, *On nonlinear scattering of states which are close to a soliton*, Astérisque **210** (1992), 6, 49–63. Méthodes semi-classiques, Vol. 2 (Nantes, 1991). MR1221351

[BP92b] V. S. Buslaev and G. S. Perel′man, *Scattering for the nonlinear Schrödinger equation: states that are close to a soliton* (Russian, with Russian summary), Algebra i Analiz **4** (1992), no. 6, 63–102; English transl., St. Petersburg Math. J. **4** (1993), no. 6, 1111–1142. MR1199635

[BP95] V. S. Buslaev and G. S. Perel′man, *On the stability of solitary waves for nonlinear Schrödinger equations*, Nonlinear evolution equations, Amer. Math. Soc. Transl. Ser. 2, vol. 164, Amer. Math. Soc., Providence, RI, 1995, pp. 75–98, DOI 10.1090/trans2/164/04. MR1334139

[BPZ98] I. V. Barashenkov, D. E. Pelinovsky, and E. V. Zemlyanaya, *Vibrations and oscillatory instabilities of gap solitons*, Phys. Rev. Lett. **80** (1998), pp. 5117–5120.

[Bre10] H. Brezis, *Functional analysis, Sobolev spaces and partial differential equations*, Universitext, Springer, New York, 2011. MR2759829

[Bro25] L. D. Broglie, *Recherches sur la théorie des quanta*, Ann. d. Phys. III **10** (1925), p. 22.

[BS00] F. Brock and A. Yu. Solynin, *An approach to symmetrization via polarization*, Trans. Amer. Math. Soc. **352** (2000), no. 4, 1759–1796, DOI 10.1090/S0002-9947-99-02558-1. MR1695019

[BS03] V. S. Buslaev and C. Sulem, *On asymptotic stability of solitary waves for nonlinear Schrödinger equations* (English, with English and French summaries), Ann. Inst. H. Poincaré Anal. Non Linéaire **20** (2003), no. 3, 419–475, DOI 10.1016/S0294-1449(02)00018-5. MR1972870

[Bur05] L. Burlando, *On nilpotent operators*, Studia Math. **166** (2005), no. 2, 101–129, DOI 10.4064/sm166-2-1. MR2109585

[Can11] T. Candy, *Global existence for an L^2 critical nonlinear Dirac equation in one dimension*, Adv. Differential Equations **16** (2011), no. 7-8, 643–666. MR2829499

[Caz03] T. Cazenave, *Semilinear Schrödinger equations*, Courant Lecture Notes in Mathematics, vol. 10, New York University, Courant Institute of Mathematical Sciences, New York; American Mathematical Society, Providence, RI, 2003. MR2002047

[CG74] J. M. Chadam and R. T. Glassey, *On certain global solutions of the Cauchy problem for the (classical) coupled Klein-Gordon-Dirac equations in one and three space dimensions*, Arch. Rational Mech. Anal. **54** (1974), 223–237, DOI 10.1007/BF00250789. MR0369952

[CGG14] A. Comech, M. Guan, and S. Gustafson, *On linear instability of solitary waves for the nonlinear Dirac equation* (English, with English and French summaries), Ann. Inst. H. Poincaré Anal. Non Linéaire **31** (2014), no. 3, 639–654, DOI 10.1016/j.anihpc.2013.06.001. MR3208458

[CGM78] S. Coleman, V. Glaser, and A. Martin, *Action minima among solutions to a class of Euclidean scalar field equations*, Comm. Math. Phys. **58** (1978), no. 2, 211–221. MR0468913

[CGNT08] S.-M. Chang, S. Gustafson, K. Nakanishi, and T.-P. Tsai, *Spectra of linearized operators for NLS solitary waves*, SIAM J. Math. Anal. **39** (2007/08), no. 4, 1070–1111, DOI 10.1137/050648389. MR2368894

[CL82] T. Cazenave and P.-L. Lions, *Orbital stability of standing waves for some nonlinear Schrödinger equations*, Comm. Math. Phys. **85** (1982), pp. 549–561.

[CM08] S. Cuccagna and T. Mizumachi, *On asymptotic stability in energy space of ground states for nonlinear Schrödinger equations*, Comm. Math. Phys. **284** (2008), no. 1, 51–77, DOI 10.1007/s00220-008-0605-3. MR2443298

[CMKS$^+$16] J. Cuevas-Maraver, P. G. Kevrekidis, A. Saxena, A. Comech, and R. Lan, *Stability of solitary waves and vortices in a 2D nonlinear Dirac model*, Phys. Rev. Lett. **116** (2016), no. 21, 214101, 6, DOI 10.1103/PhysRevLett.116.214101. MR3613280

[Cot55] M. Cotlar, *A combinatorial inequality and its applications to L^2-spaces*, Rev. Mat. Cuyana **1** (1955), 41–55 (1956). MR0080263

[CP03] A. Comech and D. Pelinovsky, *Purely nonlinear instability of standing waves with minimal energy*, Comm. Pure Appl. Math. **56** (2003), no. 11, 1565–1607, DOI 10.1002/cpa.10104. MR1995870

[CP06] M. Chugunova and D. Pelinovsky, *Block-diagonalization of the symmetric first-order coupled-mode system*, SIAM J. Appl. Dyn. Syst. **5** (2006), no. 1, 66–83, DOI 10.1137/050629781. MR2217129

[CPS16] A. Contreras, D. E. Pelinovsky, and Y. Shimabukuro, *L^2 orbital stability of Dirac solitons in the massive Thirring model*, Comm. Partial Differential Equations **41** (2016), no. 2, 227–255, DOI 10.1080/03605302.2015.1123272. MR3462129

[CPS17] A. Comech, T. V. Phan, and A. Stefanov, *Asymptotic stability of solitary waves in generalized Gross-Neveu model*, Ann. Inst. H. Poincaré Anal. Non Linéaire **34** (2017), no. 1, 157–196, DOI 10.1016/j.anihpc.2015.11.001. MR3592683

[CPV05] S. Cuccagna, D. Pelinovsky, and V. Vougalter, *Spectra of positive and negative energies in the linearized NLS problem*, Comm. Pure Appl. Math. **58** (2005), no. 1, 1–29, DOI 10.1002/cpa.20050. MR2094265

[CS18] A. Comech and D. Stuart, *Small amplitude solitary waves in the Dirac-Maxwell system*, Commun. Pure Appl. Anal. **17** (2018), no. 4, 1349–1370, DOI 10.3934/cpaa.2018066. MR3842865

[CT16] S. Cuccagna and M. Tarulli, *On stabilization of small solutions in the nonlinear Dirac equation with a trapping potential*, J. Math. Anal. Appl. **436** (2016), no. 2, 1332–1368, DOI 10.1016/j.jmaa.2015.12.049. MR3447012

[Cuc01] S. Cuccagna, *Stabilization of solutions to nonlinear Schrödinger equations*, Comm. Pure Appl. Math. **54** (2001), no. 9, 1110–1145, DOI 10.1002/cpa.1018. MR1835384

[Cuc03] S. Cuccagna, *On asymptotic stability of ground states of NLS*, Rev. Math. Phys. **15** (2003), no. 8, 877–903, DOI 10.1142/S0129055X03001849. MR2027616

[Cuc14] S. Cuccagna, *On asymptotic stability of moving ground states of the nonlinear Schrödinger equation*, Trans. Amer. Math. Soc. **366** (2014), no. 6, 2827–2888, DOI 10.1090/S0002-9947-2014-05770-X. MR3180733

[CV86] T. Cazenave and L. Vázquez, *Existence of localized solutions for a classical nonlinear Dirac field*, Comm. Math. Phys. **105** (1986), pp. 35–47.

[Dav95] E. B. Davies, *Spectral theory and differential operators*, Cambridge Studies in Advanced Mathematics, vol. 42, Cambridge University Press, Cambridge, 1995. MR1349825

[DBGRN15] S. De Bièvre, F. Genoud, and S. Rota Nodari, *Orbital stability: analysis meets geometry*, in *Nonlinear optical and atomic systems*, vol. 2146 of *Lecture Notes in Math.*, pp. 147–273, Springer, Cham, 2015.

[DBRN19] S. De Bièvre and S. Rota Nodari, *Orbital stability via the energy-momentum method: the case of higher dimensional symmetry groups*, Arch. Ration. Mech. Anal. **231** (2019), no. 1, 233–284, DOI 10.1007/s00205-018-1278-5. MR3894551

[Der64] G. H. Derrick, *Comments on nonlinear wave equations as models for elementary particles*, J. Mathematical Phys. **5** (1964), 1252–1254, DOI 10.1063/1.1704233. MR174304

[Der12] J. Derezinski, *Open problems about many-body Dirac operators*, Bulletin of International Association of Mathematical Physics (2012).

[DF07] P. D'Ancona and L. Fanelli, *Decay estimates for the wave and Dirac equations with a magnetic potential*, Comm. Pure Appl. Math. **60** (2007), no. 3, 357–392, DOI 10.1002/cpa.20152. MR2284214

[Dir28] P. Dirac, *The quantum theory of the electron*, Proc. R. Soc. Lond. Ser. A **117** (1928), pp. 610–624.

[DJKM17] T. Duyckaerts, H. Jia, C. Kenig, and F. Merle, *Soliton resolution along a sequence of times for the focusing energy critical wave equation*, Geom. Funct. Anal. **27** (2017), no. 4, 798–862, DOI 10.1007/s00039-017-0418-7. MR3678502

[DR79] D. Darby and Th. W. Ruijgrok, *A noncompact gauge group for the Dirac equation*, Acta Phys. Polon. B **10** (1979), no. 11, 959–973. MR570119

[DS58] N. Dunford and J. T. Schwartz, *Linear Operators. I. General Theory*, With the assistance of W. G. Bade and R. G. Bartle. Pure and Applied Mathematics, Vol. 7, Interscience Publishers, Inc., New York; Interscience Publishers, Ltd., London, 1958. MR0117523

[DS63] N. Dunford and J. T. Schwartz, *Linear operators. II. Spectral Theory: self adjoint operators in Hilbert space*, Interscience Publishers, Inc., New York, 1963.

[DS06] L. Demanet and W. Schlag, *Numerical verification of a gap condition for a linearized nonlinear Schrödinger equation*, Nonlinearity **19** (2006), no. 4, 829–852, DOI 10.1088/0951-7715/19/4/004. MR2214946

[Ear42] S. Earnshaw, *On the nature of the molecular forces which regulate the constitution of the luminiferous ether*, Trans. Camb. Phil. Soc **7** (1842), pp. 97–112.

[EE87] D. E. Edmunds and W. D. Evans, *Spectral theory and differential operators*, Oxford Mathematical Monographs, The Clarendon Press, Oxford University Press, New York, 1987. Oxford Science Publications. MR929030

[EGS96] M. J. Esteban, V. Georgiev, and E. Séré, *Stationary solutions of the Maxwell-Dirac and the Klein-Gordon-Dirac equations*, Calc. Var. Partial Differential Equations **4** (1996), no. 3, 265–281, DOI 10.1007/BF01254347. MR1386737

[ELZ83] W. D. Evans, R. T. Lewis, and A. Zettl, *Nonselfadjoint operators and their essential spectra*, in *Ordinary differential equations and operators (Dundee, 1982)*, vol. 1032 of *Lecture Notes in Math.*, pp. 123–160, Springer, Berlin, 1983.

[Enf73] P. Enflo, *A counterexample to the approximation problem in Banach spaces*, Acta Math. **130** (1973), 309–317, DOI 10.1007/BF02392270. MR0402468

[ES95] M. J. Esteban and É. Séré, *Stationary states of the nonlinear Dirac equation: a variational approach*, Comm. Math. Phys. **171** (1995), pp. 323–350.

[ES02] M. J. Esteban and E. Séré, *An overview on linear and nonlinear Dirac equations*, Discrete Contin. Dyn. Syst. **8** (2002), no. 2, 381–397, DOI 10.3934/dcds.2002.8.381. Current developments in partial differential equations (Temuco, 1999). MR1897689

[ES05] M. J. Esteban and É. Séré, *Dirac-Fock models for atoms and molecules and related topics*, in *XIVth International Congress on Mathematical Physics*, pp. 21–28, World Sci. Publ., Hackensack, NJ, 2005.

[EV97] M. Escobedo and L. Vega, *A semilinear Dirac equation in $H^s(\mathbf{R}^3)$ for $s > 1$*, SIAM J. Math. Anal. **28** (1997), pp. 338–362.

[Eva98] L. C. Evans, *Partial differential equations*, Graduate Studies in Mathematics, vol. 19, American Mathematical Society, Providence, RI, 1998. MR1625845

[Fed96] B. V. Fedosov, *Index theorems [MR1135117 (92k:58262)]*, Partial differential equations, VIII, Encyclopaedia Math. Sci., vol. 65, Springer, Berlin, 1996, pp. 155–251, DOI 10.1007/978-3-642-48944-0_3. MR1401125

[FFK56] R. Finkelstein, C. Fronsdal, and P. Kaus, *Nonlinear spinor field*, Phys. Rev. **103** (1956), pp. 1571–1579.

[FLR51] R. Finkelstein, R. LeLevier, and M. Ruderman, *Non-linear spinor fields*, Physical Rev. (2) **83** (1951), 326–332. MR0042204

[Gal77] A. Galindo, *A remarkable invariance of classical Dirac Lagrangians*, Lett. Nuovo Cimento (2) **20** (1977), no. 6, 210–212. MR0503135

[Gil84] P. B. Gilkey, *Invariance theory, the heat equation, and the Atiyah-Singer index theorem*, Mathematics Lecture Series, vol. 11, Publish or Perish, Inc., Wilmington, DE, 1984. MR783634

[GJY04] A. Galtbayar, A. Jensen, and K. Yajima, *Local time-decay of solutions to Schrödinger equations with time-periodic potentials*, J. Statist. Phys. **116** (2004), no. 1-4, 231–282, DOI 10.1023/B:JOSS.0000037203.79298.ec. MR2083143

[GK57] I. C. Gohberg and M. G. Kreĭn, *Fundamental aspects of defect numbers, root numbers and indexes of linear operators* (Russian), Uspehi Mat. Nauk (N.S.) **12** (1957), no. 2(74), 43–118. MR0096978

[GK69] I. C. Gohberg and M. G. Kreĭn, *Introduction to the theory of linear nonselfadjoint operators*, Translated from the Russian by A. Feinstein. Translations of Mathematical Monographs, Vol. 18, American Mathematical Society, Providence, R.I., 1969. MR0246142

[Gla77] R. T. Glassey, *On the blowing up of solutions to the Cauchy problem for nonlinear Schrödinger equations*, J. Math. Phys. **18** (1977), no. 9, 1794–1797, DOI 10.1063/1.523491. MR0460850

[GM74] J. Ginibre and M. Moulin, *Hilbert space approach to the quantum mechanical three-body problem*, Ann. Inst. H. Poincaré Sect. A (N.S.) **21** (1974), 97–145. MR0368656

[GN74] D. J. Gross and A. Neveu, *Dynamical symmetry breaking in asymptotically free field theories*, Phys. Rev. D **10** (1974), pp. 3235–3253.

[GNNN81] B. Gidas, W. M. Ni, and L. Nirenberg, *Symmetry of positive solutions of nonlinear elliptic equations in* $\mathbf{R}^n$, Mathematical analysis and applications, Part A, Adv. in Math. Suppl. Stud., vol. 7, Academic Press, New York-London, 1981, pp. 369–402. MR634248

[GNT04] S. Gustafson, K. Nakanishi, and T.-P. Tsai, *Asymptotic stability and completeness in the energy space for nonlinear Schrödinger equations with small solitary waves*, Int. Math. Res. Not. **66** (2004), 3559–3584, DOI 10.1155/S1073792804132340. MR2101699

[GO12] V. Georgiev and M. Ohta, *Nonlinear instability of linearly unstable standing waves for nonlinear Schrödinger equations*, J. Math. Soc. Japan **64** (2012), no. 2, 533–548. MR2916078

[Gol66] S. Goldberg, *Unbounded linear operators: Theory and applications*, McGraw-Hill Book Co., New York-Toronto, Ont.-London, 1966. MR0200692

[Gri88] M. Grillakis, *Linearized instability for nonlinear Schrödinger and Klein-Gordon equations*, Comm. Pure Appl. Math. **41** (1988), no. 6, 747–774, DOI 10.1002/cpa.3160410602. MR948770

[Gro66] L. Gross, *The Cauchy problem for the coupled Maxwell and Dirac equations*, Comm. Pure Appl. Math. **19** (1966), 1–15, DOI 10.1016/S0079-8169(08)61684-0. MR0190520

[GRS46] I. M. Gel′fand, D. A. Raĭkov, and G. E. Šilov, *Commutative normed rings* (Russian), Uspehi Matem. Nauk (N. S.) **1** (1946), no. 2(12), 48–146. MR0027130

[GS71] I. C. Gohberg and E. I. Sigal, *An operator generalization of the logarithmic residue theorem and Rouché's theorem* (Russian), Mat. Sb. (N.S.) **84(126)** (1971), 607–629. MR0313856

[GSS87] M. Grillakis, J. Shatah, and W. Strauss, *Stability theory of solitary waves in the presence of symmetry. I*, J. Funct. Anal. **74** (1987), no. 1, 160–197, DOI 10.1016/0022-1236(87)90044-9. MR901236

[GT83] D. Gilbarg and N. S. Trudinger, *Elliptic partial differential equations of second order*, 2nd ed., Grundlehren der Mathematischen Wissenschaften [Fundamental Principles of Mathematical Sciences], vol. 224, Springer-Verlag, Berlin, 1983. MR737190

[GT01] D. Gilbarg and N. S. Trudinger, *Elliptic partial differential equations of second order*, Classics in Mathematics, Springer-Verlag, Berlin, 2001. Reprint of the 1998 edition. MR1814364

[Gua08] M. Guan, *Solitary wave solutions for the nonlinear Dirac equations*, ArXiv e-prints (2008), arXiv:0812.2273.

[GV92] J. Ginibre and G. Velo, *Smoothing properties and retarded estimates for some dispersive evolution equations*, Comm. Math. Phys. **144** (1992), no. 1, 163–188. MR1151250

[Hei57] W. Heisenberg, *Quantum theory of fields and elementary particles*, Rev. Mod. Phys. **29** (1957), 269–278, DOI 10.1103/revmodphys.29.269. MR0090429

[Hes69] P. Hess, *Zur Störungstheorie linearer Operatoren: Relative Beschränktheit und relative Kompaktheit von Operatoren in Banachräumen* (German), Comment. Math. Helv. **44** (1969), 245–248, DOI 10.1007/BF02564526. MR0246156

[Hes70] P. Hess, *Zur Störungstheorie linearer Operatoren in Banachräumen* (German), Comment. Math. Helv. **45** (1970), 229–235, DOI 10.1007/BF02567328. MR0264432

[His00] P. D. Hislop, *Exponential decay of two-body eigenfunctions: a review*, Proceedings of the Symposium on Mathematical Physics and Quantum Field Theory (Berkeley, CA, 1999), Electron. J. Differ. Equ. Conf., vol. 4, Southwest Texas State Univ., San Marcos, TX, 2000, pp. 265–288. MR1785381

[HM15] H. Huh and B. Moon, *Low regularity well-posedness for Gross-Neveu equations*, Commun. Pure Appl. Anal. **14** (2015), no. 5, 1903–1913, DOI 10.3934/cpaa.2015.14.1903. MR3359550

[Hob63] R. H. Hobart, *On the instability of a class of unitary field models*, Proc. Phys. Soc. **82** (1963), 201–203. MR0155591

[Hör83] L. Hörmander, *The analysis of linear partial differential operators. I*, Grundlehren der Mathematischen Wissenschaften [Fundamental Principles of Mathematical Sciences], vol. 257, Springer-Verlag, Berlin, 1983. Differential operators with constant coefficients. MR705278

[Huh11] H. Huh, *Global strong solution to the Thirring model in critical space*, J. Math. Anal. Appl. **381** (2011), no. 2, 513–520, DOI 10.1016/j.jmaa.2011.02.042. MR2802088

[Huh13a] H. Huh, *Global solutions to Gross-Neveu equation*, Lett. Math. Phys. **103** (2013), no. 8, 927–931, DOI 10.1007/s11005-013-0622-9. MR3063953

[Huh13b] H. Huh, *Global strong solutions to some nonlinear Dirac equations in super-critical space*, Abstr. Appl. Anal., posted on 2013, Art. ID 602753, 8, DOI 10.1155/2013/602753. MR3070190

[Huh14] H. Huh, *On the growth rate of solutions to Gross-Neveu and Thirring equations*, Commun. Korean Math. Soc. **29** (2014), no. 2, 263–267, DOI 10.4134/CKMS.2014.29.2.263. MR3206413

[Hur98] A. Hurwitz, *Ueber die Composition der quadratischen Formen von belibig vielen Variablen*, Nachrichten von der Gesellschaft der Wissenschaften zu Göttingen, Mathematisch-Physikalische Klasse **1898** (1898), pp. 309–316.

[Hur22] A. Hurwitz, *Über die Komposition der quadratischen Formen* (German), Math. Ann. **88** (1922), no. 1-2, 1–25, DOI 10.1007/BF01448439. MR1512117

[Ily61] V. P. Ilyin, *Some integral inequalities and their applications in the theory of differentiable functions of several variables*, Mat. Sb. (N.S.) **54 (96)** (1961), pp. 331–380.

[IM99] A. Iftimovici and M. Măntoiu, *Limiting absorption principle at critical values for the Dirac operator*, Lett. Math. Phys. **49** (1999), no. 3, 235–243, DOI 10.1023/A:1007625918845. MR1743451

[Iva38] D. D. Ivanenko, *Notes to the theory of interaction via particles*, Zh. Éksp. Teor. Fiz **8** (1938), pp. 260–266.

[Jen80] A. Jensen, *Spectral properties of Schrödinger operators and time-decay of the wave functions results in* $L^2(\mathbf{R}^m)$, $m \geq 5$, Duke Math. J. **47** (1980), no. 1, 57–80. MR563367

[Jer86] D. Jerison, *Carleman inequalities for the Dirac and Laplace operators and unique continuation*, Adv. in Math. **62** (1986), no. 2, 118–134, DOI 10.1016/0001-8708(86)90096-4. MR865834

[JK79] A. Jensen and T. Kato, *Spectral properties of Schrödinger operators and time-decay of the wave functions*, Duke Math. J. **46** (1979), pp. 583–611.

[JN01] A. Jensen and G. Nenciu, *A unified approach to resolvent expansions at thresholds*, Rev. Math. Phys. **13** (2001), no. 6, 717–754, DOI 10.1142/S0129055X01000843. MR1841744

[Jör61] K. Jörgens, *Das Anfangswertproblem im Grossen für eine Klasse nichtlinearer Wellengleichungen* (German), Math. Z. **77** (1961), 295–308, DOI 10.1007/BF01180181. MR0130462

[JSS91] J.-L. Journé, A. Soffer, and C. D. Sogge, *Decay estimates for Schrödinger operators*, Comm. Pure Appl. Math. **44** (1991), no. 5, 573–604, DOI 10.1002/cpa.3160440504. MR1105875

[Kat58] T. Kato, *Perturbation theory for nullity, deficiency and other quantities of linear operators*, J. Analyse Math. **6** (1958), 261–322, DOI 10.1007/BF02790238. MR0107819

[Kat76] T. Kato, *Perturbation theory for linear operators*, 2nd ed., Springer-Verlag, Berlin-New York, 1976. Grundlehren der Mathematischen Wissenschaften, Band 132. MR0407617

[Kel51] M. V. Keldyš, *On the characteristic values and characteristic functions of certain classes of non-self-adjoint equations* (Russian), Doklady Akad. Nauk SSSR (N.S.) **77** (1951), 11–14. MR0041353

[Kel71] M. V. Keldyš, *The completeness of eigenfunctions of certain classes of nonselfadjoint linear operators* (Russian), Uspehi Mat. Nauk **26** (1971), no. 4(160), 15–41. MR0300125

[Kel83] C. Keller, *Large-time asymptotic behavior of solutions of nonlinear wave equations perturbed from a stationary ground state*, Comm. Partial Differential Equations **8** (1983), no. 10, 1073–1099, DOI 10.1080/03605308308820296. MR709163

[Kes61] H. Kestelman, *Anticommuting linear transformations*, Canadian J. Math. **13** (1961), 614–624, DOI 10.4153/CJM-1961-050-2. MR148679

[KK07] A. Komech and A. Komech, *Global attractor for a nonlinear oscillator coupled to the Klein-Gordon field*, Arch. Ration. Mech. Anal. **185** (2007), no. 1, 105–142, DOI 10.1007/s00205-006-0039-z. MR2308860

[KK12] A. Komech and E. Kopylova, *Dispersion decay and Scattering Theory*, John Wiley & Sons Inc., Hoboken, NJ, 2012.

[KLLS15] C. Kenig, A. Lawrie, B. Liu, and W. Schlag, *Stable soliton resolution for exterior wave maps in all equivariance classes*, Adv. Math. **285** (2015), 235–300, DOI 10.1016/j.aim.2015.08.007. MR3406501

[KO90] H. Kozono and T. Ozawa, *Relative bounds of closable operators in nonreflexive Banach spaces*, Hokkaido Math. J. **19** (1990), no. 2, 241–247, DOI 10.14492/hokmj/1381517358. MR1059168

[Kol73] A. A. Kolokolov,*Stability of the dominant mode of the nonlinear wave equation in a cubic medium*, J. Appl. Mech. Tech. Phys. **14** (1973), pp. 426–428.

[Kom03] A. I. Komech, *On attractor of a singular nonlinear* U(1)*-invariant Klein-Gordon equation*, Progress in analysis, Vol. I, II (Berlin, 2001), World Sci. Publ., River Edge, NJ, 2003, pp. 599–611. MR2032730

[Kom12] A. Komech, *Quantum mechanics: genesis and achievements*, Springer, Dordrecht, 2013. MR3012343

[KOY15] H. Kalf, T. Okaji, and O. Yamada, *The Dirac operator with mass $m_0 \geq 0$: non-existence of zero modes and of threshold eigenvalues*, Doc. Math. **20** (2015), 37–64. MR3398708

[KS02] T. Kapitula and B. Sandstede, *Edge bifurcations for near integrable systems via Evans function techniques*, SIAM J. Math. Anal. **33** (2002), no. 5, 1117–1143, DOI 10.1137/S0036141000372301. MR1897705

[KS07] P. Karageorgis and W. A. Strauss, *Instability of steady states for nonlinear wave and heat equations*, J. Differential Equations **241** (2007), no. 1, 184–205, DOI 10.1016/j.jde.2007.06.006. MR2356215

[KT98] M. Keel and T. Tao, *Endpoint Strichartz estimates*, Amer. J. Math. **120** (1998), no. 5, 955–980. MR1646048

[Kur78] S. T. Kuroda, *An introduction to scattering theory*, Lecture Notes Series, vol. 51, Aarhus Universitet, Matematisk Institut, Aarhus, 1978. MR528757

[Kwo89] M. K. Kwong, *Uniqueness of positive solutions of $\Delta u - u + u^p = 0$ in $\mathbf{R}^n$*, Arch. Rational Mech. Anal. **105** (1989), no. 3, 243–266, DOI 10.1007/BF00251502. MR969899

[KY89] T. Kato and K. Yajima, *Some examples of smooth operators and the associated smoothing effect*, Rev. Math. Phys. **1** (1989), no. 4, 481–496, DOI 10.1142/S0129055X89000171. MR1061120

[Lak18] T. I. Lakoba, *Numerical study of solitary wave stability in cubic nonlinear Dirac equations in 1D*, Phys. Lett. A **382** (2018), no. 5, 300–308, DOI 10.1016/j.physleta.2017.11.032. MR3739696

[Lev14] A. Levitt, *Solutions of the multiconfiguration Dirac-Fock equations*, Rev. Math. Phys. **26** (2014), no. 7, 1450014, 25, DOI 10.1142/S0129055X14500147. MR3246964

[LG75] S. Y. Lee and A. Gavrielides, *Quantization of the localized solutions in two-dimensional field theories of massive fermions*, Phys. Rev. D **12** (1975), pp. 3880–3886.

[LR47] W. E. Lamb and R. C. Retherford, *Fine structure of the hydrogen atom by a microwave method*, Phys. Rev. **72** (1947), pp. 241–243.

[LS75] B. M. Levitan and I. S. Sargsjan, *Introduction to spectral theory: selfadjoint ordinary differential operators*, American Mathematical Society, Providence, R.I., 1975. Translated from the Russian by Amiel Feinstein; Translations of Mathematical Monographs, Vol. 39. MR0369797

[Mar88] A. S. Markus, *Introduction to the spectral theory of polynomial operator pencils*, Translations of Mathematical Monographs, vol. 71, American Mathematical Society, Providence, RI, 1988. Translated from the Russian by H. H. McFaden; Translation edited by Ben Silver; With an appendix by M. V. Keldysh. MR971506

[McL93] K. McLeod, *Uniqueness of positive radial solutions of $\Delta u + f(u) = 0$ in $\mathbf{R}^n$. II*, Trans. Amer. Math. Soc. **339** (1993), no. 2, 495–505, DOI 10.2307/2154282. MR1201323

[Meh01] J. Mehra, *The Golden Age of Theoretical Physics*, The Golden Age of Theoretical Physics, World Scientific, 2001.

[Mer88] F. Merle, *Existence of stationary states for nonlinear Dirac equations*, J. Differential Equations **74** (1988), no. 1, 50–68, DOI 10.1016/0022-0396(88)90018-6. MR949625

[Mer90] F. Merle, *Construction of solutions with exactly k blow-up points for the Schrödinger equation with critical nonlinearity*, Comm. Math. Phys. **129** (1990), pp. 223–240.

[Miz08] T. Mizumachi, *Asymptotic stability of small solitary waves to 1D nonlinear Schrödinger equations with potential*, J. Math. Kyoto Univ. **48** (2008), no. 3, 471–497, DOI 10.1215/kjm/1250271380. MR2511047

[MM03] R. Mennicken and M. Möller, *Non-self-adjoint boundary eigenvalue problems*, North-Holland Mathematics Studies, vol. 192, North-Holland Publishing Co., Amsterdam, 2003. MR1995773

[MNNO05] S. Machihara, M. Nakamura, K. Nakanishi, and T. Ozawa, *Endpoint Strichartz estimates and global solutions for the nonlinear Dirac equation*, J. Funct. Anal. **219** (2005), no. 1, 1–20, DOI 10.1016/j.jfa.2004.07.005. MR2108356

[MNO03] S. Machihara, K. Nakanishi, and T. Ozawa, *Small global solutions and the non-relativistic limit for the nonlinear Dirac equation*, Rev. Mat. Iberoamericana **19** (2003), no. 1, 179–194, DOI 10.4171/RMI/342. MR1993419

[MNT10] S. Machihara, K. Nakanishi, and K. Tsugawa, *Well-posedness for nonlinear Dirac equations in one dimension*, Kyoto J. Math. **50** (2010), no. 2, 403–451, DOI 10.1215/0023608X-2009-018. MR2666663

[MO07] S. Machihara and T. Omoso, *The explicit solutions to the nonlinear Dirac equation and Dirac-Klein-Gordon equation*, Ric. Mat. **56** (2007), no. 1, 19–30, DOI 10.1007/s11587-007-0002-9. MR2330348

[Moo94] W. Moore, *A life of Erwin Schrödinger*, Canto, Cambridge University Press, Cambridge, 1994. MR1302170

[MS70] A. S. Markus and E. I. Sigal, *The multiplicity of the characteristic number of an analytic operator function* (Russian), Mat. Issled. **5** (1970), no. 3(17), 129–147. MR0312306

[MS72] C. S. Morawetz and W. A. Strauss, *Decay and scattering of solutions of a nonlinear relativistic wave equation*, Comm. Pure Appl. Math. **25** (1972), 1–31, DOI 10.1002/cpa.3160250103. MR0303097

[Nai72] M. Naimark, *Normed algebras*, Wolters-Noordhoff, 1972, 1 edn.

[NW73] L. Nirenberg and H. F. Walker, *The null spaces of elliptic partial differential operators in* $\mathbf{R}^n$, J. Math. Anal. Appl. **42** (1973), 271–301, DOI 10.1016/0022-247X(73)90138-8. Collection of articles dedicated to Salomon Bochner. MR0320821

[Oku91] S. Okubo, *Real representations of finite Clifford algebras. I. Classification*, J. Math. Phys. **32** (1991), no. 7, 1657–1668, DOI 10.1063/1.529277. MR1112690

[Oun00] H. Ounaies, *Perturbation method for a class of nonlinear Dirac equations*, Differential Integral Equations **13** (2000), no. 4-6, 707–720. MR1750047

[OY04] T. Ozawa and K. Yamauchi, *Structure of Dirac matrices and invariants for nonlinear Dirac equations*, Differential Integral Equations **17** (2004), no. 9-10, 971–982. MR2082456

[Pau36] W. Pauli, *Contributions mathématiques à la théorie des matrices de Dirac* (French), Ann. Inst. H. Poincaré **6** (1936), no. 2, 109–136. MR1508031

[Pel11] D. Pelinovsky, *Survey on global existence in the nonlinear Dirac equations in one spatial dimension*, Harmonic analysis and nonlinear partial differential equations, RIMS Kôkyûroku Bessatsu, B26, Res. Inst. Math. Sci. (RIMS), Kyoto, 2011, pp. 37–50. MR2883845

[Pok65] S. I. Pohožaev, *On the eigenfunctions of the equation* $\Delta u + \lambda f(u) = 0$ (Russian), Dokl. Akad. Nauk SSSR **165** (1965), 36–39. MR0192184

[PS05] S. Puntanen and G. Styan, *Historical introduction: Issai Schur and the early development of the Schur complement*, in F. Zhang, editor, *The Schur Complement and its Applications*, vol. 4 of *Numerical Methods and Algorithms*, pp. 1–16, Springer, 2005.

[PS12] D. E. Pelinovsky and A. Stefanov, *Asymptotic stability of small gap solitons in nonlinear Dirac equations*, J. Math. Phys. **53** (2012), no. 7, 073705, 27, DOI 10.1063/1.4731477. MR2985264

[PS14] D. E. Pelinovsky and Y. Shimabukuro, *Orbital stability of Dirac solitons*, Lett. Math. Phys. **104** (2014), no. 1, 21–41, DOI 10.1007/s11005-013-0650-5. MR3147657

[PSK04] D. E. Pelinovsky, A. A. Sukhorukov, and Y. S. Kivshar, *Bifurcations and stability of gap solitons in periodic potentials*, Phys. Rev. E (3) **70** (2004), no. 3, 036618, 17, DOI 10.1103/PhysRevE.70.036618. MR2130334

[PW97] C.-A. Pillet and C. E. Wayne, *Invariant manifolds for a class of dispersive, Hamiltonian, partial differential equations*, J. Differential Equations **141** (1997), no. 2, 310–326, DOI 10.1006/jdeq.1997.3345. MR1488355

[Rad22] J. Radon, *Lineare Scharen orthogonaler Matrizen* (German), Abh. Math. Sem. Univ. Hamburg **1** (1922), no. 1, 1–14, DOI 10.1007/BF02940576. MR3069384

[Rañ83a] A. F. Rañada, *Classical nonlinear Dirac field models of extended particles*, in A. O. Barut, editor, *Quantum theory, groups, fields and particles*, vol. 4 of *Mathematical Physics Studies*, pp. 271–291, D. Reidel Publishing Co., Dordrecht-Boston, Mass., 1983.

[Rañ83b] A. F. Rañada, *On nonlinear classical Dirac fields and quantum physics*, in *Old and new questions in physics, cosmology, philosophy, and theoretical biology*, pp. 363–376, Plenum, New York, 1983.

[Rau78] J. Rauch, *Local decay of scattering solutions to Schrödinger's equation*, Comm. Math. Phys. **61** (1978), no. 2, 149–168. MR0495958

[RN10a] S. Rota Nodari, *Perturbation method for particle-like solutions of the Einstein-Dirac-Maxwell equations* (English, with English and French summaries), C. R. Math. Acad. Sci. Paris **348** (2010), no. 13-14, 791–794, DOI 10.1016/j.crma.2010.06.003. MR2671162

[RN10b] S. Rota Nodari, *Perturbation method for particle-like solutions of the Einstein-Dirac equations*, Ann. Henri Poincaré **10** (2010), no. 7, 1377–1393, DOI 10.1007/s00023-009-0015-x. MR2593110

[RS78] M. Reed and B. Simon, *Methods of modern mathematical physics. IV. Analysis of operators*, Academic Press [Harcourt Brace Jovanovich, Publishers], New York-London, 1978. MR0493421

[RS80] M. Reed and B. Simon, *Methods of modern mathematical physics. I*, 2nd ed., Academic Press, Inc. [Harcourt Brace Jovanovich, Publishers], New York, 1980. Functional analysis. MR751959

[RS04] I. Rodnianski and W. Schlag, *Time decay for solutions of Schrödinger equations with rough and time-dependent potentials*, Invent. Math. **155** (2004), no. 3, 451–513, DOI 10.1007/s00222-003-0325-4. MR2038194

[RSN56] F. Riesz and B. Sz.-Nagy, *Functional Analysis*, Blackie & Son Limited, 1956.

[Rud73] W. Rudin, *Functional analysis*, McGraw-Hill Book Co., New York-Düsseldorf-Johannesburg, 1973. McGraw-Hill Series in Higher Mathematics. MR0365062

[Sch26] E. Schrödinger, *Quantisierung als Eigenwertproblem*, Ann. Phys. **384** (1926), pp. 361–376.

[Sch49] L. I. Schiff, *Quantum Mechanics*, McGraw-Hill, New York, 1949.

[Sch51a] L. I. Schiff, *Nonlinear meson theory of nuclear forces. I. Neutral scalar mesons with point-contact repulsion*, Phys. Rev. **84** (1951), pp. 1–9.

[Sch51b] L. I. Schiff, *Nonlinear meson theory of nuclear forces. II. Nonlinearity in the meson-nucleon coupling*, Phys. Rev. **84** (1951), pp. 10–11.

[Sch66] M. Schechter, *On the essential spectrum of an arbitrary operator. I*, J. Math. Anal. Appl. **13** (1966), 205–215, DOI 10.1016/0022-247X(66)90085-0. MR0188798

[Sch86] M. Schechter, *Spectra of partial differential operators*, 2nd ed., North-Holland Series in Applied Mathematics and Mechanics, vol. 14, North-Holland Publishing Co., Amsterdam, 1986. MR869254

[Seg63] I. E. Segal, *The global Cauchy problem for a relativistic scalar field with power interaction*, Bull. Soc. Math. France **91** (1963), 129–135. MR0153967

[Seg66] I. Segal, *Quantization and dispersion for nonlinear relativistic equations*, Mathematical Theory of Elementary Particles (Proc. Conf., Dedham, Mass., 1965), M.I.T. Press, Cambridge, Mass., 1966, pp. 79–108. MR0217453

[Ser77] J.-P. Serre, *Linear representations of finite groups*, Springer-Verlag, New York-Heidelberg, 1977. Translated from the second French edition by Leonard L. Scott; Graduate Texts in Mathematics, Vol. 42. MR0450380

[Sha77] D. B. Shapiro, *Spaces of similarities. I. The Hurwitz problem*, J. Algebra **46** (1977), no. 1, 148–170, DOI 10.1016/0021-8693(77)90397-0. MR0453636

[Sha83] J. Shatah, *Stable standing waves of nonlinear Klein–Gordon equations*, Comm. Math. Phys. **91** (1983), pp. 313–327.

[Sha85] J. Shatah, *Unstable ground state of nonlinear Klein-Gordon equations*, Trans. Amer. Math. Soc. **290** (1985), no. 2, 701–710, DOI 10.2307/2000308. MR792821

[Sha00] D. B. Shapiro, *Compositions of quadratic forms*, De Gruyter Expositions in Mathematics, vol. 33, Walter de Gruyter & Co., Berlin, 2000. MR1786291

[Shi11] D. S. Shirokov, *Extension of Pauli's theorem to the case of Clifford algebras* (Russian), Dokl. Akad. Nauk **440** (2011), no. 5, 607–610, DOI 10.1134/S1064562411060329; English transl., Dokl. Math. **84** (2011), no. 2, 699–701. MR2963685

[Shi13] D. S. Shirokov, *Pauli theorem in the description of n-dimensional spinors in the Clifford algebra formalism*, Theoret. and Math. Phys. **175** (2013), no. 1, 454–474, DOI 10.1007/s11232-013-0038-9. Translation of Teoret. Mat. Fiz. **175** (2013), no. 1, 11–34. MR3172130

[Sig93] I. M. Sigal, *Nonlinear wave and Schrödinger equations. I. Instability of periodic and quasiperiodic solutions*, Comm. Math. Phys. **153** (1993), no. 2, 297–320. MR1218303

[Sim76] B. Simon, *The bound state of weakly coupled Schrödinger operators in one and two dimensions*, Ann. Physics **97** (1976), no. 2, 279–288, DOI 10.1016/0003-4916(76)90038-5. MR0404846

[Slo09] V. A. Sloushch, *Generalizations of the Cwikel estimate for integral operators*, Proceedings of the St. Petersburg Mathematical Society. Vol. XIV, Amer. Math. Soc. Transl. Ser. 2, vol. 228, Amer. Math. Soc., Providence, RI, 2009, pp. 133–155, DOI 10.1090/trans2/228/06. MR2584397

[Slo13] V. A. Sloushch, *A Cwikel-type estimate as a consequence of some properties of the heat kernel* (Russian, with Russian summary), Algebra i Analiz **25** (2013), no. 5, 173–201, DOI 10.1090/S1061-0022-2014-01318-0; English transl., St. Petersburg Math. J. **25** (2014), no. 5, 835–854. MR3184610

[Sof06] A. Soffer, *Soliton dynamics and scattering*, in *International Congress of Mathematicians. Vol. III*, pp. 459–471, Eur. Math. Soc., Zürich, 2006.

[Sog93] C. D. Sogge, *Fourier integrals in classical analysis*, Cambridge Tracts in Mathematics, vol. 105, Cambridge University Press, Cambridge, 1993. MR1205579

[Sol70] M. Soler, *Classical, stable, nonlinear spinor field with positive rest energy*, Phys. Rev. D **1** (1970), pp. 2766–2769.

[SS94] C. M. D. Sterke and J. Sipe, *Gap solitons*, Progress in Optics **33** (1994), pp. 203–260.

[SS00] J. Shatah and W. Strauss, *Spectral condition for instability*, in *Nonlinear PDE's, dynamics and continuum physics (South Hadley, MA, 1998)*, vol. 255 of *Contemp. Math.*, pp. 189–198, Amer. Math. Soc., Providence, RI, 2000.

[ST10] S. Selberg and A. Tesfahun, *Low regularity well-posedness for some nonlinear Dirac equations in one space dimension*, Differential Integral Equations **23** (2010), no. 3-4, 265–278. MR2588476

[Ste70] E. M. Stein, *Singular integrals and differentiability properties of functions*, Princeton Mathematical Series, No. 30, Princeton University Press, Princeton, N.J., 1970. MR0290095

[Ste93] E. M. Stein, *Harmonic analysis: real-variable methods, orthogonality, and oscillatory integrals*, Princeton Mathematical Series, vol. 43, Princeton University Press, Princeton, NJ, 1993. With the assistance of Timothy S. Murphy; Monographs in Harmonic Analysis, III. MR1232192

[Str77] J. Strutt, *The theory of sound*, no. v. 1 in The theory of sound, Macmillan and Co., London, 1877.

[Str78] J. Strutt, *The theory of sound*, no. v. 2 in The theory of sound, Macmillan and Co., London, 1878.

[Str68] W. A. Strauss, *Decay and asymptotics for $cmu = F(u)$*, J. Functional Analysis **2** (1968), 409–457, DOI 10.1016/0022-1236(68)90004-9. MR0233062

[Str77a] W. A. Strauss, *Existence of solitary waves in higher dimensions*, Comm. Math. Phys. **55** (1977), no. 2, 149–162. MR0454365

[Str77b] R. S. Strichartz, *Restrictions of Fourier transforms to quadratic surfaces and decay of solutions of wave equations*, Duke Math. J. **44** (1977), no. 3, 705–714. MR512086

[Str89] W. A. Strauss, *Nonlinear wave equations*, CBMS Regional Conference Series in Mathematics, vol. 73, Published for the Conference Board of the Mathematical Sciences, Washington, DC; by the American Mathematical Society, Providence, RI, 1989. MR1032250

[Stu10] D. Stuart, *Existence and Newtonian limit of nonlinear bound states in the Einstein-Dirac system*, J. Math. Phys. **51** (2010), no. 3, 032501, 13, DOI 10.1063/1.3294085. MR2647868

[SW90] A. Soffer and M. I. Weinstein, *Multichannel nonlinear scattering for nonintegrable equations*, Comm. Math. Phys. **133** (1990), pp. 119–146.

[SW92] A. Soffer and M. I. Weinstein, *Multichannel nonlinear scattering for nonintegrable equations. II. The case of anisotropic potentials and data*, J. Differential Equations **98** (1992), no. 2, 376–390, DOI 10.1016/0022-0396(92)90098-8. MR1170476

[SW99] A. Soffer and M. I. Weinstein, *Resonances, radiation damping and instability in Hamiltonian nonlinear wave equations*, Invent. Math. **136** (1999), no. 1, 9–74, DOI 10.1007/s002220050303. MR1681113

[Tao07] T. Tao, *A (concentration-)compact attractor for high-dimensional non-linear Schrödinger equations*, Dyn. Partial Differ. Equ. **4** (2007), no. 1, 1–53, DOI 10.4310/DPDE.2007.v4.n1.a1. MR2304091

[Tao09] T. Tao, *Why are solitons stable?*, Bull. Amer. Math. Soc. (N.S.) **46** (2009), no. 1, 1–33, DOI 10.1090/S0273-0979-08-01228-7. MR2457070

[Tha92] B. Thaller, *The Dirac equation*, Texts and Monographs in Physics, Springer-Verlag, Berlin, 1992. MR1219537

[Thi58] W. E. Thirring, *A soluble relativistic field theory*, Ann. Physics **3** (1958), 91–112, DOI 10.1016/0003-4916(58)90015-0. MR0091788

[Tho97] J. Thomson, *Cathode rays*, Philosophical Magazine **44** (1897), pp. 293–316.

[Tho04] J. Thomson, *On the structure of the atom: an investigation of the stability and periods of oscillation of a number of corpuscles arranged at equal intervals around the circumference of a circle; with application of the results;to;the theory of atomic structure*, Philosophical Magazine **7** (1904), pp. 237–265.

[TY02a] T.-P. Tsai and H.-T. Yau, *Asymptotic dynamics of nonlinear Schrödinger equations: resonance-dominated and dispersion-dominated solutions*, Comm. Pure Appl. Math. **55** (2002), no. 2, 153–216, DOI 10.1002/cpa.3012. MR1865414

[TY02b] T.-P. Tsai and H.-T. Yau, *Classification of asymptotic profiles for nonlinear Schrödinger equations with small initial data*, Adv. Theor. Math. Phys. **6** (2002), no. 1, 107–139, DOI 10.4310/ATMP.2002.v6.n1.a2. MR1992875

[TY02c] T.-P. Tsai and H.-T. Yau, *Relaxation of excited states in nonlinear Schrödinger equations*, Int. Math. Res. Not. (2002), pp. 1629–1673.

[TY02d] T.-P. Tsai and H.-T. Yau, *Stable directions for excited states of nonlinear Schrödinger equations*, Comm. Partial Differential Equations **27** (2002), no. 11-12, 2363–2402, DOI 10.1081/PDE-120016161. MR1944033

[Vaz77] L. Vazquez, *Localised solutions of a non-linear spinor field*, J. Phys. A **10** (1977), no. 8, 1361–1368. MR0456046

[vdW74] B. L. van der Waerden, *Group theory and quantum mechanics*, Springer-Verlag, New York-Heidelberg, 1974. Translated from the 1932 German original; Die Grundlehren der mathematischen Wissenschaften, Band 214. MR0479090

[VK73] N. G. Vakhitov and A. A. Kolokolov, *Stationary solutions of the wave equation in the medium with nonlinearity saturation*, Radiophys. Quantum Electron. **16** (1973), pp. 783–789.

[vNW29] J. von Neumann and E. Wigner, *Uber merkwürdige diskrete Eigenwerte. Uber das Verhalten von Eigenwerten bei adiabatischen prozessen*, Physikalische Zeitschrift **30** (1929), pp. 467–470.

[Wak66] M. Wakano, *Intensely localized solutions of the classical Dirac-Maxwell field equations*, Progr. Theoret. Phys. **35** (1966), pp. 1117–1141.

[Wei82] J. Weidmann, *Absolut stetiges Spektrum bei Sturm-Liouville-Operatoren und Dirac-Systemen* (German), Math. Z. **180** (1982), no. 3, 423–427, DOI 10.1007/BF01214182. MR664527

[Wei83] M. I. Weinstein, *Nonlinear Schrödinger equations and sharp interpolation estimates*, Comm. Math. Phys. **87** (1982/83), no. 4, 567–576. MR691044

[Wei85] M. I. Weinstein, *Modulational stability of ground states of nonlinear Schrödinger equations*, SIAM J. Math. Anal. **16** (1985), no. 3, 472–491, DOI 10.1137/0516034. MR783974

[Wei86] M. I. Weinstein, *Lyapunov stability of ground states of nonlinear dispersive evolution equations*, Comm. Pure Appl. Math. **39** (1986), no. 1, 51–67, DOI 10.1002/cpa.3160390103. MR820338

[Wei17] S. Weinberg, *The Trouble with Quantum Mechanics*, The New York Review of Books **64** (2017), pp. 51–53.

[Wey09] H. Weyl, *Über beschränkte quadratische Formen, deren Differenz vollstetig ist*, Rendiconti del Circolo Matematico di Palermo **27** (1909), pp. 373–392.

[Wol59] F. Wolf, *On the essential spectrum of partial differential boundary problems.*, Comm. Pure Appl. Math. **12** (1959), 211–228, DOI 10.1002/cpa.3160120202. MR0107750

[Yaf10] D. R. Yafaev, *Mathematical scattering theory*, Mathematical Surveys and Monographs, vol. 158, American Mathematical Society, Providence, RI, 2010. Analytic theory. MR2598115

[Yaj87] K. Yajima, *Existence of solutions for Schrödinger evolution equations*, Comm. Math. Phys. **110** (1987), no. 3, 415–426. MR891945

[Yam73] O. Yamada, *On the principle of limiting absorption for the Dirac operator*, Publ. Res. Inst. Math. Sci. **8** (1972/73), 557–577, DOI 10.2977/prims/1195192961. MR0320547

[Yos80] K. Yosida, *Functional analysis*, 6th ed., Grundlehren der Mathematischen Wissenschaften [Fundamental Principles of Mathematical Sciences], vol. 123, Springer-Verlag, Berlin-New York, 1980. MR617913

[Zak67] V. Zakharov, *Instability of self-focusing of light*, Zh. Éksp. Teor. Fiz **53** (1967), pp. 1735–1743.

[Zei84] E. Zeidler, *Nonlinear functional analysis and its applications III. Variational methods and optimization*, Springer, 1984, 1 edn.

[Zet05] A. Zettl, *Sturm-Liouville theory*, Mathematical Surveys and Monographs, vol. 121, American Mathematical Society, Providence, RI, 2005. MR2170950

[ZS75] V. Zakharov and V. Synakh, *The nature of the self-focusing singularity*, Zh. Éksp. Teor. Fiz **41** (1975), pp. 465–468, Russian original - ZhETF, Vol. 68, No. 3, p. 940, June 1975.

[ZSS71] V. E. Zakharov, V. V. Sobolev, and V. S. Synakh, *Character of singularity and stochastic phenomena in self focusing*, ZhETF Letters **14** (1971), pp. 564–568.

[ZZ15] Y. Zhang and Q. Zhao, *Global solution to nonlinear Dirac equation for Gross-Neveu model in* $1 + 1$ *dimensions*, Nonlinear Anal. **118** (2015), 82–96, DOI 10.1016/j.na.2015.02.007. MR3325607

Index

List of symbols

Symbol	Meaning	Page
$1 \le i,\, j \le n$	$1 \le i \le n$ and $1 \le j \le n$.	
$\mathbb{1}_S$	A characteristic function of $S \subset T$: $\mathbb{1}_S(x) = \begin{cases} 1, & x \in S; \\ 0, & x \in T \setminus S. \end{cases}$	
$\lvert\alpha\rvert$	$\lvert\alpha\rvert = \alpha_1 + \dots + \alpha_n, \quad \alpha \in \mathbb{N}_0^n$.	
$(a_i)_{i\in\mathbb{N}}$	A sequence (ordered set) of elements $a_1,\, a_2,\, a_3,\, \dots$.	
$\{a_i\}_{i\in\mathbb{N}}$	A set of elements $a_1,\, a_2,\, a_3,\, \dots$.	
$\boldsymbol{\alpha}^i,\ \boldsymbol{\beta},\ \mathbf{J}$	$\boldsymbol{\alpha}^i = \begin{bmatrix} \operatorname{Re}\alpha^i & -\operatorname{Im}\alpha^i \\ \operatorname{Im}\alpha^i & \operatorname{Re}\alpha^i \end{bmatrix},\ \boldsymbol{\beta} = \begin{bmatrix} \operatorname{Re}\beta & -\operatorname{Im}\beta \\ \operatorname{Im}\beta & \operatorname{Re}\beta \end{bmatrix},\ \mathbf{J} = \begin{bmatrix} 0 & I_N \\ -I_N & 0 \end{bmatrix}.$	Page 178
$\mathbb{B}_R(\mathbf{X})$	$\{x \in \mathbf{X}\colon\ \lvert x\rvert < R\},\ R > 0.$	
$\mathbb{B}^n_R,\ \mathbb{B}^n_R(x_0)$	$\{x \in \mathbb{R}^n\colon\ \lvert x\rvert < R$ or $\lvert x - x_0\rvert < R$ if x_0 is provided$\},\ R > 0.$	
$\mathscr{B}(\mathbf{X}),\ \mathscr{B}(\mathbf{X},\mathbf{Y})$	Vector space of bounded linear operators $\mathbf{X} \to \mathbf{X}$ and $\mathbf{X} \to \mathbf{Y}$.	Page 25
$\mathscr{B}_0(\mathbf{X}),\ \mathscr{B}_0(\mathbf{X},\mathbf{Y})$	Vector space of compact linear operators $\mathbf{X} \to \mathbf{X}$ and $\mathbf{X} \to \mathbf{Y}$.	Page 26
$\mathscr{C}(\mathbf{X}),\ \mathscr{C}(\mathbf{X},\mathbf{Y})$	Set of closed linear operators $\mathbf{X} \to \mathbf{X}$ and $\mathbf{X} \to \mathbf{Y}$.	Page 26
$\mathbb{C}_+,\ \mathbb{C}_-$	$\{z \in \mathbb{C}\colon\ \operatorname{Im} z > 0\},\quad \{z \in \mathbb{C}\colon\ z < 0\}.$	
$C^k_{\mathrm{b}}(\mathbb{R}^n)$	$\{f \in C(\mathbb{R}^n)\colon\ \max\limits_{\alpha\in\mathbb{N}_0^n,\,\lvert\alpha\rvert\le k} \sup_{x\in\mathbb{R}^n} \lvert\partial_x^\alpha f(x)\rvert < \infty\}$	
$\mathscr{C}_{\lambda,\omega}(M,\mathcal{N},\rho,\nu)$	The set of admissible functions for the Carleman estimates for the linearization operator $J(D_m + V - \omega)$.	Page 131
$\mathscr{C}_{\lambda}(M,\mathcal{N},\rho,\nu)$	The set of admissible functions for the Carleman estimates for the operator $D_m + V$.	Page 117
$\mathbf{Cl}_\ell(\mathbb{C})$	Clifford algebra over $\mathbb{C}$ formed by ℓ generators.	Page 143

$\mathbf{coker}(A)$	$\mathbf{Y}/\mathfrak{R}(A)$, the cokernel of a linear operator $A : \mathbf{X} \to \mathbf{Y}$. Page 38
$\mathfrak{D}(A)$	Domain of a linear operator A. Page 25
∂_x^α	$\partial_x^\alpha = \partial_{x^1}^{\alpha_1} \dots \partial_{x^n}^{\alpha_n} = \left(\frac{\partial}{\partial x^1}\right)^{\alpha_1} \dots \left(\frac{\partial}{\partial x^n}\right)^{\alpha_n}$ for $x = (x^1, \dots, x^n) \in \mathbb{R}^n$, $\alpha \in \mathbb{N}_0^n$.
D_m, D_0	The Dirac operator: $D_m = -\mathrm{i}\boldsymbol{\alpha} \cdot \nabla + \beta m$, $D_0 = -\mathrm{i}\boldsymbol{\alpha} \cdot \nabla$. Page 62
$\mathbb{D}_R$, $\mathbb{D}_R(z_0)$	$\{z \in \mathbb{C}: \ \lvert z\rvert < R \ \text{ or } \ \lvert z - z_0\rvert < R \ \text{ if } z_0 \text{ is provided}\}$, $R > 0$.
$E^\perp$	$\{\xi \in \mathbf{X}^*: \ \langle \xi, x\rangle = 0 \ \ \forall x \in E\}$, where E is a subset of a Banach space $\mathbf{X}$.
$\mathrm{End}(\mathbb{C}^N)$	The endomorphism ring of $\mathbb{C}^N$; $\mathrm{End}(\mathbb{C}^N) \cong M_N(\mathbb{C}) \cong \mathbb{C}^{N\times N}$.
Φ_A, $\Phi_A^\pm$, Ψ_A	Fredholm domains of a linear operator A. Page 38
$\mathcal{G}(A)$	Graph of a linear operator A. Page 26
$\gtrapprox$	$a \gtrapprox b$ if $a \in (b, b + \varepsilon)$ for $\varepsilon > 0$ small enough.
H_r^1, L_r^2, etc.	Corresponding subspaces of spherically symmetric functions. Page 17
I, $I_{\mathbf{X}}$	I or $I_{\mathbf{X}} : \mathbf{X} \to \mathbf{X}$, $x \mapsto x \ \forall x \in \mathbf{X}$, where $\mathbf{X}$ is a Banach space.
I_N	For $N \in \mathbb{N}$, $I_N = I_{\mathbb{C}^N} : \mathbb{C}^N \to \mathbb{C}^N$.
$\mathbf{ker}(A)$	Kernel (null space) of a linear operator A. Page 27
$\lessapprox$	$a \lessapprox b$ if $a \in (b - \varepsilon, b)$ for $\varepsilon > 0$ small enough.
$\mathfrak{L}(A)$, $\mathfrak{L}_\lambda(A)$	Root lineal of A corresponding to eigenvalue zero (λ if specified). Page 28
L^p_{loc}, L^∞_{loc}	$\{ u \in \mathscr{D}': \ \ \varphi u \in L^p, \ \ \forall \varphi \in \mathscr{D} \}$ Page 12
$\mathbb{N}$, $\mathbb{N}_0$	$\mathbb{N} = \{1, 2, 3, \dots\}$, $\mathbb{N}_0 = \{0, 1, 2, 3, \dots\}$.
Ω_R^n	$\mathbb{R}^n \setminus \overline{\mathbb{B}_R^n} = \{x \in \mathbb{R}^n: \ \lvert x\rvert > R\}$, $R > 0$.
$\mathbf{O}(N, \mathbb{F})$	Orthogonal matrices with coefficients in the field $\mathbb{F}$. Page 150
$\bar\psi = \psi^*\beta$	The Dirac adjoint of $\psi \in \mathbb{C}^N$. Page 165
$\mathbb{R}_+$, $\mathbb{R}_-$	$\{t \in \mathbb{R}: \ \ t > 0\}$, $\{t \in \mathbb{R}: \ \ t < 0\}$.

$r(A)$	Spectral radius of a bounded linear operator A Page 35
r, $\langle r\rangle$	$r = \|x\|$ for $x \in \mathbb{R}^n$, $n \in \mathbb{N}$. The operators of multiplication by $\|x\|$ and $\langle x\rangle = (1 + \|x\|^2)^{1/2}$ are denoted by by r and $\langle r\rangle$, respectively.
$\mathfrak{R}(A)$	Range of a linear operator A. Page 25
$\rho(A)$	Resolvent set of a linear operator A. Page 33
$\mathbb{S}_R^{n-1}$	The sphere in $\mathbb{R}^n$ of radius $R > 0$.
$(\sigma_i)_{1\le i\le n}$	Generalized Pauli matrices $\sigma_i \in \mathrm{End}(\mathbb{C}^N)$ which satisfy the relations $\sigma_i\sigma_j^* + \sigma_j\sigma_i^* = 2\delta_{ij}I_N$, $1 \le i, j \le n$, with $n, N \in \mathbb{N}$. Page 142
σ_{p}, σ_{c}, σ_{res}	Point, continuous, and residual spectra of a linear operator. Page 36
$\sigma_{\mathrm{ess},k}$	Different definitions of the essential spectra of a linear operator. Page 50
σ_{ap}	Approximate point spectrum of a linear operator. Page 37
$W^{s,p}(\mathbb{R}^n)$, $W_0^{s,p}(\Omega)$	L^p-based Sobolev spaces of order s. Page 12
$\mathbb{Z}$	$\{\dots, -2, -1, 0, 1, 2, 3, \dots\}$

For a complete list of titles in this series, visit the
AMS Bookstore at **www.ams.org/bookstore/survseries/**.